Areas under the Normal Curve from 0 to Z

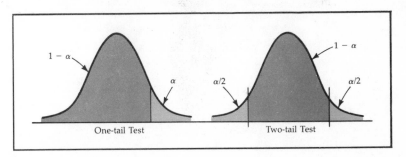

One-tail Test Two-tail Test

Z	0.0	0.01	0.02	0.03	0.04	0.05	0.06	0.07	0.08	0.09
0.0	.0000	.0040	.0080	.0120	.0160	.0199	.0239	.0279	.0319	.0359
0.1	.0398	.0438	.0478	.0517	.0557	.0596	.0636	.0675	.0714	.0753
0.2	.0793	.0832	.0871	.0910	.0948	.0987	.1026	.1064	.1103	.1141
0.3	.1179	.1217	.1255	.1293	.1331	.1368	.1406	.1443	.1480	.1517
0.4	.1554	.1591	.1628	.1664	.1700	.1736	.1772	.1808	.1844	.1879
0.5	.1915	.1950	.1985	.2019	.2054	.2088	.2123	.2157	.2190	.2224
0.6	.2257	.2291	.2324	.2357	.2389	.2422	.2454	.2486	.2517	.2549
0.7	.2580	.2611	.2642	.2673	.2704	.2734	.2764	.2794	.2823	.2852
0.8	.2881	.2910	.2939	.2967	.2995	.3023	.3051	.3078	.3106	.3133
0.9	.3159	.3186	.3212	.3238	.3264	.3289	.3315	.3340	.3365	.3389
1.0	.3413	.3438	.3461	.3485	.3508	.3531	.3554	.3577	.3599	.3621
1.1	.3643	.4665	.4686	.3708	.3729	.3749	.3770	.3790	.3810	.3830
1.2	.3849	.3869	.3888	.3907	.3925	.3944	.3962	.3980	.3997	.4015
1.3	.4032	.4049	.4066	.4082	.4099	.4115	.4131	.4147	.4162	.4177
1.4	.4192	.4207	.4222	.4236	.4251	.4265	.4279	.4292	.4306	.4319
1.5	.4332	.4345	.4357	.4370	.4382	.4394	.4406	.4418	.4429	.4441
1.6	.4452	.4463	.4474	.4484	.4495	.4505	.4515	.4525	.4535	.4545
1.7	.4554	.4564	.4573	.4582	.4591	.4599	.4608	.4616	.4625	.4633
1.8	.4641	.4649	.4656	.4664	.4671	.4678	.4686	.4693	.4699	.4706
1.9	.4713	.4719	.4726	.4732	.4738	.4744	.4750	.4756	.4761	.4767
2.0	.4772	.4778	.4783	.4788	.4793	.4798	.4803	.4808	.4812	.4817
2.1	.4821	.4826	.4830	.4834	.4838	.4842	.4846	.4850	.4854	.4857
2.2	.4861	.4864	.4868	.4871	.4875	.4878	.4881	.4884	.4887	.4890
2.3	.4893	.4896	.4898	.4901	.4904	.4906	.4909	.4911	.4913	.4916
2.4	.2918	.4920	.4922	.4925	.4927	.4929	.4931	.4932	.4934	.4936
2.5	.4938	.4940	.4941	.4943	.4945	.4946	.4948	.4949	.4951	.4952
2.6	.4953	.4955	.4956	.4957	.4959	.4960	.4961	.4962	.4963	.4964
2.7	.4965	.4966	.4967	.4968	.4969	.4970	.4971	.4972	.4973	.4974
2.8	.4974	.4975	.4976	.4977	.4977	.4978	.4979	.4979	.4980	.4981
2.9	.4981	.4982	.4982	.4983	.4984	.4984	.4985	.4985	.4986	.4986
3.0	.4987	.4987	.4987	.4988	.4988	.4989	.4989	.4989	.4990	.4990
3.1	.4990	.4991	.4991	.4991	.4992	.4992	.4992	.4992	.4993	.4993
3.2	.4993	.4993	.4994	.4994	.4994	.4994	.4994	.4995	.4995	.4995
3.3	.4995	.4995	.4995	.4996	.4996	.4996	.4996	.4996	.4996	.4997
3.4	.4997	.4997	.4997	.4997	.4997	.4997	.4997	.4997	.4997	.4998
3.6	.4998	.4998	.4999	.4999	.4999	.4999	.4999	.4999	.4999	.4999
3.9	.5000	.5000	.5000	.5000	.5000	.5000	.5000	.5000	.5000	.5000

Source: Reprinted by permission from *Statistical Methods* by George W. Snedecor and William G. Cochran, sixth edition © 1967 by Iowa State University Press, Ames, Iowa.

Basic Statistics
A Real World Approach

Basic Statistics
A Real World Approach

Third Edition

Vincent E. Cangelosi
Louisiana State University

Phillip H. Taylor
University of Arkansas

Philip F. Rice
Louisiana Tech Univeristy

WEST PUBLISHING COMPANY
St. Paul   New York   Los Angeles   San Francisco

A study guide has been developed to assist you in mastering concepts presented in this text. The study guide reinforces concepts by presenting them in condensed concise form. Additional illustrations and examples are also included. The study guide is available from your local bookstore under the title *Study Guide to Accompany Basic Statistics: A Real World Approach,* Third Edition, prepared by Phillip H. Taylor and John Thanapolous.

COPYRIGHT © 1976, 1979, 1983 by
WEST PUBLISHING CO.
50 West Kellogg Boulevard
P.O. Box 3526
St. Paul, Minnesota 55165

Printed in the United States of America

Library of Congress Cataloging in Publication Data

Cangelosi, Vincent E.
  Basic statistics.

  Includes index.
  1. Social sciences—Statistical methods.
  2. Commercial statistics. 3. Statistics.
  I. Taylor, Phillip H. II. Rice, Philip F.
  III. Title.
  HA29.C27   1983   519.5   82-23913
  ISBN 0-314-69637-7   INTL. ED. ISBN 0-314-68852-8
  1st reprint 1983   1st reprint 1983

To Jean, Gloria, and Jane

Contents

Chapter 13
Correlation and Regression Analysis: The Multivariate Case 367

Chapter 14
Time-Series Analysis— Trend 391

CONTENTS

Appendixes

APPENDIXES

Preface

In writing this text for students in business administration, economics, and other social sciences, we attempted to accomplish the following objectives:

1. To write a book that will be understandable to students—one from which students can learn statistics.

2. To make learning statistical techniques interesting.

3. To present the subject of statistics in such a way that it has relevance to *real-world* applications.

4. To design a text that gives the instructor maximum flexibility in presenting a course suitable to student needs.

The first objective is accomplished by making a realistic evaluation of the mathematics background of the types of students who will most likely take a statistics course for which this text will be used. Not only do we consider that most of these students will probably have not taken any more than one course in algebra, but we also acknowledge that the mathematics retention factor is reasonably low for many students. Moreover we feel that this book is written in a comprehensible, explanatory, and personal fashion.

We also feel that learning statistics does not have to be dull and monotonous. Since most everyday, real-world affairs involve statistics of some sort, a text dealing with statistical techniques can draw upon many real-world experiences for applications of these techniques. We have attempted to use the real world as our laboratory to help accomplish objectives two and three.

While the text may serve as the basis for a one-semester course in statistics, the topical coverage is rather extensive. Therefore the text is designed so that the instructor may eliminate chapters without disrupting the continuity of the course or running out of material. For example, one instructor may not want to include the multiple regression analysis or the nonparametric chapter in a course for business students; yet another may want to exclude time-series analysis and index numbers for other social science students. In either case these eliminations are possible with sufficient material in the remainder of the book to cover a semester's work.

This book was conceived several years ago when the authors felt the need for a text in statistics which would help students with real-world problems. It is based upon what we have learned in our many years as classroom teachers and practitioners, as well as on the education and experience provided by our many friends, colleagues, and professors.

The second edition of this text was published in 1979, and the acceptance of the text by colleagues and their students encouraged us to continue to strive for further improvement. The third edition contains a new chapter, substantial rewriting of several chapters, and as always, attempts to improve and enhance the presentation of the material. Definitions are provided for the end-of-chapter equation summaries, and the problem sets have been changed substantially. Additionally, a case study is provided for each chapter.

While we recognize that the final responsibility for this text, good or bad, rests with us, we are thankful for the many useful suggestions from colleagues, students, and editors. Special appreciation is given to Professor Jim Maas who provided an insightful and comprehensive review of the second edition. Also, we express appreciation to Bruce Lewis and Richard Ford for writing a SAS-based book, *Basic Statistics Using SAS*, to accompany our text; to John Thanopoulos for coauthoring the study guide/workbook; and to Professors William and Vicky Crittenden for preparing the solutions manual. However, we are most indebted to the members of our respective families for their encouragement when we were discouraged; for their patience during the many days and nights we were unable to be with them; and foremost of all, their continued love when we were unlovable.

Basic Statistics
A Real World Approach

1   Processing and Presenting Data

This text focuses on statistics as a research tool for decision making. Accordingly, we shall define the term *statistics* as the collection, presentation, analysis, and interpretation of quantitative information. In collecting such information, one usually searches for existing data, takes a sample survey, or runs an experiment. Almost anyone who collects data does so for a purpose, usually because it helps to measure and evaluate. Through analyzing and evaluating statistical data, public opinion has been determined; young athletes have been tendered enormous contracts; and economic programs have been initiated.

Evaluating or analyzing data, however, usually does not complete the researcher's task. Frequently the researcher must communicate this knowledge to others. To provide this communication, the researcher may incorporate tables, charts, or empirical frequency distributions in the presentation of the research report. The purpose of this chapter is to introduce you to some of these tools. Subsequent chapters will deal with the collection, analysis, and interpretation techniques.

Measurement

The term *data* often implies that *numbers* are being used in some fashion, but we seldom ask where and how the numbers originated. Obviously, someone or some piece of equipment created any numbers that we perceive as data. For example, the amount of electric energy consumed by a house may be shown by the kilowatt hours utilized, with the amount (some number) determined by a meter. Another, and quite different, type of number can originate with the judges of some contest where the entrants in the contest are *ranked*, 1, 2, 3, and so forth. Thus numbers can be, and are, used to represent many different things. How these numbers are created determines how they may be properly used.

Assigning numbers to observations is called *measurement*. There are various ways, or *levels*, of measurement, and the level of measurement determines what you can do with the numbers. In other words, how you may analyze or manipulate data to acquire new information is limited by the *level of measurement*.

There are four levels of measurement, each with its own characteristics. The levels are also sometimes referred to as *scales of measurement*. In order, from lowest to highest, they are nominal, ordinal, interval, and ratio.

Nominal Scale

The nominal scale is the lowest, or weakest, level of measurement. With this level of measurement, numbers, and sometimes other symbols, are used to identify groups to which various observations, or objects, belong. The numbers or symbols identifying the groups make up the scale, and the measurement consists of placing the observation in the correct group or classification. Thus this level of classification is also referred to as the *classificatory scale*.

An example of the use of the nominal scale is the two-digit numbering system used by many football teams. Under this system a number beginning with 1, such as 13 or 19, denotes a quarterback; a number beginning with 5 is assigned to centers, and so forth. Another example is the zip code scheme of the U.S. Postal Service whereby a portion of the zip code designates a specific geographic area such as a city.

A characteristic of the nominal scale is that the numbers or symbols can be interchanged. Thus the numbers for all quarterbacks could begin with 5 and that of all centers with a 1. It really makes no difference what number or symbol you use to identify a class so long as you are consistent.

With the examples given for this level of measurement, it may be obvious to you that two or more nominal scale numbers should not be added, subtracted, or multiplied, or divided by one another. You may count them and arrange them into groups, and, in later sections of this text, you will be introduced to a few other forms of analysis that can be applied to nominal scale data. By comparison with other levels of measurement, however, there are few ways to analyze data derived using this level of measurement.

Ordinal Scale

The ordinal level of measurement is the second weakest of the scales. A measurement of this type exists when the categories into which objects are grouped have some *order* or ranking. In fact it is also called the *ranking scale*. The grades students earn in a class—A, B, C, D, and F—are an example of an ordinal scale. Similarly, income levels such as low, middle, and high are another example of an ordinal or ranking scale of measurement.

It is permissible to assign different names to categories in an ordinal scale, but the new names must not alter the order of the classes. As an illustration you might elect to identify a grade of F as a 1 and an A as a 5, or just the reverse with F as a 5 and A as a 1. In both cases the order would be preserved. You cannot identify an A as 1, F as 2, and C as 3, however.

The operations that can be performed with data developed using the various levels of measurement are *cumulative* from the weakest to the strongest. Thus the permissible operations for ordinal data include those for nominal data as well as a large group of techniques generally associated with the order of the various observations.

Interval Scale

If in addition to some order of ranking of categories the distance between categories or numbers is known, then an *interval scale* exists. An interval scale requires a constant unit of measurement, but both the zero point and the unit of measurement are arbitrary.

Perhaps the best known example of the interval scale is the measurement of temperature. Two scales are commonly used, centigrade and Fahrenheit, and neither has a true zero point. That is, both the scale and zero points are arbitrary.

Since the distances between categories as points are known, it is permissible to utilize methods of analysis that deal with the difference in numbers when using an interval scale. In this sense measurements derived using the interval scale are sufficiently strong that this scale can be properly termed a quantitative scale, whereas weaker scales are thought of as qualitative ones. A large portion of the statistical techniques presented in this text require a level of measurement at least of this strength.

Ratio Scale

Some measurement methods have all the characteristics of the interval scale and in addition have an origin or true zero point. For example, measures of distance, speed, and weight fall into this group. Measurements of this type are said to be made with a *ratio scale* and are considered to be the strongest and most quantitative of the four scales.

Only with the ratio scale can arithmetic operations be performed on the numeric values themselves. (Remember that with an interval scale you can work with the *differences in numbers but not the numbers themselves*.) As was previously mentioned, the forms of analysis are cumulative through the scales from the weakest to the strongest. Thus only with values derived using the ratio scale can *all* forms of statistical analysis be utilized.

Tables

A *table* is an orderly presentation of data. It should have a *title* that accurately and clearly identifies the contents and *column headings* that categorize the data within the table. The first column usually identifies for whom, what, when, or where the other columns of data were collected. That is, the first column usually represents years, months, states, countries, nationality, religion, or some such breakdown for which the same data are to be presented for comparison. And finally, a good table will identify the *source* of the original data.

For an illustration of a well-prepared table, take a look at table 1.1. You should notice that the table has a title and a source. The first column identifies what the information in the other columns is about. Notice also that the other columns have headings so that the data underneath are understandable. This table exhibits how data can be differentiated using what is called qualitative classes (see column 1). Table 1.2, on the other hand, presents its data on the basis of quantitative classes (see column 1).

If you examine table 1.2, you will find it is a good example of a complete table, for it is titled, sourced, and in a clear manner

Table 1.1
Projections of population and personal income by place

	Population		Personal income		Per capita income (U.S. average = 100)	
	Distribution per 1,000 persons, 1990	Percent change, 1980–1990	Distribution per $1,000, 1990	Real percent change, 1980–1990	1980	1990
Total U.S.	1000.0	9.4%	$1,000.0	46.5%	100	100
New England	54.0	6.2	52.9	39.3	101	99
Middle Atlantic	151.8	0.5	153.5	32.2	104	104
East North Central	181.4	6.3	188.3	41.6	105	105
West North Central	75.7	6.9	73.7	43.2	97	98
South Atlantic	165.2	13.0	158.0	53.9	94	95
East South Central	65.2	11.2	56.4	57.8	81	87
West South Central	107.6	14.5	105.6	57.7	94	95
Mountain	54.1	19.9	53.3	65.8	95	96
Pacific	145.0	14.1	158.3	48.7	112	108

Source: The Conference Board, *Across the Board*, vol. 18 (May 1981), p. 56.

Table 1.2
Civilian labor force by age and sex, 1990 (in thousands)

Age	Males	Females	Total
16–19	4,606	4,447	9,053
20–24	7,397	7,280	14,677
25–34	18,287	16,570	34,857
35–44	16,555	13,916	30,471
45–54	10,511	8,236	18,747
55–64	5,635	4,406	10,041
64+	1,609	1,146	2,755
16+	64,600	56,001	120,601

Source: Reprinted with permission from *American Demographics*, February, 1981.

provides data for the groups delineated. Study this table carefully to see if you can read it.

Frequency Distributions

Take a look at table 1.3. What do you see? At best, you see numbers for fifty states. Now suppose it has become your job to present this set of data in some other orderly and meaningful manner. A tabular listing of the numbers in descending or ascending order (an ordered array) would help somewhat, but it would certainly be a long table. So let's investigate a way to aggregate the data.

A *frequency distribution* is developed when the data are divided into two or more *groups* called *classes*, with some identification

Table 1.3
State and local taxes as percent of personal income, 1978–1979

State	Percent	State	Percent	State	Percent	State	Percent	State	Percent
Ala	9.4	Hawaii	14.0	Mass	13.7	NM	10.1	SD	11.1
Alaska	14.0	Ida	10.6	Mich	11.1	NY	15.8	Tenn	9.5
Ariz	13.5	Ill	10.7	Minn	12.7	NC	10.0	Tex	8.9
Ark	8.9	Ind	9.4	Miss	10.8	ND	9.0	Utah	12.2
Calif	10.9	Iowa	10.7	Mo	9.6	Ohio	9.2	Vt	13.1
Colo	11.7	Kans	10.4	Mont	11.2	Okla	8.6	Va	10.2
Conn	10.5	Ky	9.9	Nebr	15.6	Oreg	11.1	Wash	11.8
Del	11.2	La	9.4	Nev	12.5	Pa	11.0	W Va	11.6
Fla	9.8	Me	12.1	NH	8.9	RI	12.0	Wis	12.7
Ga	10.4	Md	12.0	NJ	11.5	SC	10.1	Wyo	13.4

Source: The Conference Board, "The Two-Way Squeeze 1981," *Economic Road Maps* (April 1981), p. 4.
Note: State and local taxes of all types, except severance and corporate income, for fiscal year ending in the twelve months preceding June 1979. Personal income in calendar 1978.

given to each class. The identifying nature of the classes can be either qualitative or quantitative, just the same as the first column of a complete table. Thus a frequency distribution is the presentation of data according to the number of observations appearing in the given intervals or categories. In order to construct a frequency distribution, however, we need to examine the concept of grouping data.

Grouping the Data

To arrange a set of raw data into groups, it is necessary to establish classes. The classes represent the groups (intervals for quantitative classes and categories or attributes for qualitative classes). Each class that is established must be unique. That is, there must be no overlap in the classes so that each individual item in the data set will belong to only one class. The number of crimes classified according to type, as shown in table 1.4, is an example of *qualitative* classes. The population data, classified according to age and presented in table 1.5, are an illustration of *quantitative* classes.

Table 1.5 needs some explanation, however, regarding its construction. The first column, age, in the table represents the classes. All but one of the classes have *limits*, upper and lower, that make the classes unique, but the placing of the data into

Table 1.4
Victimization rates against persons, 1978 (per 1,000 persons, 12 years old and over)

Crime	Total[a]	White	Black	Victim–offender relationship	
				Stranger	Nonstranger
Total[b]	34	33	41	21	12
Rape	1	1	2	1	(Z)
Robbery:					
With injury	2	2	3	1	(Z)
Without injury	4	3	8	3	1
Aggravated assult	10	9	13	6	3
Simple assult	17	18	14	10	7
Personal larceny	97	98	90	(NA)	(NA)

Source: U.S. Department of Commerce, Bureau of the Census, *Reflections of America* (December 1980), p. 71.
Note: Based on National Crime Survey; includes attempted crimes. NA = not available; Z = less than 0.5.
[a]Includes other races not shown separately.
[b]Excludes personal larceny.

Table 1.5
Estimated and projected population, by age for selected years, 1950–2010 (in thousands)

| Age | 1950 | 1960 | 1970 | 1975 | 1978 | 1979 | Projections | | | |
							1985	1990	2000	2010
All ages	152,271	180,671	204,878	213,559	218,717	220,584	232,880	243,513	260,378	275,335
Under 5 years	16,410	20,341	17,101	15,879	15,378	15,649	18,803	19,437	17,852	19,221
5–13 years	22,423	32,965	36,636	33,440	31,397	30,647	29,098	32,568	35,080	33,067
14–17 years	8,444	11,219	15,910	16,934	16,651	16,276	14,392	12,771	16,045	15,439
18–21 years	8,947	9,555	14,707	16,484	17,101	17,148	15,442	14,507	14,990	16,319
22–24 years	7,129	6,573	9,980	11,120	11,871	12,136	12,411	10,642	9,663	12,043
25–34 years	24,036	22,919	25,294	30,919	33,971	35,024	39,859	41,086	34,450	36,246
35–44 years	21,637	24,221	23,142	22,816	24,409	25,136	31,376	36,592	41,344	34,685
45–54 years	17,453	20,578	23,310	23,769	23,199	22,957	22,457	25,311	35,875	40,551
55–64 years	13,396	15,625	18,664	19,777	20,677	20,952	21,737	20,776	23,257	32,926
65 years and over	12,397	16,675	20,087	22,420	24,064	24,658	27,305	29,824	31,822	34,837
Median age	30.2	29.4	27.9	28.8	29.7	30.0	31.5	32.8	35.5	36.6

Source: U.S. Department of Commerce, Bureau of the Census, *Reflections of America* (December 1980), p. 29.
Note: Includes Armed Forces abroad; base date for projections is 1976.

classes may still require some interpretation. Consider a person who was 17 years and 8 months old. Into which class would 17⅔ go?

Where 17⅔ resides is decided in the following manner. If the data require rounding, more definite *boundaries* are needed than the class limits given in table 1.5. That is, it may be decided to include all items less than 18 but equal to or more than 14 in the third class. As a matter of fact, when dealing with ages, this is the usual procedure. Had the data, however, involved weights, dollars, inches, or other similar units, another rounding procedure might be used. For example, the boundary of a class might be halfway between the upper limit of one class and the lower limit of the next class. The point is that a decision has to be made about the boundaries so that you will know where to place values that fall between the upper limit of one class and the lower limit of another.

Now notice in table 1.5 that the class limits are unique. That is, the upper limit of one class must be less than the lower limit of the next class. Another way to present the class limits for this table is as follows:

under 5
5 and under 14
14 and under 18

Stating limits in this fashion will sometimes produce a slightly different grouping, but it will also eliminate any doubt as to where a particular item belongs.

So far we have looked at class limits without discussing the class interval. The *interval* is the width of the class or the difference between the upper boundary and the lower boundary of a class. Thus in table 1.5 the boundaries, and not the class limits, must be used to establish the class interval. Consequently, using the boundaries for the second class (5 and 14), subtraction yields a class intervals of nine.[1]

The next item in table 1.5 that we need to examine is the *class midpoint*. The class midpoint is the value represented by the point equidistant between the upper and lower boundaries of the class. The midpoint of any class can be established by adding the two boundaries and dividing their sum by two. Thus the midpoint of the second class of table 1.5 is $(5 + 14)/2 = 9.5$.

Finally, it should be pointed out that not all the class intervals in table 1.5 are equal. While equal intervals for all classes are frequently desired, they are not essential. In this instance there were also some extreme values that required the last class (65 years and over) to be left without an upper limit. (Otherwise we might have had several classes in which no observations fell.) When a class is missing a limit, either upper or lower, it is called an *open end* class.

Frequency Distribution Construction

Now that you know something about grouping data, let us proceed to the construction of a frequency distribution for the data shown in table 1.3. To facilitate the grouping, first arrange the tax/income data in order of magnitude. (It makes no difference whether you begin with the smallest and proceed to the largest or vice versa.) Data arranged in this way are called an *ordered array*. Table 1.6 shows the per capita tax data arrayed in ascending order.

It is now necessary to face the question: How many classes are needed? Unfortunately, there are no rules that will fix the number of classes that should be in a given frequency distribution. There are, however, two guidelines that will aid you in frequency distribution construction. They are (1) keep the number of intervals between 5 and 20 and (2) avoid empty classes.

---

1. Remember that in this instance the upper boundary of the second class approaches the lower limit of the third class. Since this depends on the rounding method used, sometimes the upper boundary and the lower limit will be different.

Table 1.6
Array of state and local taxes as percent of personal income, 1978–79 data

8.6	9.6	10.6	11.2	12.7
8.9	9.8	10.7	11.5	12.7
8.9	9.9	10.7	11.6	13.1
8.9	10.0	10.8	11.7	13.4
9.0	10.1	10.9	11.8	13.5
9.2	10.1	11.0	12.0	13.7
9.4	10.2	11.1	12.0	14.0
9.4	10.4	11.1	12.1	14.0
9.4	10.4	11.1	12.2	15.6
9.5	10.5	11.2	12.5	15.8

Source: Table 1.3.

With these two guidelines in mind, a good starting place in constructing a frequency distribution is to examine the difference in the extreme values in our data set. That is, subtract the lowest data item from the highest (15.8 − 8.6 = 7.2). To assure that there is a place in the frequency distribution for every item in the data set, the product of the number of classes and the class interval must be equal to or greater than this difference (7.2).

At this point either experience or experimentation is the only help available.[2] In our particular case class intervals of 0.75, 1.00, or 1.50 might do the job adequately. Selecting the interval of 1.00 indicates the need for eight classes. That is, dividing the selected class interval into the difference between the largest and smallest values in the data set yields 7.2 ÷ 1.00 = 7.2. Then, the number of classes is rounded to eight. Table 1.7 shows the tax data grouped into eight classes with 8.0 chosen as the lower limit of the first class.

It should be noted that the boundaries of the first class are 7.95 and 8.95, which gives the class interval of 1.00. The midpoint for the first class is (7.95 + 8.95) ÷ 2 = 8.45. This method of stating the class limits, however, has the disadvantage that the class limits shown are not the actual boundaries of the classes. Table 1.8

2. There is a method, called Sturges's rule, that may assist you in determining the number of classes. Sturges's rule is stated as follows:

$k = 1 + 3.3 \ln n$

where k = the number of classes when rounded to the nearest whole number.
n = the number of observations in the data set.

It should be remembered, however, that this method is a general guide and not a substitute for experienced judgment.

Table 1.7
Frequency distribution for tax/income data

Class interval (percent)	Frequency
8.0– 8.9	4
9.0– 9.9	9
10.0–10.9	12
11.0–11.9	10
12.0–12.9	7
13.0–13.9	4
14.0–14.9	2
15.0–15.9	2
	50

Source: Table 1.6.

Table 1.8
Frequency distribution for tax/income data

Class interval (percent)	Frequency
8.0 and under 9.0	4
9.0 and under 10.0	9
10.0 and under 11.0	12
11.0 and under 12.0	10
12.0 and under 13.0	7
13.0 and under 14.0	4
14.0 and under 15.0	2
15.0 and under 16.0	2
	50

Source: Table 1.6.

presents the data grouped using class limits that correspond to the boundaries. As stated before, this type of grouping eliminates uncertainty about the placement of any observations into classes. Consequently, this is the method of stating class intervals that will be used throughout this text. Since it is not intended that this is the only frequency distribution that might be established from the data set—nor is it necessarily the "best"—why not take a few minutes and make your own!

The frequency distribution in table 1.8 is an *absolute frequency distribution*. That is, it shows the actual number of data items falling within each class. Thus in the first class there are 4 items, in the second class there are 9 items, and so on. In contrast to this absolute reporting is the type of distribution known as a *relative frequency distribution*. The relative frequency distribution states

Table 1.9
Relative frequency distribution for tax/income data

Class interval (percent)	Relative frequency (percent)
8.0 and under 9.0	8.0
9.0 and under 10.0	18.0
10.0 and under 11.0	24.0
11.0 and under 12.0	20.0
12.0 and under 13.0	14.0
13.0 and under 14.0	8.0
14.0 and under 15.0	4.0
15.0 and under 16.0	4.0
	100.0

Source: Table 1.8.

the frequency of each class as a percent of the total frequencies. Table 1.9 presents the tax data as a relative frequency distribution. The first class contains 8 percent of the observations in the distribution or $(4 \div 50)100 = 8$ percent.

Cumulative Frequency Distributions

At times you may find it convenient to present the data in a manner that will show the number of values that are greater than or less than some quantity. Such a presentation is called a *cumulative frequency distribution*. For an example of a "more than" distribution, take a look at table 1.10. In that table you will notice that the class limits have been changed to allow the frequencies to accumulate. That is, the data are shown as 8.0 or more," "9.0 or more," and so on, rather than each class interval being exclusive. The cumulative frequencies are obtained by summing down the

Table 1.10
"More than" cumulative frequency distribution

Taxes as a percent of income	Frequency
8.0 or more	50
9.0 or more	46
10.0 or more	37
11.0 or more	25
12.0 or more	15
13.0 or more	8
14.0 or more	4
15.0 or more	2
16.0 or more	0

Source: Table 1.8.

Table 1.11
"Less than" cumulative frequency distribution

Taxes as a percent of income	Frequency
Less than 8.0	0
Less than 9.0	4
Less than 10.0	13
Less than 11.0	25
Less than 12.0	35
Less than 13.0	42
Less than 14.0	46
Less than 15.0	48
Less than 16.0	50

Source: Table 1.8.

frequency column in table 1.8. That is, the first cumulative frequency is 50 because all items are "8.0 or more." For the second cumulative frequency the summation begins with 9 and takes in all frequencies down the column (9 + 12 + 10 + 7 + 4 + 2 + 2 = 46). This procedure is continued until all data items are accounted for.

Reversing the manner in which the frequencies are accumulated produces a "less than" cumulative frequency distribution as shown in table 1.11. Notice that the frequency begins with 0 and ends with 50 in the "less than" distribution, and that the frequency begins with 50 and ends with 0 in the "more than" distribution.

Now these cumulative frequency distributions can be used to establish additional information about the data set. For example, table 1.10 indicates that all 50 selected areas have taxes of 8.0 percent or more of income, and 15 of them have taxes of 12 or more percent. According to these data, no area had taxes in excess of 16 percent of income. Similarly, table 1.11 points out that none of the areas has taxes below 8.0 percent and that all are below 16 percent.

Frequency Array

Sometimes arranging data into classes is not convenient because you may wish to retain the individual identity of each item of data. Other times, the data set may not contain enough different values to justify constructing the type of frequency distribution shown in table 1.8, or still another possibility is that there are only a few different values in the data set and these values occur in whole number form. In this latter case we simply mean that no fractional units occur.

Table 1.12
Frequency array of the number of banks in Arkansas counties, 1977

Number of banks	Number of counties (frequency)
1	9
2	22
3	12
4	16
5	5
6	3
7	5
8	1
9	1
10	0
11	1
	75

Source: 1977 Report of the Bank Commissioner of Arkansas, Office of State Bank Department, Little Rock.

Look at table 1.12. This is an example of a *frequency array*. Rather than tabulating the data by classes, a frequency array displays the data as a tabulation of the occurrence of specific values. In the example in table 1.12, the number of banks in any county is discrete (a whole number). Thus the table is simply a tabulation of the number of counties with 1, 2, and so on, up to 11 banks.

Charts

Charts, as opposed to tables, resort to a *pictorial* presentation of data. And, because of the sense of familiarity with pictures, a reader usually feels more at ease with charts. Consequently, charts are very popular data presentation aids. Two widely used charts are the *bar chart* and the *pie chart*.

Bar Charts

Just as you would expect, *bar charts* communicate information through the use of bars. Two examples of bar charts are shown in figures 1.1 and 1.2. Notice that in one of the charts the bars are vertical and in the other they are horizontal. In both cases the width of the bars is constant, and the length of each bar indicates the magnitude of the data.

Pie Charts

As you probably guessed, a *pie chart* is simply a division of a circle (a pie) into "pie slices." This circular shaped chart is particularly useful when data are stated in percentage terms. The "pie" represents the whole (100 percent) and is subdivided into "slices" that represent the respective parts. Of course the "slices" ex-

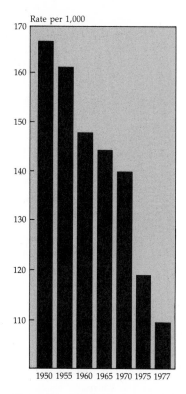

Figure 1.1 Marriage rate per 1,000 unmarried women, 15–44 years old for selected years, 1950–1977. Source: U.S. Department of Commerce, Bureau of the Census, *Reflections of America* (December 1980), p. 162.

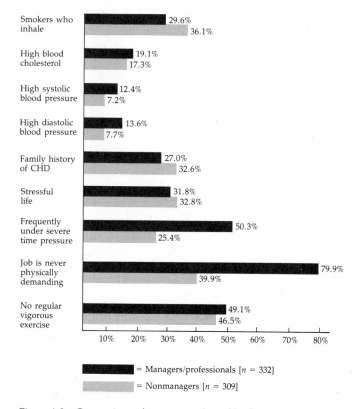

= Managers/professionals [*n* = 332]

= Nonmanagers [*n* = 309]

Figure 1.2 Comparison of coronary-risk profiles for managerial/professional and nonmanagerial employees.
Source: Reprinted by permission from *Business* Magazine. "Warning: Your Job May Be Killing You," by Robert Krietner, Steven D. Wood, and Glenn M. Friedman, January–February 1981 Copyright © 1981, by the College of Business Administration: Georgia State University, Atlanta.

pressed as percentages must total 100 percent. Figure 1.3 illustrates the use of a pie chart.

Graphs

A *graph* is a line connecting points representing the coordinated values of two variables. One variable is represented on the vertical axis (the ordinate). Because it relates the variables in a precise way, the graph form is frequently utilized to present statistical data when the main objective is to stress the relationship between the two variables. Figures 1.4 and 1.5 are examples of graphs using two different methods of scaling the vertical axis.

In figure 1.4 an arithmetic scale is used along the vertical axis. This means that the distance used to represent 100 million (or any amount) is constant up and down the axis. If data are displayed in this manner, the graph is called *arithmetic*.

14

Billions of dollars

Figure 1.3 School expenditures by source of funds, 1979.
Source: U.S. Department of Commerce, Bureau of the Census, *Reflections of America* (December 1980), p. 93.

Using an arithmetic scale, however, can create a problem if the data to be displayed are very different in size. For example, you might find it necessary to plot the numbers 1, 10, 150, 1,000, and 10,000 on the ordinate scale of a graph. To show this set of five numbers using an arithmetic scale would require a graph that would be either very tall or one that would not show much, if any, difference in the numbers 1, 10, and 150. The problem can be avoided by using a graph similar to that shown in figure 1.5. This type of graph differs from an arithmetic graph in that the scale on the vertical axis is logarithmic rather than an arithmetic. Since a logarithmic scale is used only on the ordinate, this type of graph is sometimes called a *semilog graph*.

Figure 1.4 Domestic telegraph messages and telephone conversations, 1880–1977.
Source: U.S. Department of Commerce, Bureau of the Census, *Reflections of America* (December 1980), p. 54.

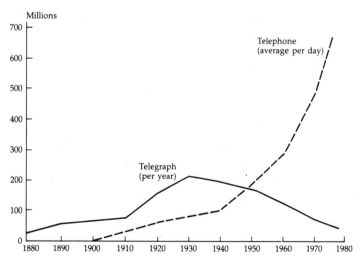

Figure 1.5 Prophecy vs. reality, forecasts of U.S. crude oil production.
Source: The Conference Board, *Across the Board*, vol. 15 (January 1978), p. 20.

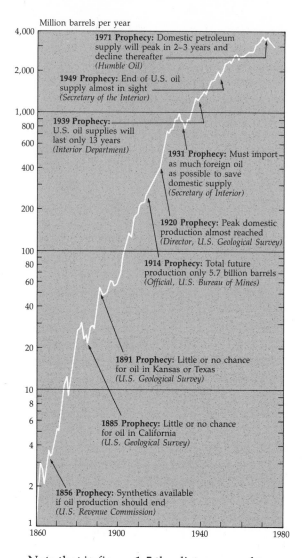

Million barrels per year

1971 Prophecy: Domestic petroleum supply will peak in 2–3 years and decline thereafter
(Humble Oil)

1949 Prophecy: End of U.S. oil supply almost in sight
(Secretary of the Interior)

1939 Prophecy: U.S. oil supplies will last only 13 years
(Interior Department)

1931 Prophecy: Must import as much foreign oil as possible to save domestic supply
(Secretary of Interior)

1920 Prophecy: Peak domestic production almost reached
(Director, U.S. Geological Survey)

1914 Prophecy: Total future production only 5.7 billion barrels
(Official, U.S. Bureau of Mines)

1891 Prophecy: Little or no chance for oil in Kansas or Texas
(U.S. Geological Survey)

1885 Prophecy: Little or no chance for oil in California
(U.S. Geological Survey)

1856 Prophecy: Synthetics available if oil production should end
(U.S. Revenue Commission)

Note that in figure 1.5 the distance used to represent 10 million barrels of oil diminishes as you move away from the intersection of the axes. This occurs because distances used to represent values are determined by the logarithms of the values. The logarithm of 1 is 0.00000; that of 2 is 0.30103; and the logarithmic value for 10 is 1.00000. Thus on a log scale the value 2 is shown as 30.103 percent of the way between 1 and 10. The logarithmic scale in figure 1.5 is called a "four-phase" scale. So long as the largest number to be plotted on an axis does not exceed ten times the smallest number, only a single phase is needed. However, another phase must be added to accommodate numbers larger than ten times the minimum number. That is, for the numbers 1

Table 1.13
Stock prices (dollars per share)

Year	Cherry Computers	Little Computer, Inc.
1979	2.00	5.00
1980	2.40	6.00
1981	2.88	7.20
1982	3.46	8.64

and 12 two phases are required; for 1 and 150, three phases; and so on. Finally, one disadvantage of this type of graph is that neither 0 nor negative numbers can be plotted on a logarithmic scale.

Semilog graphs make it possible to compare the rate of change of two or more series of data. Specifically, the slope of a line connecting data points plotted on a semilog graph reflects the rate of change. Thus, looking at the data in table 1.13, you can easily see that the prices of the two stocks are changing by differing absolute amounts. When the two data series are plotted in semilog form as in figure 1.6, however, the lines connecting the respective data points have identical slopes. Thus it is apparent that the prices of the two stocks have equal rates of growth.

Histograms, Frequency Polygons, and Ogives

The data given in table 1.8 could be displayed graphically to present the information pictorially. One form this presentation could take is known as a histogram. A *histogram* is a bar chart that

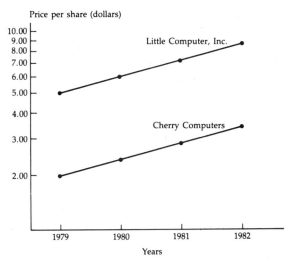

Figure 1.6   Semilog graph of stock prices.
Source: Table 1.13.

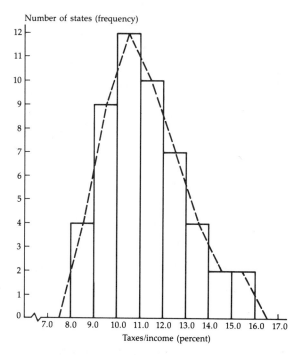

Figure 1.7   Tax/income data displayed as a histogram and a frequency polygon. Source: Table 1.8.

shows a frequency distribution with data presented along both axes. The class interval is represented by the width of the bar on one axis, and the frequency of the class is given by the height of the bar on the other axis. As with other bar charts this conveys to anyone viewing the histogram both the absolute frequency of each class and the size of the one class frequency relative to the frequencies of all other classes. The data in table 1.8 are shown as a histogram in figure 1.7.

In presenting a frequency distribution as a histogram, there is an implicit assumption that the data items in each class are evenly distributed throughout the class. If this even distribution is not a realistic presentation of the data, it may be more accurate to display the data using a *frequency polygon*. To construct a frequency polygon, a series of points, whose coordinates are the midpoints and frequencies of each class, is plotted. A frequency polygon, then, is a line joining these points. The dashed line in figure 1.7 is a frequency polygon showing the same information as the histogram and also having the same meaning. That is, it should provide pictorially both a medium for viewing the shape of the frequency distribution and a look at the frequency of occurrence for each class.

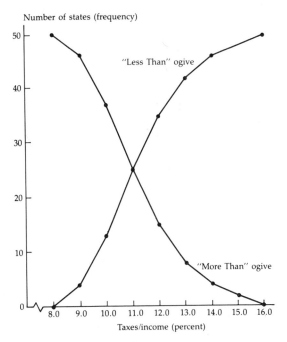

Figure 1.8 Tax/income data displayed as ogives.
Source: Tables 1.10 and 1.11.

There is one final point that you should notice concerning the frequency polygon in figure 1.7. The dashed line does not terminate at the midpoint of the last class, nor does it originate at the midpoint of the first class. The polygon is "closed in" by extending the dashed line to the abscissa. In practice the line is made to intersect the axes at points that correspond to the midpoints of what would be the next class at each end of the frequency distribution.

Finally, the information shown in table 1.10, as well as that shown in table 1.11, can be displayed graphically. Such a graph is called an *ogive*. In the case of an ogive using the data in the form shown in table 1.10, a series of points with coordinates of the lower limits and respective frequencies is plotted. These points are then connected to form the ogive. Figure 1.8 presents, in ogive form, the cumulative distributions of the tax/income data as shown in tables 1.10 and 1.11.

Symbolic Notation and Arithmetic Operations

Before launching your studies of the many statistical techniques presented in the chapters to follow, you should become familiar with some of the symbolic notation used in statistics. This notation is a form of shorthand that enables those familiar with it to communicate through the use of symbols.

Generally speaking, statistical analysis is conducted on the values of some variable. In other words, statistical analysis may be carried out on such items as the incomes of families living in some neighborhood, the daily high temperature in a city for a year, the number of points scored by each basketball team in the NBA, and the numeric grade earned by each student on some statistics examination. Rather than writing out something like "the income of families in Blue Eye, Missouri" each time the variable is identified, statisticians will give the variable a new name, perhaps X.

Since a variable may take on several values, symbolic notation also provides a way in which to identify each value. This is done by a method of notation known as subscripting so that instead of just using X we use X_i. The i is the subscript, and the notation is read as "X sub i" or "the ith value of X." As an example, suppose the following data represent the number of towboats passing a given point on the Intracoastal Canal each day for five days: 26, 31, 22, 18, and 35. The number of towboats (the variable) can be renamed as X_i so that $X_1 = 26$, $X_2 = 31$, $X_3 = 22$, $X_4 = 18$, and $X_5 = 35$. In this way we can identify each value of X separately by using the subscript values of 1 through 5.

Another symbol you will frequently encounter is Σ (pronounced *sigma*). This is a notation that is used to represent *summation*. In other words, the expression ΣX_i tells you to sum, or add, the values of X_i. Using the numbers of towboats, $\Sigma X_i = 26 + 31 + 22 + 18 + 35 = 132$.

The expression ΣX_i may be altered slightly to provide some additional information. Specifically, it may appear in the form $\Sigma_{i=1}^{n} X_i$, and, when written in this manner, it means *sum the* n *number of values of* X_i *beginning with i = 1*. This is a much more complete set of symbolic notation and is the form required if you attempt to write a computer program to perform this function.

Since many students are not experienced with arithmetic operations using Σ, it is worthwhile to take note of two common sources of errors. The first of these is

$$\sum_{i=1}^{n} X_i^2 \neq \left(\sum_{i=1}^{n} X_i \right)^2 .$$

The left-hand side of the above expression instructs you to *sum the* n *number of squared values of* X_i *beginning with i = 1*. The right-hand side of the expression tells you to *sum the* n *number of values of* X_i *beginning with i = 1 and then square that sum*. These two expressions are *not* algebraically equal.

A second common source of error is related to the following:

$$\sum_{i=1}^{n} X_i Y_i \neq \left(\sum_{i=1}^{n} X_i\right)\left(\sum_{i=1}^{n} Y_i\right).$$

The first part of the expression says *determine the products of* n *pairs of numbers beginning with* i = 1 *and then sum the products.* The second, or right-hand part, instructs you to *sum the* n *values of* X_i *and* Y_i *beginning with* i = 1 *and then determine the product of the sum of* X_i *and the sum of* Y_i. Just as in the first case these two expressions are *not* algebraically equal.

There are, however, some operations you can perform that may not be initially obvious to you. If *a* is the symbol for some constant, then

$$\sum_{i=1}^{n} a X_i = a \sum_{i=1}^{n} X_i.$$

In a similar, but slightly more complicated, expression involving the constants *a* and *b*,

$$\sum_{i=1}^{n} (a + b X_i) = na + b \sum_{i=1}^{n} X_i.$$

Finally, the following expressions are also equal:

$$\sum_{i=1}^{n} (X_i + Y_i) = \sum_{i=1}^{n} X_i + \sum_{i=1}^{n} Y_i.$$

There are many more symbols that go to make up the formulas you will encounter in this text. This brief introduction to statistical shorthand should help you to understand most of the notations and, at the same time, keep you from making some of the more common errors.

1. Define
(a) statistics
(b) measurement
(c) table
(d) frequency distribution
(e) frequency array
(f) graph
(g) class interval
(h) class midpoint
(i) class limits
(j) class boundaries

2. Name and discuss the levels of measurement.

3. List the components of a good table.

4. Distinguish between a relative and an absolute frequency distribution.

5. What are the advantages and disadvantages of using a semilog graph?

Problems

1. The Phillips Petroleum Company's *1980 Annual Report—Special Edition* contains the following information on the number of thousands of barrels of petroleum products the company sells each day in the United States:

	Year	
Type of product	1975	1980
Automotive gasoline	304	214
Aviation fuels	24	19
Distillates	99	81
Liquified petroleum gas	97	106
Other products	31	40
Total	555	460

(a) Present these data in bar chart form for each year.
(b) Present these data in pie chart form for each year.

2. The *Survey of Current Business,* January 1982, contains the following information on the unemployment rate in the United States during 1981:

Month	Percent of civilian labor force unemployed (seasonally adjusted)
January	7.4
February	7.3
March	7.3
April	7.3
May	7.6
June	7.3
July	7.0
August	7.2
September	7.5
October	8.0
November	8.4
December	8.9

(a) Construct a bar chart showing the monthly unemployment rates.

(b) Construct an arithmetic scale graph showing the monthly unemployment rates.

3. In its *1982 Annual Report* the Commercial National Bank provided the following information on its assets:

Item	1981	1982
Cash and due from banks	$ 48,531,000	$ 45,707,000
Investments	21,393,000	31,091,000
Funds loaned	58,700,000	61,575,000
Net loans	115,853,000	142,708,000
Other assets	9,906,000	10,542,000
Total assets	$254,382,000	$291,623,000

(a) Present these data in bar chart form for each year.

(b) Present these data in pie chart form for each year.

4. Several alternative projections of natural gas demand and supply for the Ozarks region have been made. The following one for 1985 assumes a high economic growth rate and low levels of supply. Demand will be: Oklahoma—710, Missouri—564, Louisiana—2176, Kansas—530, Arkansas—530. The supply available in each state will be: Louisiana—941, Kansas—192, Missouri—147, Arkansas—114, Oklahoma—561. (All demand and supply data are in billions of cubic feet of natural gas. The source of this projection is *The Ozarks Regional Commission Regional Energy Alternatives Study: Arkansas Summary* published in August 1977.)

(*a*) Present these data in table form. Include in your table the shortage or surplus expected in each state and the region as a whole.

(*b*) Present these data in bar chart form for each state.

(*c*) Present these data in pie chart form, using one chart for demand and one for supply.

5. One of the breakdowns for gross national product is by major type of product: durable goods, nondurable goods, services, and structures. Durable goods were $458.6 billion in 1980 and $507.0 billion in 1981; nondurable goods were evaluated at $671.9 billion in 1980 and $764.2 billion in 1981; services were $1,229.6 billion in 1980 and $1,370.3 billion in 1981; and structures were $266.0 billion in 1980 and $280.7 billion in 1981. The source of these data is tables 1.3 and 1.4 on page 9 of the January 1982 issue of the *Survey of Current Business*.

(*a*) Present these data in table form.

(*b*) Present these data in a bar chart for each year.

(*c*) Present these data in a pie chart for each year.

6. On page 29 of the January 1982 issue of the *Survey of Current Business*, the following personal income data are given:

State	Personal income
Connecticut	$39,751
Maine	9,812
Massachusetts	64,146
New Hampshire	9,354
Rhode Island	9,811
Vermont	4,497

Actually, these data are for the third quarter of 1981 and are stated in millions of dollars, seasonally adjusted at annual rates. The states as shown in the publication are the New England region.

(*a*) Present these data in a bar chart.

(*b*) Use a pie chart to present the information.

(*c*) Comment on the two methods of presenting the data.

7. A small toy company recorded the following information concerning last year's operating expenses:

Item	Expenditures
Cost of products sold	$150,000
Administrative expenses	25,000
Commissions	75,000
Interest	10,000
Depreciation	60,000
Income taxes	40,000

(a) Present these data in a bar chart.

(b) Present these data in a pie chart.

(c) Comment on the two presentations.

8. Construct a bar chart of the following employment data for Washington County in 1982. The source of these data is the Bureau of Business and Economic Research, University of Arkansas.

Occupation	Number of employed
Professional and technical	8,200
Farm workers	6,800
Clerical and kindred	4,500
Sales personnel	6,000
Craftsmen and foremen	14,000
Operative workers	9,700
Household workers	200
Service workers	7,300
Laborers	5,200
Other	2,500

9. S and S News Service has recorded the following five-year earnings pattern. Present the data in a bar chart.

Year	Earnings per share
1978	$0.90
1979	0.85
1980	0.70
1981	0.75
1982	0.95

10. Convert the data in problem 8 to percentages and present these percentages in a pie chart.

11. Present the data in problem 9 on a graph.

12. The following data are for West County for 1982:

Age	Number of females	Number of males
15	100	95
16	125	130
17	130	120
18	160	150
19	190	180
20	205	200
21	200	210
22	240	225
23	250	260

(a) Construct a graph for females.

(b) Construct a graph for males.

13. The following data are federal government receipts and outlays (in $ billions) for 1974 through 1981 (estimated), as reported in the *Statistical Abstract of the United States*, 1981, pp. 247–248:

Year	Receipts	Outlays
1974	$264.9	$269.6
1975	281.0	326.2
1976	300.0	366.4
1977	367.8	402.7
1978	402.0	450.8
1979	465.9	493.6
1980	520.1	579.6
1981	607.5	655.2

(a) Construct a graph showing both receipts and outlays.

(b) Construct a bar chart with the bars for receipts and outlays for each year adjacent to one another.

14. The Commercial National Bank lists the following information concerning borrowed funds in its *1982 Annual Report:*

Year	Amount	Year	Amount
1973	$ 591,000	1978	$34,007,000
1974	1,740,000	1979	21,501,000
1975	9,650,000	1980	44,716,000
1976	12,110,000	1981	51,655,000
1977	42,655,000	1982	62,315,000

Present this information in semilog graph form.

15. The Sunflower Nursing Home began operations in 1973 and the following is its record of the number of residents the home has had:

Year	Number
1973	5
1974	12
1975	55
1976	75
1977	120
1978	125
1979	140
1980	242
1981	136
1982	248

(a) Construct an arithmetic graph with these data.
(b) Construct a semilog graph with these data.
(c) Comment on your results using the two graph forms.

16. At the end of its first year of operations, a new credit union has 60 accounts as given below:

$ 50	$300	$510	$720	$ 950	$1,250
50	350	550	750	1,000	1,250
70	370	560	750	1,000	1,260
120	380	560	800	1,000	1,270
150	390	570	800	1,000	1,280
150	400	580	800	1,010	1,290
200	500	580	810	1,050	1,300
250	500	600	820	1,080	1,350
260	500	600	820	1,090	1,380
280	500	630	830	1,090	1,400

(a) Using these data, construct a frequency distribution.
(b) Convert the frequency distribution to a "More Than" cumulative frequency distribution.

17. Use the following data on ages of women working at the Northwest Egg Breaking Plant to prepare the five presentations listed below:

43	58	21	24	31	49	40	51	55	28
50	33	62	30	25	39	59	29	36	42
38	46	42	18	50	41	37	35	40	52
47	35	57	55	50	36	45	32	45	42
36	52	64	34	32	47	41	44	39	38

(a) Present these data as an array.
(b) Prepare a frequency distribution with about seven or eight classes. (Save your work for use later.)
(c) Plot the data in part (b) as a histogram.
(d) Convert the frequency distribution to a relative frequency distribution.
(e) Convert the histogram of part (c) to a frequency polygon.

18. Use the following frequency distribution to answer the questions given below:

Data	Frequency
50 and under 60	5
60 and under 70	11
70 and under 80	18
80 and under 90	14
90 and under 100	6

(a) How many classes are there?

(b) What is the class interval?

(c) What are the class midpoints?

(d) Convert the above frequency distribution to a relative frequency distribution.

(e) Construct a histogram.

(f) Construct a frequency polygon.

(g) What percent of the data are 70 or above?

(h) What percent of the data are less than 80?

19. Use your frequency distribution in problem 17 to answer the following questions:

(a) Convert the frequency distribution to a "More Than" cumulative frequency distribution.

(b) Construct a "More Than" ogive.

(c) Convert the frequency distribution to a "Less Than" cumulative frequency distribution.

(d) Construct a "Less Than" ogive.

20. A service station owner has the following information on customer purchases of gasoline:

Gallons purchased	Frequency
5 and under 8	17
8 and under 11	25
11 and under 14	29
14 and under 17	52
17 and under 20	39
20 and under 23	28
	190

(a) What is the class interval?

(b) What are the midpoints of the first and last class?

(c) Construct a "More Than" and "Less Than" cumulative frequency distribution.

(d) How many of the customers purchased 14 or more gallons? What percentage of the customers is this?

(e) How many customers purchased at least 8 but less than 17 gallons?

21. Using the following charts, answer the following questions:

(a) What were the administrative expenses in 1980?

(b) How much was the insurance in 1980?

(c) What was the dollar increase in expenditures from 1979 to 1981?

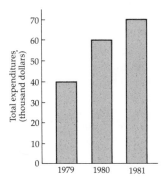 

Disposition of expenditures in 1980

(d) What was the percentage increase in expenditures from 1979 to 1981?

Case Study

Calories Unlimited, an ice cream parlor specializing in unusual flavors, has been in business for several years. Actually, the business has been very successful and its owners, Sarah and Craig Sullivan, want to expand their operation. They must convince their local banker to grant them a loan to finance the expansion and have assembled some data from their business records.

How could the following data be accurately and attractively presented in graph or chart form? Prepare the appropriate graphs and chart.

Flavors as a percent
of total ice cream sales

Flavor	Percent
Pistachio Pineapple Nut	15
Mostly Mint Chocolate Chip	26
Chunky Chewy Pecan Fudge	29
Coconut Peanut Butter Cream	12
Honey Almond Apple	10
All others	8

Average number of customers
per month

Month	Customers	Month	Customers
January	8,000	July	10,900
February	7,900	August	11,300
March	8,200	September	10,100
April	9,000	October	9,600
May	9,500	November	9,000
June	10,200	December	8,700

Relative frequency distribution
of order sizes

Class interval	Percent
Less than $0.50	4.8
$0.50 and under 1.00	20.2
1.00 and under 1.50	35.8
1.50 and under 2.00	21.4
2.00 and under 2.50	9.5
2.50 and more	8.3

Case study contributed by Zoe Teague, Interactive Data Corporation, Chicago, Ill.

2 Methods for Measuring and Describing Data

In chapter 1 we examined various ways of presenting quantitative data. As you might have guessed, however, data presentation alone does not provide much explicit information about the data under investigation. The data must be described and analyzed. It is time, then, to introduce methods that can be used to measure and to describe certain features of the data. In this chapter we are going to become familiar with certain *measures of central tendency*, which are often referred to as *averages*. We will also examine measures for describing the dispersion, or spread of data, along with ways to measure its symmetry and peakedness.

When we attempt to describe data, we usually do so using parameters. *A parameter is a general characteristic of a population.* If we do not know the population parameter, we often estimate the parameter using a statistic. *A statistic is an estimator of a population parameter obtained from a sample.*

Measures of Central Tendency

When describing a set of data, whether the set is individual data items or a frequency distribution such as the one constructed for taxes in table 1.8, a *measure of central tendency* is often given to identify the center of the data. In the following discussion, you will be introduced to five measures of central tendency sometimes referred to as averages. They are the *arithmetic mean*, the *weighted arithmetic mean*, the *median*, the *mode*, and the *geometric mean*. As you will discover, we have more than one measure of central tendency because there is more than one way to describe the "middle." Their meanings are different, and often there is a need to have more than one way to describe the "middle."

The Arithmetic Mean

Perhaps the best known average (measure of central tendency) is the *arithmetic mean*, or simply the "mean," which is the "center" of the values of a set of observations. That is, the *arithmetic mean* of a set of observations is the sum of their values divided by the number of observations. Algebraically, the equation for determining the arithmetic mean of n observations is

$$\overline{X} = \frac{\sum\limits_{i=1}^{n} X_i}{n} = \frac{X_1 + X_2 + X_3 + \cdots + X_n}{n}. \tag{2.1}$$

As a matter of accepted practice, the symbol $\overline{X}$ (pronounced X bar or bar X) will be used in this text to refer to the mean of sample data. This value, $\overline{X}$, is a statistic and is an estimate of the population mean (a parameter). A sample consists of n observations drawn from a larger group called the population whose number of observations is represented by the symbol N. For the population, the symbol representing the arithmetic mean is the Greek letter μ (mu). Thus for a population we have

$$\mu = \frac{\sum\limits_{i=1}^{N} X}{N} = \frac{X_1 + X_2 + X_3 + \cdots X_N}{N}. \tag{2.2}$$

To illustrate the computation, suppose that the ages of five students in a particular class are 19, 20, 21, 18, and 22. If these observations represent a population of five (the whole class), then the arithmetic mean (parameter) is

$$\mu = \frac{\sum\limits_{i=1}^{N} X_i}{N} = \frac{\sum\limits_{i=1}^{5} X_i}{5} = \frac{19 + 20 + 21 + 18 + 22}{5} = 20.$$

Now, let's suppose that a sample of three students is drawn from the population of five. The three ages drawn are 20, 21, and 22. Since the three ages represent a sample from the population, the arithmetic mean (statistic) is

$$\overline{X} = \frac{\sum\limits_{i=1}^{n} X_i}{n} = \frac{\sum\limits_{i=1}^{3} X_i}{3} = \frac{20 + 21 + 22}{3} = 21.$$

The arithmetic mean may be thought of as the point of balance with regard to the values of the data items. That is, when the mean is subtracted from each observation, the sum of these differences will equal 0, $\sum_{i=1}^{n}(X_i - \overline{X}) = 0$. In other words, the negative differences equal the positive ones. Additionally, if these differences are squared and then summed, $\sum_{i=1}^{n}(X_i - \overline{X})^2$, this sum is the *minimum value* that can be attained from summing the squared differences. No other number can be used in place of the mean to give a smaller sum of squared differences. For any given data set, no other number demonstrates these two properties, and in this respect the arithmetic mean is the most representative measure of central tendency.

As a method of locating the middle, the arithmetic mean has both advantages and disadvantages. The fact that a mean exists for every set of data and that every value of a variable in a set of data influences the size of the arithmetic mean are considered as advantages. Among its disadvantages is the fact that the mean can be overly influenced by extreme values of the variable. (For example, the mean of 2, 4, 6, 8, 10, and 900 is 155, although five of the six observations are values of 10 or less.) In addition, since the mean is a *calculated* number, it may not be an actual number in the data set.

Perhaps the most significant advantages of the arithmetic mean in relation to other measures of central tendency are those related to other statistical measures and analysis techniques. As you cover the material in the following chapters, you will see repeated reference to the arithmetic mean. Overall, it is used in more ways than the other measures of central tendency, and thus it may be thought of as being more popular than either the median or the mode.

The Median

A second average (measure of central tendency) is the median. It is the *positional middle* of the data. In other words, in an ordered array of the data, one-half of the values of the variable precede the median and one-half follow it.

Procedurally, the first step in calculating the median is to locate its position in the ordered array. For ungrouped data the median is located at the middle position in the array. Equation (2.3) is used to locate the median position:

$$\text{Median position} = \frac{n + 1}{2}. \tag{2.3}$$

For example, in the ordered array 10, 9, 8, 7, 6, there are 5 values. Thus the median is located at $(5 + 1)/2 = 3$. Referring to the array, the third number is 8. Therefore 8 is the median.

If the data set happens to have an even number of observations, the median is assumed to be halfway between the two central values in the ordered array. Thus in table 2.1 the median is the $(50 + 1)/2 = 25.5$th number in the array. Since the 25th value

Table 2.1
Array of state and local taxes as percent of personal income, 1978–79 data

8.6	9.6	10.6	11.2	12.7
8.9	9.8	10.7	11.5	12.7
8.9	9.9	10.7	11.6	13.1
8.9	10.0	10.8	11.7	13.4
9.0	10.1	10.9	11.8	13.5
9.2	10.1	11.0	12.0	13.7
9.4	10.2	11.1	12.0	14.0
9.4	10.4	11.1	12.1	14.0
9.4	10.4	11.1	12.2	15.6
9.5	10.5	11.2	12.5	15.8

Source: Table 1.6.

is 10.9 and the 26th is 11.0, the median is (10.9 + 11.0)/2 = 10.95.

As an average, the median is not affected by extreme values of the variable, only by the position of values. This characteristic of the median provides its primary advantage. For instance, if Charlie Harris's sales are exceptionally high compared to those of other salesmen, it is probably best to use a median in computing the average sales of the salesmen. In that way Charlie's sales will not influence the average because only those values of the variable that are near the positional middle of an ordered array influence the median.

The median is also used when you desire to know the positional middle. An illustration of this might be the average number of employees absent each day. In this case a positional average would indicate that a certain number, or more, of employees were absent every day. Furthermore, when the median is subtracted from each of the values of the variable, the sum of the absolute differences is the minimum that can be obtained. That is, $(\Sigma_{i=1}^{n}|X_i - Md| = \text{minimum})$.[1]

Like the arithmetic mean, the median is sometimes an artificial number. That is, *no value* in the set of observations might actually be the median. An added disadvantage is the fact that it cannot be used with many statistical or analytical procedures.

The Mode

A third measure of central tendency is the *mode*. By definition, the mode is the observed value that occurs most frequently. It is a

---

1. Notice that for the median the sum of the *absolute differences* is a minimum and that for the arithmetic mean the sum of the differences, *paying attention to the sign,* is 0.

measure of central tendency in that it locates the point where the variable values occur with the greatest density. For ungrouped data, the mode is determined by counting the frequency of each value and defining the mode as the value with the highest frequency of occurrence.

To illustrate, consider the following eight values: 2, 4, 5, 7, 7, 9, 10, 12. All values occur once except the 7 which occurs twice. Thus 7 is the mode of these data.

In Example 2.1, a real-world problem is presented that illustrates a situation in which the mode is a more meaningful measure of central tendency than either the arithmetic mean or the median.

Example 2.1

▷ An architect has been employed to design a series of model homes for a large housing development. One of the questions the architect must resolve is the size of the garage. Should it hold one, two, or more cars? After doing some research, he discovered the following averages with respect to the automobile ownership of the families that are in the potential market for houses of this type:

$\overline{X} = 2.25,$
$Md = 1.5,$
$Mo = 2.$

Of these three averages the mode is the one the architect decided to use. Why? First, it is rather obvious that the mean and median may not be the actual number of cars owned by any family. But more important, the mode indicates the garage size that would meet the needs of the largest numbers of families. ◁

The mode is always equal to a value of one of the observations. This characteristic is not necessarily true for the mean or the median. Thus at times the mode is the best of these three measures of central tendency.

Finally, the mode is not influenced by extreme values but rather by the frequency of occurrence. However, the mode can be located at one extreme of the data. Consequently, its use could result in misleading conclusions.

The Weighted Arithmetic Mean

When calculating the arithmetic mean using equation (2.1), each value of the variable was assumed to be of equal importance. In some situations, however, such an assumption might be misleading. In fact, if the values are not of the same importance, it is better to calculate a *weighted arithmetic mean* using equation (2.4):

$$\overline{X}_w = \frac{\displaystyle\sum_{i=1}^{n} (W_i X_i)}{\displaystyle\sum_{i=1}^{n} W_i},$$ (2.4)

where
$\overline{X}_w$ = weighted arithmetic mean,
W_i = weight for each ith value,
X_i = the ith individual values,
n = the number of values.

To illustrate, let us consider calculating the semester grade point average for a college student taking a sixteen-hour load. This sixteen-hour load is composed of four three-hour courses and two two-hour courses. Now, let us assume that our student earns the grade of A in two three-hour courses and one two-hour course, the grade of B in one two-hour course, the grade of C in a three-hour course, and a grade of D in a three-hour course. What is his semester grade point average? As you already know, it would be incorrect just to assign A = 4, B = 3, C = 2, and D = 1 to the grades and divide by the number of courses, (4 + 4 + 4 + 3 + 2 + 1)/6 = 3.000, because each course does not carry the same hour credit. Consequently, the correct procedure is to calculate the weighted arithmetic mean as follows:

$$\overline{X}_w = \frac{\displaystyle\sum_{i=1}^{6} (W_i X_i)}{\displaystyle\sum_{i=1}^{6} W_i},$$

$$\overline{X}_w = \frac{3(4) + 3(4) + 2(4) + 2(3) + 3(2) + 3(1)}{3 + 3 + 2 + 2 + 3 + 3},$$

$$\overline{X}_w = 2.9375.$$

Thus the correct grade point average is 2.9375 rather than 3.000.

The arithmetic mean of a frequency array can also be found using equation (2.4). Table 2.2 contains an illustration of the use of the equation in this manner. In making this computation, the number of banks per county is X_i, and the number of counties is W_i.

The Geometric Mean

Another average is the *geometric mean* which is the nth root of the product of n numbers. It is calculated as follows:

$$G = \sqrt[n]{X_1 \cdot X_2 \cdot X_3 \cdot \ \cdots \ \cdot X_n.}$$ (2.5)

Table 2.2
Computing the arithmetic mean of a frequency array

Number of banks (X_i)	Number of counties (W_i)	$W_i X_i$
1	9	9
2	22	44
3	12	36
4	16	64
5	5	25
6	3	18
7	5	35
8	1	8
9	1	9
10	0	0
11	1	11
	75	259

$$\bar{X}_w = \frac{\sum_{i=1}^{n} (W_i X_i)}{\sum_{i=1}^{n} W_i}$$

$$= \frac{259}{75}$$

$$\bar{X}_w = 3.453$$

Actually, this mean has a special use in business and economic problems, for frequently the situation arises where there is a need for an *average percentage change*.

Example 2.2

▷ To illustrate, let us consider the sales of Mom's Red Hot shown in table 2.3. In the recent annual report the president claimed that the sales had shown an average annual increase of 106.37 percent. This percentage was determined as follows:

$$\bar{X} = \frac{\sum_{i=1}^{3} X_i}{3} = \frac{104 + 203.8 + 311.3}{3} = 206.37,$$

$206.37 - 100 = 106.37.$

The 100 was subtracted to convert to an annual percentage increase. If the average increase of 106.37 percent is applied to the original sales figure, however, the resulting sales at the end of the four-year period would be $439,450.23.[2] You should note that

2. ($50)(2.0637) = $103.1850; ($103.1850)(2.0637) = $212.9428845; (212.9428845) (2.0637) = $439.45023; ($439.45023)(1,000) = $439,450.23.

Table 2.3
Sales of Mom's Red Hot

Year	Sales ($1,000)	Percent of previous year
1972	50	
1973	52	104
1974	106	203.8
1975	330	311.3

this is much larger than the actual sales figure of $330,000. So, what happened?

What happened is that we used the wrong measure of central tendency. We overlooked the effect of compounding. That is, we forgot that the increase also increases. If we had used the geometric mean, the problem would have been solved as follows using equation (2.5):

$$G = \sqrt[3]{104 \cdot 203.8 \cdot 311.3},$$
$$\log G = 1/3 \,(\log 104 + \log 203.8 + \log 311.3)$$
$$= 1/3 \,(2.01703 + 2.30920 + 2.49318)$$
$$= 1/3 \,(6.81941)$$
$$\log G = 2.27314,$$
$$G = 187.6\%.$$

And if this percentage is applied to the original sales data, the sales increase is as follows:

$$50 \times 1.876 = \$93.8,$$
$$93.8 \times 1.876 = 175.97,$$
$$175.97 \times 1.876 = 330.12 \cong \$330 \text{ thousand.}$$

Thus the sales averaged an annual increase of 87.6 percent (187.6 percent − 100 percent). ◁

Measures of Dispersion, Skewness, and Kurtosis

Earlier in this chapter measures of central tendency were discussed in an effort to familiarize you with averages used to describe the "middle" of the data. But central tendency doesn't tell the whole story. In describing and analyzing data, we must also consider its dispersion, skewness, and kurtosis. *Dispersion* relates to the *spread* or *variability* of the data. *Skewness* refers to the *symmetry* or lack of symmetry of the data, and *kurtosis* describes its *peakedness*. Therefore this chapter is concerned with measuring the extent to which the values of a set of observations are equal and how they are spread about a central value. To illustrate

Table 2.4
Three data sets with different amounts of dispersion

A	B	C
50	25	0
50	25	0
50	50	0
50	50	100
50	75	100
50	75	100

dispersion, consider the sets of data A, B, and C shown in table 2.4. Each set has the same arithmetic mean (50) and median (50), but they are not alike in dispersion or variability.

In set A the values of the six items are identical. There is no dispersion; they are perfectly homogeneous. In set B the items vary over three values; and in set C the items vary over two values, but there is greater spread in this set. No item in set C is close to the mean or the median, so you can see that dispersion (spread) is an important characteristic of data description.

Our discussion of dispersion will include the following measures: the *range*, the *quartiles*, the *standard deviation*, and the *coefficient of variation*. First, the range and the quartiles will be discussed.

The Range and Quartiles

The *range* of a set of data, as you probably already realize, is the difference between the highest and lowest values in the data set. In table 2.4 the range of set A is 0; the range of set B is 50; and the range of set C is 100. Thus the range is a very easy measure of dispersion to compute. It has the disadvantage, however, that it can be influenced by extreme values, thereby providing misleading information about the true spread of the majority of the data.

Consequently, the spread of data is sometimes expressed in terms of quartiles so that segments of the data can be observed. In figure 2.1 the lower or first quartile is the value Q_1 below which 25 percent of data can be found and above which 75 percent of the data lie. The second quartile (Q_2 and also the median) is the value above which and below which can be found 50 percent of the data. And finally, the third quartile is the value Q_3 above which 25 percent of the data lie and below which 75 percent of the data occur.

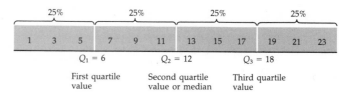

Figure 2.1   Presentation of quartiles

The quartile positions can be determined using the following equations:

$$Q_1 \text{ position} = \frac{n + 2}{4}, \qquad (2.6)$$

$$Q_2 \text{ position} = \frac{n + 1}{2}, \qquad (2.7)$$

$$Q_3 \text{ position} = \frac{3n + 2}{4}. \qquad (2.8)$$

Remember, though, that these equations identify the location and not the value of the quartile. Thus in figure 2.1 the first quartile position is

$$Q_1 \text{ position} = \frac{12 + 2}{4} = \frac{14}{4} = 3.5.$$

Numerically, Q_1 is the average of the third and fourth values or $(5 + 7)/2 = 6 = Q_1$.

Use of either the range or quartiles results in representing dispersion in terms of a few selected values. Thus this type of dispersion measure fails to make use of most of the values of the observations in the data set. To overcome this, another type of dispersion measure is available. This second type deals with measuring dispersion from one of the measures of central tendency. There are several of this second type of dispersion measure, but we will investigate only one, the *standard deviation*.

The Standard Deviation

In order to measure dispersion from some measure of central tendency, it is of course necessary to choose one of the averages. To compute the standard deviation, the average used is the arithmetic mean. If you recall our earlier discussion, one property of the arithmetic mean is that $\Sigma_{i=1}^{n}(X_i - \overline{X})^2$ is the minimum obtainable from summing the squared differences about any of the averages. And if we are dealing with a population, $\Sigma_{i=1}^{N}(X_i - \mu)^2$ is the minimum also. Now if we take this minimum sum of squared differences, divide it by the number of observations, and then take the square root, we have calculated the standard deviation.

For a population the standard deviation σ (pronounced sigma) is

$$\sigma = \sqrt{\dfrac{\sum\limits_{i=1}^{N}(X_i - \mu)^2}{N}}, \tag{2.9}$$

where
μ = the population mean,
N = the number of items in the population,
X_i = the *i*th individual values.

If you should happen to be working with sample data, and you want the sample standard deviation to provide a good estimate of the population standard deviation, use equation (2.10):

$$s = \sqrt{\dfrac{\sum\limits_{i=1}^{n}(X_i - \overline{X})^2}{n - 1}}. \tag{2.10}$$

There are two differences in equation (2.9) and (2.10). First, $\overline{X}$ is used rather than μ. Second, the denominator in the equation for s is $n - 1$, while for σ it is N. This second difference looks like a small one, but it warrants an additional explanation.

The earlier discussion of the arithmetic mean mentions that the sum of the variations about the mean is 0 or $\sum_{i=1}^{n}(X_i - \overline{X}) = 0$. With this in mind, suppose you have five numbers, 2, 4, 6, 8, and 10. Their arithmetic mean is 6. If you select any four of them, the value of the fifth number is fixed. In other words, it is not free to vary. For example, if you select the first four numbers (2, 4, 6, and 8), the fifth number *must* be 10 if $\sum_{i=1}^{n}(X_i - \overline{X}) = 0$:

$$(2 - 6) + (4 - 6) + (6 - 6) + (8 - 6) + (X - 6) = 0$$
$$-4 \quad - \quad 2 \quad + \quad 0 \quad + \quad 2 \quad + (X - 6) = 0$$
$$-4 \quad + (X - 6) = 0$$
$$X - 10 = 0$$
$$X = 10.$$

In computing s from a sample containing n values of the variable X, you have only $n - 1$ *degrees of freedom*. Put in a more general form, *degrees of freedom is the number of values that can vary freely in a given set of variables.*

As you can see from equations (2.9) and (2.10), the standard deviation is the square root of the mean of the squared deviations. Thus it is sometimes referred to as the root mean square deviation. If we remove the radical, that is, leave the result as σ^2 or s^2, we have what is known as the *variance*. The *variance* is the *standard deviation squared*, or the *standard deviation* is the *square root* of the *variance*.

Shown in table 2.5 are the necessary computations for the determination of s. Notice that the arithmetic mean $\overline{X}$ is first determined to be $500, and then $500 is subtracted from each sales observation X_i to find the individual deviations from the mean $(X_i - \overline{X})$ shown in the second column. These deviations are then squared and summed and placed in equation (2.10), which yields $s = \$63.25$.

Consistent with the definition of the arithmetic mean, the sum of these variations is 0. These deviations are then squared and summed. Next, the sum of the squared deviations is divided by the degrees of freedom $(n - 1)$ to determine the variance (4,000). The square root of the variance is $63.25 and is the sample standard deviation s. Notice that the variance is in terms of "dollars squared" or, in other words, it is not in the same units as the original data. The standard deviation, however, is in the same units as the original data and is thus a more practical measure of variation than is the variance.

It has probably already occurred to you that, if there were 100 sales figures rather than six as shown in table 2.5, then a much greater computational effort would be necessary to calculate the

Table 2.5
Computation of the standard deviation (long method)

Daily sales for Market's market

Sales (X_i)	$(X_i - \overline{X})$	$(X_i - \overline{X})^2$
$ 525	+ 25	625
475	− 25	625
425	− 75	5,625
450	− 50	2,500
525	+ 25	625
600	+100	10,000
$3,000	0	20,000

$$\overline{X} = \frac{\sum_{i=1}^{n} X_i}{n} \qquad s = \sqrt{\frac{\sum_{i=1}^{n} (X_i - \overline{X})^2}{n - 1}}$$

$$\overline{X} = \frac{\$3,000}{6} \qquad s = \sqrt{\frac{20,000}{6 - 1}}$$

$$\overline{X} = \$500 \qquad s = \sqrt{4,000}$$

$$s = \$63.25$$

standard deviation. Your observation would be correct, and you would have identified a need for a shortcut procedure. Equation (2.11) is such a shortcut method for entire populations, and equation (2.12) is the shortcut method when using sample data:[3]

$$\sigma = \sqrt{\frac{\sum_{i=1}^{N} X_i^2}{N} - \left(\frac{\sum_{i=1}^{N} X_i}{N}\right)^2} \qquad (2.11)$$

$$s = \sqrt{\frac{\sum_{i=1}^{n} X_i^2 - \frac{\left(\sum_{i=1}^{n} X_i\right)^2}{n}}{n - 1}} \qquad (2.12)$$

To illustrate how much easier it is to use equation (2.12) rather than equation (2.10), take a look at table 2.6. Now compare the work involved against the effort necessary in table 2.5. Don't you agree that the shortcut method is easier?

Table 2.6
Computation of the standard deviation (short method)

Market's market

Sales (X_i)	X_i^2
$ 525	275,625
475	225,625
425	180,625
450	202,500
525	275,625
600	360,000
$3,000	1,520,000

$$s = \sqrt{\frac{\sum_{i=1}^{n} X_i^2 - \frac{\left(\sum_{i=1}^{n} X_i\right)^2}{n}}{n - 1}}$$

$$s = \sqrt{\frac{1,520,000 - \frac{(3,000)^2}{6}}{6 - 1}} \qquad s = \sqrt{\frac{1,520,000 - 1,500,000}{5}}$$

$$s = \sqrt{\frac{20,000}{5}} \qquad s = \sqrt{4,000} \qquad s = \$63.25$$

3. Both of these so-called shortcut equations are really mathematical derivations of the so-called long-method equations. Therefore the solutions from equations (2.9) and (2.11) are equal. The same is true of equations (2.10) and (2.12).

Coefficient of Variation

Up until now two types of measures of dispersion have been discussed. The first type was of the nature of the range and quartiles where only a few observations are used, and the second type was of the nature of the standard deviation (or variance) where all observations are used to measure dispersion about the arithmetic mean. The third and last type of measure of dispersion we will discuss shows *relative dispersion;* the particular one we are concerned with is called the *coefficient of variation.*

An *absolute measure* of dispersion (for example, the standard deviation) is converted to a relative measure (the coefficient of variation) by stating the absolute measure as a percent of a measure of central tendency, usually the arithmetic mean. Equations (2.13) and (2.14) are used to calculate the coefficient of variation, which is the ratio of the standard deviation to the mean:

$$V = \left(\frac{s}{\overline{X}}\right) 100, \tag{2.13}$$

$$V = \left(\frac{\sigma}{\mu}\right) 100. \tag{2.14}$$

Since the coefficient of variation expresses the standard deviation as a percentage of the arithmetic mean, it can be used to compare the dispersion of data series whose values are substantially different in size. This coefficient is also useful in comparing the dispersion of data expressed in different units. These in fact are the important characteristics of any measure of relative dispersion. The following example illustrates the use of this statistical technique.

Example 2.3

▷ The Big Pickle Company has been in operation for several years. Up until twelve months ago average daily output of pickles was 126 cases, with a standard deviation (σ) of 14.97 cases. Last year the owners of the company invested additional capital in plant and equipment. Consequently, over the past twelve months the mean (μ) daily output increased to 150 cases, but the standard deviation of daily output rose to 16.00 cases. Since it was hoped that the added investment would cause the level of output to be more consistent, as well as larger, the production manager feels he may have a problem and undertakes the analysis shown in table 2.7.

After completing the calculations shown in table 2.7, the production manager realized that the capital expansion had in fact achieved its objective. Even though the standard deviation in-

Table 2.7
Calculation of a coefficient of variation

Period	μ	σ	$V = \left(\dfrac{\sigma}{\mu}\right)100$
Before expansion	126	14.97	11.9
After expansion	150	16.00	10.7

creased from 14.97 cases to 16.00 cases, *relative variation* actually decreased. This would be good news to the production manager for daily output is now both relatively more homogeneous as well as greater than it was prior to the expansion. ◁

Skewness and the Relationship of the Mean, Median, and Mode

Skewness refers to the shape of a frequency distribution; specifically, skewness can be related to symmetry. If a frequency distribution is bell shaped as shown in figure 2.2, it is said to be symmetrical. That is, 50 percent of the values lie on each side of the mode. Figure 2.2 is also a *unimodal* distribution. That is, it has a single mode.

Figure 2.2 demonstrates another property of the symmetrical distribution. In addition to being unimodal, the median and the mean are equal to the mode. Now turning to figure 2.3, we see the relationship of the mean, median, and mode for *skewed* distributions. If the distribution is skewed right, the value of the mode is less than the median, which is less than the mean. But if the distribution is skewed left, the value of the mode is greater than the median, which is greater than the mean. In both situations the median value lies between the value of the mean and the mode. Also the mean is always in the direction of the skewness relative to the mode. That is, if the data are skewed right, the

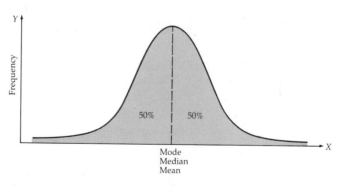

Figure 2.2   Symmetrical frequency distribution

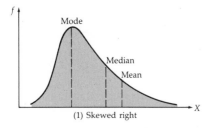

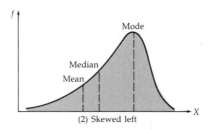

Figure 2.3   Skewed frequency distributions

extreme values are to the right, so the mean is to the right of, or greater than, the mode. The converse is true for data that are skewed left.

There are several methods available for measuring skewness. The ones presented here, in equations (2.15) and (2.16), are known as the *Pearsonian coefficient of skewness:*

$$Sk = \frac{3(\overline{X} - Md)}{s}, \tag{2.15}$$

$$Sk = \frac{3(\mu - Md)}{\sigma}. \tag{2.16}$$

The Pearsonian coefficient of skewness ranges from -3 to $+3$, with $Sk = 0$ meaning a perfectly symmetrical distribution.[4] A coefficient with a positive sign indicates the data are skewed to the right, while a negative sign indicates the data are skewed to the left.

The use of this coefficient is illustrated in example 2.4.

Example 2.4

▷ The Big Pickle production manager has been evaluating the daily output data for his plant and is beginning to suspect he has a problem with production. While he knows that the mean (u) daily output recently was 150 cases, he feels this may be the result of having a few days of high productivity, with output many other days being lower than the mean. Since he also knows that the standard deviation of daily output is 16.00 cases and the median daily output is 138 cases, he decides to calculate a measure of skewness:

---

4. Equations (2.15) and (2.16) are based on the observation that the median will lie approximately one-third of the way between the arithmetic mean and the mode. Since this is not always true, you should refer to a more advanced text if you need a more precise measure of skewness.

BASIC STATISTICS

$$Sk = \frac{3(150 - 138)}{16.00}$$

$$= \frac{3(+12)}{16.00} = \frac{+36}{16.00},$$

$$Sk = +2.25.$$

With a Pearsonian coefficient of skewness of +2.25, the plant manager has his suspicions confirmed. Daily output is indeed positively skewed by a relatively large amount. Apparently plant output is high only a relatively few days, and on a larger number of days it is less than the arithmetic mean output. ◁

Kurtosis

The extent to which the data contained in a distribution tend to be concentrated at its midpoint is referred to as the *kurtosis* of the distribution. Thus kurtosis is another term for *peakedness*.

In figure 2.4 three distributions are shown, each with different amounts of kurtosis. The curve labeled A is called *leptokurtic* and represents a distribution with values concentrated around its midpoint. The second, curve B, is *mesokurtic*. It has intermediate kurtosis but less concentration at the midpoint than does curve A. Finally, curve C is relatively flat and is called *platykurtic*. In other words, the values in the distribution are relatively uniformly scattered so that there is little peakedness.

Grouped Data

You have learned how to calculate the arithmetic mean, median, mode, quartiles, and standard deviation when you know each individual value in a data set. However, sometimes you do not know each individual value; instead, your data set is in the form of a frequency distribution or grouped data. Should this be the

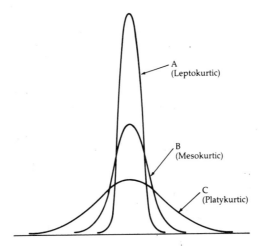

Figure 2.4   Three distributions with differing amounts of kurtosis

case, it is frequently possible to calculate the arithmetic mean, median, quartiles, and standard deviation if you are willing to make certain assumptions concerning the frequency distribution. Let's look at how these calculations can be made. At the same time we'll examine when, and with what assumptions, they can be made.

It is possible to compute an arithmetic mean of a data set presented as a frequency distribution *if you assume that the midpoint of each class is representative of all the values in that class.* If that assumption is made, the following equations may be used to compute the arithmetic mean of a frequency distribution:[5]

$$\mu = \frac{\Sigma fx}{N} \, , \tag{2.17}$$

$$\overline{X} = \frac{\Sigma fx}{\Sigma f} \, , \tag{2.18}$$

where f = frequency of each class,
$\quad\quad x$ = midpoint of each class.

Since you must know the midpoint (x) of each class in order to use equations (2.17) and (2.18), it follows that this equation cannot be used if a frequency distribution has an open end class.

Equations (2.17) and (2.18) are also equations for calculating a weighted arithmetic mean. If you examine them closely, you will see they are identical with equation (2.4). Thus the arithmetic mean of a frequency distribution is simply a weighted arithmetic mean of the class midpoints.

The median of a frequency distribution can be determined *if you assume that the values of the variable falling in the median class are evenly spaced throughout the class.* (The *median class* is the class containing the median.) In that case equation (2.19) can be used:

$$Md = L + i\left(\frac{n/2 - F}{f_{md}}\right) , \tag{2.19}$$

where
$\quad L$ = the lower boundary of the class in which the median is located,
$\quad i$ = the class interval,
$\quad n$ = the number of values of the variable in the distribution,
$\quad F$ = the cumulated frequency of the classes preceding the median class,

---

5. The subscripts for these equations and others that follow in this chapter have been omitted for simplicity of presentation.

f_{md} = the frequency of the median class (the median class is the class where the cumulative frequency is equal to, or exceeds for the first time, $n/2$).

In determining the quartile values of a frequency distribution, essentially the same assumptions must be made as was used in determining the median. That is, *the values of the variable are evenly spaced in each quartile class.* With this assumption in mind, the following equations can be utilized:[6]

$$Q_1 = L + i\left(\frac{n/4 - F}{f_{Q_1}}\right), \tag{2.20}$$

$$Q_2 = L + i\left(\frac{n/2 - F}{f_{Q_2}}\right), \tag{2.21}$$

$$Q_3 = L + i\left(\frac{3n/4 - F}{f_{Q_3}}\right), \tag{2.22}$$

where

L = the lower boundary of the class in which the quartile is located,

n = number of observations,

F = the cumulative frequencies of the classes preceding the quartile class,

f_Q = the frequency of the quartile class,

i = the class interval.

Calculating the standard deviation of grouped data requires the same assumption as does computing the arithmetic mean since it is a measure of dispersion about the mean. That is, the midpoint of each class is representative of all the values in that class. Equations (2.23) and (2.24) can be used to compute the standard deviation of a population, whereas equations (2.25) and (2.26) are appropriate for calculating the standard deviation of a sample. Finally, you may find equations (2.24) and (2.26) somewhat easier to use than equations (2.23) and (2.25):

$$\sigma = \sqrt{\frac{\Sigma f(x - \mu)^2}{N}}, \tag{2.23}$$

$$\sigma = \sqrt{\frac{\Sigma fx^2 - \frac{(\Sigma fx)^2}{N}}{N}}, \tag{2.24}$$

---

6. Notice that equation (2.21) is the same as the equation to determine the median.

$$s = \sqrt{\frac{\Sigma f(x - \overline{X})^2}{n - 1}}, \tag{2.25}$$

$$s = \sqrt{\frac{\Sigma fx^2 - \frac{(\Sigma fx)^2}{n}}{n - 1}}, \tag{2.26}$$

where

x = midpoint of each class,
f = frequency of each class,
N = total number of observations in the population,
n = total number of observations in the sample.

Example 2.5 provides an illustration that includes the calculation of all the measures of central tendency and dispersion discussed in this section. Examine it carefully, and you will see how the various equations can be utilized.

Example 2.5

▷ The Executive Committee of National Credit Card has for some time been concerned with lost production time associated with employee absenteeism. The company has several hundred employees, and on any given day many of them are absent from their jobs. Because of the serious nature of the problem the committee requested that the Personnel Department prepare a report that would answer the following three questions:

(*a*) What is the average number of employees absent each day?

(*b*) Are there large differences in the numbers of employees absent from day to day?

(*c*) Could a few days in which there are an extremely large number of employees absent be causing the average absentee level to be what it is?

The Personnel Manager, upon receiving the request, prepared a study of the number of employees absent using a sample of 365 days. The results of his study, along with the analysis he performed, are given in table 2.8. His answers to each question the Executive Committee asked are presented below:

(*a*) The average number of employees absent is either 116.11 or 114.71.

(*b*) There are no less than 75 nor more than 165 employees absent on any given day. The relative variability in absenteeism is less than 15 percent. On 50 percent of the days between 104.01 and 127.75 employees miss work.

(c) There is only a relatively small impact on the arithmetic mean absentee level. ◁

Look at table 2.8 closely and see why the Personnel Manager answered each question in the way that he did.

Table 2.8
Employee absenteeism in National Credit Card Company

Number of employees	Frequency f	Class midpoint x	fx	fx^2
75 and under 85	10	80	800	64,000
85 and under 95	20	90	1,800	162,000
95 and under 105	68	100	6,800	680,000
105 and under 115	87	110	9,570	1,052,700
115 and under 125	75	120	9,000	1,080,000
125 and under 135	50	130	6,500	845,000
135 and under 145	39	140	5,460	764,400
145 and under 155	11	150	1,650	247,500
155 and under 165	5	160	800	128,000
	365		42,380	5,023,600

$$\overline{X} = \frac{\Sigma fx}{\Sigma f}$$

$$\overline{X} = \frac{42,380}{365}$$

$$\overline{X} = 116.11$$

$$Md = L + i\left(\frac{n/2 - F}{f_{md}}\right)$$

$$Md = 105 + 10\left(\frac{365/2 - 98}{87}\right)$$

$$= 105 + 10\left(\frac{84.5}{87}\right)$$

$$Md = 114.71$$

$$s = \sqrt{\frac{\Sigma fx^2 - \frac{(\Sigma fx)^2}{n}}{n - 1}}$$

$$s = \sqrt{\frac{5,023,600 - (42,380)^2/365}{365 - 1}}$$

$$= \sqrt{\frac{5,023,600 - 4,920,724.38}{364}}$$

$$= \sqrt{282.63}$$

$$s = 16.81$$

$$Q_1 = L + i\left(\frac{n/4 - F}{f_{Q_1}}\right)$$

$$Q_1 = 95 + 10\left(\frac{365/4 - 30}{68}\right)$$

$$= 95 + 10\left(\frac{61.25}{68}\right)$$

$$Q_1 = 104.01$$

$$Q_2 = L + i\left(\frac{n/2 - F}{f_{Q_2}}\right)$$

$$Q_2 = 105 + 10\left(\frac{365/2 - 98}{87}\right)$$

$$= 105 + 10\left(\frac{84.5}{87}\right)$$

$$Q_2 = 114.71$$

$$Q_3 = L + i\left(\frac{3n/4 - F}{f_{Q_3}}\right)$$

$$Q_3 = 125 + 10\left(\frac{3(365)/4 - 260}{50}\right)$$

$$= 125 + 10\left(\frac{13.75}{50}\right)$$

$$Q_3 = 127.75$$

$$V = \frac{s}{\overline{X}}(100)$$

$$V = \frac{16.81}{116.11}(100)$$

$$V = 14.48$$

$$Sk = \frac{3(\overline{X} - Md)}{s}$$

$$Sk = \frac{3(116.11 - 114.71)}{16.81}$$

$$Sk = +0.25$$

Summary of Equations

2.1   Arithmetic mean of a sample:

$$\overline{X} = \frac{\sum\limits_{i=1}^{n} X_i}{n}$$

2.2   Arithmetic mean of a population:

$$\mu = \frac{\sum\limits_{i=1}^{N} X_i}{N}$$

2.3   Location of the median:

$$\text{Median position} = \frac{n + 1}{2}$$

2.4   Weighted arithmetic mean of a sample:

$$\overline{X}_W = \frac{\sum\limits_{i=1}^{n} (W_i X_i)}{\sum\limits_{i=1}^{n} W_i}$$

2.5   Geometric mean of a sample:

$$G = \sqrt[n]{X_1 \cdot X_2 \cdot X_3 \cdot \ \cdots \ \cdot X_n}$$

2.6   Location of the first quartile:

$$Q_1 \text{ position} = \frac{n + 2}{4}$$

2.7   Location of the second quartile (median):

$$Q_2 \text{ position} = \frac{n + 1}{2}$$

2.8   Location of the third quartile:

$$Q_3 \text{ position} = \frac{3n + 2}{4}$$

2.9   Standard deviation of a population:

$$\sigma = \sqrt{\frac{\sum\limits_{i=1}^{N} (X_i - \mu)^2}{N}}$$

2.10   An estimate of σ using sample data:

$$s = \sqrt{\frac{\sum\limits_{i=1}^{n}(X_i - \overline{X})^2}{n-1}}$$

2.11   Standard deviation of a population—a shortcut method:

$$\sigma = \sqrt{\frac{\sum\limits_{i=1}^{N} X_i^2}{N} - \left(\frac{\sum\limits_{i=1}^{N} X_i}{N}\right)^2}$$

2.12   An estimate of σ using sample data—a shortcut method:

$$s = \sqrt{\frac{\sum\limits_{i=1}^{n} X_i^2 - \dfrac{\left(\sum\limits_{i=1}^{n} X_i\right)^2}{n}}{n-1}}$$

2.13   Coefficient of variation of a sample:

$$V = \left(\frac{s}{\overline{X}}\right)100$$

2.14   Coefficient of variation of a population:

$$V = \left(\frac{\sigma}{\mu}\right)100$$

2.15   Pearsonian coefficient of skewness of a sample:

$$Sk = \frac{3(\overline{X} - Md)}{s}$$

2.16   Pearsonian coefficient of skewness of a population:

$$Sk = \frac{3(\mu - Md)}{\sigma}$$

2.17   Arithmetic mean of a population—grouped data:

$$\mu = \frac{\Sigma fx}{N}$$

2.18 Arithmetic mean of a sample—grouped data:

$$\bar{X} = \frac{\Sigma fx}{\Sigma f}$$

2.19 Median—grouped data:

$$Md = L + i\left(\frac{n/2 - F}{f_{md}}\right)$$

2.20 Value of the first quartile—grouped data:

$$Q_1 = L + i\left(\frac{n/4 - F}{f_{Q_1}}\right)$$

2.21 Value of the second quartile (median)—grouped data:

$$Q_2 = L + i\left(\frac{n/2 - F}{f_{Q_2}}\right)$$

2.22 Value of the third quartile—grouped data:

$$Q_3 = L + i\left(\frac{3n/4 - F}{f_{Q_3}}\right)$$

2.23 Standard deviation of a population—grouped data:

$$\sigma = \sqrt{\frac{\Sigma f(x - \mu)^2}{N}}$$

2.24 Standard deviation of a population—grouped data, shortcut method:

$$\sigma = \sqrt{\frac{\Sigma fx^2 - \frac{(\Sigma fx)^2}{N}}{N}}$$

2.25 An estimate of σ using grouped sample data:

$$s = \sqrt{\frac{\Sigma f(x - \bar{X})^2}{n - 1}}$$

2.26   An estimate of σ using grouped sample data—shortcut method:

$$s = \sqrt{\frac{\Sigma fx^2 - \dfrac{(\Sigma fx)^2}{n}}{n-1}}$$

Review Questions

1. Define the terms *parameter* and *statistic,* and distinguish between them.

2. As measures of central tendency, what are the advantages and disadvantages of the

(a) arithmetic mean
(b) median

3. Define

(a) arithmetic mean   *(e)* quartiles
(b) median   *(f)* standard deviation
(c) mode   *(g)* skewness
(d) range   *(h)* kurtosis

4. Distinguish between the arithmetic mean and the weighted arithmetic mean.

5. Why is the coefficient of variation useful?

6. The arithmetic mean of grouped data is simply a weighted arithmetic mean of class midpoints. Comment.

7. Explain the significance of the assumption that the items in the median class are evenly spaced when calculating the median from grouped data.

Problems

1. The Underhanded Company has developed, and is about to market, a new "long-life" light bulb. They would like to make a statement about the "average life" of their bulbs. Since Underhanded is mostly short on everything, including money, they have decided to base any statement they make on a sample of 10 bulbs. As Underhanded's statistician you are given the following data on the "life," in hours, of the 10 bulbs:

106,   99,   78,   110,   97,   82,   15,   86,   108,   97.

Compute the:

(a) arithmetic mean
(b) median
(c) mode
(d) standard deviation

2. A fruit market is considering wrapping their onions in packages. They have decided to make the average number purchased by the next six customers the standard package size. The six purchases are as follows: 2, 4, 3, 10, 4, and 7. What is the number of onions which should be placed in a package if the average used is the

(a) arithmetic mean
(b) median
(c) mode

3. A recent driver's education class took their driving exam and scored as follows: 98, 63, 92, 88, 92, 70, 98, 92, 75, and 70. For the exam scores, compute the

(a) arithmetic mean
(b) median
(c) mode
(d) standard deviation
(e) range

4. In a recent survey a sample of consumers was asked how many bottles of Burp Cola they drink each week. The following answers were given by 11 respondents included in the sample:

2, 0, 36, 4, 2, 6, 5, 8, 1, 10, 3.

(a) Compute the arithmetic mean.
(b) Find the mode.
(c) Determine the median.
(d) Compute the standard deviation.

5. A sample of 10 workers in Walnut Ridge was asked how many miles they drove to work each day with the following results:

25, 8, 1, 2, 3, 7, 4, 3, 9, 5.

(a) Compute the arithmetic mean.
(b) Compute the median.
(c) Compute Q_3.
(d) What is the range?
(e) Compute the standard deviation (s).
(f) What is the mode?
(g) Comment on the use of the arithmetic mean in this instance.

6. The following data represent the monthly electric utility charges for a local resident for one year. Compute the arithmetic mean, median, and variance for these charges.

$63, $24, $80, $175, $90, $150,
$32, $39, $79, $102, $103, $85.

7. Listed are the number of overtime hours per worker last week at Rooster's Heaven Poultry Company. Determine the arithmetic mean, median, and mode of the number of overtime hours. What can you tell management about the average amount of overtime?

2	6	8	10	5
5	6	2	3	3
5	8	2	3	6
9	5	4	6	10
2	6	6	5	1
1	6	9	3	6

8. Sun Dial Manufacturing Co. has determined that the average number of overtime hours per employee per year was 300 for the previous year. The company currently has 25 employees. Based on a 40-hour week and a 50-week operating year, does Sun Dial need any additional employees? If so, how many? What assumption did you make?

9. Mr. Rent owns seven houses, which are rented on a monthly basis. If he receives monthly rents of $125, $150, $275, $100, $110, $200, and $80, what is

(a) the mean monthly rent?
(b) the median monthly rent?
(c) the modal monthly rent?

10. Fast Sam is a senior finance major at Quick Buck College. Sam is working his way through his last semester by selling Sanskrit dictionaries between classes. In order to stay in school, Sam must average 24 sales per day. His last 24 days' sales are listed below. Do you think he will make it? Find the mean, median, and mode to help you make up your mind.

34	50	32	42	14	26	33	34	11	14	5
38	5	39	7	13	26	26	23	1	37	2
37	47									

11. In the small community of Robinsonville there are eleven families—Mr. Robinson, and his ten employees. The following is a list of the family incomes in the community last year: $5,000, $7,600, $5,500, $9,000, $7,000, $100,000, $7,800, $6,250, $10,000, $15,000, and $8,500.

(a) Determine the best measure of central tendency to represent the average yearly family income of Robinsonville.
(b) Calculate the value of this best measure.

12. The Dean of Students wants to include a statement in the University catalog concerning the cost of off-campus housing. Since he cannot get this information from every student living off-campus, he decided to ask 25 students how much they spend and use their average expenditure in his statement. The following data are the monthly expenditures reported by this group of students:

23	33	76	90	38
25	40	61	92	54
27	59	45	63	71
27	73	51	52	93
29	93	62	37	34

Help the Dean by determining the
(a) arithmetic mean
(b) median
(c) mode
(d) standard deviation
(e) coefficient of variation
(f) Q_1
(g) Q_3
(h) range

13. Suppose you have $13,000 to invest, and are considering two types of investments, A and B. Your investment counselor gives you the following information based on the past experience of other clients:

	A	B
Arithmetic mean value after 10 years	$18,000	$22,350
Standard deviation	1,100	3,200

Under what criteria would you opt for A? For B?

14. An examination of ice cream prices (per gallon) in this SMSA yielded the following results:

Price	Number of brands
$2.10 but less than $2.20	1
$2.20 but less than $2.30	5
$2.30 but less than $2.40	20
$2.40 but less than $2.50	10
$2.50 but less than $2.60	5
$2.60 but less than $2.70	4

Compute the following assuming the data are a sample:

(a) arithmetic mean
(b) median
(c) variance

15. The following is a tabulation of the number of drownings, by age of the victim, in hotel swimming pools in 1977:

Age of victim	Number of deaths
Less than 5	16
5 but less than 10	9
10 but less than 15	11
15 but less than 20	7
20 but less than 25	5
25 but less than 30	8
30 but less than 35	4

(a) Determine the median age of the victims.
(b) Compute the arithmetic mean of the ages.
(c) Compute the variance of the ages.
(d) Compute the coefficient of variation.
(e) Determine both Q_1 and Q_3.

16. From a sample ($n = 100$) of the records of hospital patients, the following frequency distribution was developed:

Length of stay (days)	Number of patients
1 but less than 2	20
2 but less than 3	25
3 but less than 4	30
4 but less than 5	10
5 but less than 6	8
6 or more	7
	100

Note: The total number of "patient days" in the last class (6 or more) is 50.

(a) Compute the arithmetic mean of the length of stay.
(b) Compute the median for the length of stay.
(c) Compute the standard deviation of the length of stay.

17.

Price of used textbooks	Number
$0 but less than $5.00	10
$5.00 but less than $10.00	35
$10.00 but less than $15.00	50
$15.00 but less than $20.00	30
$20.00 but less than $25.00	20
$25.00 but less than $30.00	5

Given the above sample information, compute the following:
(a) arithmetic mean
(b) median
(c) variance
(d) Pearsonian coefficient of skewness

18. A recent study of television viewing habits produced, in part, the following data. Compute the coefficient of variation and the coefficient of skewness for these data.

Number of hours viewing per day	Frequency
0 but less than 2	10
2 but less than 4	38
4 but less than 6	34
6 but less than 8	10
8 but less than 10	7
10 but less than 12	1

19. A sample of 25 charge customers of Deferred Payments Unlimited revealed the following frequency distribution of unpaid balances:

Unpaid balance	Frequency
$30 and under 40	3
40 and under 50	8
50 and under 60	7
60 and under 70	4
70 and under 80	3

(a) Determine the mean unpaid balance.
(b) Determine the median unpaid balance.

20. At the annual convention of Life Insurance, Inc., the following distribution of whole life insurance policies was presented:

Dollar value of policy	Number sold
$ 0 and under 10,000	15
10,000 and under 20,000	18
20,000 and under 30,000	30
30,000 and under 40,000	10
40,000 and under 50,000	7
50,000 and over[a]	20

[a]The average value of each policy in this group is $90,000.

(a) Estimate the total dollar value of whole life policies sold.
(b) Calculate the mean value of each policy.
(c) Calculate the median value of the policies.

21. Hazel bought 10 pounds of sugar for 40¢ per pound, 5 pounds of potatoes for 15¢ per pound, 3 pounds of tomatoes for 60¢ per pound, and 8 pounds of bananas for 30¢ per pound. What is the average price per pound of her purchases?

22. Mr. Logan has a large apartment complex which has one-, two-, and three-bedroom apartments. The one bedroom rents for $135 per month; the two bedroom rents for $205 per month; and the three bedroom rents for $265 per month. If there are 50 one-bedroom apartments, 40 two-bedroom apartments, and 10 three-bedroom apartments, what is the mean monthly rental of the apartments?

23. At a major state university assistant professors have an arithmetic mean salary of $15,000 with a standard deviation of $5,000, while professors have an arithmetic mean salary of $40,000 with a standard deviation of $10,000. Which group has the greater relative dispersion? What is it?

24. Two growers of grapefruit have determined the following statistics for their current crops.

Grower M   Grower N
$\overline{X}$ = 15 ounces $\overline{X}$ = 14 ounces
x = 1 ounce s = 2 ounces

(a) Which grower has grapefruit that are more uniform in weight?
(b) Which grower probably has the larger grapefruit?

25.

Classes	Frequency
5 and under 15	6
15 and under 25	10
25 and under 35	24
35 and under 45	12
45 and under 55	8
	60

(a) Determine the arithmetic mean.
(b) Determine the standard deviation.
(c) Determine the variance.
(d) Determine the first quartile value.
(e) Determine the second quartile value.
(f) Determine the third quartile value.
(g) Determine the coefficient of variation.
(h) Compute the Pearsonian coefficient of skewness.

26. A southern state established public school kindergartens six years ago. In successive years enrollment grew 50, 30, 20, 10, and 5 percent. Using the geometric mean, what is the average annual rate of growth?

27. Mr. and Mrs. Savings have four savings accounts. In account A they have $4,000; in account B they have $6,000; in account C they have $3,000; and in account D they have $10,000. If account A earns 5½ percent per year, account B earns 4 percent per year, account C earns 6 percent per year, and account D earns 7 percent per year, what is the average annual percent earnings of the Savings?

28. Over five years State University recorded the following enrollments:

Year	Enrollment
1974	20,000
1975	22,500
1976	24,600
1977	27,500
1978	29,000

(a) What has been the average annual rate of increase in enrollment?
(b) If the percentage increase had been maintained exactly each year, what would have been the enrollments for each year?

29. An organization has the following distribution of ages for its current employees:

Ages	Number of employees
18 and under 28	2
28 and under 38	10
38 and under 48	30
48 and under 58	25
58 and under 68	17

(a) Determine the first quartile value.
(b) Determine the third quartile value.
(c) Comment on the age situation in this organization.
(d) Compute the Pearsonian coefficient of skewness.

30. The following frequency distribution was compiled from the records of Feline, Unlimited, a local firm providing cat care:

Number of daily customers	Frequency
0 and under 5	6
5 and under 10	10
10 and under 15	25
15 and under 20	35
20 and under 25	20
25 and under 30	13

Compute:
(a) the arithmetic mean for the number of daily customers
(b) the median number of daily customers
(c) the standard deviation of the number of daily customers
(d) the coefficient of skewness for the distribution

Case Study

One year ago the Evan Wilson Company employed a Chicago advertising consultant to help increase the firm's average daily sales. The consultant, in a recent report, cited the following results:

	Last year	Current year
Total sales	$2,500,000	$3,000,000
Number of business days	250	252
Coefficient of skewness	−0.5	+2.0
Coefficient of variation	0.2	0.8

Evan, the owner, seems pleased since, as he puts it, "All the numbers for the current year are larger." Gwen, his wife, argues that the consultant should be fired for according to her, "If things keep going as they did this past year, Wilson's might as well not open for business about half the time."

What do you think the consultant's numbers show? How could Gwen and Evan reach such opposing conclusions using the same data?

Case study contributed by Zoe Teague, Interactive Data Corporation, Chicago, Ill.

3 Elementary Probability Theory

Before a football game, a coin is flipped to determine which team will kick off and which team will receive the football. What is going to turn up—a head, a tail? Neither we nor the teams involved can be sure until the coin is flipped and the outcome is observed. We do know, however, that it will be a head or a tail, assuming the coin cannot stand on edge. Moreover, if the coin can be assumed to be perfectly balanced in weight so that neither side is more likely to be the up side, it would seem reasonable to conclude that either side is equally likely to appear. That is, the head should occur half of the time, and the tail should occur the other half of the time, thereby giving each team an equal chance of winning the coin toss.

In the preceding paragraph, the word *likely* appeared twice and the word *chance* once with reference to the possible occurrence of a future event. Such situations where chance and likelihood are discussed represent situations where the conclusions are often stated in terms of *probability*. In fact, probability stands as the foundation upon which statistical analysis is built. Later we will define probability more specifically, but for the moment let us accept "likelihood" as a reasonable synonym for it.

For instance, one might assert, "There is a strong likelihood that it will rain" or, "There is a better chance that it will rain than that it will not rain." Yet either statement could have been phrased, "There is a higher *probability* that it will rain than that it will not."

More specifically, the statement might have assigned some numerical value to describe the likelihood of rain. Now such a statement might read, "There is a 0.60 probability of rain." To convey the same information, the television weatherman would say, "There is a 60 percent chance of rain today." So let us see what this 0.60 probability, or 60 percent chance, connotes.

Consider a scale from 0 to 1.0, where zero means there is no possibility and 1.0 means there is absolute certainty that an event will occur. Thus, if the probability of rain is 0, there is no possibility of rain, whereas, if the probability of rain is 1.0, *it will rain.* Moreover, if the probability of rain is 0.5, then it is equally likely that it will or will not rain.

Probability of rain	Interpretation
0	No possibility
0.5	Equally as likely
1.0	Absolute certainty

A probability value can be any value between and including the two extremes of zero and one. It is equivalent to expressing a relationship on a numerical scale. Thus, as a working definition, let us define probability as the prospect that an event will occur at some future time measured on a scale from zero (the event will not occur) to one (the event will occur with certainty).

Sources of Probability Numbers

So far we have established that *probability numbers* range from zero to one inclusive. Zero probability implies that there is no possibility that an event can occur (it is impossible) whereas a probability of one implies that an event is certain to occur. The next question that needs to be answered is, "Where do we get probability numbers?" The answer can be uncovered by examining three concepts of a probability number.

Classical Concept

The classical concept of probability is that probability numbers are known or can be determined *a priori* (before the fact). Thus the exact probability that a particular event might occur would be known before the fact. Illustrations of this type of probability number are

1. the probability of drawing an ace from a deck containing 52 cards;
2. the probability of drawing a red ball from a jar containing 3 red balls and 7 green balls.

To determine each of the preceding *a priori* probabilities, the following formula can be used:

$$P(A) = \frac{N(A)}{N(A) + N(\bar{A})}, \tag{3.1}$$

where

$P(A)$ = the probability of defined event A occurring,

$N(A)$ = the number of possible ways A can occur in a situation,

$N(\overline{A})$ = the number of possible ways A cannot occur in the same situation,

$N = N(A) + N(\overline{A})$ = the total number of possibilities of A and not-A involved in the situation.

Considering the illustration of drawing an ace from a deck of 52 cards, we see that by defining an ace as the event (A) of interest the following substitution results:

$N(A) = 4$ (aces),

$N(\overline{A}) = 48$ (all other cards),

$N = 52$ (all cards),

$$P(A) = \frac{N(A)}{N(A) + N(\overline{A})}$$

$$= \frac{N(A)}{N} = \frac{4}{4 + 48} = \frac{4}{52} = \frac{1}{13} = 0.077.$$

Thus the probability of drawing an ace on a single draw is 0.077. Similarly, if a ball is drawn from the aforementioned jar, the following substitution results when A, the event of interest, is defined as drawing a red ball:

$N(A) = 3$ (red balls),

$N(\overline{A}) = 7$ (green balls),

$N = 10$ (all balls),

$$P(A) = \frac{N(A)}{N(A) + N(\overline{A})} = \frac{N(A)}{N} = \frac{3}{3 + 7} = \frac{3}{10} = 0.3.$$

And so, the probability of a red ball being drawn is 0.3.

Now it should be remembered that these are examples of *a priori* probabilities. *A priori* probabilities are those that can be determined before any event occurs; however, *a priori* probabilities are subjected to a particular criticism that we should discuss.

As you may have realized when you worked through the previous two illustrations, for the $P(A)$ calculation to be correct, it is necessary to assume that all cards or all balls had an equal probability of being selected. Now notice what has happened. We have described a type of probability (*a priori*) that relies on the assumption of equal probability. Thus we have used probability to define probability, and that is the criticism.

As this is a rather technical criticism, let us not concern ourselves with the ambiguity that results from the definition but simply remember that these are situations where the possibilities (number of ways) can be determined before the occurrence of the event. And from these possibilities we can determine the classical probability (*a priori*) of an event before it happens.

Relative Frequency Concept

Another concept of probability is based on the relative frequency of occurrence. This concept utilizes the relationship of the number of times a particular event actually occurred to the total number of trials observed. To illustrate, consider that, when flipping a coin 1,000 times (1,000 trials), the head turned up 450 times. We would then determine a probability of the occurrence of a head based on relative frequency of 450/1,000 = 0.45. Since this probability is determined after the fact (*a posteriori*) and is based on observed data, it is referred to frequently as an empirical probability. The empirical probability of an event A can be determined as follows:

$$P(A) = \frac{a}{n},$$

where a = the number of times event A occurred out of n total observations or trials.

As the number of trials, n, increases infinitely, the relative frequency probability becomes stable. That is, the probability value a/n becomes the limit of the relative frequency as n approaches infinity.[1] As a further illustration of the concept of relative frequency, consider the following example.

Example 3.1

▷ The Careless Manufacturing Company has maintained records on the number of defective tires returned over the past year. Of the 200,000 tires produced, 3,000 were defective. In order to determine the probability of any one tire being defective, the relative frequency concept is utilized, and the empirical probability is

$$P(\text{defective tire}) = \frac{3,000 \ (\text{number of defectives})}{200,000 \ (\text{total number produced})} = 0.015.$$

Thus the implication is that, if a'. aspects of the manufacturing process remain unchanged, future production of defective tires will occur at a rate of 15 defectives for each 1,000 tires produced. ◁

Here again we must be reminded to take care in interpreting this probability. Remember, the concept of relative frequency is based on an infinite number (or at least a large number) of trials. Obviously, we never use an infinite number of observations, but the empirical probability can be erroneous if too few trials are

---

1. If you are famliar with limits, then $P(A) = \lim\limits_{n \to \infty} \frac{a}{n}$.

BASIC STATISTICS

observed. Certainly 10 trials would not be enough to determine the probability of a defective tire, whereas with our 200,000 trials we have a right to feel pretty secure about our estimate of the probability of a defective tire.

Finally, you might like to know that empirical probability has been found to be quite useful in setting premium rates in the insurance industry. These rates are based on the probability of the occurrence of events against which a company may issue an insurance policy—such as insurance on your car. Since there are few, if any, cases in the insurance industry where the possible number of ways events may occur can be determined, the probabilities upon which insurance rates are based are determined by the relative frequency of occurrence. Thus relative frequency probability explains, in part, why automobile insurance premiums differ in some parts of the country and for different age-groups of drivers.

Subjective Probability

At times there are situations (certainly in business) where probabilities cannot be determined by either the *a priori* or the empirical methods. In these cases one may have to resort to a probability based on intuition or judgment. This type of probability estimate is called *subjective probability*. However, do not confuse intuitive judgment with empirical observation. The latter is based on observed data, while the former relies on personal evaluation (something like shooting from the hip). Since subjective probability is based on experience, the subjective probability estimates should also get better as one's experience improves. And from time to time the probability may undergo revision based on additional experience.

Methods of Counting

In using the classical (*a priori*) concept of probability, the equation is dependent on our being able to determine the values that go into the numerator and denominator—the possibilities:

$$P(A) = \frac{N(A)}{N(A) + N(\bar{A})}, \tag{3.1}$$

where $N(A) + N(\bar{A}) = N$ (total number of possibilities).

That is, *a priori* probabilities are based on the number of possibilities. Sometimes these possibilities are reasonably easy to count; other times they are difficult to determine. When they are difficult to determine, we frequently need assistance in counting the possibilities. Four methods used to help count the possibilities are presented in this chapter.

The Multiplication Method

If a particular experiment is determined to have m possible outcomes on the first trial and n possible outcomes on the second trial, the total number of possible outcomes for the two trials is the product $m \times n$. Let us examine this method of counting using a tree diagram.[2]

Suppose you have four required and five elective courses that you could take. Further suppose that required courses are available only in the fall semester and that only electives can be taken in the spring. If you are planning to take one required course in the fall and one elective course in the spring, how many different pairs of courses are available to you?

Notice that we have two distinct groups, required courses (4 items) followed by electives (5 items). Then the solution is $4 \times 5 = 20$ as demonstrated by the tree diagram in figure 3.1. This solution was obtained by setting $m = 4$ and $n = 5$. The examples can be expanded to any number of groups, and the total number of outcomes equals the product of the number of elements of each group.

Multiple Choice Arrangement

Notice that this section deals with a type of arrangement. Such is the case for the next three methods of counting. The first method (the *multiplication method*) dealt with two or more groups, and we will now examine arrangements within one of the groups.

Suppose we identify the four required courses as A, B, C, and D; and we want to arrange them in pairs so that one is taken during one semester and the other is taken in the next semester. How many distinctly different arrangements of two courses can we get from the required four courses? At this point the question cannot be answered until we make the two assumptions required of *multiple choice arrangements:*

1. Duplication *is* permissible. That is, a single item may be duplicated in a given pair. (You might fail a given course and have to repeat it.)

2. Order *is* important. That is, AB is considered a different arrangement from BA.

The possible multiple choice arrangements of the courses are shown in figure 3.2.

From the tree diagram, we see that there are 16 arrangements

---

2. Tree diagrams can be very helpful and will be used frequently in this chapter. They are pictorial representations of the situation we are trying to evaluate.

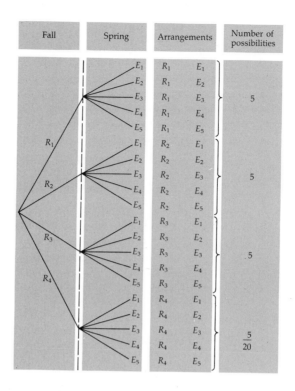

Figure 3.1 Tree diagram for the multiplication method

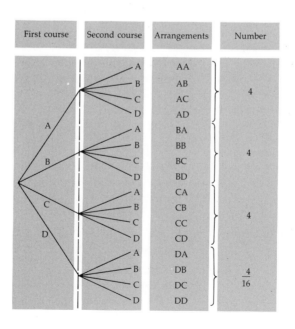

Figure 3.2 Tree diagram for multiple choice arrangements

under the assumptions of multiple choice. It is obvious, however, that for larger problems use of the tree diagram would prove cumbersome. Thus the easier way is to use the following formula:

$$M_x^n = n^x, \tag{3.2}$$

where

n = the size of the set,

x = the number of places in the arrangement.

Example 3.2

▷ Now let us repeat the illustration and demonstrate the use of the preceding equation. There are four courses ($n = 4$), and you are going to take one each semester for two semesters ($x = 2$). How many possible pairs do you have? If we assume that you might have to repeat a course (duplication is permissible) and that sequence AB is different from BA (order is important), then our number of pairs is $M_2^4 = 4^2 = 16$. ◁

Permutation Arrangement

Suppose we continue with your schedule plans for subsequent semesters and that we decide to change the condition regarding duplication. That is, let us now assume:

1. Duplication *is not* permissible. (You do not expect to fail any courses.)

2. Order *is* important. That is, AB is considered a different arrangement from BA.

Then figure 3.3 shows the possible pairs of courses.

Now instead of 16 arrangements, we have 12 arrangements. These are the *permutation arrangements*. The permutation formula for finding the number of possible arrangements under the assumptions of (1) no duplication and (2) order is important is

$$P_x^n = \frac{n!}{(n - x)!} . \tag{3.3}$$

This equation states that the number of permutations of n items taken x at a time is equal to n factorial ($n!$) divided by ($n - x$) factorial, $(n - x)!$.[3] As another example of counting with permutations, consider the following situation.

Example 3.3

▷ Let us use equation (3.3) to determine the number of arrangements shown in figure 3.3. Remember, the situation is to sched-

---

3. The symbol, !, is the notation for factorial. $n! = n \cdot (n - 1) \cdot (n - 2) \cdot \cdots \cdot (1)$. Thus $5! = 5 \cdot 4 \cdot 3 \cdot 2 \cdot 1 = 120$, and, by definition, $0! = 1$.

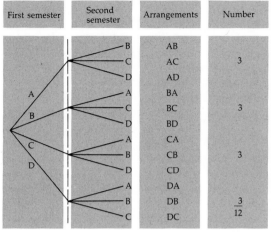

Figure 3.3 Tree diagram for permutation arrangements

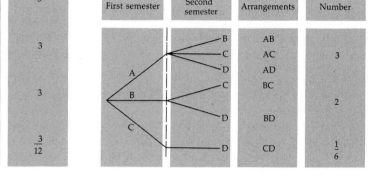

Figure 3.4 Tree diagram for combinatorial arrangements

ule one of four courses (A, B, C, D) in each of the next two semesters under the assumptions of (1) no duplication and (2) order is important. Setting $n = 4$ and $x = 2$,

$$P_2^4 = \frac{4!}{(4-2)!} = \frac{4 \cdot 3 \cdot 2 \cdot 1}{2 \cdot 1} = 12. \triangleleft$$

Combination Arrangement

Suppose we go back to planning your schedule of courses and decide that in addition to the restriction of no duplication we will consider the arrangement AB to be indistinguishable from the arrangement BA. That is, we are now assuming the following:

1. Duplication *is not* permissible.
2. Order *is not* important.

As far as your schedule is concerned, this means that no course may be taken twice; and that, if A is taken first and B second, the reverse order will not be considered. The tree diagram of figure 3.4 depicts the possible combinatorial arrangements.

Thus, under the conditions of (1) no duplication and (2) order is not important, we get only six distinct arrangements. The combination of n items taken x at a time is

$$C_x^n = \frac{n!}{x!(n-x)!}. \tag{3.4}$$

This formula is called the *combinatorial formula.* As an example of the use of this formula, consider the following situation.

Example 3.4

▷ Let us use equation (3.4) to determine the number of arrangements shown in figure 3.4. The situation is to schedule one of four courses (A, B, C, D) in each of the next two semesters under the assumptions of (1) no duplication and (2) order is not important. Setting $n = 4$ and $x = 2$,

$$C_2^4 = \frac{4!}{2!(4-2)!} = \frac{4 \cdot 3 \cdot 2 \cdot 1}{2 \cdot 1 \cdot 2 \cdot 1} = 6. ◁$$

Probability: Characteristics and Terms

The general formulas,

$$P(A) = \frac{N(A)}{N} \quad (a\ priori),$$

$$P(A) = \frac{a}{n} \quad \text{(relative frequency)},$$

can be used to determine probabilities of simple events. There are many times, however, when probabilities will be determined by combining simple probabilities. In this type of circumstance, we will be investigating compound events, and, to do this, we need to become familiar with certain characteristics and terms associated with probability numbers.

Characteristics of Probability Numbers

Let's suppose that an experiment has several possible outcomes which we'll call events—$A_1, A_2, A_3, \ldots, A_k$. If a group of numbers is to be identified as probability numbers, they must possess the following characteristics:

1. The probability associated with any event cannot be any smaller than 0 or greater than 1. That is, $0 \leq P(A_i) \leq 1$.
2. The sum of the event probabilities in the experiment $(A_1 + A_2 + A_3 + \cdots + A_k)$ must equal 1.

Thus

$$\sum_{i=1}^{k} P(A_i) = 1.$$

The first characteristic states that for all outcomes $(A_1, A_2, A_3, \ldots, A_k)$ of an experiment, the probability of any particular outcome, denoted $P(A_i)$, must be within the range 0 and 1. The second characteristic states that the probability numbers assigned to all outcomes $[P(A_1), P(A_2), \ldots, P(A_k)]$ must sum to one, $P(A_1) + P(A_2) + \cdots + P(A_k) = 1$. Since all possible events are considered, one of them *must* occur.

Consider, as an illustration, the flipping of a coin. The possible outcomes are a head and a tail. If the coin is fair, and each outcome is equally likely, then, using equation (3.1), $P(H) = N(H)/[N(H) + N(T)] = 1/(1 + 1) = 1/2$, and similarly $P(T) = 1/2$. Now notice that each probability is in the 0–1 range, and that the sum of the probabilities is 1. Thus the numbers have the necessary characteristics of probability numbers.

Now that we know something about the characteristics of probability numbers, we are almost ready to consider the so-called *laws of probability*. Before we do this, though, let us take a quick look at some terms we need to know.

Complement

Let us understand the term *complement*. The complement of heads is tails; or the complement of tails is heads. More generally, the complement of any outcome A is all outcomes other than A or simply "not-A". Symbolically, the expression denoting not-A is often written as $\overline{A}$. And for any outcome A, the probability of not-A is

$$P(\overline{A}) = 1 - P(A). \tag{3.5}$$

In our example, if the probability of heads is $P(A)$, then $P(\overline{A}) = P(T) = 1 - P(H)$. When the experiment has several possible outcomes, $P(\overline{A})$ represents the probability that some outcome other than A will occur.

Mutually Exclusive Events

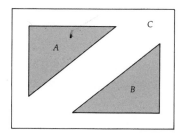

Figure 3.5   Mutually exclusive events

What does it mean to say that two events (outcomes), A and B, are *mutually exclusive*? Events A and B are mutually exclusive if they cannot occur simultaneously. Figure 3.5 diagrams an example of mutually exclusive events. In the figure the area designated A refers to the number of possible ways event A might occur, and the area marked B refers to the number of possible ways event B might occur. The area outside A and B (call it C) depicts the possible ways for neither event A nor B to occur. You should notice that there is no area indicating possible ways for events A and B to occur together. So events A are B are mutually exclusive.

Independent and Dependent Events

In addition to mutually exclusive events, it is important that we understand the difference between independent and dependent events. If two events are such that the occurrence of one event does not affect or change the probability of the other event occurring, these events are said to be *independent*. But if the events are so related that, if one event occurs, the probability of the other event occurring is changed, these events are *dependent*. The probability of the second event would be considered *conditional* on the occurrence of the first event.

Example 3.5

▷ Consider a jar that contains four red balls and six green balls. Then $P(R) = 4/10$ and $P(G) = 6/10$. Now if a red ball is drawn and not replaced, the probability that a green ball would be drawn on the second draw is no longer 6/10. It has been changed to 6/9, conditional that a red ball was drawn on the first draw. Conditional probabilities are written as follows:

$P(G|R) = 6/9.$

In this case it is read as "the probability of a green ball, given that a red has occurred, equals 6/9." If the events had been independent (the red ball was replaced after the first draw) rather than dependent, then the probability of a green ball on the second draw would have remained 6/10. ◁

You should notice that the conditional probability was determined from a reduced population. That is, we started with 10 balls but considered only 9 when defining the conditional probability. This is always the case when dealing with dependent events. If the population from which the second event could occur was not reduced after the occurrence of the first event, then the events would have to be independent because one event is not affected by the occurrence of the other event.

Laws of Probability

Now that we have learned certain terms in probability, we are ready to examine and use the *laws of probability* as they relate to *compound events*.

Multiplication Law

One particular compound event involves the occurrence of *all* the individual events under consideration. These events are considered either to occur all at the same time or successively. Such a compound event requires the calculation of a *joint* probability using the following general expression:

$$P(A \text{ and } B) = P(A) \cdot P(B|A). \tag{3.6}$$

Notice that the joint probability is a product of the probabilities of each event required to occur. Equation (3.6) is for *dependent* events; if the events are *independent*, then the expression for joint probabilities becomes

$$P(A \text{ and } B) = P(A) \cdot P(B). \tag{3.7}$$

Example 3.6

▷ Consider tossing a fair coin and rolling a fair die. What would be the probability of a head occurring on the coin toss and a six occurring on the die roll? Since the events are independent, then

$$P(\text{Head and Six}) = P(H) \cdot P(S) = 1/2 \cdot 1/6 = 1/12. \triangleleft$$

The outcomes and probabilities of the experiment described in example 3.6 are shown in the tree diagram, figure 3.6. The figure shows all four possible outcomes when a coin is tossed and a die rolled. They are: Head and Six; Head and No Six; No Head and Six; No Head and No Six; and the sum of all of these joint occurrences is one, just as we expected. The occurrence we are interested in is "Head and Six" with a probability of 1/12.

Example 3.7

▷ Now let us apply the *multiplication law* to *dependent events*. Suppose we return to our jar that has four red balls and six green balls. What would be the joint probability of a red on the first of two draws and a green on the second draw, given that the red ball is not replaced?

$$P(\text{Red and Green}) = P(\text{Red}) \cdot P(\text{Green} | \text{Red}),$$
$$P(\text{Red and Green}) = 4/10 \cdot 6/9,$$
$$P(\text{Red and Green}) = 4/15. \triangleleft$$

This example also can be illustrated using a tree diagram as shown in figure 3.7. Again there are four possible outcomes: Red and Green; Red and Not Green; Not Red and Green; Not Red and Not Green; and the probabilities sum to one as expected. The outcome of interest to us is "Red and Green" which is equal to 24/90 or 4/15 as demonstrated earlier. This also illustrates that the use of the tree diagram is very helpful when calculating joint probabilities.

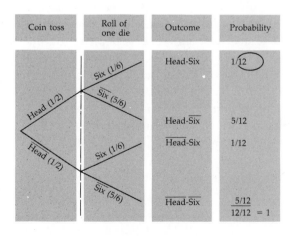

Figure 3.6   Tree diagram for coin toss and roll of one die

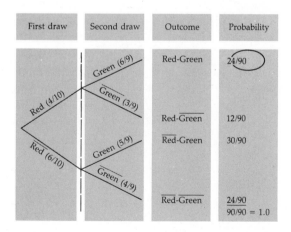

Figure 3.7   Tree diagram for red and green balls

Remember that joint probabilities require the multiplication law which is

$$P(A \text{ and } B) = P(A) \cdot P(B|A), \tag{3.6}$$

and, if A and B are independent events, the law is

$$P(A \text{ and } B) = P(A) \cdot P(B). \tag{3.7}$$

Addition Law

Suppose you wanted to know the probability of A or B occurring. In this case, if either A or B occurs, then we say the event occurs. Thus for two events, A or B, the probability of either A or B occurring is determined by the following equation:

$$P(A \text{ or } B) = P(A) + P(B) - P(A \text{ and } B). \tag{3.8}$$

In equation (3.8), if events A and B are mutually exclusive, then the last term on the right side, $P(A \text{ and } B)$, is equal to zero, and the addition law for mutually exclusive events reduces to

$$P(A \text{ or } B) = P(A) + P(B). \tag{3.9}$$

Example 3.8

▷ Let us determine the probability of drawing a red or green ball from a jar containing three red, four green, and five blue balls. Since the events are mutually exclusive, then

$$P(\text{Red or Green}) = P(\text{Red}) + P(\text{Green}) = 3/12 + 4/12 = 7/12.$$

Also notice that the probability of drawing a red or a green or a blue ball is

$$P(\text{R or G or B}) = 3/12 + 4/12 + 5/12 = 12/12 = 1.0.$$

That conforms to our understanding that something will occur with a probability of one if all possible outcomes are defined. ◁

Now let's move to an illustration where the events are *not mutually exclusive*. If the events are not mutually exclusive, they have a common or like area of occurrence as depicted by the shaded area in figure 3.8.

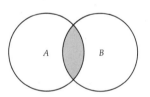

Figure 3.8 Diagram of events that are not mutually exclusive

For instance, if we define event A as the drawing of a heart (there are 13 hearts) from a deck of 52 cards and event B as the drawing of a king (there are four kings) from the same deck, then the shaded area of figure 3.8 represents the king of hearts. Thus the king of hearts is an outcome that would mean both events occurred. This identifies them as events that are not mutually exclusive. Consequently, if you want to determine the probability of a heart or king on one draw, you must be careful to avoid double counting the king of hearts. That is, if you consider that there are thirteen hearts (including the king of hearts), you must pretend that there are only three kings; or conversely, if you con-

BASIC STATISTICS

sider that there are four kings, you must pretend that there are only twelve hearts. Either way, it should be apparent that the number of cards available to make your draw successful is sixteen—either 12 + 4 or 13 + 3. The probability of drawing a heart or king is 16/52.

Now, utilizing the addition law,

$$P(H \text{ or } K) = P(H) + P(K) - P(H \text{ and } K)$$
$$= 13/52 + 4/52 - 1/52,$$
$$P(H \text{ or } K) = 16/52.$$

Generalizing the Laws

In the preceding discussion of the multiplication and addition laws, two events are considered. These laws, however, are valid for as many events as you care to consider as long as you are careful.

To illustrate, if the addition law is applied to three events (A, B, C), then

$$P(A \text{ or } B \text{ or } C) = P(A) + P(B) + P(C) - P(A \text{ and } B)$$
$$- P(A \text{ and } C) - P(B \text{ and } C) + P(A \text{ and } B \text{ and } C).$$

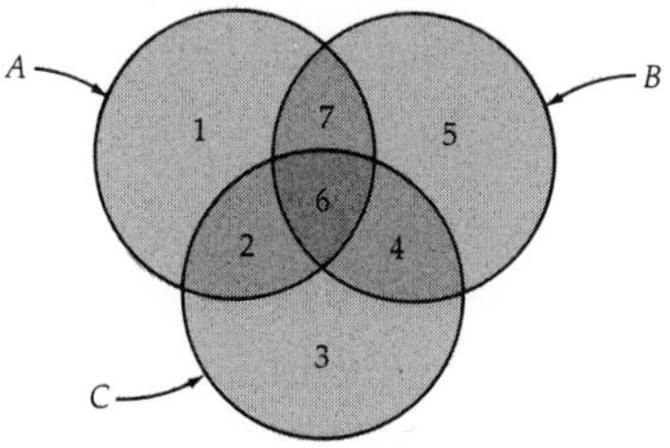

Figure 3.9 can be used to show how this equation accomplishes the desired result. Each numbered area corresponds to a specific probability so that $P(A)$ = area 1 + area 2 + area 6 + area 7. Similarly,

$$P(B) = 4 + 5 + 6 + 7,$$
$$P(C) = 2 + 3 + 4 + 6,$$
$$P(A \text{ and } B) = 6 + 7,$$
$$P(A \text{ and } C) = 2 + 6,$$
$$P(B \text{ and } C) = 4 + 6,$$
$$P(A \text{ and } B \text{ and } C) = 6.$$

Figure 3.9 Compound events with three events

Then

$$P(A \text{ or } B \text{ or } C) = (1 + 2 + 6 + 7) + (4 + 5 + 6 + 7) + (2 + 3 + 4 + 6) - (6 + 7) - (2 + 6) - (4 + 6) + 6,$$
$$P(A \text{ or } B \text{ or } C) = 1 + 2 + 3 + 4 + 5 + 6 + 7,$$

which are the actual areas within the three circles.

Examples Using the Multiplication and Addition Laws

There are instances where the determination of a desired probability involves combining the laws of multiplication and addition as illustrated in the following examples.

Example 3.9

▷ The Go-Save Shopping Center, in celebrating its first anniversary, is giving away two automobiles by a drawing. The rules are simple. There are two barrels—one red and one green—in the shopping center mall. Every time a purchase is made, a green ticket is given to a male customer for deposit in the green barrel, and a red ticket is given to a female customer for deposit in the red barrel.

Mr. and Mrs. Armstrong are not regular customers of the shopping center, but they have been shopping there lately because they both would like to win a new automobile. Mr. Armstrong has deposited 10 tickets in the green barrel, and Mrs. Armstrong has deposited 20 tickets in the red barrel. On the night of the drawing, there are 1,000 tickets in the green barrel and 5,000 tickets in the red barrel.

Now let us determine the probability that Mr. and Mrs. Armstrong win cars. First let us caution you that there are three ways they might win a car; either Mr. Armstrong or Mrs. Armstrong or both might win a car. Perhaps the easiest approach to this type of problem is through a tree diagram as shown in Figure 3.10.

Assuming that a ticket is drawn from the green barrel first and the red barrel second (the order can be reversed if you desire), the four outcomes are

$P(HS)$ (Both win) $= 200/5,000,000,$
$P(H\overline{S})$ (He wins, She loses) $= 49,800/5,000,000,$
$P(\overline{H}S)$ (He loses, She wins) $= 19,800/5,000,000,$
$P(\overline{H}\overline{S})$ (Neither wins) $= 4,930,200/5,000,000.$

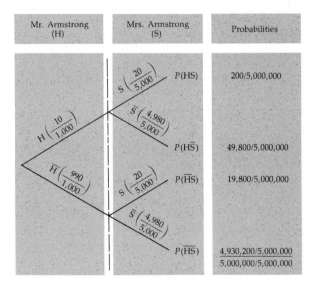

Figure 3.10 Tree diagram for the Armstrongs winning a car

Now notice the outcomes are joint probabilities and that the *multiplication law* was used to determine the probability of each; but to determine the probability of winning a car, it is necessary to use the addition law. Thus

$P(\text{car}) = P(\text{HS}) + P(\text{H}\overline{\text{S}}) + P(\overline{\text{H}}\text{S})$
$= 200/5{,}000{,}000 + 49{,}800/5{,}000{,}000 + 19{,}800/5{,}000{,}000$
$= 69{,}800/5{,}000{,}000,$
$P(\text{car}) = 349/25{,}000 = 0.01396.$ ◁

Example 3.10

▷ Suppose there are 100 stores that stock the same product. Table 3.1 presents a cross classification of the stores according to product sales and the height of shelf display. Then the following probabilities, among others, might be determined:

$P(\text{Average Sales}) = 36/100 = 0.36,$
$P(\text{Eye-Level Display}) = 51/100 = 0.51,$
$P(\text{High Sales} \mid \text{Eye-Level Display}) = 32/51 = 0.63.$

And, the question of independence can be answered. Notice that the probability of high sales and high shelf display is 4/100. If product sales and shelf display are independent, then

$P(\text{High Sales and High Display}) = P(\text{HS}) \cdot P(\text{HD}),$

that is,

$4/100 = (41/100)(20/100),$

but

$4/100 \neq 820/10{,}000.$

Thus they are not independent. There is a relationship between product sales and shelf display, and the actual relationship becomes

Table 3.1
Product sales for 100 stores

Product sales	Shelf display			
	High	Eye-level	Low	Total
High	4	32	5	41
Average	10	15	11	36
Low	6	4	13	23
	20	51	29	100

$$P(\text{High Sales and High Display}) = P(\text{HS}) \cdot P(\text{HD}\,|\,\text{HS})$$
$$= (41/100)(4/41)$$
$$= 4/100.$$

And this result agrees with our earlier determination. ◁

Revised Probabilities and Bayes's Formula

From time to time a situation arises where the probability of an event can be reevaluated, or revised, based on information obtained about a subsequent event. This situation is an example of another type of *a posteriori* (after the fact) probability. It differs from the empirical probability that was also determined after the fact in that empirical probability, in a sense, attempts to estimate the *a priori* (before the fact) or classical probability.

The type of posterior probability that we are now interested in is a revision of the original priors, whether they be *a priori* or *empirical* or *subjective* probabilities. To determine these revised possibilities, we will employ Bayes's formula.

Around two hundred years ago the Reverend Thomas Bayes discovered an interesting relationship that permits the calculation of posterior (revised) probabilities from prior probabilities. This discovery is called Bayes's theorem; in essence, Bayes's theorem states: If there are two events A and B, such that the $P(A)$ and $P(B\,|\,A)$ are known, it is possible to determine $P(A\,|\,B)$ from the original priors. Notice that $P(A\,|\,B)$ is a revision of the original prior $P(A)$. To demonstrate how this can be done, it is necessary to observe that

$$P(A \text{ and } B) = P(A) \cdot P(B\,|\,A)$$

and

$$P(A \text{ and } B) = P(B) \cdot P(A\,|\,B),$$

and then, equating the right sides so that

$$P(B)P(A\,|\,B) = P(A)P(B\,|\,A),$$

thus

$$P(A\,|\,B) = \frac{P(A) \cdot P(B\,|\,A)}{P(B)}. \qquad (3.10)$$

As you can see, the numerator of the right side contains $P(A)$ and $P(B\,|\,A)$, the two original known probabilities. The denominator, however, still must be determined. As the denominator is the probability of the event B occurring, it is necessary that *all* possible ways that B might occur be included in this probability, $P(B)$. Thus $P(B) = P(A) \cdot P(B\,|\,A) + P(\overline{A}) \cdot P(B\,|\,\overline{A})$, and there may be more than one $P(\overline{A}) \cdot P(B|\overline{A})$ since there may be several events identified as not-A. Consider the following example.

Example 3.11

▷ A plant has two machines. Machine A produces 60 percent of the output with a fraction defective of 0.02. Machine B produces 40 percent of the output with a fraction defective of 0.04. If a single unit of output is observed to be defective, determine the revised probabilities that should be assigned to each machine (see table 3.2).

Even though Machine A produces a larger proportion of the output, the revised probability of 0.571 indicates that the quality control investigation should begin with Machine B.

Another approach to the preceding problem could have been the use of a tree diagram as shown in figure 3.11. Notice again in figure 3.11 that the sum of the probabilities for all outcomes is 1.00. Now from Bayes's formula, we know that

Table 3.2
Probability revision example calculations

Event (Ei)	Prior P(Ei)	Conditional P(D|Ei)	Joint P(Ei)·P(D|Ei)	Posterior P(Ei|D)
Machine A	0.60	0.02	0.012	0.012/0.028 = 0.429
Machine B	0.40	0.04	0.016	0.016/0.028 = 0.571
			0.028	

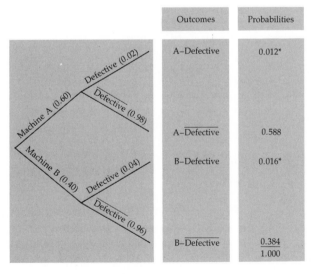

Figure 3.11   Tree diagram for Bayes's formula (example 3.11)

$$P(A|D) = \frac{P(A \text{ and } D)}{P(D)} = \frac{P(A \text{ and } D)}{P(A) \cdot P(D|A) + P(B) \cdot P(B|A)}.$$

All we need to do is select the probabilities from the tree diagram needed to make the revision calculation. The ones we need are denoted by asterisks. They are the outcomes involving a defective. Each of these probabilities appears in the denominator, and the probability involving Machine A and a defective appears in the numerator so that

$$P(A|D) = \frac{0.012}{0.012 + 0.016}$$

$$= \frac{0.012}{0.028},$$

$P(A|D) = 0.429.$

And, similarly, the probability associated with Machine B can be revised as follows:

$$P(B|D) = \frac{P(B \text{ and } D)}{P(D)}$$

$$= \frac{0.016}{0.012 + 0.016}$$

$$= \frac{0.016}{0.028},$$

$P(B|D) = 0.571.$

Notice on the tree diagram (figure 3.11) that the first branches represent outcomes for the original event, and the second group of branches represent the observable outcomes. Now let's look at another example where the original probabilities are more subjective than empirical. ◁

Example 3.12

▷ A marketing manager believes the market demand potential of a new product to be high with a probability of 0.30, or to be average with a probability of 0.50, or to be low with a probability of 0.20. From a sample of 20 employees, 14 indicated a very favorable reception to the new product. In the past such an employee response (14 of 20 favorable) has occurred with the following probabilities: if the actual demand is high, the probability of a favorable reception is 0.80; if the actual demand is average, the probability of a favorable reception is 0.55; and if the actual demand is low, the probability of the favorable reception is 0.30.

Table 3.3
Probability revision example calculations

Event (Ei)	Prior P(Ei)	Conditional P(F|Ei)	Joint P(Ei) · P(F|Ei)	Posterior P(Ei|F)
High demand	0.3	0.80	0.240	0.24/0.575 = 0.4174
Average demand	0.5	0.55	0.275	0.275/0.575 = 0.4783
Low demand	0.2	0.30	0.060	0.06/0.575 = 0.1043
			0.575	

Thus, given a favorable reception, revise the prior probabilities estimated by the manager (table 3.3).

Thus the probability of high demand is greater (0.4174), whereas the probability of low demand is reduced to 0.1043.

As in example 3.11, another approach to example 3.12 could have been the use of a tree diagram as shown in figure 3.12.

As before, we select the probabilities from the tree to make the desired revision. For instance, if we want the revised probability associated with high demand, then

$$P(H|F) = \frac{P(H \text{ and } F)}{P(F)}$$

$$P(H|F) = \frac{P(H \text{ and } F)}{P(H) \cdot P(F|H) + P(A) \cdot P(F|A) + P(L) \cdot P(F|L)}$$

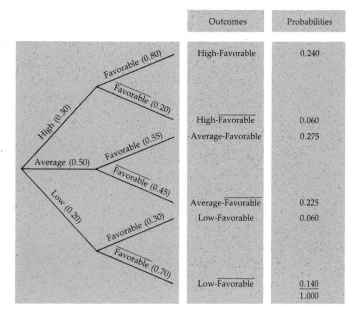

Figure 3.12   Tree diagram for Bayes's formula

$$P(H|F) = \frac{0.24}{0.24 + 0.275 + 0.060}$$

$$= \frac{0.24}{0.575}$$

$P(H|F) = 0.4174.$

In a like manner $P(A|F)$ and $P(L|F)$ can be determined. Why don't you try to calculate them using the answers in table 3.3 to check your work? ◁

Summary of Equations

3.1   *A priori* probability:

$$P(A) = \frac{N(A)}{N(A) + N(\overline{A})}$$

3.2   Multiple choice:

$$M_x^n = n^x$$

3.3   Permutations:

$$P_x^n = \frac{n!}{(n - x)!}$$

3.4   Combinations:

$$C_x^n = \frac{n!}{x!(n - x)!}$$

3.5   Probability of a complement:

$$P(\overline{A}) = 1 - P(A)$$

3.6   Joint probability dependent events:

$$P(A \text{ and } B) = P(A) \cdot P(B \,|\, A)$$

3.7   Joint probability independent events:

$$P(A \text{ and } B) = P(A) \cdot P(B)$$

3.8   Probability of either of two events:

$$P(A \text{ or } B) = P(A) + P(B) - P(A \text{ and } B)$$

3.9   Probability of either of two mutually exclusive events:

$$P(A \text{ or } B) = P(A) + P(B)$$

3.10   Bayes's formula:

$$P(A \,|\, B) = \frac{P(A) \cdot P(B \,|\, A)}{P(B)}$$

1. What are the sources of probability numbers?

2. Define
(a) *a priori* probability
(b) combination
(c) permutation
(d) mutually exclusive events
(e) complement

3. State the characteristics by which a group of numbers can be identified as probability numbers.

4. Explain the purpose of the multiplication law.

5. Explain the purpose of the addition law.

Problems

1. A local bachelor always eats lunch downtown and dinner in one of the restaurants near his apartment. There are 10 places to eat downtown and 6 restaurants near his apartment. How many pairs does he have from which to choose when deciding where to eat his meals in a given day?

2. A dietician has 5 meats and 12 vegetables to select from in preparing a menu. Assuming each menu contains 1 meat and 2 vegetables, how many menus are possible?

3. Richard went in to buy a new car and has everything picked out except for the body style, color, and the radio. If he has four different body styles, eight different colors (he is only considering a one color car), and three different radios, how many different choices does he have to select from?

4. The Hot Air Company is introducing a new logo to be placed on its products. It is necessary to select three different colors for the three parts of the logo. If they have eight colors to select from and each of the three parts must be a different color, in how many ways can they complete the design of the logo?

5. There are six flavors of candy available at the Fat-Is-Fun Shop. The Manager of FIF wants to create gift-packs that contain three different flavors of candy. How many types of gift-packs can be created?

6. A car company is thinking of introducing a new sports car and the marketing department needs to come up with a name for this car. They have already determined that a three-character name

with at least one letter and one number will be used. Furthermore they will only select from six letters and from four numbers (only the letters E, L, S, R, X, and Z and the numbers 2, 4, 6, and 9 seem to work best for car names). They will allow a number to be repeated but not a letter in the name; how many different names do they have to select from?

7. A florist has two show windows at the front of her store. As the Christmas season approaches, she must decide on the window decorations to use. If there are six possible window decorations, how many possible arrangements are available? (Assume no duplication and order is not important.)

8. A distinguished economics professor teaches a course in monetary theory. He has a library on the subject that consists of twenty books, but he always selects four of them, at random, to use for a given class. As a student the previous semester, you own four books in his library. What is the probability that you will have to buy one or more new books when you repeat the course this year?

9. From a container that has 10 identical units, 8 of which are defective, 2 units are drawn in succession without replacement; determine

(a) the probability that both units are not defective.
(b) the probability that the second unit drawn is defective.

10. On a slot machine there are three reels with the ten digits, 0, 1, 2, 3, 4, 5, 6, 7, 8, 9, plus a star on each reel. When a coin is inserted and the handle is pulled, the three reels revolve independently several times, and come to rest. What is the probability of

(a) two stars?
(b) one star?
(c) no stars?

11. In the Big Mac Burger Company there are five economists and seven engineers. A committee of two economists and three engineers is to be formed. In how many ways can this be done if

(a) one particular engineer must be on the committee.
(b) two particular economists cannot be on the same committee.

12. If a die is rolled twice, what is the probability that the first roll yields a 5 or 6 and the second roll anything but a 3?

13. Fat-Is-Fun has two clerks besides the Manager and whoever arrives first opens the shop. (They are supposed to be there at 10 A.M.) Actually, one clerk arrives late 25 percent of the time; the other clerk arrives late 10 percent of the time; and the Manager arrives late 50 percent of the time. (a) If you arrive at the store at 10 A.M., what is the probability you will find it open? (b) Open, but with only one person there?

14. Three people are working independently at solving a statistics problem. The respective probabilities that they will solve the problem are 1/4, 1/2, and 1/3. What is the probability that the problem will be solved?

15. Big Al, Susan, and Steve are applying for three different jobs at Smash, Inc. The probability is 1/2 that Big Al will get his job, 1/3 that Susan will get hers, and 3/4 that Steve will be employed.

(a) What is the probability that all three get their respective jobs?
(b) None will be employed?
(c) Only one will get his/her job?

16. Jean and Phyllis practice shooting at a target. If, on the average, Jean hits the target 3 times out of 5, and Phyllis hits it 4 times out of 8, what is the probability, when both shoot simultaneously, that the target is not hit?

17. A certain professor owns four automobiles. The probability of each one of them being operational at any given time is presented below. What is the probability that on a given morning

(a) none will be operational?
(b) only the Volkswagen will be operational?

Car	Probability of operating
Chrysler	0.95
Trans Am	0.97
Corvette	0.85
Volkswagen	0.60

18. A nervous student has ten ink pens in his brief case, but three of them will not write. As he hurries to a statistics test, he selects two pens at random to take to the exam. What is the probability that neither will write?

19. What is the probability of a bridge player having 13 cards of one suit in a given hand. (Note for nonbridge players: There are 4 suits of 13 cards each in a deck. A bridge hand consists of 13 cards. The entire deck is dealt so that each of four players has 13 cards.)

20. One urn contains 6 red discs and 3 blue discs, and a second urn contains 7 red discs and 4 blue ones. A disc is selected from each urn, and they are placed in a bag containing 4 red and 4 blue discs. What is the probability of drawing a red disc from the bag?

21. An urn contains 7 red discs and 3 blue discs. After two draws with replacement, what is the probability of

(a) 2 red discs?

(b) only 1 blue disc?

22. Repeat problem 21 without replacement after the first draw.

23. Two cards are drawn successively at random from a 52-card deck, without replacement, and reshuffling after each draw. What is the probability that one or the other, but not both, will be a heart?

24. From 5 men and 4 women, how many committees can be selected consisting of

(a) 3 men and 2 women?

(b) 5 people of which at least 3 are men?

25. If a machine produces to specification 98 percent of the time, what is the probability of two defective items in a row? (Assume independent events.)

26. An airline has a daily flight between two cities that is on schedule 80 percent of the time. When the flight is on schedule, the weather has been rated good 90 percent of the time, and, when the flight is not on schedule, the weather has been rated poor 70 percent of the time. If on a given day, the weather is rated good, what is the revised probability that the flight will be on schedule?

27. In a given geographic area it rains on 20 percent of the days in a year. When it rains, the local weather forecaster has predicted rain 70 percent of the time. When it does not rain, she (the weather forecaster) has predicted no rain 90 percent of the time.

(a) Given a prediction of no rain, what is the probability the forecast is correct?

(b) For a given day, what is the probability the forecaster will predict rain?

28. The Safety Council did a traffic survey of 1,000 drivers to determine whether age made a difference in the usage of safety belts. Based on the following results,

	Safety belts	No safety belts	Total
Under 40	275	200	475
40 and over	325	200	525
	600	400	1,000

(a) what is the probability that a driver under 40 uses the safety belt?

(b) what is the probability of a driver over 40 not using a safety belt?

(c) is the use of a safety belt independent of age?

29. Use the information concerning 400 individuals in the following table to answer the questions:

	Checking account	No checking account	Total
Savings account	175	50	225
No savings account	100	75	175
	275	125	400

(a) What is the probability of selecting an individual who has a checking account?

(b) What is the probability of selecting an individual who has a savings account?

(c) What is the probability of selecting an individual who has both a savings account and checking account?

(d) Is having a checking account independent of having a savings account?

30. Suppose that 100 stores have been cross-classified according to management and sales.

Sales	Management			Total
	A	B	C	
High	15	3	2	20
Average	5	50	15	70
Low	1	2	7	10
	21	55	24	100

Use the data to determine
(a) the probability of high sales.
(b) the probability of B management.
(c) the probability of average sales given A management.
(d) the probability of B management given high sales.
(e) if sales and management are independent.

31. The owner of a ladies' specialty shop has cross-classified her customers in the following manner:

	Purchase	No purchase	Total
Female	50	160	210
Male	70	20	90
	120	180	300

(a) What is the probability that a customer is a female?

(b) If a purchase is made, what is the probability that the purchaser was a male?

(c) Is the likelihood of a purchase independent of who makes the purchase?

32. A candidate for political office runs spot announcements on both TV and radio. A recent survey of registered voters revealed that 50 percent of those interviewed had seen at least one of the TV announcements; of this 50 percent, 60 percent had heard the candidate on radio. Of all the people interviewed, however, 70 percent had heard the candidate on radio.

(a) What is the probability that a registered voter has seen the candidate on TV and heard him on radio?

(b) What is the probability that a voter has seen or heard the candidate?

33. If in problem 32 a voter is selected who has heard the candidate on radio, what is the probability that the voter has also seen the candidate on TV?

34. Professor Taylor wins 30 percent of the racquetball games he plays. When he wins, 80 percent of the time he is playing one of his graduate students. When he loses, 10 percent of the time he is playing a graduate student. Given that he is playing a graduate student, what is the probability that Dr. Taylor will win his next racquetball game?

35. Extensive research has shown that 70 percent of all applicants for a particular graduate program possess the ability to succeed in graduate study. The University requires that all applicants take a battery of tests. Past experience has shown that 80 percent of the students who score well on the test are successful in graduate school. However, 40 percent of the students who score poorly on the exam also do well in graduate school. Given that a student does well in graduate school, what is the probability he scored well on the test?

36. When Jim Leadfoot drives home, he has three different roads to take, the interstate, the state road, and a county road. He can be late or on time regardless of the road he takes, however, if he goes by interstate he is late 10 percent of the time. He is late and takes the state road 30 percent of the time. Of all trips home, he is late 37 percent of the time. On any given day when he starts home, he has a 20 percent chance of taking the county road and a 50 percent chance of going on the state road. If he comes home on time one day, what is the probability that he used the interstate?

Case Study

Mr. Spice of Spice Sweets, Inc., has been looking over the sales records in an attempt to understand better the marketing strategy of his firm. The three different products sold by his company (chocolate candy, sugarless candy, and gum) are distributed through two different types of stores (regular and discount) and displayed in two different locations within the stores (in candy departments and at the checkout registers). Last year his company sold $1,352,000 worth of candy.

Spice Sweets, Inc. do not market any of their chocolate candies through discount stores nor do any of the regular stores display chocolate candy at the checkout registers. Half of last year's revenue was derived from sugarless candies, of which 80 percent was distributed through discount stores. Mr. Spice has records to indicate that discount stores display 70 percent of his sugarless candy items by the cash registers. He also knows that discount stores do not display any of his gum items on the shelves in their candy departments.

Last year's records also showed that $608,400 of candy revenue came from discount stores while $540,800 was the result of chocolate sales. Fifty-nine percent of the revenue for last year resulted from candy sold through candy departments while 4 percent of the total revenue was from sugarless candy sold by regular stores displayed at their checkout registers.

Draw a tree diagram to illustrate the relationships and probabilities associated with the different events in this case:

1. Mr. Spice would like to know how many dollars worth of gum was sold through regular stores from the candy department.

2. What percent of total sales comes from the sale of gum?

3. What other analyses, if any, can you make from the data given in this case?

Case study contributed by Richard Ford, University of Arkansas at Little Rock.

BASIC STATISTICS

4

The Random Variable and Probability Distributions

In the previous chapter we were concerned with defining and calculating probability numbers. In this chapter we define the term *random variable* and show that there are various ways for the occurrence of a random variable to be described. These ways to describe random variables are known as *probability distributions*.

The Random Variable

A random variable is a variable whose value is determined by the chance outcome of an experiment. To illustrate, let us consider a box that contains a red and a green ball. We shall define the outcomes of interest to be the results of two draws, with replacement, from the box. The possible outcomes of the experiment are shown in figure 4.1. Now if we are interested in how many times the green ball occurs for each outcome, then let the random variable be the *number* of green balls per outcome.

Table 4.1 shows the possible values of the random variable. Notice that the first column in the table represents the outcomes of the experiment and that the second column is the value of the random variable resulting from each outcome.

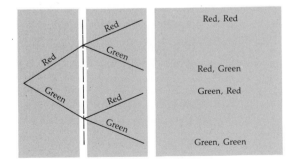

Table 4.1
Random variable: the number of green balls

Outcomes	Number of green balls (the random variable)
Red Red	0
Red Green	1
Green Red	1
Green Green	2

Classification of Random Variables

A random variable will display one of two classifications. It will either be a *discrete* random variable or a *continuous* random variable.

A discrete random variable is a random variable that may assume a countable or limited number of quantitative values. Table 4.1 is an example of a discrete random variable. The numerical values 0, 1, 1, 2 are certainly countable as there are only three values possible. Thus these numerical values represent a discrete random variable.

In contrast to the discrete random variable is the continuous random variable. *If, over a range, the random value may assume any numerical value in the range, the random variable is called a continuous random variable.* Notice that the number of values that the random variable might assume would not be countable or finite. That is, the random variable might take on any one of an infinite number of numerical values; thus both the scale and the random variable are said to be continuous.

Probability Distributions

If we know all possible outcomes of a random experiment, and if we know all values that the random variable can assume, the probability that a particular random variable might occur can be determined. That is, we could determine the probability that the random variable X would take on one of the possible values, namely x_i. Symbolically, this would be written $P(X = x_i)$. Further, if we determine all the probability numbers for the random variable so that

$$0 \leq P(X = x_i) \leq 1$$

and

$$\sum_{i=1}^{n} P(X = x_i) = 1,$$

then we have established a probability distribution for the random variable X. Thus, very generally, we can think of a probability distribution as the *listing* of the probabilities for each numerical value that the random variable might assume. There are differences, however, between *discrete* and *continuous* probability distributions. We shall discuss discrete probability distributions first.

Discrete Probability Distributions

To illustrate the general concepts of a probability distribution and of a cumulative probability distribution, we shall employ the discrete random variable. Consider, if you will, the data presented in table 4.1. Looking at the table, we observe that there are four outcomes (Red Red, Red Green, Green Red, Green Green) that determine the values of the random variable of interest. Further we observe that each outcome occurs only once. Since each outcome occurs only once, and there are four outcomes, the probability of any outcome can be determined *a priori* to be 0.25; thus a third column concerning probability might be added to table 4.1 yielding table 4.2.

In table 4.2 we see that probabilities have been assigned for each value of the random variable (number of green balls). This column of probabilities coincides with our definition of a probability distribution, which is the listing of the probability numbers for each numerical value that the random variable might assume. And if the sum of these probabilities is 1, then we know that we have a complete probability distribution. In our illustration, the probabilities do sum to 1, so we have constructed the probability distribution for the number of green balls that occur in two draws, with replacement, from a container with a red and green ball.

Notice in table 4.2 that the probability associated with one green ball is really 0.50. This is obtained by adding the probability associated with Red Green to the probability associated with Green Red. Since we are interested in only the random variable

Table 4.2
Outcomes, random variables, and probabilities

Outcomes	Number of green balls	Probabilities
Red Red	0	0.25
Red Green	1	0.25 ⎫
Green Red	1	0.25 ⎬ 0.50
Green Green	2	0.25

Table 4.3
Cumulative probability distribution

Number of green balls (x_i)	Probability $P(X \leq x_i)$
0	0.25
1	0.75
2	1.00

(the number of green balls), there are only three numerical values of the random variable in this case.

Table 4.2 can be converted easily into a cumulative probability distribution such as that displayed in table 4.3. In this table the probability column is interpreted as the probability that the random variable will be equal to or less than a given numerical value x. Symbolically, the probability column is

$P(X \leq x_i)$

Thus table 4.3 can be developed in the following manner:

(1) $P(X \leq 0) = P(X = 0)$
    $P(X \leq 0) = 0.25,$

(2) $P(X \leq 1) = P(X = 0) + P(X = 1)$
    $= 0.25 + 0.50$
    $P(X \leq 1) = 0.75,$

(3) $P(X \leq 2) = P(X = 0) + P(X = 1) + P(X = 2)$
    $P(X \leq 2) = 0.25 + 0.50 + 0.25$
    $P(X \leq 2) = 1.0.$

These results are presented in table 4.3.

Now that we have developed the concepts of a probability distribution and of a cumulative probability distribution, we are ready to investigate certain specific probability distributions. First, we will examine four well-known discrete probability distributions.

The Uniform Distribution

If all possible numerical values that a random variable may assume have an equal probability of occurring, then the resulting probability distribution is said to be a *uniform* distribution. The general equation for a uniform distribution is[1]

$$P(X = x_i) = \frac{1}{N}. \tag{4.1}$$

---

1. For discrete distributions, such equations are called probability mass functions (*pmf*). Notice we are using the notation $P(X = x_i)$.

This equation calculates the probability that the random variable X will assume the numerical value x_i where there are N values.

A common illustration of the uniform distribution is the single roll of a fair die. Since there are six surfaces on a die and each side has an equal chance to occur, the probability mass function would be

$$P(X = x_i) = 1/6 \quad \text{for } x_i = 1, 2, \ldots, 6.$$

By defining $N = 6$, we have described a specific uniform distribution. $P(X = x_i) = 1/6$ tells us that the random variable X is uniformly distributed, and the probability that X equals any of the possible values is 1/6. Such a distribution as this is called a *one-parameter distribution*, because there is only one parameter, N, in the probability equation.

The Binomial Distribution

Suppose that, in tossing three fair coins, we want to determine the probability that exactly two heads will show. The question that arises is how to proceed.

One approach would be to use the counting schemes learned in chapter 3. Using the multiple choice procedure, we can determine that there are eight different arrangements possible when a coin is tossed three times: $M_x^n = n^x = M_3^2 = 2^3 = 8$. These eight arrangements are shown in figure 4.2.

Next, we need to determine the number of ways that two heads might occur. In figure 4.2 these are the circled outcomes. Thus the probability of two heads in three tosses of a fair coin is

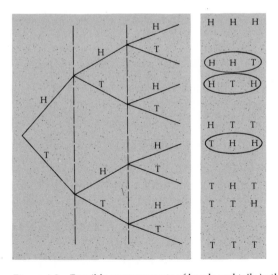

Figure 4.2   Possible arrangements of heads and tails in three tosses

$$P(\text{two heads, three tosses}) = \frac{\text{Number of ways to}}{\text{Number of outcomes}} \atop {\text{get two heads}}$$

$$P(\text{two heads, three tosses}) = \frac{\text{Number of ways to get two heads}}{\text{Number of outcomes from three tosses}}$$

$$= 3/8 = 0.375.$$

Obviously, for large experiments with many outcomes the preceding procedure would not be very effective. Fortunately, however, there is an expression for the binomial distribution just as there was for the uniform distribution. The general equation for a binomial distribution is

$$P(X = x) = C_x^n \pi^x (1 - \pi)^{n-x} \tag{4.2}$$

which calculates the probability of x occurrences in n trials. The term π is the probability that x occurs on any one trial. Thus, for our preceding probability of two heads in three tosses of a fair coin, equation (4.2) would yield

$$P(X = 2) = C_2^3 (0.50)^2 (0.50)^1$$

$$= \frac{3!}{2!(3 - 2)!} (0.50)^2 (0.50)$$

$$= \frac{3!}{2!} (0.25)(0.50)$$

$$= \frac{3 \cdot 2 \cdot 1}{2 \cdot 1} (0.125)$$

$$P(X = 2) = 0.375.$$

Equation 4.2, the binomial equation, is valid under three conditions. They are

1. *Each trial is dichotomous; there are only two possible outcomes such as head–tail or success–failure.*

2. *On each trial, the probability of success (head or tail) is equal to π and remains constant.*

3. *There are n independent trials.*

Example 4.1

▷ Consider a jar containing four green chips and six red ones. Suppose four chips are drawn *with replacement*.

(*a*) What is the probability that exactly three of the four chips drawn are green?

(*b*) What is the probability that at least one chip is green?

(c) What is the probability that at most two of the four chips are green?

The solution to this probability problem can be found easily by using the binomial formula, where

$$\pi = 4/10 = 0.40,$$
$$1 - \pi = 6/10 = 0.60,$$
$$n = 4.$$

Thus the specific binomial equation for this situation is

$$P(X = x) = C_x^4(0.4)^x(0.6)^{4-x}.$$

(a) To answer the first part of the question, we simply substitute 3 for x. Therefore

$$P(X = 3) = C_3^4(0.4)^3(0.6)^1$$
$$= (4)(0.064)(0.6)$$
$$P(X = 3) = 0.1536.$$

(b) The solution to the second part of the question requires that we find the probabilities for $X = 1$, $X = 2$, $X = 3$, and $X = 4$ and sum them. That is,

$$P(X > 0) = P(1) + P(2) + P(3) + P(4).$$

However, since X can be only 0, 1, 2, 3, or 4, we may simplify the calculations somewhat by taking advantage of the complementarity characteristic of probability numbers. Specifically,

$$P(X > 0) = 1 - P(X = 0)$$
$$= 1 - C_0^4(0.40)^0(0.60)^4$$
$$= 1 - 0.1296$$
$$P(X > 0) = 0.8704.$$

(c) To say "at most two chips are green" is to say there can be no more than two green chips. In other words, there can be 0, 1, or 2 green ones. Thus we want $P(X = 0, 1, \text{or } 2) = P(0) + P(1) + P(2)$, or

$$P(X \le 2) = C_0^4(0.4)^0(0.6)^4 + C_1^4(0.4)^1(0.6)^3 + C_2^4(0.4)^2(0.6)^2$$
$$= 0.1296 + 0.3456 + 0.3456,$$
$$P(X \le 2) = 0.8208. \triangleleft$$

Example 4.2

▷ A particular telephone system for Public Surety Insurance Agency is busy 30 percent of the time. Suppose five separate calls are made to Public Surety. What is the probability that one of the five calls gets a busy signal? What assumption are you making regarding the different calls?

Here again we are faced with a probability problem that suggests the binomial. Let X equal the number of busy calls; X can possibly be 0, 1, 2, 3, 4, or 5 in this case. Also $\pi = 0.3$, and $1 - \pi = 0.7$. Using the binomial function, we find that

$$P(X = 1) = C_1^5 (0.3)^1 (0.7)^4$$
$$P(X = 1) = 0.3602.$$

The underlying assumption made here is that each call coming to Public Surety is independent of all other calls, which may not really be a good assumption. ◁

After studying the preceding examples, you have probably realized that computations using equation (4.2) could become quite time-consuming. For this reason tables have been developed to make using the binomial much easier. Appendix D and appendix E are provided to assist in obtaining binomial probabilities for various values of π, n, and x. Let us demonstrate the use of these tables by determining the answers for example 4.1.

In example 4.1(a), appendix D (additional discussion on the use of the appendix is provided at the beginning of appendix D) can be used to determine $P(X = 3)$. Across the top of each page in appendix D are the various values of π that are provided in the table, and down the left side of each page are the various values of n and x. Locating $n = 4$ and then $x = 3$, we go across the row until we are under the column headed $\pi = 0.40$. At the intersection of the row and column is the value, 0.1536, which we previously calculated as $P(X = 3)$.

In example 4.1(c), it is possible to determine $P(X \le 2)$ from appendix E (additional discussion on the use of the appendix is provided at the beginning of appendix E). In appendix E, across the top of each group of numbers is the value of n. Locating $n = 4$, we then locate the intersection of $\pi = 0.40$ and $x = 2$ resulting in the value, 0.8208. Thus $P(X \le 2) = 0.8208$.

Finally, in example 4.1(b), we have to be a little more careful in our use of appendix E. Appendix E determines $P(X \le x)$, so we will have to subtract as was done in the example to determine $P(X > 0)$. Specifically, we locate the value for $P(X \le 0)$ in the table and subtract the value from 1. Thus, using appendix E, when $n = 4$, $\pi = 0.40$, and $x = 0$, then $P(X \le 0) = 0.1296$, and $P(X > 0) = 1.000 - 0.1296 = 0.8704$.

In the case of the binomial distribution, the events relating to the random variable were independent. Recall that in the binomial example the chips were drawn *with replacement*.

The *hypergeometric* mass function is an appropriate model for estimating probabilities of a random variable when selection is made *without replacement* from a finite number of items. In other words, if it is *not reasonable* to assume that the events are independent, the hypergeometric distribution is appropriate for the random variable. As in the case of the binomial, the dichotomy (either–or situation) condition applies to the hypergeometric distribution.

Stated alternatively, the hypergeometric distribution is appropriate when a sample of size n is drawn, without replacement, from a finite population of size N. Under these conditions the probability that the sample contains exactly x items of the first type and $n - x$ items of the second type can be determined given that the population N contains N_1 items of the first type and N_2 items of the second type.

The formula for the hypergeometric probability mass function (pmf) is shown as

$$P(X = x) = \frac{C_x^{N_1} C_{n-x}^{N_2}}{C_n^N},\qquad(4.3)$$

where

    x = value of the random variable X,

    N_1 = the number of one kind of outcome (say, a success) from which the selection is made,

    N_2 = the number of the other kind of outcome (say, failure) from which the selection is made,

    n = the number of items in the random sample,

    $N = N_1 + N_2$.

Example 4.3

▷ Consider a jar containing four green chips and six red ones. Suppose four chips are drawn *without* replacement. (They may be drawn all at once.)

(*a*) What is the probability that exactly three of the four chips drawn are red?

(*b*) What is the probability that at least one chip is red?

A hypergeometric pmf is the appropriate one for making these probability estimates. We have a dichotomy, but the events are not independent, because the color of the chips drawn each time influences the probabilities of obtaining a red or green chip on the next draw. To answer both questions, let us first define and assign values to the terms of the hypergeometric pmf:

N_1 = number of red chips in the jar = 6,
N_2 = number of green chips in the jar = 4,
n = number of chips to be selected or drawn = 4,
x = value of the random variable, the number of red chips drawn = 3,
$N = N_1 + N_2 = 10$.

(a) $P(X = 3) = \dfrac{C_3^6 C_1^4}{C_4^{10}}$

$$= \dfrac{\dfrac{6!}{3!3!} \dfrac{4!}{1!3!}}{\dfrac{10!}{4!6!}}$$

$$P(X = 3) = \dfrac{(20)(4)}{210} = \dfrac{8}{21} = 0.381.$$

(b) Here we can find the probability of getting 0 red chips and subtract this probability from 1. This difference will give us the probability of getting 1, 2, 3, or 4 red chips. Thus let $x = 0$, and using equation (4.3),

$$P(X = 0) = \dfrac{C_0^6 C_4^4}{C_4^{10}}$$

$$= \dfrac{\dfrac{6!}{0!6!} \dfrac{4!}{4!0!}}{\dfrac{10!}{4!6!}}$$

$$P(X = 0) = \dfrac{1}{210}.$$

Therefore

$$P(X \geq 1) = 1 - \dfrac{1}{210} = \dfrac{209}{210} = 0.9952. \lhd$$

The Poisson Distribution

Whereas the binomial distribution can be used to determine the probability for a certain number of occurrences in a given number of trials, it is sometimes desirable to be able to determine the probability of the number of occurrences per unit of time or space. When the frame of reference is a unit of time or a unit of space rather than the number of trials, it is appropriate to use the Poisson distribution. The Poisson probability function is shown by the following equation:

$$P(X = x) = \frac{\mu^x e^{-\mu}}{x!},$$ (4.4)

where

x = value of the random variable X,
μ = the average number of occurrences per unit of time or space,
e = the base of the natural log system (2.71828).

Examples of situations where the Poisson distribution is particularly useful for computing probabilities include events within a specified time frame: how many machines break down, how many vehicles arrive at a toll booth, how many customers need service, or how many calls come in at a switchboard. Other examples that do not necessarily involve time include, number of defects per yard of woven material, number of typeset errors per page of print, and number of fatalities per passenger mile of driving. In each of the preceding situations, the Poisson distribution can be employed to calculate the probabilities associated with the particular number of occurrences per unit of measurement if the following conditions are met:

1. μ, the average number of occurrences per unit of time or space, remains constant;

2. the number of occurrences in one time period is independent of previous occurrences;

3. μ is proportional to the length of time or space and decreases as the length decreases so that the probability of more than one occurrence in a very short length is practically zero.

The following examples demonstrate the use of the Poisson distribution.

Example 4.4

▷ The Key Insurance Agency receives an average of two calls per minute through its switchboard. Use the Poisson distribution and determine

(a) the probability of three calls in any one minute,

(b) the probability of five calls in two minutes.

(a) Since $\mu = 2$, the solution is

$$P(X = x) = \frac{\mu^x e^{-\mu}}{x!},$$

$$P(X = 3) = \frac{2^3 e^{-2}}{3!}$$

$$= \frac{(8)(2.71828)^{-2}}{6}$$

$$P(X = 3) = 0.1804.$$

(b) In this case $\mu = 2 \cdot 2 = 4$ (two calls per minute and there are two minutes) since the time measurement associated with μ must be the same as the time measurement associated with x:

$$P(X = 5) = \frac{4^5 e^{-4}}{5!}$$

$$= \frac{(1024)(2.71828)^{-4}}{120}$$

$$P(X = 5) = 0.1563. \triangleleft$$

Solution Note:

The preceding probabilities could have been obtained directly from appendix F. Use of the appendix is explained in a section at its beginning. Why don't you rework example (4.4) using appendix F now!

Example 4.5

$\triangleright$ On a particular thruway there has been an average of three accidents per day. Use appendix F to determine

(a) the probability of no accidents on a given day,

(b) the probability of three or four accidents on a given day.

(a) Using appendix F and $\mu = 3$, the column heading for $\mu = 3$ is located, and the row heading for $x = 0$ is located. The intersection of this particular row and column yields the answer. Thus

$$P(X = 0) = 0.0498.$$

(b) Again using appendix F and $\mu = 3$, the column heading for $\mu = 3$ is located, and then the row headings for $x = 3$ and $x = 4$ are located. From the appendix the two required probabilities are read as

$$P(X = 3) = 0.2240$$
$$P(X = 4) = 0.1680.$$

And the solution is

$$P(X = 3 \text{ or } 4) = 0.2240 + 0.1680 = 0.3920. \triangleleft$$

The Poisson distribution can also be used as an approximation for the binomial when μ is small. In this situation the *mean* of the Poisson distribution μ is determined as $n\pi$.

Example 4.6

▷ Consider the situation of a very large box of electrical fuses, 4 percent of which are known to be defective. Determine the probability of selecting 5 fuses at random and obtaining 1 defective fuse. The problem clearly suggests use of the binomial distribution which yields

$$P(X = 1) = C_1^5(0.04)^1(0.96)^4$$
$$= 5(0.04)(0.8493)$$
$$P(X = 1) = 0.1699.$$

However, using the Poisson approximation, we let

$$\mu = n\pi = 5(0.04) = 0.20,$$

and, then, using appendix F and $x = 1$,

$$P(X = 1) = 0.1637.$$

Thus the approximation is accurate through the second decimal place and should be considered quite satisfactory. ◁

Continuous Probability Distributions

The distributions discussed so far have all been discrete distributions. Before ending our discussion of probability distributions, however, it is important to introduce continuous distributions. In particular, we need to become familiar with the normal distribution and the exponential distribution. First, let us discuss the exponential distribution.

The Exponential Distribution

The Poisson distribution can be used to calculate the probability of the number of occurrences per unit of time or space. Suppose, however, that rather than determining the number of occurrences per unit of time, it is desirable to determine the length of time or space between occurrences. We can do this with the exponential distribution. In particular, if a random process follows the Poisson distribution, the probabilities of the length of time between two occurrences can be determined using the exponential distribution.

The exponential distribution is a continuous distribution, and the function which determines the shape is[2]

$$f(x) = \frac{1}{\lambda}e^{-x/\lambda}, \tag{4.5}$$

where
$f(x) =$ the ordinate value (height) when $X = x$,

---

2. For continuous variables, the equations are called probability density functions (pdf). Notice for the continuous distribution we are using $f(x)$.

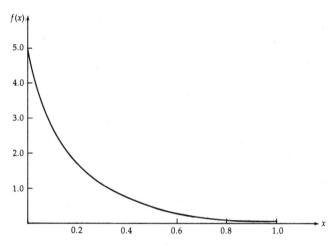

Figure 4.3   Exponential curve ($\lambda = 0.2$)

λ = average time between occurrences (the reciprocal of the Poisson parameter, μ),

x = any value of the random variable X,

e = 2.71828, the base of the natural log system.

Figure 4.3 shows an exponential curve developed using equation (4.5). In figure 4.3, λ was assumed to be 0.2 minutes between occurrences, which would correspond to $\mu = 5$ in a Poisson distribution. That is, $\lambda = 0.2$ minutes (12 seconds) is associated with a Poisson process that averaged 5 occurrences per minute.

Equation (4.6) can be used to calculate the cumulative probability in the right end of the exponential curve. Equation (4.6) will be employed in examples 4.7 and 4.8:

$$P(X > x) = e^{-x/\lambda}, \tag{4.6}$$

where x = time before an occurrence.

Example 4.7

▷ An electronic part has a performance history that follows the exponential curve so that average time between failures is 150 hours. Thus, for any part, what is the probability that it will operate more than 300 hours?

$P(X > x) = e^{-x/\lambda}$,
$P(X > 300) = e^{-300/150} = e^{-2.0}$,
$P(X > 300) = 0.13534$ (from appendix G). ◁

Example 4.8

▷ At one of the large eastern airports, FAA records show that the average interval between flight arrivals is 4 minutes between 7 A.M. and 9 A.M. The FAA records also indicate that the intervals between arrivals are exponentially distributed. From time to time

BASIC STATISTICS

the control tower suffers equipment failure. When this happens, it takes no less than 4 but not more than 8 minutes to repair. What is the probability that a plane will arrive during any given equipment failure?

Using equation (4.6), we can answer this question by determining $P(X > 4)$ and $P(X > 8)$ and then subtracting the two probabilities:

$P(X > 4) = e^{-4/4}$
$P(X > 4) = 0.36788$ (from appendix G),
$P(X > 8) = e^{-8/4}$
$P(X > 8) = 0.13534$ (from appendix G).

Thus

$P(4 < X < 8) = 0.36788 - 0.13534$
$P(4 < X < 8) = 0.23254.$ ◁

The Normal Distribution

Distributions of data are somewhat like people—they come in many shapes. Figure 4.4 contains a few of the more common ones, but the one that is of particular significance is the normal distribution. Why? Because the normal distribution can be used to approximate the distribution of many physical phenomena, and forthcoming chapters deal with several topics based on the knowledge of the normal distribution.

With all the possible shapes—both human and statistical—how do you recognize a normal distribution when you see one? Knowing a few characteristics of normal distributions might help:

1. Continuous—the values of the variable are infinitely divisible and infinite in number.

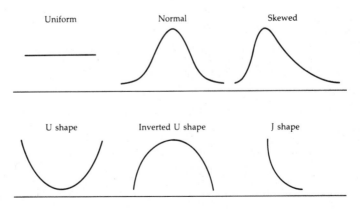

Figure 4.4   Various distribution shapes

2. Symmetrical—the mean, median, and mode are equal.

3. Asymptotic—the left- and right-hand tails of the distribution approach the abscissa but reach it only at infinity.

4. Unimodal—there is only one mode.

These characteristics will help you to recognize many distributions as being nonnormal, but they will not let you be *certain* that a particular distribution is a normal distribution. The precise shape of a normal distribution is outlined by the following equation:

$$f(x) = \frac{1}{\sqrt{2\pi\sigma^2}} e^{-(x-\mu)^2/2\sigma^2},\tag{4.7}$$

where

$f(x)$ = the ordinate value (height) when $X = x$,

x = any value of the random variable X,

μ = the mean of X,

σ^2 = the variance of X,

π = 3.1416 (a constant),

e = 2.71828 (base of natural log system—a constant).

Using equation (4.7), you can determine the coordinates $(x, f(x))$ for a sufficiently large set of points to produce a normal curve (a graph of a normal distribution) for any arithmetic mean and variance.

The statement of the exact shape of a normal curve enables the statistician to rely upon the existence of a definite relationship between the arithmetic mean, the variance (or standard deviation), and the area under the curve. Specifically, for any normal distribution the arithmetic mean plus one standard deviation ($\mu + 1\sigma$) will always encompass 34.13 percent of the area under the curve, as shown in figure 4.5. The mean plus two standard deviations ($\mu + 2\sigma$) includes 47.72 percent of the area, and the mean plus three standard deviations ($\mu + 3\sigma$) bounds 49.86 percent of the area. And, since the normal curve is symmetrical, these values also hold when the standard deviation is subtracted from the mean.

There are an infinite number of possible normal distributions, each dependent upon the values of μ and σ. However, we can convert any normal distribution (with parameters μ and σ) into a standard form. This standardized distribution is referred to as the *standard normal distribution*.

The standard normal distribution is a normal distribution with parameters $\mu = 0$ and $\sigma = 1$. Appendix H contains the areas under

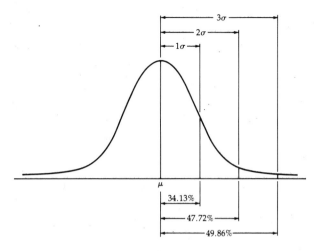

Figure 4.5 Areas under a normal curve

the curve for the standard normal distribution. In appendix H we see that areas under the normal curve are listed for various values of Z, where Z represents the numer of standard deviations away from μ. That is, if $Z = 1$, then the area between μ and $(\mu + 1\sigma)$ can be read from appendix H. Thus we can read in the table the area between the mean and one standard deviation from the mean, which is 0.3413. Look back at figure 4.5, and you can see the area that we are discussing. Consequently, we use appendix H to determine areas under the normal curve for any normally distributed random variable.

Fortunately, it is not necessary to convert an entire normal distribution to a standard normal distribution in order to use appendix H. Rather it requires only that we measure the number of standard deviations from the mean associated with that portion of the distribution that is of interest. This can be accomplished using equation (4.8):

$$Z = \frac{X - \mu}{\sigma},$$
(4.8)

where
Z = the number of standard deviates,
X = some value of the random variable—a point on the abscissa,
μ = the mean of the normal distribution,
σ = the standard deviation of the normal distribution.

Thus we express the distance from the mean μ along the abscissa in standard deviates (Z's) rather than in units of the variable (dollars, bushels, etc.). Look at the following examples in order to see how appendix H is used.

Example 4.9

▷ A small town has maintained records showing that the daily high temperature during the spring has averaged 15° Celsius with a standard deviation of 4° Celsius. Assume the random variable (high temperature) is normally distributed and determine the following:

(*a*) The percentage of days that will have a high temperature between 15° C and 19° C.

(*b*) The percentage of days that the high temperature will exceed 21° C.

(*c*) The probability that the high temperature will be less than 12° C.

(*d*) The probability that the high temperature will be between 13° C and 14° C.

(*e*) The high temperature that will be exceeded only 10 percent of the time.

(*a*) In order to determine the area under a normal curve (expressed as a percentage) between the mean (15) and a value of X(19), it is necessary to determine the appropriate Z and then use appendix H. Look at figure 4.6 to see the area that is being determined:

$$Z = \frac{X - \mu}{\sigma}$$

$$= \frac{19 - 15}{4},$$

$$Z = 1.0.$$

Referring to appendix H, when Z = 1, the area is 0.3413. Thus 34.13 percent of the days have temperatures between 15° and 19°.

Figure 4.6 Area under normal curve in example 4.9(a)

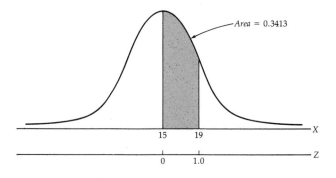

BASIC STATISTICS

Figure 4.7 Area under normal
curve in example 4.9(b)

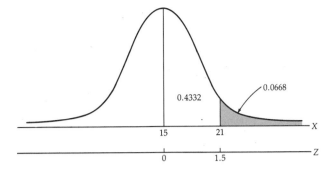

(b) In order to determine the area under a normal curve that is above a particular value of X, such as above 21, as shown in figure 4.7, it is first necessary to determine the area between 15 and 21. Then the area between 15 and 21 must be subtracted from 0.5000:

$$Z = \frac{X - \mu}{\sigma}$$

$$= \frac{21 - 15}{4}$$

$Z = 1.5.$

Using appendix H, when $Z = 1.5$, the area $= 0.4332$. Thus the area above 21 is $0.5000 - 0.4332 = 0.0668$; or the percentage of days with high temperature above 21° is 6.68 percent.

(c) To determine the area under a normal curve less than a particular value of X, such as below 12, as shown in figure 4.8, initially we determine the area between 12 and 15. Then the area between 12 and 15 must be subtracted from 0.5000:

$$Z = \frac{X - \mu}{\sigma}$$

$$= \frac{12 - 15}{4}$$

$Z = -0.75.$

From appendix H, if $Z = -0.75$, the area $= 0.2734$. When using appendix H, ignore the sign (plus or minus) of Z because the sign identifies only whether we are looking on the left (minus sign) side of the mean or the right (plus sign) side of the mean. Thus the area below 12 is $0.5000 - 0.2734 = 0.2266$, and the probability that the high temperature will be less than 12° C is 0.2266.

(d) When a problem requires that we find the area under a normal curve between two values of X that are both either less than

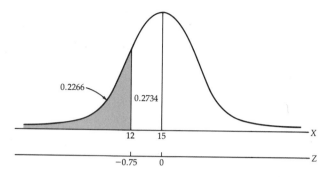

Figure 4.8   Area under normal curve in example 4.9(c)

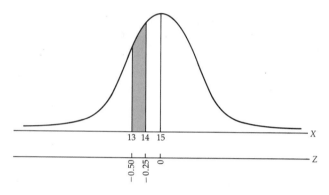

Figure 4.9   Area under the curve in example 4.9(d)

or more than μ, as illustrated by the shaded area between 13 and 14 in figure 4.9, we must solve the problem somewhat indirectly. First, we find the area between 13 and 15. Next, we determine the area between 14 and 15. Finally, we subtract the second area from the first so that what remains is the area between 13 and 14:

$$Z = \frac{X - \mu}{\sigma},$$

$$Z = \frac{13 - 15}{4} \qquad Z = \frac{14 - 15}{4}$$

$$Z = -0.50, \qquad Z = -0.25.$$

Referring to appendix H, when $Z = -0.50$, the area $= 0.1915$ and, when $Z = -0.25$, the area $= 0.0987$. Therefore the probability that the high temperature will be between 13° C and 14° C is $0.1915 - 0.0987 = 0.0928$.

(e) Sometimes we know the probability and must determine the value of X associated with that probability. Look at figure 4.10 to

BASIC STATISTICS

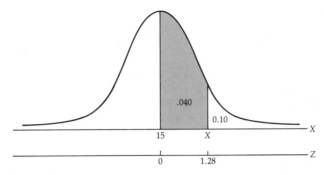

Figure 4.10   Area under normal curve in example 4.9(e)

get the idea. You are already familiar with equation (4.8), and we can restate that equation so that you can solve for X:

$$Z = \frac{X - \mu}{\sigma}$$

$$Z\sigma = X - \mu$$

$$X = \mu + Z\sigma.$$

Using this variation of equation (4.8), we must first determine the value of Z before we can solve for X. Looking again at figure 4.10, the area between 15 and X is 0.4000, or 0.5000 − 0.1000. Thus we use appendix H in a reverse procedure. That is, we begin with areas rather than Z values. Looking in appendix H, the area closest to 0.4000 is 0.3997 which corresponds to $Z = 1.28$. Consequently, the unknown value of X is 1.28 standard deviations to the right of μ. Now we can solve for X:

$$X = \mu + Z\sigma$$
$$= 15 + (1.28)4 = 15 + 5.12,$$
$$X = 20.12.$$

Thus the temperature that should be exceeded only 10 percent of the time is 20.12° C. ◁

At this point let us emphasize a few things about the normal distribution:

1. The area under a normal curve contains all (100 percent) of the values of the variable in the normal distribution.

2. One-half of the area of the curve, and thus one-half (50 percent) of the values of the variable, are less than and one-half are more than the arithmetic mean.

3. The ranges along the abscissa are infinitely divisible, and the area under the curve at any specific point is 0.

4. Standard deviates are not additive; that is, it is necessary to add or subtract areas (not deviates) to determine the area under the curve over some specific range of the variable. Let us consider another example.

Example 4.10

▷ A gasoline distributor has maintained records that indicate sales of Special gasoline average 3,500 gallons per week with a standard deviation of 250 gallons. Assuming that sales are normally distributed, determine the probability of weekly sales between 3,300 and 3,800 gallons. Look at figure 4.11 to see the areas that represent the solution to the problem.

Area 1 is determined as follows:

$$Z_1 = \frac{X - \mu}{\sigma}$$

$$Z_1 = \frac{3,300 - 3,500}{250} = -0.80.$$

From appendix H, area 1 = 0.2881.

Area 2 is determined in the same manner as area 1:

$$Z_2 = \frac{X - \mu}{\sigma}$$

$$Z_2 = \frac{3,800 - 3,500}{250} = 1.2.$$

From appendix H, area 2 = 0.3849.

Now, to obtain the area between 3,300 and 3,800, it is necessary to add area 1 and area 2 together. Thus the probability of sales between 3,300 and 3,800 gallons is

$P(3,300 < X < 3,800) = 0.2881 + 0.3849$
$P(3,300 < X < 3,800) = 0.6730.$ ◁

Try some of the problems at the end of this chapter that involve normal distributions. If you cannot work them, or do not under-

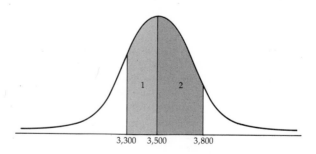

Figure 4.11   Areas under normal curve in example 4.10

116

stand why a particular procedure should be used, reread the normal distribution section of this chapter. An understanding of the normal distribution will help you comprehend more clearly some of the material presented in subsequent chapters.

Normal Approximation of the Binomial

The expression $P(X = x) = C_x^n \pi^x (1 - \pi)^{n-x}$ is increasingly cumbersome to work with as n becomes larger. Fortunately, though, when n is large and π is neither very large nor very small, the binomial distribution can be approximated by the normal distribution.

In order to use the normal distribution as an approximation to the binomial, it is necessary to calculate the mean and standard deviation of the binomial distribution. To illustrate this calculation, suppose we investigate the probabilities associated with obtaining either 0, 1, 2, or 3 heads on the three tosses of a fair coin ($\pi = 0.50$) by using appendix D to determine $P(X = x)$. The four outcomes and their respective probabilities are shown in table 4.4. With the information in table 4.4 it is now possible to calculate μ and σ in the same fashion as presented in chapter 2. Specifically, we can do so by using the following equations:

$$\mu = \frac{\Sigma fx}{N},$$

$$\sigma = \sqrt{\frac{\Sigma f(x - \mu)^2}{N}}.$$

Table 4.4
Calculation of the mean and standard deviation for a binomial distribution

Number of heads, x	Probability, $P(X = x)$	$P(X = x) \cdot x$	$(x - \mu)$	$(x - \mu)^2$	$P(X = x) \cdot (x - \mu)^2$
0	0.1250	0.0000	−1.5	2.25	0.28125
1	0.3750	0.3750	−0.5	0.25	0.09375
2	0.3750	0.7500	0.5	0.25	0.09375
3	0.1250	0.3750	1.5	2.25	0.28125
	1.0000	1.5000			0.75000

$$\mu = \frac{\Sigma[P(X = x) \cdot x]}{\Sigma P(X = x)} \qquad \sigma = \sqrt{\frac{\Sigma[P(X = x) \cdot (x - \mu)^2]}{\Sigma P(X = x)}}$$

$$= \frac{1.5}{1.0} \qquad\qquad = \sqrt{\frac{0.75}{1.00}}$$

$$\mu = 1.5 \qquad\qquad \sigma = 0.866$$

However, before using these equations, it is necessary to redefine f so that, instead of meaning frequency, it means $P(X = x)$. This new definition of f can be interpreted to mean that we are weighting each outcome x by the probability that outcome will occur, and the resulting values of μ and σ are then referred to as the expected values of μ and σ. The equations that result from this redefinition of f are

$$\mu = \frac{\Sigma[P(X = x) \cdot x]}{\Sigma P(X = x)}, \tag{4.9}$$

$$\sigma = \sqrt{\frac{\Sigma[P(X = x) \cdot (x - \mu)^2]}{\Sigma P(X = x)}}. \tag{4.10}$$

The actual calculations for the μ and σ of the binomial distribution are shown in table 4.4.

The computations in table 4.4 are rather time-consuming for large values of n. Fortunately, μ and σ can be determined much easier using equations (4.11) and (4.12). Then this mean and standard deviation can be used to provide a normal approximation to binomial probabilities. Generally, the normal approximation is reasonably satisfactory when both $n\pi$ and $n(1 - \pi) > 5$:

$$\mu = n\pi \tag{4.11}$$
$$= 3(0.5)$$
$$\mu = 1.5,$$

$$\sigma = \sqrt{n\pi(1 - \pi)} \tag{4.12}$$

$$= \sqrt{3(0.50)(0.50)}$$

$$= \sqrt{0.75}$$

$$\sigma = 0.866.$$

Example 4.11

▷ Suppose that 45 percent of a telephone operator's service requests are long distance calls. What is the probability of having to place 11 long distance calls in 20 service requests? Using the binomial table in appendix D, and setting $\pi = 0.45$ and $n = 20$, $P(X = 11) = 0.1185$.

Now let us approximate this binomial situation by using the normal distribution. First, we must calculate the mean and standard deviation as follows:

$$\mu = n\pi \qquad \sigma = \sqrt{n\pi(1 - \pi)}$$

$$= 20(0.45) \qquad = \sqrt{20(0.45)(0.55)}$$

$$\mu = 9.0, \qquad = \sqrt{9(0.55)}$$

$$\sigma = \sqrt{4.95} = 2.22.$$

Now look at figure 4.12 to see how the normal distribution (a continuous distribution) is divided up to approximate the binomial distribution (a discrete distribution). As you know, we would ordinarily calculate only the probability of more than or less than 11. Thus, in order to calculate the probability of exactly 11, we must make adjustments so that the normal distribution can be used to determine exact probabilities. As you can see in figure 4.12, the adjustment is to place boundaries (10.5 and 11.5) around the value of interest (11). The purpose is to create an area under the normal curve that can serve as the approximation of $P(X = 11)$. Once the boundaries are established, we then calculate the approximate value for $P(X = 11)$ by determining $P(9 < X < 10.5)$ and subtracting it from $P(9 < X < 11.5)$:

$$Z = \frac{X - \mu}{\sigma} \qquad\qquad Z = \frac{X - \mu}{\sigma}$$

$$= \frac{10.5 - 9}{2.22} \qquad\qquad = \frac{11.5 - 9}{2.22}$$

$$= \frac{1.5}{2.22} \qquad\qquad = \frac{2.5}{2.22}$$

$$Z = 0.68, \qquad\qquad Z = 1.13,$$
$$\text{Area} = 0.2517; \qquad \text{Area} = 0.3708.$$

Thus $P(X = 11) = 0.3708 - 0.2517 = 0.1191$, which is very close to the value (0.1185) from appendix D. ◁

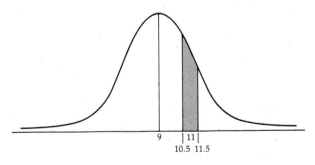

Figure 4.12   Use of the normal distribution to approximate the binomial

Summary of Equations

4.1   Uniform distribution:

$$P(X = x_i) = \frac{1}{N}$$

4.2   Binomial distribution:

$$P(X = x) = C_x^n \pi^x (1 - \pi)^{n-x}$$

4.3   Hypergeometric distribution:

$$P(X = x) = \frac{C_x^{N_1} C_{n-x}^{N_2}}{C_n^N}$$

4.4   Poisson distribution:

$$P(X = x) = \frac{\mu^x e^{-\mu}}{x!}$$

4.5   Exponential distribution:

$$f(x) = \frac{1}{\lambda} e^{-x/\lambda}$$

4.6   Cumulative probability in the right tail of an exponential curve:

$$P(X > x) = e^{-x/\lambda}$$

4.7   Ordinate value of a normal curve:

$$f(x) = \frac{1}{\sqrt{2\pi\sigma^2}} e^{-(x-\mu)^2/2\sigma^2}$$

4.8   The number of standard deviates:

$$Z = \frac{X - \mu}{\sigma}$$

4.9   Arithmetic mean of a binomial distribution:

$$\mu = \frac{\Sigma[P(X = x) \cdot x]}{\Sigma P(X = x)}$$

4.10 Standard deviation of a binomial distribution:

$$\sigma = \sqrt{\frac{\Sigma[P(X = x) \cdot (x - \mu)^2]}{\Sigma P(X = x)}}$$

4.11 Arithmetic mean of a binomial distribution—shortcut method:

$$\mu = n\pi$$

4.12 Standard deviation of a binomial distribution—shortcut method:

$$\sigma = \sqrt{n\pi(1 - \pi)}$$

Review Questions

1. Define
(a) random variable
(b) discrete random variable
(c) continuous random variable
(d) standard normal distribution
(e) probability distribution

2. What are the two characteristics that identify a group of numbers as being a probability distribution?

3. Name the continuous probability distributions presented in this chapter.

4. Name the discrete probability distributions presented in this chapter.

5. State the three characteristics of a binomial distribution.

6. Identify when the Poisson distribution is appropriate.

7. Identify when the exponential distribution is appropriate.

Problems

1. Identify which of the following are probability distributions:
(a) $P(X = x) = x$ for $x = 1/6, 1/3, 1/2$.
(b) $P(X = x) = 2/x$ for $x = 3, 4, 5$.

(c) $P(X = x) = \dfrac{x - 2}{12}$ for $x = 1, 2, 3, 4$.

(d) $P(X = x) = \dfrac{x^2}{14}$ for $x = 1, 2, 3$.

(e) $P(X = x) = \dfrac{x - 3}{9}$ for $x = 3, 4, 5, 6, 7$.

(f) $P(X = x) = \dfrac{x}{3}$ for $x = \dfrac{1}{2}, 1, \dfrac{3}{2}$.

(g) $P(X = x) = \dfrac{x^2 - 1}{50}$ for $x = 2, 3, 4, 5$.

(h) $P(X = x) = \dfrac{1}{x}$ for $x = 2, 4, 6, 12$.

2. Determine the outcomes that could result if a coin is flipped three times. Using these outcomes, develop the probability distribution that results if the random variable is defined as the number of heads that occur.

3. Develop the cumulative probability distribution for problem 2.

4. A company's records yield the following information on the mean monthly sales of its fifty outlets:

Mean monthly sales ($1,000)	Number of outlets
50	2
100	15
150	17
200	13
250	3

(a) Determine the probability distribution for mean monthly sales of outlets.
(b) Construct a cumulative probability distribution for mean monthly sales of outlets.

5. There are equal numbers of red, green, blue, and yellow lights on a Christmas tree. If a light is defective, what is the probability it is a blue one?

6. If there is a 0.32 probability of getting a bill in the mail each day, what is the probability of not getting a bill all week? (Mail is not delivered on Saturday and Sunday.)

7. A given hunter hits the target 80 percent of the time. If he shoots at a target 8 times, what is the probability that he will hit it no more than 2 times?

8. Professor Zippo is late to 20 percent of the classes he teaches. If the Dean checks 10 of his classes, what is the probability that Zippo will be late to 4 or more of them?

9. If 40 percent of all consumers have hospitalization insurance, what is the probability that in a sample of 10 consumers,

(a) exactly 4 will be insured?
(b) at least 4 will be insured?
(c) 8 or more will be insured?
(d) no more than 1 will be insured?

10. Piggy U. has a very large number of classrooms, 50 percent of which are air conditioned. If classes are randomly assigned to classrooms, what is the probability that a student taking 6 classes will have less than 3 in air conditioned rooms?

11. The probability of passing a given test is 0.90. (Consider each test an independent event.) If you take four tests, what is the probability of passing exactly two.

12. In the manufacture of no-deposit-no-return bottles, 5 percent of the bottles are defective. What is the probability in a sample of 7 bottles that there are

(a) no defectives.
(b) 3 defectives.
(c) more than 3 defectives.

13. If actuarial tables indicate that 85 percent of all 10-year olds will survive 40 years, what is the probability that only one of three 10-year olds will survive 40 years?

14. Use appendix D or appendix E to determine the following probabilities;

(a) $P(X = 4)$ if $n = 15$ and $\pi = 0.40$.
(b) $P(X \le 4)$ if $n = 12$ and $\pi = 0.10$.
(c) $P(5 \le X \le 11)$ if $n = 20$ and $\pi = 0.35$.
(d) $P(X = 6)$ if $n = 14$ and $\pi = 0.65$.
(e) $P(X \ge 9)$ if $n = 19$ and $\pi = 0.60$.
(f) $P(5 \le X < 8)$ if $n = 14$ and $\pi = 0.75$.

15. If vehicles arrive at a toll booth at an average of 2 per minute:
(a) What is the probability of no arrivals in one minute?
(b) What is the probability of 1 arrival in one minute?
(c) What is the probability of 5 arrivals in three minutes?

16. The transportation system at a new large airport is designed so that it will have one failure every ten days. What is the probability that it will not fail on the "Grand Opening Day"?

17. The long distance operator receives an average of 3 calls per minute.

(a) What is the probability of no calls in one minute?

(b) What is the probability of 2 calls in one minute?

(c) What is the probability of 4 calls in three minutes?

18. A particular type of cloth averages a defect every two yards. Given that it requires four yards of the cloth to make a piece of clothing, what is the probability that it will contain zero defects?

19. If computer requests from terminals are Poisson distributed with an average of 4 requests per minute, what is the probability of

(a) 2 requests in one minute?

(b) 6 requests in two minutes?

(c) no more than 5 requests in one minute?

20. The coffee machines of a certain vendor break down on the average of 2 per week.

(a) What is the probability of no breakdowns in a week?

(b) What is the probability of 1 breakdown in 1/4 of a week?

(c) What is the probability of 6 breakdowns in 2 weeks?

21. A box contains 100 marbles. Five of the marbles are purple, and the rest are green. If 8 marbles are drawn without replacement, what is the probability of obtaining exactly 2 purple marbles?

22. A retailer has 80 cans of vacuum packed tennis balls. If 3 of the cans contain yellow balls, what is the probability that sample of 10 cans will contain 1 can of yellow balls?

23. A machine produces its output with a fraction defective of 0.06. If a lot consists of 100 items, what is the probability that a sample of 5 items will contain exactly 1 defective?

24. Assume you have a bag containing 20 jelly beans, 5 of which are lemon flavored and 15 of which are strawberry flavored. Assuming you eat 6 of them (randomly selected), what is the probability that less than 3 are lemon flavored?

25. Records indicate that in a small Arkansas community of 50 families, 10 families cheated on their U.S. tax returns. If the IRS selects 5 families to audit, what is the probability none will have cheated.

26. Assume a normal distribution with $\mu = 45$ and $\sigma = 10$, and determine the following probabilities:

(a) $P(X < 65)$ *(d)* $P(X > 50)$
(b) $P(X > 35)$ *(e)* $P(X < 42)$
(c) $P(35 < X < 55)$ *(f)* $P(37 < X < 44)$

27. Assume a normal distribution with $\mu = 30$ and $\sigma = 5$, and determine the following probabilities:

(a) $P(X > 34)$ *(d)* $P(X > 40)$
(b) $P(X < 27)$ *(e)* $P(X < 24)$
(c) $P(31 < X < 33)$ *(f)* $P(27 < X < 32)$
                      (g) $P(25 < X < 29)$

28. A local trout hatchery is preparing to place its latest fish "crop" on the market. However, from past experience they know potential buyers will have certain questions, examples of which are given below. If the weights of the trout are assumed to be normally distributed about a mean of 2.5 pounds, with a standard deviation of 0.5 pounds, provide the answers to the questions. Incidentally, there are 200,000 trout in the "crop."
(a) What proportion of the trout weigh more than 3.0 pounds?
(b) What proportion of the trout weigh between 2.75 and 2.10 pounds?
(c) If only those trout weighing more than 1.80 pounds are marketable, how many trout meet this condition?
(d) One potential buyer likes to know how many trout weigh between 3.00 and 3.50 pounds. What is the answer to her question?

29. Average annual rainfall in our state is 40 inches, with a standard deviation of 5 inches. (Annual rainfall is normally distributed.)
(a) If it takes at least 36 inches of rain to make a "good" rice crop, what is the probability that rice farmers will have enough rain for a "good" crop?
(b) The apple crop will suffer if it rains less than 38 inches or more than 50 inches. What is the probability that the crop will suffer because of rain?
(c) In a recent speech, a political figure said that rainfall of the amount we had this year happened, on the average, only once in a hundred years. (Actually, the politician meant this amount or less.) How much rainfall—in inches—must we have had this year?

30. Research has proven that the life of Christmas tree lights is normally distributed with an arithmetic mean of 4 years and a standard deviation of 0.8 years. What percentage of all Christmas tree lights last more than 3 years?

31. A manufacturer of batteries has determined the mean life of their battery in a smoke detector is 15 months with a standard deviation of 1.5 months. If battery lives are normally distributed, what should be the length of the guarantee if they want to replace a maximum of (a) 5 percent? (b) 2 percent?

32. If $\pi = 0.30$, use the normal approximation for the binomial to determine the probability of 140 defectives in a lot size of 500 items.

33. A manufacturing process ordinarily produces 35 defective bottles per 100 bottles manufactured. If a lot size of 1,000 bottles is manufactured, determine the probability of (a) exactly 370 defectives. (b) more than 370 defectives.

34. A very expensive lightbulb has been designed so that the average time between failures is 1,000 hours. What is the probability a particular bulb will last more than 1,500 hours? (Assume an exponential distribution.)

35. A study has determined that long-distance, operator-assisted calls occur at the rate of 1 every 5 minutes. Assuming that the time interval between occurrences is exponentially distributed, determine the probability that more than 8 minutes will elapse before service is needed.

36. A recent study by a store manager established the following frequency distribution for customers checking out of the store during a very busy afternoon. Assume the time interval between checkouts is exponentially distributed and determine
(a) the probability of more than 1 minute elapsing before the next customer arrives to check out.
(b) the probability of more than 3 minutes elapsing before the next customer arrives to check out.

Number of customers arriving at checkout per minute	Frequency
0	2
1	7
2	15
3	12
4	11
5	10
6	6

37. One of the major tire manufacturers is testing the quality of its tires. While the company is not yet aware of it, 1 percent of the tires manufactured are defective.

(a) Assuming that the number of defectives is Poisson distributed, in a sample of 100 tires what is the probability of there being less than 2 defective tires?

(b) Assume the number of defective tires to be binomially distributed, and use the normal curve approximation of the binomial to determine the probabilty of less than 2 defective tires in a sample of 100.

38. Between the hours of 4 and 5 P.M. on a "snowy" afternoon cars arrive at the rate of 2 per minute at a particular intersection.

(a) If the arrival of cars is Poisson distributed, what is the probability that zero cars (none) will arrive during any given minute?

(b) Since the street is slick, it will take you a full one-half minute to cross the intersection in your car. There are no stop signs or traffic lights, so if another car arrives at the intersection while you are crossing, you will be "hit." What is the probability that you will be hit one or more times?

(c) There are two ways part *(b)* can be worked. However you solved it the first time, use another way and rework the problem.

Case Study

For the past fifteen years, Main State University has required all new freshmen to take the American College Test (ACT) as part of its application for admission. However, the scores are used only for placement into classes and *not* for selective admission into the university. A frequency array of composite ACT scores for men and women for the freshmen class of 1982 are given in the following table.

ACT composite scores for women and men, Main State University, fall semester 1982

Score	Number women	Number men
36	0	0
35	0	0
34	0	0
33	1	1
32	3	4
31	11	26
30	12	39
29	35	56
28	48	85
27	66	76
26	83	119

Score	Number women	Number men
25	84	117
24	94	122
23	103	140
22	97	121
21	106	111
20	114	111
19	120	120
18	108	102
17	101	112
16	113	88
15	94	75
14	92	73
13	72	45
12	50	44
11	53	38
10	37	27
9	20	20
8	25	15
7	8	9
6	7	6
5	1	2
4	2	0
3	1	0
2	0	1
1	0	0

1. How would you describe the two distributions of ACT scores? What are the differences?

2. Assuming both distributions approach normality, what is the probability that a freshman woman has a composite score of 15 or less? What about the male freshman?

3. Why would you consider either of these distributions as normal? Why would you not?

4. If the administration of Main State were to consider a cutoff score of the ACT for selective admission, would you recommend that the same score be used for men and women?

5. Prepare a brief report summarizing the characteristics of each distribution and your recommendations.

5

Random Sampling and the Sampling Distribution of Sample Means and Sample Proportions

The method of *inductive reasoning* involves drawing a conclusion or making some generalization based upon limited or less than complete (specific) information. In a vague sense it is often stated that inductive reasoning involves "reasoning from the specific to the general." In statistics, using "limited information," obtained through *sampling*, to make generalizations is a form of inductive reasoning.

Elements of Statistical Sampling

Have you ever seen a large jar of jelly beans or peanuts used in a store window or a shopping mall as a promotion stunt? Shoppers may be asked to guess something about the contents of the jar to win a prize. The contest might involve guessing something like how many beans are in the jar, their weight, or the proportion of them that are red in color.

Suppose that each jelly bean in the jar has a number on it; the shopper might be required to determine the mean, median, or modal averages of these numbers to win the prize. Besides merely winning prizes, it might also be important to learn something about the characteristics of the jelly beans in the jar for some other reason. In any event you could examine the entire jar of candy beans and thus gain the information desired, but this procedure could be costly as well as time-consuming. If the jelly beans had to be tasted or eaten to be tested, there would be no jelly beans left if all of them were tested.

A reasonable alternative to examining, measuring, or testing *all* the jelly beans is through the use of sampling. *Sampling*, using *inductive reasoning*, is often employed to determine certain characteristics of things like the jelly beans in the large jar. Let us see what is involved in this process called *sampling*.

The purpose of *sampling* is usually to make an *inference*. To infer, in this sense, means to use limited information from a sample to generalize about one or more characteristics of some larger body of data. In our example with the jar of jelly beans then, we might test a smaller number than the total number of beans in the jar (the sample) in order to learn something about the characteristics of all the beans in the entire jar (the *population*).

Now let us examine the jelly bean illustration in terms of the practical aspects of sampling. The entire collection of all jelly beans in the jar is called the *universe*. This universe, consisting of any number of jelly beans, or other items, may generate several variables. One variable might be the number of jelly beans of various colors; another variable might be the proportion of the jelly beans of various flavors; and another variable might be the numbers on each jelly bean. Each variable generates a *population* distribution. General characteristics of the population (color, flavor, numbers as in our illustration) are called *parameters*. The *population parameters* are what we try to make inferences about.

The components in the universe from which a selection is drawn are called the *sample units*. In our example each individual jelly bean is a *sample unit* because that is the basic component from which the selection is made. The basic components providing the values of the variables or data generating the various populations are called *sample elements*. In other words, *sample units* actually selected make up the sample, and *sample elements* provide the quantitative data.

A *sample*, then, is any number of components drawn or selected that does not include the entire group of the units in the universe. A characteristic measure of the sample is called an *obtained statistic*, or an *estimator*. *Estimators* (obtained from the sample) estimate *parameters* (characteristics of the entire population).

Random Sampling

As stated previously, the purpose of sampling is to make an inference or generalization about a population characteristic. In order to make valid generalizations, it is necessary to select random (probability) samples. The concept of random sampling implies that no external factors affect the selection of the sample units; that is, the person collecting the sample has no influence over the specific units that either appear or do not appear in the sample.

A *random* or *probability sample* is a sample in which the probability of any item in the population being selected is known. All samples used in this text will be assumed to be random samples.

BASIC STATISTICS

To be quite specific, in this text all samples will be assumed to be *simple random samples.*[1] Since this is the case, it is necessary to investigate the concept of a *simple random sample* selected from a *finite population* and from an *infinite population.*

Simple Random Sampling: A Finite Population

If a sample containing n elements is selected from a finite population of N elements when $n < N$, it is a *simple random sample if each sample of size n has an equal probability of being selected.*

For example, let us consider a finite population containing four elements, A, B, C, and D. Suppose we select a sample of two elements from this finite population. Figure 5.1 shows the possible samples and the probability that any of the 12 possible samples will be selected is equal to 1/12.

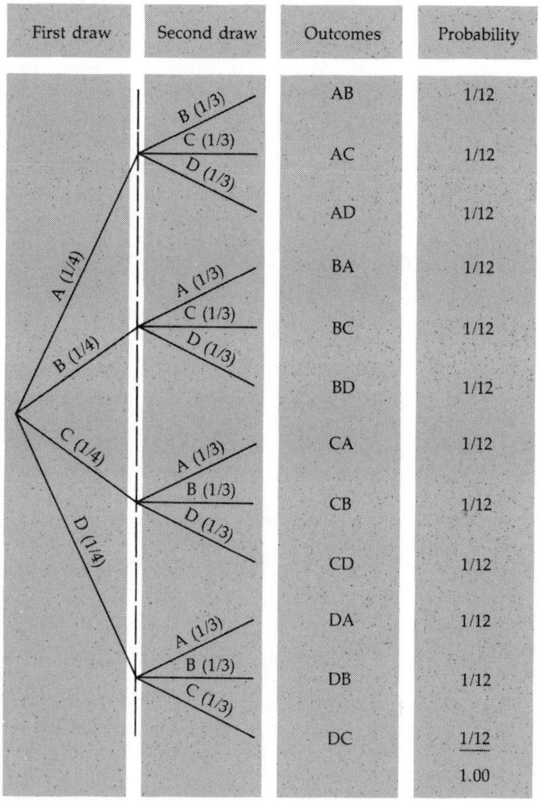

First draw	Second draw	Outcomes	Probability
A (1/4)	B (1/3)	AB	1/12
	C (1/3)	AC	1/12
	D (1/3)	AD	1/12
B (1/4)	A (1/3)	BA	1/12
	C (1/3)	BC	1/12
	D (1/3)	BD	1/12
C (1/4)	A (1/3)	CA	1/12
	B (1/3)	CB	1/12
	D (1/3)	CD	1/12
D (1/4)	A (1/3)	DA	1/12
	B (1/3)	DB	1/12
	C (1/3)	DC	1/12
			1.00

Figure 5.1   Simple random sample from a finite population

1. Other types of random samples (cluster, stratified, and systematic) are discussed in chapter 7.

A closer examination of figure 5.1 reveals that there are only six individual samples if viewed from the eyes of the individual taking the sample. That is, the outcome AB is distinguishable from BA only by noting the order of selection. If the order of selection is ignored, then there are only six outcomes (AB or BA), (AC or CA), (AD or DA), (BC or CB), (BD or DB), and (CD or DC). And, in fact, as far as the sample is concerned, it is the elements in the sample and not the order of the elements that is important. Thus any of the six outcomes has a probability of selection of $1/12 + 1/12 = 1/6$.

Now, realizing that the number of sample outcomes possible when order is not considered is 6, it is apparent that the total number of outcomes could have been determined using the *combinational equation*. That is, from a finite population of 4 elements, the number of possible samples of size 2 is

$$C_2^4 = \frac{4!}{2!(4-2)!} = \frac{4 \cdot 3 \cdot 2 \cdot 1}{2 \cdot 2} = 6,$$

and, extending our knowledge to generate the expressions for the probability of any sample of size 2 from a finite population of 4, the probability becomes

$$\frac{1}{C_2^4} = \frac{1}{6}.$$

Thus we see that from a finite population of size N simple random samples of size n have equal probability which can be calculated as

$$\frac{1}{C_n^N}.$$

Finally, you should notice that the preceding illustration relied on sampling *without replacement*, the most common type of sampling.

Simple Random Sampling: An Infinite Population

Sampling from an infinite population is the same as sampling from a finite population *with replacement*. A simple random sample for an infinite population is a sample of size n selected so that each element of the universe has an equal chance of being chosen as a sample unit. This results in each sample of n items having an equal chance of being selected. Thus it is apparent that this is the same definition of a simple random sample as that given for a finite population.

Under the infinite population, however, the assumption of independence is met. Obviously, this is the case with replacement.

The Sampling Distribution of Sample Means

As stated earlier, the usual purpose of sampling is to obtain information that can be used to describe a large body of data. The purpose might be to estimate the population mean of a particular universe. The procedure would be to select a sample of size n, determine $\overline{X}$, and use $\overline{X}$ as a point estimate of μ.

Now, realizing that the sample from which the estimator, $\overline{X}$, would be obtained is only one of many possible samples (as demonstrated in figure 5.1), it is reasonable to raise the question, "How good is the estimate of μ?" To answer such a question, we must investigate what is known as the sampling distribution of $\overline{X}$. A sampling distribution is the probability distribution of a sample statistic.

First, let us state (without proof) that if the random samples were selected from a population that was known to be normally distributed with a finite mean μ and standard deviation σ, then the sampling distribution of sample means ($\overline{X}$'s) will be normally distributed. We should recognize, however, that on many occasions the population may not be normally distributed. Then, we have to use the central limit theorem to help determine the shape of the sampling distribution of sample means.

The *central limit theorem,* as you will soon begin to realize, is a very important theorem in statistical sampling. In fact, the central limit theorem will form the basis from which most of the interval estimation and inferences in this text will be made. So what does it say?

Central Limit Theorem: The central limit theorem states that, if all samples of size n are selected from a population with a finite mean μ and standard deviation σ, then as n is increased in size, the distribution of sample means ($\overline{X}$'s) will tend toward a normal distribution with a mean of μ (the population mean) and a standard deviation equal to $\sigma/\sqrt{n}$ (called the standard error of the mean or $\sigma_{\overline{X}}$).

In other words, it says that sample means, $\overline{X}$'s, tend to have a frequency distribution that is normally distributed, and this occurrence is generally the case regardless of the shape of the population frequency distribution. Usually, a sample size of $n > 30$ is considered large enough for the central limit theorem to be effective. As you can imagine, this theorem is quite powerful.

Sampling from Infinite Populations

Let us investigate further this prediction of normality by the central limit theorem. To do so, we will use a hypothetical distribution having as members of the population the numbers 1, 2, 3, and 4. Notice in figure 5.2 that this hypothetical distribution is a

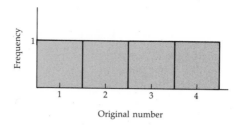

Original number

Figure 5.2   Histogram of original data

uniform distribution. Notice also that the hypothetical distribution is a finite distribution with the following parameters:

$$\mu = \frac{\sum\limits_{i=1}^{N} X_i}{N} = \frac{10}{4} = 2.5,$$

$$\sigma = \sqrt{\frac{\sum\limits_{i=1}^{N} (X_i - \mu)^2}{N}} = \sqrt{\frac{5.00}{4}} = 1.12.$$

Now let us take all possible samples of size 2 from this population under the assumption of replacement which, as stated previously, is equivalent to sampling from an infinite population. Table 5.1 shows all sixteen samples and the mean of each sample.

Using the data in table 5.1 to plot a histogram yields figure 5.3.

Table 5.1
Samples and sample means

Sample number	Observed values	Sample mean
1	1, 1	1.0
2	1, 2	1.5
3	1, 3	2.0
4	1, 4	2.5
5	2, 1	1.5
6	2, 2	2.0
7	2, 3	2.5
8	2, 4	3.0
9	3, 1	2.0
10	3, 2	2.5
11	3, 3	3.0
12	3, 4	3.5
13	4, 1	2.5
14	4, 2	3.0
15	4, 3	3.5
16	4, 4	4.0

Note: Number of samples $= n' = M_x^n = M_2^4 = 4^2 = 16$.

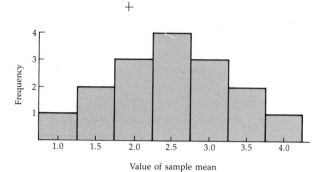

Figure 5.3 Histogram for sampling distribution of means when $n = 2$

You should find it interesting to compare figures 5.2 and 5.3. Why? Because from a uniform distribution of only four numbers (figure 5.2) came a symmetrical distribution of sample means using sample size 2 (figure 5.3). Thus you can see that the distribution of sample means has started toward a normal one as the sample size increases. (If you think of figure 5.2 as a histogram of sample means when $n = 1$, it may help you to see the symmetry developing.)

The central limit theorem suggests that the parameters of the sampling distribution will be a mean equal to μ and a standard deviation equal to $\sigma/\sqrt{n}$. To make sure, we can calculate the mean and standard deviation of the sampling distribution.

First, let us determine the mean of the sampling distribution, which we will denote as $\mu_{\bar{X}}$. Now $\mu_{\bar{X}}$ is equal to the sum of the sample means shown in table 5.1 divided by n', where n' is the number of samples selected (16):

$$\mu_{\bar{X}} = \frac{\sum_{i=1}^{n'} \bar{X}_i}{n'} = \frac{40}{16} = 2.5. \tag{5.1}$$

And we see that $\mu_{\bar{X}} = 2.5$ which is also the value of μ, the population mean.[2] *In other words, $\mu_{\bar{X}} = \mu$.*

Now let us learn more about the standard deviation of the sampling distribution of $\bar{X}$'s. Call this measure of dispersion the *standard error of the mean* and denote it $\sigma_{\bar{X}}$. In fact $\sigma_{\bar{X}}$, the standard deviation of the distribution of sample means, is computed just like any other standard deviation. Compare the following formulas to those you learned in chapter 2.

---

2. This computation will hold for any set of data. That is,
$$\frac{\sum \bar{X}}{\text{all samples of size } n} = \mu.$$

Ungrouped data

$$\sigma_{\bar{X}} = \sqrt{\frac{\sum (\bar{X} - \mu_{\bar{X}})^2}{n'}}.$$ (5.2)

Grouped data

$$\sigma_{\bar{X}} = \sqrt{\frac{\sum f(\bar{X} - \mu_{\bar{X}})^2}{n'}}.$$ (5.3)

Or, from the central limit theorem we know that the standard deviation of the sampling distribution is

$$\sigma_{\bar{X}} = \frac{\sigma}{\sqrt{n}}.$$ (5.4)

The information in table 5.1 is presented again in table 5.2. But this time it is shown as a frequency distribution. The calculations included in the table show the same numerical result using either equation (5.3) or (5.4).

To summarize, we have seen that the central limit theorem is very important. Because of the central limit theorem we will expect the distribution of sample means to tend toward normality regardless of the shape of the original population distribution. We will expect the tendency toward normality to be stronger as the sample size, n, is increased. And, finally, the parameters of

Table 5.2
Calculation of standard error of the mean

Sample mean ($\bar{X}$)	Frequency	$\bar{X} - \mu_{\bar{X}}$	$f(\bar{X} - \mu_{\bar{X}})^2$
1.0	1	−1.5	2.25
1.5	2	−1.0	2.00
2.0	3	−0.5	0.75
2.5	4	0	0
3.0	3	+0.5	0.75
3.5	2	1.0	2.00
4.0	1	1.5	2.25
	16		10.00

$$\mu_{\bar{X}} = \frac{\sum \bar{X}}{n'} = 2.5, \qquad \sigma_{\bar{X}} = \sqrt{\frac{\sum f(\bar{X} - \mu_{\bar{X}})^2}{n'}} = \sqrt{\frac{10}{16}} = 0.79$$

eq. 5.3

$$\sigma_{\bar{X}} = \frac{\sigma}{\sqrt{n}} = \frac{1.12}{\sqrt{2}} = \frac{1.12}{1.41} = 0.79$$

eq. 5.4

BASIC STATISTICS

Table 5.3
Sampling without replacement

Sample number	Observed values	Sample mean
~~1~~	~~1, 1~~	~~1.0~~
2	1, 2	1.5
3	1, 3	2.0
4	1, 4	2.5
5	2, 1	1.5
~~6~~	~~2, 2~~	~~2.0~~
7	2, 3	2.5
8	2, 4	3.0
9	3, 1	2.0
10	3, 2	2.5
~~11~~	~~3, 3~~	~~3.0~~
12	3, 4	3.5
13	4, 1	2.5
14	4, 2	3.0
15	4, 3	3.5
~~16~~	~~4, 4~~	~~4.0~~

the sampling distribution are estimated as mean $= \mu$ and standard deviation (called standard error of the mean) $= \sigma/\sqrt{n}$.

Sampling from Finite Populations

The preceding discussion was for sampling from an infinite population. However, sampling from finite populations is more likely to occur in business applications. Thus we must investigate the effect of the finite population on the estimate of the parameters for the sampling distribution. To avoid overcomplicating the situation, it will suffice to say that there is an effect on the estimation procedure due to the finite population, but this effect can be corrected.

First, to demonstrate the effect, let us repeat table 5.1 in table 5.3. Notice, however, that four of the sample means have been eliminated. That is because sampling without replacement eliminates sample numbers 1, 6, 11, and 16.

It may be a good idea to remind you that the original population is the same. It is still the four numbers 1, 2, 3, and 4. It is still a uniform distribution as shown in figure 5.2, and it still has $\mu = 2.5$ and $\sigma = 1.12$. Remember, it is the distribution of $\overline{X}$'s that has changed, because the number of possible samples has been reduced from 16 to 12. *Thus the mean and standard error of the mean must be determined for sampling distributions of finite populations.* These calculations are shown in table 5.4.

Table 5.4
Sampling distribution from finite population

Sample mean	Frequency	$f\overline{X}$	$\overline{X} - \mu_{\overline{X}}$	$f(\overline{X} - \mu_{\overline{X}})^2$
1.5	2	3.0	−1.0	2.00
2.0	2	4.0	−0.5	0.50
2.5	4	10.0	0	0
3.0	2	6.0	0.5	0.50
3.5	2	7.0	1.0	2.00
	12	30.0		5.00

$$\mu_{\overline{X}} = \frac{\sum_{i=1}^{n'} f\overline{X}_i}{n'} = \frac{30}{12} = 2.5 \qquad \sigma_{\overline{X}} = \sqrt{\frac{\sum f(\overline{X} - \mu_{\overline{X}})^2}{n'}} = \sqrt{\frac{5}{12}} = 0.645$$

As you can see from table 5.4, the mean of the sampling distribution of $\overline{X}$'s has remained the same $\mu_{\overline{X}} = 2.5$; however, the standard deviation of the sampling distribution has changed. This is because there are only 12 sample means, and the dispersion is not the same as the previous 16 sample means (sampling with replacement).

Direct calculation of the standard error of mean using the population standard deviation can be accomplished using equation (5.5):

$$\sigma_{\overline{X}} = \frac{\sigma}{\sqrt{n}} \sqrt{\frac{N - n}{N - 1}}. \tag{5.5}$$

For our example, the results using equation (5.5) are the same as shown in table 5.4 where equation (5.3) was used:

$$\sigma_{\overline{X}} = \frac{1.12}{\sqrt{2}} \sqrt{\frac{4 - 2}{4 - 1}} = (0.79)(0.816) = 0.645.$$

Using the Sampling Distribution of Sample Means

Let us reiterate the guidelines for establishing the shape of the sampling distribution of sample means obtained from a population with a finite mean μ and standard deviation σ. First, note that, if the sample size is large ($n > 30$), the sampling distribution of sample means is normally distributed, as explained by the central limit theorem; and second, remember that, if the random samples were taken from a normal population, the distribution of sample means will be normal regardless of the sample size. Under either of these situations the sampling distribution will be as shown in figure 5.4.

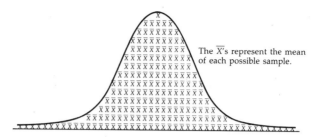

The $\overline{X}$'s represent the mean of each possible sample.

Figure 5.4   Sampling distribution of sample means

Now that the shape of the sampling distribution of sample means has been established, we need to know how to make use of this information. Since the sampling distribution of sample means will be normally distributed under the conditions we stated, it is reasonable to suspect that we can make the same type of computations and probability statements for $\overline{X}$ that we did for a normally distributed random variable X in chapter 4. Recall from chapter 4 that any normal distribution could be converted to the standard normal distribution by using equation (4.8) which is repeated here:

$$Z = \frac{X - \mu}{\sigma},$$

where

Z = distance (measured in standard deviations) that the random variable is from the true population mean,

X = random variable,

μ = population mean,

σ = population standard deviation.

The same approach can be extended to a sampling distribution of sample means to yield

$$Z = \frac{\overline{X} - \mu}{\sigma_{\overline{X}}}. \qquad (5.6)$$

Equation (5.6) results from putting $\overline{X}$, which is also a random variable, in place of X, and putting $\sigma_{\overline{X}}$, the standard deviation of the sampling distribution (standard error of the mean), in place of σ. Figure 5.5 shows how the normally distributed $\overline{X}$'s are dispersed about the population mean. And, because the sampling distribution is normally distributed, we can relate the probabilities associated with any normal distribution to our sampling distribution. For example, 68.27 percent of the sample means lie within the range of $+1\sigma_{\overline{X}}$ and $-1\sigma_{\overline{X}}$ from the mean, μ.

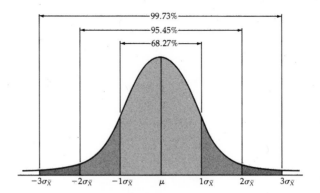

Figure 5.5   Sampling distribution of sample means

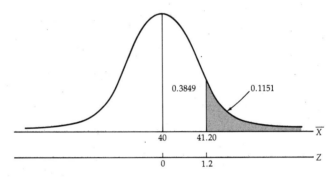

Figure 5.6   Area under normal curve in example 5.1

Example 5.1

▷ If a random sample of 25 items was selected from a normally distributed population with $\mu = 40$ and $\sigma = 5$, what is the probability that the sample mean will exceed 41.20? (See figure 5.6.)

$$Z = \frac{\overline{X} - \mu}{\sigma_{\overline{X}}} \quad \text{and} \quad \sigma_{\overline{X}} = \frac{\sigma}{\sqrt{n}},$$

so
$$Z = \frac{41.2 - 40}{5/\sqrt{25}}$$

$$= \frac{1.2}{1}$$

$$Z = 1.2,$$
$$\text{Area} = 0.3849 \text{ (from appendix H)},$$
$$P(X > 41.20) = 0.5000 - 0.3849 = 0.1151. ◁$$

Example 5.2

▷ Suppose the Account Receivable balances at a large finance company had $\mu = \$925$ and $\sigma = \$72$. What is the probability that a

BASIC STATISTICS

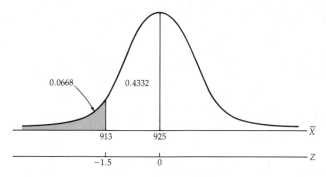

Figure 5.7   Area under normal curve in example 5.2

sample of 81 Accounts Receivables would have an $\overline{X}$ less than $913? (See figure 5.7.)

$$Z = \frac{\overline{X} - \mu}{\sigma_{\overline{X}}} \quad \text{and} \quad \sigma_{\overline{X}} = \frac{\sigma}{\sqrt{n}}$$

so

$$Z = \frac{913 - 925}{72/\sqrt{81}} = \frac{-12}{8},$$

$$Z = -1.5,$$

$$\text{Area} = 0.4332 \text{ (from appendix H)},$$

$$P(\overline{X} < \$913) = 0.5000 - 0.4332 = 0.0668. \triangleleft$$

The Student _t_ Distribution

As you recall, the central limit theorem states that the sampling distribution of $\overline{X}$'s tends to be normally distributed with a mean of μ and standard deviation (called _standard error_ of the _mean_) of $\sigma/\sqrt{n}$. It has been pointed out, however, that the σ of a population is frequently unknown. Thus in many sampling situations we estimate $\sigma_{\overline{X}}$ from sample data so that for an infinite population

$$s_{\overline{X}} = \frac{s}{\sqrt{n}}. \tag{5.7}$$

And when the population is finite, the estimator of $\sigma_{\overline{X}}$ is

$$s_{\overline{X}} = \frac{s}{\sqrt{n}} \sqrt{\frac{N - n}{N - 1}}. \tag{5.8}$$

Recognizing that this estimation of $\sigma_{\overline{X}}$ will take place whenever σ is unknown, the following question has to be raised: "What is the effect on the ratio $[(\overline{X} - \mu)/\sigma_{\overline{X}}]$ when it is changed to $[(\overline{X} - \mu)/s_{\overline{X}}]$?" And the answer is that the result is quite significant since $[(\overline{X} - \mu)/s_{\overline{X}}]$ is not normally distributed. In fact, when

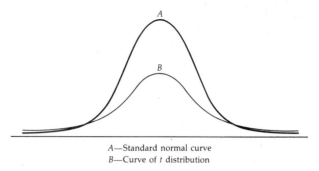

A—Standard normal curve
B—Curve of t distribution

Figure 5.8   Comparison of normal curve (t distribution with ∞ degrees of freedom) with curve of t distribution whose degrees of freedom are relatively small

σ is unknown, the distribution of $[(\overline{X} - \mu)/s_{\overline{X}}]$ is a sampling distribution known as the *Student t distribution*, provided the *original population was normally distributed*.[3]

The *t distribution* is similar to the standard normal distribution. Its curve resembles the physical characteristics of the standard normal in many ways. It is bell shaped, symmetrical, and universal. The two main differences are (1) the t distribution has greater dispersion (the curve is flatter and its tails extend farther) and (2) there is a particular t distribution for varying degrees of freedom (degrees of freedom, df, are one less than the sample size or $n - 1$).

This second difference requires a little explanation. Unlike the normal distribution, there is not a single t distribution. There is a specific t distribution for each value of $n - 1$ (degrees of freedom). The smaller the degrees of freedom ($n - 1$), the flatter, or more dispersed, the curve becomes. Conversely, as ($n - 1$) becomes large, the t distribution approaches the normal distribution. Figure 5.8 demonstrates the idea of a flatter, more dispersed curve as the t distribution.

Appendix I shows values of the t distribution for various degrees of freedom at selected probability levels. Notice that, where the degrees of freedom (df) $= \infty$, the *t distribution* is equivalent to the *standard normal distribution*. This table must be read for the appropriate degrees of freedom each time you want to determine a particular t value.

In converting to Student t distributions, the procedure is essentially the same as when converting to a standard normal dis-

---

3. Credit for developing the theoretical standard t distribution goes to William S. Gosset. This notion was first published in 1908 in a paper written by Gosset in which he used the pen name, "A. Student."

tribution. Look at equation (5.9). We will make use of this relationship in the next chapter:

$$t = \frac{\overline{X} - \mu}{s_{\overline{X}}} \qquad (5.9)$$

Except for using the values in the t table instead of those in the standard normal distribution table, equation (5.9) is used just as you use the standard normal distribution (equation 5.6).

The Sampling Distribution of Sample Proportions

So far we have discussed sampling distributions of the arithmetic mean. There will be times, however, when the statistic of interest will not be the arithmetic mean. That is, the statistic might be a percentage sometimes referred to as a *proportion*. Since this is the case, we must investigate how *sample proportions (p)* are distributed, where p is defined as the ratio of the number of successes to the total number of events in a binomial situation. Equation (5.10) shows how sample proportions are calculated:

$$p = \frac{X}{n}, \qquad (5.10)$$

where

p = sample proportion,
X = number of occurrences of the random variable,
n = number of trials.

Let us recall that proportions (when there are two responses such as yes-no) are actually distributed according to the *binomial distribution*. As the sample size increases, however, sample proportions will approach a normal distribution under the central limit theorem, provided the population proportion, π, is not very large or very small.[4]

Figure 5.9 demonstrates the effect of the population proportion, π, on the distributions of sample proportions, p. As you can see, if the population proportion is quite small ($\pi < 0.5$), the p's tend to be skewed to the right, and if the population proportion is quite large ($\pi > 0.5$), the p's tend to be skewed left. Also you must remember that, for symmetry to develop, the sample size must be large.

Now that we know something of the sampling distribution of sample proportions, we need to know how to calculate the mean and standard deviation for this type of data. Let us consider the

---

4. A reasonable guide is that $n\pi$ and $n(1 - \pi)$ are both >5.

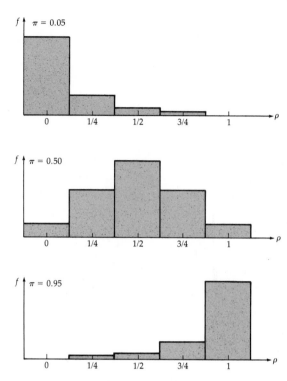

Figure 5.9   Sampling distributions of p's with different values of π and $n = 4$

same finite population composed of the four numbers 1, 2, 3, and 4, which we used earlier in this chapter. Again let us take all possible samples of size 2, and let us count the number of 3's in each sample. Table 5.5 shows these results. (To help understand the entries in table 5.5, consider sample number 7, which has observed values (2, 3). The number of 3's in the sample is 1, and the sample size is 2; so the sample proportion is $p = X/n = 1/2$.)

Now let us use the results in table 5.5 to calculate the average of the sample proportions μ_p so that

$$\mu_p = \frac{\text{sum of sample proportions}}{\text{number of samples}} = \frac{\sum p}{n'} \qquad (5.11)$$

$$= \frac{1/2 + 1/2 + 1/2 + 1/2 + 1 + 1/2 + 1/2}{16}$$

$$= 4/16$$

$$\mu_p = 1/4.$$

Notice that $\mu_p = 1/4$ is equal to the true proportion of 3's in the population (1, 2, 3, 4), which is 1/4. The average of possible sam-

Table 5.5
Samples of size 2 and the number of 3's observed

Sample number	Observed values	Number of threes	Proportion of 3's in sample (p)
1	1, 1	0	0
2	1, 2	0	0
3	1, 3	1	1/2
4	1, 4	0	0
5	2, 1	0	0
6	2, 2	0	0
7	2, 3	1	1/2
8	2, 4	0	0
9	3, 1	1	1/2
10	3, 2	1	1/2
11	3, 3	2	1
12	3, 4	1	1/2
13	4, 1	0	0
14	4, 2	0	0
15	4, 3	1	1/2
16	4, 4	0	0

ple proportions (p's) is equal to the true population proportion, π. Thus $\mu_p = \pi$.

Continuing on, let us calculate the standard deviation of the sample proportions, which is usually called the *standard error of the proportion*. The calculation in table 5.6 shows the standard error of the proportion, σ_p, to be 0.306.

The calculations in table 5.6 are rather long, however, particularly when the number of samples is large. Fortunately, a shorter way exists to determine the standard error of the proportion. Using equation (5.12) with our example, we find that the answer in table 5.6 could have been found much more readily:

$$\sigma_p = \sqrt{\frac{\pi(1 - \pi)}{n}}, \qquad (5.12)$$

$$\pi = 1/4,$$
$$1 - \pi = 3/4,$$
$$n = 2;$$

$$\sigma_p = \sqrt{\frac{(1/4) \cdot (3/4)}{2}}$$

$$= \sqrt{3/32}$$
$$\sigma_p = 0.306.$$

Table 5.6
Calculation of standard error of the proportion

Sample number	Observed values	Proportion of 3's in sample, p	$(p - \pi)$	$(p - \pi)^2$
1	1, 1	0	−1/4	1/16
2	1, 2	0	−1/4	1/16
3	1, 3	1/2	1/4	1/16
4	1, 4	0	−1/4	1/16
5	2, 1	0	−1/4	1/16
6	2, 2	0	−1/4	1/16
7	2, 3	1/2	1/4	1/16
8	2, 4	0	−1/4	1/16
9	3, 1	1/2	1/4	1/16
10	3, 2	1/2	1/4	1/16
11	3, 3	1	3/4	9/16
12	3, 4	1/2	1/4	1/16
13	4, 1	0	−1/4	1/16
14	4, 2	0	−1/4	1/16
15	4, 3	1/2	1/4	1/16
16	4, 4	0	−1/4	1/16

$$\sum(p - \pi)^2 = \frac{24}{16}$$

$$\sigma_p = \sqrt{\frac{\sum(p - \pi)^2}{n'}}$$

$$= \sqrt{\frac{24/16}{16}}$$

$$\sigma_p = \sqrt{3/32} = 0.306$$

As before, if we sample from a finite population, we should use the finite correction factor as shown in equation (5.13):

$$\sigma_p = \sqrt{\frac{\pi(1 - \pi)}{n}} \sqrt{\frac{N - n}{N - 1}}. \tag{5.13}$$

Now that you are familiar with the sampling distributions of sample proportions, you need to learn to make use of it. Since the *central limit theorem applies to sample proportions*, we have an equation for converting the sampling distribution to a standard normal distribution. And equation (5.14),

$$Z = \frac{p - \pi}{\sigma_p}, \tag{5.14}$$

can be used just as you have done previously.

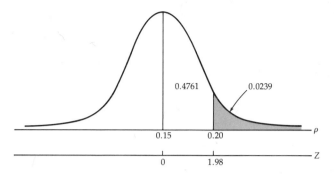

Figure 5.10 Area under normal curve in example 5.3

Example 5.3

▷ Angell Greeting Cards, Inc. produces approximately 250,000 greeting cards annually. During a recent production run, it was determined that 15 percent (π) of the cards were defective. What is the probability that a sample of 200 cards would contain a sample proportion exceeding 20 percent? Note that the finite population correction factors may be ignored since the universe is quite large. (See figure 5.10.)

$$Z = \frac{p - \pi}{\sigma_p}, \quad where \quad \sigma_p = \sqrt{\frac{\pi(1 - \pi)}{n}},$$

$$= \frac{0.20 - 0.15}{\sqrt{\dfrac{(0.15)(1 - 0.15)}{200}}}$$

$$= \frac{0.20 - 0.15}{\sqrt{\dfrac{(0.15)(0.85)}{200}}}$$

$$= \frac{0.05}{\sqrt{0.0006375}}$$

$$= \frac{0.05}{0.02525},$$

$$Z = 1.98,$$

Area = 0.4761 (from appendix H),
$P(p > 0.20) = 0.5000 - 0.4761 = 0.0239.$ ◁

You should recognize that the same situation exists for proportions that is true for arithmetic means; that is, we do not often know the population parameters, π and $1 - \pi$. Accordingly, we must frequently compute an estimate of σ_p (labeled s_p) as follows:

$$s_p = \sqrt{\frac{p(1 - p)}{n}}. \tag{5.15}$$

Equation (5.15) is appropriate if the sampling is from an infinite population. If the population is finite, the equation to use is

$$s_p = \sqrt{\frac{p(1-p)}{n}} \sqrt{\frac{N-n}{N-1}}. \qquad (5.16)$$

Example 5.4

▷ If a random sample of 200 individuals in a very large city showed 26 persons with the flu, estimate the following:

(a) The proportion of people with the flu.

(b) The standard error of the proportion.

(a) $p = X/n = \dfrac{26}{200} = 0.13.$

(b) $s_p = \sqrt{\dfrac{p(1-p)}{n}}$

$$= \sqrt{\frac{(0.13)(1-0.13)}{200}} = \sqrt{\frac{(0.13)(0.87)}{200}} = \sqrt{0.000566}$$

$$= 0.0238. \ ◁$$

Summary of Equations

5.1 Mean of the sampling distribution of $\overline{X}$, equal to the population mean:

$$\mu_{\overline{X}} = \frac{\sum\limits_{i=1}^{n'} \overline{X}_i}{n'}$$

5.2 Standard error of the mean—ungrouped data:

$$\sigma_{\overline{X}} = \sqrt{\frac{\sum (\overline{X} - \mu_{\overline{X}})^2}{n'}}$$

5.3 Standard error of the mean—grouped data:

$$\sigma_{\overline{X}} = \sqrt{\frac{\sum f(\overline{X} - \mu_{\overline{X}})^2}{n'}}$$

5.4 Standard error of the mean—using σ:

$$\sigma_{\overline{X}} = \frac{\sigma}{\sqrt{n}}$$

5.5 Standard error of the mean—using finite population correction factor:

$$\sigma_{\overline{X}} = \frac{\sigma}{\sqrt{n}} \sqrt{\frac{N - n}{N - 1}}$$

5.6 The number of standard deviates:

$$Z = \frac{\overline{X} - \mu}{\sigma_{\overline{X}}}$$

5.7 Standard error of the mean—using s:

$$s_{\overline{X}} = \frac{s}{\sqrt{n}}$$

5.8 Standard error of the mean—using s and the finite population correction factor:

$$s_{\overline{X}} = \frac{s}{\sqrt{n}} \sqrt{\frac{N - n}{N - 1}}$$

5.9   The *t* distribution:

$$t = \frac{\overline{X} - \mu}{s_{\overline{X}}}$$

5.10   Sample proportion:

$$p = \frac{X}{n}$$

5.11   Average of the sampling distribution of *p*—equal to the population proportion (π):

$$\mu_p = \frac{\Sigma p}{n'}$$

5.12   Standard error of the proportion:

$$\sigma_p = \sqrt{\frac{\pi(1 - \pi)}{n}}$$

5.13   Standard error of the proportion—using the finite population correction factor:

$$\sigma_p = \sqrt{\frac{\pi(1 - \pi)}{n}} \sqrt{\frac{N - n}{N - 1}}$$

5.14   Number of standard deviates:

$$Z = \frac{p - \pi}{\sigma_p}$$

5.15   Standard error of the proportion—using *p*:

$$s_p = \sqrt{\frac{p(1 - p)}{n}}$$

5.16   Standard error of the proportion—using *p* and the finite population correction factor:

$$s_p = \sqrt{\frac{p(1 - p)}{n}} \sqrt{\frac{N - n}{N - 1}}$$

1. Identify what each equation at the end of this chapter calculates.

2. Define

(a) inductive reasoning
(b) inference
(c) sample units
(d) random sample
(e) simple random sample
(f) standard error of the mean
(g) t distribution
(h) universe
(i) sample proportion
(j) standard error of the proportion

3. Distinguish between s and $s_{\bar{x}}$.
4. What is a sampling distribution?
5. State the central limit theorem.
6. When is the finite population correction factor needed?

Problems

1. From a finite population of four items (A, B, C, D), determine how many samples of size 2 are possible:

(a) If duplication is allowed and order is observed.
(b) If duplication is not allowed and order is observed.
(c) If duplication is not allowed and order is not observed.

2. Write out the possible samples for each situation in problem 1.

3. Using replacement, determine all possible samples of size 2 and the sample means from a population composed of the numbers 3, 5, 7, and 9.

4. Plot the sample means determined in problem 3 in the form of a histogram.

5. Use your results in problem 3 to

(a) determine μ by calculating $\mu_{\bar{x}}$.
(b) determine the standard error of the mean using equation (5.2) or (5.3).
(c) duplicate the results of part (b) using equation (5.4).

6. Using the population in problem 3, determine all possible samples of size 2 that can result when sampling without replacement.

7. Determine the sample means in problem 6, and plot these sample means as a histogram.

8. Determine the standard error of the mean for problem 6:

(a) Use equation (5.2) or (5.3).

(b) Duplicate the results of part (a) using equation (5.5).

9. Given a population composed of 8, 11, 5, 14:

(a) Determine the number of samples of size 2 that could be obtained if the sampling were without replacement.

(b) List all possible samples of size 2 obtained when sampling without replacement.

(c) Determine μ by calculating $\mu_{\bar{X}}$.

(d) Determine the standard error of the mean using equation (5.3).

(e) Determine the standard error of the mean using equation (5.5).

10. From the results of problem 3, determine the sample proportions that result by counting the number of 5's in each sample. (Hint: see table 5.5)

11. Determine the population proportion π, using the sample proportions from problem 10.

12. Determine the standard error of the proportion using the π from problem 10:

(a) Use equation (5.12).

(b) Duplicate this result by following the procedure shown in table 5.6.

13. If random samples of size 49 are taken from a large population with $\mu = 80$ and $\sigma = 21$, what percent of the sample means

(a) lie between 77 and 83?

(b) would be greater than 82?

(c) lie between 74 and 86?

(d) would be less than 76?

14. In a population of 21,000 management executives, the mean (μ) salary of the population is $48,000 with standard deviation (σ) of $6,000. Determine the standard deviation of the distribution of sample means for a sample size of 600.

15. If the population size is 3,149, the population mean is 185, the population standard deviation is 18, and the sample size is 144, determine the standard deviation of the distribution of sample means:

(a) Use equation (5.4).

(b) Use equation (5.5).

(c) Compare the two answers.

16. In a sample of 50 stocks, 15 declined over the period of a week. Estimate the

(a) population proportion of stocks that declined.

(b) standard error of the proportion of stocks that declined.

17. A department store has 7,000 charge accounts. The comptroller takes a random sample of 49 of the account balances and calculates the standard deviation to be $42.00. If the actual mean (μ) of the account balances is $175.00, what is the probability that the sample mean would be

(a) between $164.50 and $185.50?

(b) greater than $180.00?

(c) less than $168.00?

18. If a sample of size 16 from a normal population produces a sample mean of 22 and a standard deviation of 6,

(a) determine the standard error of the mean.

(b) Are the sample means normally distributed?

19. Determine the standard error of the proportion if $\pi = 0.65$ and $n = 225$. (Assume infinite population.)

20. Solve problem 19 if $N = 8000$.

21. If a sample of size 200 yields a sample proportion of 0.20, estimate the standard error of the proportion.

22. A distributor has 5,000 customers. If a random sample of 150 customers showed 20 delinquent accounts, estimate the

(a) population proportion of delinquent accounts.

(b) standard error of the proportion of delinquent accounts.

23. If a sample of size 36 from a normal population yields $\overline{X} = 19$ and $s = 4.8$,

(a) determine the standard error of the mean.

(b) are the $\overline{X}$'s t distributed?

(c) explain your answer to part *(b)*.

24. Historically, 0.15 of the active accounts of VESACHARGE are delinquent. What is the probability that a random sample of 200 accounts would have 40 delinquent accounts?

25. An organization with 4,000 employees has established that Monday absenteeism has been 8 percent for several years. If a random sample of 80 employees was taken last Monday, what is the probability that the sample would contain more than 15 absentees?

26. If 40 percent of the homeowners in a very large city also own two or more televisions, what is the probability that a random sample of 300 homeowners will contain more than 145 homeowners who own at least two televisions?

27. The mean vase life of cut roses with preservative is 192 hours with a standard deviation of 24 hours. If a random sample of 36 roses were tested for vase life, determine the probability that the sample mean would exceed 199.

28. A local TV talk show is generally watched by 35 percent of the local viewing public. What is the probability that a sample of 100 families would find less than 25 families watching the talk show?

Case Study

Schlitz Brewing Co. reformulated its brewing procedures in 1977 so that once again it produced a premium beer. However, after three years of producing a quality product, Schlitz's market share had not increased. The chief executive officer knew that the key to gaining market share in the premium beer industry rests with image rather than quality since only a negligible portion of beer drinkers have a sufficiently trained palate to differentiate between premium American beers. When presented with a new advertising campaign that would feature a *live* TV taste test between his beer and one of his competitor's beers, he was intrigued. If his beer could be demonstrated to be preferred over another brand, the public's perception of Schlitz beer would be enhanced and market share would increase.

The proposed TV taste test would be conducted using 100 tasters, all of whom are self-professed Budweiser drinkers. The Great American Beer Test (as it was later named) consisted of giving each of the tasters two identical mugs, each filled with a different, fresh, chilled beer, with which they would indicate their preference. If 40 or more of the tasters selected Schlitz over Budweiser, you should consider the advertising campaign a success. What is the probability that the campaign is successful?

Case study contributed by Richard Ford, University of Arkansas at Little Rock.

6

Estimation:
Means and Proportions

The major reason for taking a sample is to draw conclusions about population parameters. For instance, the management of a local television station may want to know the station's listening rate for its local newscast; or your congressman may want to know what percent of the people in his congressional district agree with the position he has taken on a certain controversial issue. These conclusions about populations are usually in the form of estimates. In fact the values of the sample are frequently called *estimators*. Short of an enumeration of the entire population, we depend on sample estimators to approximate population parameters.

In sampling, there are usually two types of estimates—*point estimates* and *interval estimates*. A point estimate is just what it implies, a single estimate of the *population parameter* made with a sample statistic. By contrast an interval estimate is a range constructed about the single point estimate by using the appropriate sampling distribution. You studied these sampling distributions in the previous chapter.

In this chapter the emphasis will be on learning to construct interval estimates for certain population parameters. The reason for stressing interval estimates over point estimates is that point estimates are somewhat fragile. That is, it is rather unlikely that a single statistic (say the sample mean, $\overline{X}$) is going to equal exactly the population parameter under investigation. Thus, to avoid being wrong most of the time with a single point estimate, interval estimates are usually made.

The Confidence Interval   Interval estimates are referred to as *confidence intervals* and are constructed in a manner that a probability value can be assigned giving the likelihood that the population parameter falls within the boundaries of the interval. This probability value is called the *confidence level*. In essence, a confidence interval is really a range, probabilistically constructed, about a point estimate to provide an interval estimate of a population parameter. The confidence level indicates the likelihood that the *statement* concerning the population parameter is true.

Let's look at a likely situation. Manufacturers of dry-cell batteries often make claims in advertisements about the life of one of their particular batteries. Suppose a representative of one such manufacturer states that she is 95 percent confident that the average life of her company's "special" batteries is between 125 and 131 hours. Assuming that this statement is based on sound statistical analysis, we could conclude that the manufacturer's representative has constructed a 95 percent confidence interval for the average life of this particular type dry-cell battery. The *implication* is that the average battery life of all such batteries made by this firm is between 125 and 131 hours, with a probability of 0.95.

A more correct interpretation of the confidence interval, however, is to apply the probability level to the statement. That is, the *statement* that the true mean life of the battery is between 125 and 131 hours is *correct with a probability of 0.95*. A moment's reflection and this statement makes better sense because the true mean is either within the interval bounds or outside of them. Thus the probability here is really 0 or 1.0, not 0.95. In other words, the true mean cannot jump in the interval 95 times out of 100 and out of the interval 5 times out of 100. Rather the statement about the interval should be correct 95 times out of 100. However, the interval itself is either right or wrong!

Confidence Intervals for the Population Means

In the previous chapter two sampling distributions with $\overline{X}$ were introduced. These two sampling distributions are the normal and the t distributions. At that time it was pointed out that, when dealing with arithmetic means, sometimes the normal distribution is appropriate and sometimes the t distribution is more appropriate. So, before constructing confidence limits for the arithmetic mean, it is important to analyze the underlying conditions dictating which of the two possible sampling distributions to use.

Situations under Which Samples Might Be Selected

In constructing a confidence interval from information obtained from a random sample, we should raise three fundamental questions about the conditions under which the sample was collected. These three questions are as follows:

1. Is the population known to be normally distributed?
2. Is the population standard deviation, σ, known or unknown?
3. Is the sample size large, $n > 30$, or small?

Each of these conditions—normal or nonnormal population distribution, known or unknown population standard deviation, and large or small sample size—can affect the construction of the confidence interval. The tree diagram in figure 6.1 considers each of these three questions in providing the expected sampling distribution for each situation. By following the branches as they relate to each question, a terminal position is reached which indicates the appropriate $\overline{X}$ sampling distribution to be used in the given situation. For instance, the branches leading to situation A are (1) population known to be normal, (2) population standard deviation, σ, known, and (3) the sample size, n, is large.

A close inspection of the appropriate $\overline{X}$ sampling distribution shows that four different categories are presented. The first category is comprised of situations A and B in which the sample was taken from a normal population with a known standard deviation. In this case the appropriate sampling distribution is the normal distribution regardless of whether or not the sample size is large.

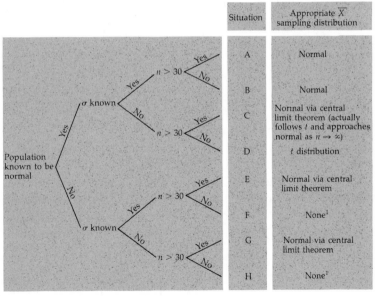

1. Various approaches to handling situations F and H appear in chapter 11.

Figure 6.1   Tree diagram of sampling situations

The second category is made up of those situations where the appropriate sampling distribution is the normal distribution as a result of the central limit theorem—situations C, E, and G. In each situation either the population distribution or the population standard deviation, or both, is unknown, and the sample size is large, $n > 30$. This large sample size allows the use of the central limit theorem, thereby establishing the normality of the sampling distribution.

The third category in which the t distribution must be used is situation D. The t distribution is necessary because the unknown population standard deviation and the small sample size prevents us from using the central limit theorem. Notice, finally, that the population distribution is normal and that this condition is necessary in order to use the t distribution.

The fourth category contains situations F and H for which there is no appropriate sampling distribution and thus no way to make interval estimates. What has happened in these two cases is that a small sample is combined with an unknown population distribution that prevents the use of the central limit theorem or the t distribution.[1]

Summarizing, we can say that there are three categories of sampling situations that permit the construction of confidence intervals. They are (1) normal population and known population standard deviation, (2) large samples, and (3) small samples from a normal population with unknown population standard deviation. The first two categories employ the normal distribution, and the third category employs the t distribution. Now, knowing the sampling situations we might encounter, let us investigate the construction of confidence intervals for μ in each of the three categories.

Confidence Intervals for the Population Mean When Population Distribution Is Normal and Population Standard Deviation Is Known

As you might have already guessed, the situation of sampling from a normal population distribution with the population standard deviation being known is rather unlikely. It is, however, a situation where the sampling distribution of $\overline{X}$'s is always normally distributed regardless of sample size. And, because the sampling distribution is normal, we can relate the probabilities associated with any normal distribution to our sampling distribution.

---

1. Various approaches to handling situations F and H appear in chapter 11.

For example, 68.27 percent of the sample means ($\overline{X}$'s) lie within the range of plus or minus one $\sigma_{\overline{X}}$ from the true mean, μ. Since this is true, there is a 0.6827 probability that a range of plus or minus one $\sigma_{\overline{X}}$ will contain the true population mean, μ. Using this same logic, it would be safe to say that there is a 0.95 probability that μ will fall within the interval of any sample mean plus or minus 1.96 $\sigma_{\overline{X}}$. Figure 6.2 presents this relationship. Notice in figure 6.2 that 95 percent of all the sample means in the sampling distribution are within a distance of $\pm 1.96\ \sigma_{\overline{X}}$ of μ. Thus any sample mean that has 1.96 $\sigma_{\overline{X}}$ added to it and 1.96 $\sigma_{\overline{X}}$ subtracted from it forms an interval that has 0.95 probability of containing the true mean, μ.

This probability level is usually referred to as the level of confidence for the interval. That is, confidence intervals may have confidence levels such as 0.80, 0.90, or 0.95, and the limits of the confidence interval are established from the expression

$$Z = \frac{\overline{X} - \mu}{\sigma_{\overline{X}}}.$$

Solving the preceding expression for μ yields

$$Z\sigma_{\overline{X}} = \overline{X} - \mu,$$
$$\mu = \overline{X} - Z\sigma_{\overline{X}},$$

and, since Z can be positive or negative,

$$\mu = \overline{X} - Z\sigma_{\overline{X}} \quad \text{when } Z \text{ is positive,}$$
$$\mu = \overline{X} + Z\sigma_{\overline{X}} \quad \text{when } Z \text{ is negative.}$$

Algebraically, this relationship may be expressed as follows:

$$(\overline{X} - Z\sigma_{\overline{X}}) \le \mu \le (\overline{X} + Z\sigma_{\overline{X}}),$$

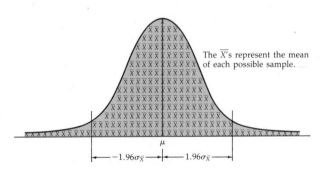

The $\overline{X}$'s represent the mean of each possible sample.

Figure 6.2 A representation of a sampling distribution showing the area under the normal curve and 95 percent of the sample means within the range $\pm 1.96\ \sigma_{\overline{X}}$

where $(\overline{X} - Z\sigma_{\overline{X}})$ is the lower confidence limit of the population mean, and $(\overline{X} + Z\sigma_{\overline{X}})$ is the upper confidence limit of the population mean. Now we can write an expression for the construction of confidence intervals as follows:

$(1 - \alpha)\%$ confidence interval for $\mu = \overline{X} \pm Z\sigma_{\overline{X}}$. (6.1)

Notice that equation (6.1) is in terms of the $(1 - \alpha)$ percentage. This alpha (α), which represents the probability that the interval is incorrect, takes as a numerical value the difference between 1.00 and the level of confidence so that for a 95 percent confidence interval, $\alpha = 0.05$. To demonstrate the construction of confidence intervals, let us consider the following example.

Example 6.1

▷ Suppose that a random sample of 25 airline tickets sold at O'Hare Airport (assume a large population so the finite population correction factor may be ignored) yielded a mean average ticket price of $250 over the past year. Now, if the population of annual tickets sold at O'Hare is *known* to be normally distributed and σ is *known* to be $100, what would be the 95 percent confidence interval for the true mean price of tickets sold at O'Hare over the past year?

From equation (6.1)

$(1 - \alpha)\%$ confidence interval for $\mu = \overline{X} \pm Z\sigma_{\overline{X}}$,
95% confidence interval for $\mu = \overline{X} \pm Z\sigma_{\overline{X}}$,

where Z must be determined from the normal curve in appendix H. To use appendix H, divide the level of confidence by $2(0.95/2 = 0.4750)$ and locate the associated Z value of 1.96 in appendix H, which leaves 2.5 percent of the area under the curve in each end of the distribution. Thus

95% confidence interval for $\mu = \$250 \pm (1.96)\left(\dfrac{\$100}{\sqrt{25}}\right)$

$= \$250 \pm (1.96)(\$20)$
$= \$250 \pm \$39.20,$
$\$210.80 \leq \mu \leq \$289.20.$

Thus we would state that the average airline ticket sold at O'Hare Airport during the past year is in the range of $210.80 to $289.20 with a probability of 0.95. You are cautioned to remember, however, that the probability is really attached to the statement about the range because the true mean is either in or out of the interval estimate. ◁

Confidence Interval for the Population Mean When Large Samples Are Selected	Large samples ($n > 30$) are necessary in order for the central limit theorem to be used as the basis of normality for a sampling distribution. The need for the large sample arises because, unlike the preceding example, either the population distribution or the population standard deviation, or both, happen to be unknown. If the population distribution is unknown and the population standard deviation is known, the confidence interval is

$$(1 - \alpha)\% \text{ confidence interval for } \mu = \overline{X} \pm Z\sigma_{\overline{X}}, \qquad (6.1)$$

where $\sigma_{\overline{X}} = \sigma/\sqrt{n}$.

However, if the standard deviation is unknown, the confidence interval is

$$(1 - \alpha)\% \text{ confidence interval for } \mu = \overline{X} \pm Zs_{\overline{X}}, \qquad (6.2)$$

where $s_{\overline{X}} = s/\sqrt{n}$

and s is the sample estimator of the population standard deviation, σ.

Keep in mind that the circumstances are generally such that equation (6.2) is used much more often than equation (6.1) as a practical matter. Also, if the sampling is from a finite population, be sure to use the finite population correction factor (fpc) when calculating $s_{\overline{X}}$ or $\sigma_{\overline{X}}$.

Example 6.2

▷ Joe Fabin's World Famous Steak House receives monthly orders for approximately 4,000 ribeye steaks. In testing the average weight of the steaks, 400 steaks were randomly selected and carefully weighed. The mean weight of the steaks in the sample was found to be 16.1 ounces.

The owner, Mr. Joe Fabin, if he so desired, could claim that his average-size ribeye steak is 16.1 ounces based on the point estimate ($\overline{X}$) of the sample taken. What would happen, though, if he were called to task on this estimate by some truth-in-advertising group, particularly by someone who knew that $\overline{X} = 16.1$ ounces is only one of many possible sample means that might have occurred? So instead of a point estimate, Mr. Fabin decided to use an interval estimate at the 90 percent level of confidence. He determined that the sample standard deviation, s, was 0.20 ounces. Then he used equation (6.2) and $\alpha = 0.10$ to determine the 90 percent confidence interval. Thus the following calculations and formulas would apply:

$$(1 - \alpha)\% \text{ confidence interval for } \mu = \overline{X} \pm Zs_{\overline{X}}.$$

In addition $s_{\bar{X}} = (s/\sqrt{n})\sqrt{(N - n)/(N - 1)}$ since the sampling was from a finite population. Thus the

90% confidence interval for $\mu = \bar{X} \pm Zs_{\bar{X}}$

$$= 16.1 \pm Z\left(\frac{0.2}{\sqrt{400}}\right)\sqrt{\frac{4000 - 400}{4000 - 1}}$$

$$= 16.1 \pm (1.64)\frac{0.2}{20}\sqrt{\frac{3600}{3999}}$$

$$= 16.1 \pm 1.64(0.01)\sqrt{0.9002}$$
$$= 16.1 \pm 1.64(0.01)(0.949)$$
$$= 16.1 \pm 0.0156,$$
$$16.0844 \le \mu \le 16.1156 \text{ ounces.}$$

Now Mr. Fabin has a range that he thinks includes the true average weight of his ribeye steaks with a probability of 0.90. Remember, though, the probability is really attached to the statement about the range or interval. ◁

Confidence Interval for the Population Mean When Small Samples are Selected from a Normal Population with Unknown Population Standard Deviation

When small samples ($n < 31$) are selected, use of the central limit theorem to imply normality of the sampling distribution is not valid. As you recall, this category was discussed in an earlier section of this chapter. It was explained that the t distribution was the proper sampling distribution for small samples taken from *normal* populations with an *unknown* population standard deviation. Thus, beginning with the expression

$$t = \frac{\bar{X} - \mu}{s_{\bar{X}}}$$

and solving for μ yields

$$\mu = \bar{X} \pm ts_{\bar{X}}.$$

So, using our knowledge of confidence intervals, we can write the expression

$$(1 - \alpha)\% \text{ confidence interval for } \mu = \bar{X} \pm ts_{\bar{X}}, \tag{6.3}$$

where the value of t must be determined from appendix I similar to the way we have used Z in the preceding confidence intervals, except we now have to consider the degrees of freedom $(n - 1)$ and the value of α. Recall that the chapter 5 discussion about the t distribution pointed out there is not a single t distribution; rather, a specific t distribution exists for each value of $n - 1$ degrees of freedom. Also, as the degrees of freedom $(n - 1)$ becomes large, say greater than 30, the normal distribution can approximate the t distribution fairly accurately.

Example 6.3

▷ Farmer's Feed and Seed, a large, regional retail feed and seed outlet in Omaha, Nebraska, was forced to purchase large quantities of a particular feed product because the manufacturer of the product was faced with a prolonged strike. Three months have elapsed since the feed purchase was made, and the general manager, Margie Winborne, is concerned about the stability of the composition of this particular type of feed, especially the loss of water which affects the weight. The net weight of the contents of a standard container of the feed is supposed to be 8 pounds. Ms. Winborne has decided to take a sample of 16 containers and weigh and test the contents in order to construct a 98 percent confidence interval to see if the weight is holding at 8 pounds. She has chosen a small sample because the product is expensive and the test involves damaging the package and destroying some of the feed. (A large sample is usually impractical in cases involving *destructive testing*.) So Ms. Winborne expects to use the t distribution because the sample is small, and she has been advised by the manufacturer of this feed that the net weight of all containers of this product is approximately normally distributed. If the sample results showed that $\overline{X} = 7.95$ pounds and $s = 0.6$ pounds, then

$(1 - \alpha)\%$ confidence interval for $\mu = \overline{X} \pm ts_{\overline{X}}$,

98% confidence interval for $\mu = 7.95 \pm t\dfrac{0.6}{\sqrt{16}}$.

From appendix I, when $\alpha = 0.02$ and df $= 15$, t is 2.602. Thus

$$98\% \text{ confidence interval for } \mu = 7.95 \pm 2.602\left(\frac{0.6}{4}\right)$$
$$= 7.95 \pm 2.602(0.15)$$
$$= 7.95 \pm 0.39,$$
$$7.56 \leq \mu \leq 8.34.$$

Since this range does include 8 pounds, Ms. Winborne should not fear that much, if any, deterioration has taken place during the three months of storage. ◁

Confidence Interval for the Population Proportion

In chapter 5 it was pointed out that the statistic of interest will not always be the mean. At times we will be interested in percentages, usually referred to as proportions. Furthermore it was pointed out in chapter 5 that even though proportions (when there are two responses, like yes-no) are binomially distributed, the normal distribution can be used as the sampling distribution of the sample proportions when both $n\pi$ and $n(1 - \pi) > 5$.

We can start with the familiar standardized normal deviate expression,

$$Z = \frac{p - \pi}{\sigma_p},$$

and replace σ_p with the estimate of the standard error of the proportion given by equation (6.4):

$$s_p = \sqrt{\frac{p(1 - p)}{n}}. \tag{6.4}$$

Using the estimated standard error of the proportion instead of σ_p, the standardized normal deviate expression becomes

$$Z = \frac{p - \pi}{s_p}.$$

Solving for π yields

$(1 - \alpha)\%$ confidence interval for $\pi = p \pm Zs_p. \tag{6.5}$

Example 6.4

▷ Each year the director of housing at State University must attempt to determine what proportion of students applying for admission to State U. will request dormitory space. This year there are 12,000 applications for admission to the fall semester. The director took a random sample of 200 applications to check the proportion of students requesting dormitory facilities. Of these 200 applications, 80 applications, or 40 percent, requested dormitory housing. For this situation a 95 percent confidence interval could be constructed using equation (6.5). The sample proportion, p, is $80/200 = 0.40$.

$(1 - \alpha)\%$ confidence interval for $\pi = p \pm Zs_p,$

95% confidence interval for $\pi = 0.40 \pm 1.96 \sqrt{\dfrac{(0.40)(1 - 0.40)}{200}}$

$$= 0.400 \pm (1.96) \sqrt{\frac{(0.40)(0.60)}{200}}$$

$$= 0.400 \pm 1.96 \sqrt{\frac{0.2400}{200}}$$

$$= 0.400 \pm 1.96 \sqrt{0.0012}$$
$$= 0.400 \pm 1.96(0.0346)$$
$$= 0.400 \pm 0.068,$$
$$0.332 \le \pi \le 0.468.$$

Thus the director can see that that at least one out of every three applications for admission will show a request for dormitory space, and he can prepare accordingly to meet this demand for the fall semester. ◁

Summary of Sampling Situations

In this chapter we discussed the various sampling situations and the appropriate sampling distribution to use in each of several cases. Table 6.1 gives a summary of the sampling situations for the sample mean ($\overline{X}$). It is a handy reference guide for sampling problems in this text as well as in the "real world."

Table 6.1
Sampling situations

Situations	Population distribution	Population standard deviation	Sample size	Appropriate $\overline{X}$ sampling distribution	Equation
A	Normal[a]	Known	Large	Normal	6.1
B	Normal[a]	Known	Small	Normal	6.1
C	Normal[a]	Unknown	Large	Normal via central limit theorem	6.2
D	Normal[a]	Unknown	Small	t	6.3
E	Unknown	Known	Large	Normal via central limit theorem	6.1
F	Unknown	Known	Small	None[c]	—
G	Unknown[b]	Unknown	Large	Normal via central limit theorem	6.2
H	Unknown	Unknown	Small	None[c]	—

a. Population distributions that differ slightly from normal are frequently treated as if they are normal distributions.
b. Populations that differ a great deal from normality or that contain extreme values will probably require samples that seem very large relative to the size of the population.
c. Approaches to these situations will be discussed in chapter 11.

Summary of Equations

6.1 Confidence interval for population mean—σ known and normal population:

$(1 - \alpha)\%$ confidence interval
for $\mu = \overline{X} \pm Z\sigma_{\overline{X}}$

6.2 Confidence interval for population mean—using s and large sample:

$(1 - \alpha)\%$ confidence interval
for $\mu = \overline{X} \pm Zs_{\overline{X}}$

6.3 Confidence interval for population mean—using s and small sample selected from a normal population:

$(1 - \alpha)\%$ confidence interval
for $\mu = \overline{X} \pm ts_{\overline{X}}$

6.4 Confidence interval for population proportion:

$(1 - \alpha)\%$ confidence interval
for $\pi = p \pm Zs_p$

Review Questions

1. Identify what each equation shown at the end of this chapter determines and when to use each one.

2. Define
(a) point estimate
(b) interval estimate
(c) confidence interval
(d) confidence level

3. Interpret the probability associated with confidence intervals.

4. List the four categories of sampling situations.

5. In figure 6.1, why is there no appropriate sampling distribution for Cases F and H?

6. What conditions are necessary to construct confidence intervals for proportions?

Problems

1. A random sample of 81 items was selected from a normal population. If the sample mean was 42 and the sample standard deviation was 5, construct a 95 percent confidence interval estimate of the population mean.

2. A random sample of 36 items yielded a sample mean of 40 and a sample standard deviation of 4. Construct a 90 percent confidence interval estimate of the population mean.

3. The owner of Flowers Beautiful desires to estimate the mean size of her outstanding accounts receivable. If a random sample of 25 accounts yielded an average unpaid balance of $60 with a standard deviation of $10, construct a 90 percent confidence interval estimate of the population mean. (Assume the balances are normally distributed.)

4. A recent random sample of 36 car owners revealed an average monthly expenditure on gasoline of $30 with a standard deviation of $12. Use the sample information and construct a 95 percent confidence interval estimate for the population mean expenditure.

5. A random sample of 49 pocket calculators yielded a mean life of 21 months with a standard deviation of 2.8 months. Construct a 95 percent confidence interval estimate for the average life expectancy.

6. The tear strength of a particular paper product is known to be normally distributed. If a random sample from 9 rolls yielded a mean tear strength of 225 pounds per square inch with a standard deviation of 15 pounds per square inch, construct a 90 percent confidence interval estimate for the average tear strength.

7. The commuting times of a very large group of workers have a standard deviation of 21 minutes. If a random sample of 49 commuting times had a mean of 52 minutes, construct a 98 percent confidence interval estimate for the mean commuting time of all workers.

8. A manufacturer of a certain type of X-ray developing fluid has recently collected information on the life expectancy of the fluid from a random sample of 64 customers. If the random sample yielded a sample mean of 16 days and a standard deviation of 4 days, construct a 90 percent confidence interval estimate for the average life expectancy of the fluid.

9. The number of shirts finished per hour by a particular production line is normally distributed. A random sample of 25 hours output had a mean of 40 shirts per hour and a standard deviation of 9 shirts. Construct a 95 percent confidence interval estimate for the mean number of shirts produced per hour.

10. It is known that the distribution of the gasoline consumption of all cars is slightly different from normal and has a standard deviation of 2.7 miles per gallon. If a sample size of 81 produced a sample mean of 19 miles per gallon, estimate with a level of confidence of 99 percent the mean gasoline consumption (μ) of cars.

11. A new gasoline additive was tested by a professional testing service. The testing service used a random sample of 50 cars that produced an average increase in gasoline mileage of 30 miles per tankful with a standard deviation of 8 miles. Use the sample information to construct a 95 percent confidence interval estimate of the population mean.

12. Tests on 26 2-by-4 steel studs yielded an average breaking strength of 580 pounds with a standard deviation of 68 pounds. Assuming that breaking strengths are normally distributed, construct a 90 percent confidence interval estimate for the average breaking strength.

13. A random sample of 25 motels in Vacation City revealed that the average room charge was $45 per day with a standard deviation of $6. Assuming that charges are normally distributed, construct a 90 percent confidence interval estimate for μ.

14. A random sample of a hundred newly designed throwaway writing pens indicated a mean writing time of 90 minutes with a standard deviation of 20 minutes. Estimate the population mean writing time by constructing a 90 percent confidence interval estimate.

15. A random sample of 25 items produced a sample mean of 80 and a standard deviation of 12. Assuming a normal distribution, estimate the population mean by constructing a 95 percent confidence interval estimate.

16. A random sample of 16 boxes of a popular fruit cereal contained a mean ($\overline{X}$) of 100 raisins per box with a standard deviation of 36 raisins per box. Assume that the number of raisins per box is normally distributed, and construct a 95 percent confidence interval estimate for μ.

17. Actually, the machine putting the raisins in the cereal boxes in problem 16 was out of adjustment long enough for 400 boxes to be processed. If the sample of 16 boxes was taken from the batch of 400, estimate the mean number of raisins per box in this batch with a 95 percent confidence interval?

18. Land, Inc., wishes to estimate the average price of an acre of highway property in the country. A random sample of 100 recent land purchases showed a mean sale price of $4,000 per acre with a standard deviation of $400. Use the sample information to construct a 98 percent confidence interval estimate for the average price.

19. A random sample of 200 accounts yielded an average unpaid balance of $75.00 with a standard deviation of $20. Use the sample information to construct a 95 percent confidence interval estimate for the average unpaid balance.

20. A concessionaire randomly sampled 200 people after the Big Game. If the sample indicated an average expenditure for concessions of $3.00 with a standard deviation of $0.50, construct a 95 percent confidence interval estimate for μ.

21. Repeat problem 20 assuming there were 6,000 people attending the game.

22. A sample size of 225 managers of Barely Fried Chicken produced a mean salary of $22,000 with a standard deviation of $1,500. Assuming there is a very large number of the Barely Fried Chicken places, construct a 90 percent confidence interval estimate for the average salary for all the managers.

23. In a large city a research analysis organization wants to estimate the average rental expense for a particular group of merchants. If a sample of 100 merchants produced $\overline{X}$ = $550 and s = $75, construct a 99 percent confidence interval estimate for the average rental expense of all merchants.

24. Rework problem 23 if there are only 500 merchants of this type in the city.

25. A random sample of 100 families had 65 families that owned one or more color television sets. Based on the sample results, construct a 99 percent confidence interval estimate of the population proportion that owns a color television set.

26. In a physical fitness test 240 out of 400 (60 percent) high school students could run a mile in less than 8 minutes. If the 400 students were a random sample, construct a 90 percent confidence interval estimate of the proportion of all high school students that can run a mile in less than 8 minutes.

27. A large bank recently studied a random sample of 80 accounts and observed that 20 of the accounts would not have to pay a service charge if a new minimum balance policy was adopted. Use the sample result to construct a 90 percent confidence interval estimate of the proportion of bank customers that would benefit from the new policy.

28. Repeat problem 27 if there are 600 customers that might be affected by the new policy.

29. If a random sample of 500 registered voters had 320 respondents who planned to vote for the Last Candidate, construct a 99 percent confidence interval estimate of the proportion of voters that Last Candidate will receive from all the registered voters?

30. The manufacturer of a high-precision product wishes to estimate the proportion of items manufactured that are defective. If a random sample of 200 items produced 10 defectives, construct a 99 percent confidence interval estimate for π.

31. A repair service owner is interested in the proportion of service calls that requires repair time exceeding one hour. If a random sample of 100 calls from last year's calls yielded 25 service calls exceeding one hour, construct a 99 percent confidence interval estimate for the proportion of service calls exceeding one hour.

32. A random sample of 100 subscribers of a daily paper indicated 40 persons who had read a particular advertisement in the paper. Construct a 90 percent confidence interval estimate for π.

33. A manufacturer of multispeed bicycles recently observed the performance of a random sample of 50 bicycles. If 15 of the bicycles were observed to perform improperly on the hill climb, construct a 99 percent confidence interval estimate of the proportion of bicycles having the malfunction.

Case Study

Lock Underwriters is a casualty insurance company specializing in liability and physical damage (collision) insurance for high-risk drivers. They currently do not sell insurance in Florida, but because of an inquiry from a general insurance agent there, Mr. Park Jackson, the president of Lock, is considering expanding his market into that state.

Before he can make a decision, Mr. Jackson feels he must know something about the market for high-risk insurance in Florida, with only a 5 percent chance of being wrong. Among other items of concern are

1. the proportion of high-risk drivers receiving traffic citations for driving while intoxicated,

2. the proportion of Florida's high-risk drivers who reside or work in Dade County (Miami),

3. the premium amounts currently being paid by high-risk drivers.

The Market Research Department of Lock acquired a list of all insured, high-risk drivers in Florida and, using that list, selected a random sample of 196 to contact by telephone. Here are some of the results of that survey:

1. Of those sampled, 42 lived or worked in Dade County.

2. Twenty-seven percent of the 196 drivers had one or more citations for driving while intoxicated.

3. The annual premiums currently being paid by those sampled were

a) for liability,
$\overline{X} = \$335, \quad s = \$85,$

b) for physical damage,
$\overline{X} = \$220, \quad s = \$30.$

Based on these results, what can you tell Mr. Jackson about the Florida market for high-risk insurance?

7

Some Practical
Aspects of Sampling

Other chapters on sampling emphasize the theoretical or conceptual aspects of sampling though attempts may be made to show applications of these theories and concepts. This chapter is directed toward the practical aspects of sampling in the real world. It considers the use, costs, and practicality of obtaining information for whatever purpose the information may have.

Why Sample?

Perhaps the most practical question one can ask about sampling is simply, "Why take a sample anyway?" Remember, samples are taken to obtain information about population characteristics. From a sample, *sample estimators* are computed; these sample data estimate the *parameters* of the population from which the sample was drawn. To extend the questions asked above, "Why work with estimators?" "Why not examine the entire population?" The answer here is rather simple.

First, a *complete enumeration* of all *sample units* in the entire *universe* is often unnecessary to obtain reasonably accurate results. Second, an examination of the entire *population* is often too costly and too time-consuming in the real world. Moreover, in cases of infinite or extremely large universes, a sample is required because an enumeration is impractical, if not impossible. Third, in the case of destructive testing situations, the *sample elements* or *units* must be destroyed or consumed in order to obtain the necessary measurements. (In example 6.3 of chapter 6 the test required the containers of feed to be broken and a portion of the product consumed as part of the measurement procedure.) Obviously, tests that destroy, damage, or consume sample elements require that the tests be conducted by some method of sampling.

Why Not Sample?

Perhaps then, the appropriate questions should be, "Why not sample? What are the risks involved?" One answer, of course, is that there are risks. For the advantage of saving time, money, effort, and product, one gives up a margin of *accuracy* and *precision*. In fact we might continue the argument that a sample is not a complete enumeration and only from a complete count can one be 100 percent sure of the accuracy of the results. Yet this observation is not entirely true either. Let's look at *precision* and *accuracy* in sampling. Perhaps then the issue can be addressed more correctly and appropriately.

Precision and Accuracy

In chapter 5 we learned that for simple random samples we measure the dispersion of the $\overline{X}$'s in the sampling distribution by the *standard error of the mean*. In a real sense this dispersion represents an error or variation of the sample means ($\overline{X}$'s) from the true mean (μ). In this same sense the *standard error* (mean or proportion) is a measure of *precision* or the lack of it. A smaller standard error, other things remaining the same, means more precision (less variance in the sampling distribution). But using the standard error as a measure of precision relates only to the chance, or sampling error, involved. In other words, this recognizes that in a sampling situation there is always *internal error* due to the fact that a sample has been taken. This idea can be seen in the discussion of sampling distributions in chapter 5 of this text. Because of this sampling error we also recognize that the sample statistic is expected to deviate from the population parameter.

However, there are two other factors that may cause a sample estimator to deviate from the parameter that is to be estimated. Often these factors are overlooked in sample design, but they should not be for they are very important. These two factors are *bias* and *external error*. Therefore there are really three factors that cause deviation between the sample statistic (the estimator) and its parameter value:

1. sampling (internal) error,
2. external error,
3. bias.

All three of these factors may contribute to a difference between the estimator and the parameter it is estimating. The closeness of the estimator to the parameter as affected by all three factors measures the *accuracy* of an estimation.

Only *sampling or internal error* is measured by the standard error, and it has been examined previously. It is caused by the

BASIC STATISTICS

fact that a sample is taken. Other things being equal, the sampling error is a function of the sample size. As the sample size increases, the value of the standard error is decreased, thereby reducing the sampling error.

External error is error related to the practical considerations in taking a sample. These errors are usually caused by the people conducting the sample and processing the data. The main components of an external error are *recording errors* (recording statistical data) and *processing errors* (editing and calculating data). This type of error usually increases as the sample size increases because individuals are dealing with larger quantities of data. For example, an interviewer who must call on 200 students in a sample survey is apt to make more mistakes in recording the information than an interviewer who calls on 100 students. Likewise, someone computing information from 100 items is likely to make fewer mistakes than when computing from 2,000 items. External error can usually be reduced by reducing the number of items in the sample. Better trained personnel will also help reduce this type error.

While *external* and *internal* errors may be in either direction, *sample bias* is a factor that causes the sample estimator to deviate from the parameter in a single direction. If you have ever seen a tree where the wind blows constantly in one direction, and the tree is bent over, you have a good idea of sample bias.

Bias, viewed from a practical consideration, can be the most insidious of the three factors causing deviation between the estimator and the parameter because it may be difficult to detect. We expect internal error, and you can repeat calculations and otherwise detect external error. Yet we may base conclusions on sample results without recognizing that bias exists.

Some of the activities in sampling that tend to lead toward biased results are

1. a poorly defined universe,

2. an inadequate sample design,

3. improperly worded questions,

4. any phenomenon providing the basis for respondents to distort the actual answer. For instance, it is quite likely that data concerned with the ages (especially the over-30 crowd) are biased downward.

There are many classic examples of bias in sampling. Perhaps the best known one is the case of the survey taken by the *Literary Digest* in connection with the Roosevelt–Landon presidential election of 1936. Based on a sample using the telephone directories and automobile registrations as the sample frame, it was concluded, "Landon wins by a landslide." Of course this result was not the case. The bias was due to a poor definition of the universe (names listed in the phone directories and automobile registers). Remember that in 1936 those who could afford to have telephones and automobiles were not the masses of unemployed, most of whom identified with the Democratic Party at that time.

If there is bias in a design or question or procedure, increasing the sample size from n to an N (a complete enumeration) will not eliminate the bias necessarily. External error increases with an increase in sample size. Thus a complete enumeration itself will eliminate the *internal error*, but it will not eliminate the *bias*, and it will actually increase *external* error. Therefore, why **not** take a sample?

Types of Random Samples

This chapter (and the previous one) is concerned with random (probability) samples. The main advantages of random samples are as follows:

1. They depend on random outcomes (not the judgment of the sampler) and are more likely to be representative.

2. The result is a probability distribution (normal or t) for which the magnitude of internal error can be estimated.

In a finite universe random samples may be drawn *with* or *without* replacement. Given N sample units in a universe, if we draw a sample of n units without replacement, the number of different possible samples is C_n^N. Therefore, if $N = 9$ and $n = 3$, the number of possible samples that can possibly be drawn is $C_3^9 = 84$.

If instead of sampling *without* replacement the samples were taken *with* replacement, the number of samples possible is N^n. In our example it is 9^3 or 729 possible samples *with* replacement, a difference of 645 when compared to a sample *without* replacement.

Simple Random Sample

A simple random sample may be drawn *without* replacement, and most often it is. Remember, however, the *finite population correction factor* should be used when the population is finite. The *simple random sample* is the basic design for most probability samples; other probability sample designs may be considered modifications of this basic one.

A simple random sample requires that the sample be taken in such a way that each *sample unit* in the universe has an equal probability of being selected. The technique for selecting a simple random sample is in fact fairly simple. Suppose that you want to select a simple random sample of 100 customer accounts from a list of 900 such credit accounts for some particular store. A list of the credit customers or account numbers (the *sample units*) is called the *sample frame*. One way to select the sample is to place credit customers' names or account numbers on slips of paper and then place them into a container so that the slips can be mixed in a random fashion. Next, select 100 slips (names or numbers) from the container without replacement. In this example we are selecting 100 units from a universe of 900 accounts. Thus, the number of possible random samples in this case is $C_{100}^{900} = 900!/(100!)(800!)$. If a simple random sample is selected, each of the possible samples has an equal probability of selection, which in this case is equal to $1/C_{100}^{900}$.

It is quite possible that writing all the terms on slips of paper and putting them into a container to select the sample will be cumbersome and unwieldy. For this reason a table of random numbers may be useful because it facilitates the procedure of a simple random selection. To use a table of random numbers (for example appendix L), we take some form of sample frame and assign numbers to each sample unit if the unit is not already in some numerical form. In our example, the accounts may be in customers' names or account numbers. If the accounts are not by number but by name, then numbers must be assigned to each account. The assignment need not be explicit so long as a unique number can be related to each sampling unit so it can be identified by the assigned number. For our illustration we might take the list (sample frame) of the 900 credit accounts and assign a number to each one, starting with the number 1 and assigning consecutive digits to the 900 customer accounts. The maximum number of digits used in the largest assigned number will determine the number of columns required in the table of random numbers. In this example, we assigned numbers from 001 to 900. Thus three columns of random digits will accommodate all possible assigned numbers to any of the accounts in the universe.

Now, suppose the following numbers were selected from three columns of the table of random numbers:

098
021
777

989
146
007
021

The item (account) corresponding to number 098 would be in the sample. Units numbered 21, 777, 146, and 7 would also be included. Since there are only 900 accounts in the universe, and we have numbered the units from 1 to 900, there is no sample unit numbered 989. We simply discard this number and replace it with a number that does correspond to a numbered account. Also, notice the number 021 appears twice. Remember, we are sampling *without* replacement; so we do not consider this account twice. Rather we replace it with another randomly selected account. Since there are circumstances in which certain numbers are discarded, it may have already occurred to you that we will probably have to select several more than 100 numbers in order to complete the selection of the sample.

Systematic Sampling

Systematic sampling offers us a modified way to *select* a random sample as opposed to the *simple random* selection technique. Because systematic sampling and simple random sampling are really both techniques for selection of sample units from the universe, either one or both methods may be employed in most *random* sample designs.

In the *systematic sample* procedure every ith sample unit in a universe is chosen for the sample. However, the very first unit is selected randomly. So we say that every ith unit is chosen but with a *random start*. Referring back to our example with credit accounts, if we selected every 100th account in our universe for the sample, we would be taking a systematic sample. Other likely examples are to select every 4th house in a series of blocks, every 50th pay voucher, every 80th car passing an intersection, or every 250th auto part produced on an assembly line.

This selection procedure requires that the sample units in the universe be ordered in some sequence. The ordering can be in explicit detail, or a rule may be specified which results in an ordering. In the case of ordering in explicit detail, the ordering may be numerical, geographical, by time interval, alphabetical, or any other such ordering approach that sequences the sample units. For example, pay vouchers may be ordered by the date on which the liability occurs. Another example of explicit ordering is the listing of names in the telephone directory.

Less explicit ordering is established by some rule. For example, if the sample frame consists of occupied dwelling units on a map for a particular city, the dwelling units (houses) could be ordered for a systematic sample by establishing a rule concerning the sequence in which the streets or perhaps city blocks will be listed in the sample frame and subsequently how the dwelling units will be counted. There is of course an administrative advantage to using the systematic method when the sample units are ordered because the selection process is convenient and requires less effort.

To maintain randomness in a systematic sample, the *start* of the sample is determined by chance. For instance, if every 50th item is to be selected, we start the sample selection between items 1 and 50. Then we select every 50th item from there on. The particular item between 1 and 50 to start the selection would be *probabilistically determined*. In this case we might select a two-column list of random digits from a table of random numbers. The first number between 1 and 50 would be the start. Or else we might simply draw from slips of paper numbered between 1 and 50; the number drawn would indicate the start. In any case a random start is necessary to involve probability concepts in a systematic sample.

In addition to a random start, another consideration in conducting a systematic sample is *the interval at which selection is to be made*. That is, how do we determine whether every 10th, 50th, 100th, or *i*th item is to be chosen? In many real-world situations the interval depends on the sample size. For instance, if we want to select a sample of 100 units from a universe consisting of only 1,000 units, we would select every 10th unit. That is,

$$i = \frac{N}{n}, \tag{7.1}$$

where
  i = the interval,
  n = sample size,
  N = universe size.

Let us consider again the universe with 9 units in it and a sample of size 3. That is, $N = 9$ and $n = 3$. How many different possible samples can we draw using the systematic sample selection procedure? In this case, $i = 9/3 = 3$.

Suppose the sample units in the universe are numbered 1–2–3–4–5–6–7–8–9. Given random a start, we might randomly choose the number 1 or 2 or 3 to begin the selection. If number 1 is randomly chosen, then the sample would consist of units 1–4–7. If 2 is picked, the sample would be 2–5–8, and the sample would consist of units 3–6–9 if 3 is taken randomly. Thus there are really only three possible samples to be selected in this example of $n = 3$ and $N = 9$. They are

Sample A	Sample B	Sample C
1	2	3
4	5	6
7	8	9

The point is that, in using the systematic method of selection, the entire sample is chosen when the random start is selected.

There is one real danger in using a systematic sample design. If there is a *pattern* in the ordering of the universe, we may get a sample with statistics that do not represent the population parameters to be estimated. Suppose we were taking a systematic sample of items coming off a production line. Let us further assume that one of the important machines used to produce the items is faulty in that every 10th item produced is defective. If we were to select every 10th item (or multiple of 10, such as 20 or 30), all the items in the sample will be defective if the first item selected is defective. Similarly, if the first item is not defective, the sample will contain 0 defectives.

Stratified Sample

Another modification of the simple random design is the *stratified sample*. This sample design gets its name from the division of sample units in the population into *strata* or *layers*. When a population is *heterogeneous* overall but within it there are *homogeneous* subpopulations (or *strata*), the population is said to be *stratified*. Under this circumstance it is usually desirable to recognize this division of the population and design the sample accordingly. Taking advantage of the homogeneity within each *stratum*, units are selected from each of the strata rather than from the universe as a whole, depending of the population to be sampled (age, GPA, height, income, weight, and so on).

Consider an illustration of taking a sample of 200 students from a universe of 5,000 undergraduates in a small college. No doubt, the student body consists of freshmen, sophomores, juniors, and seniors. Depending on the population characteristics to be measured, the four classes of students may fall into four homogeneous strata in the particular population. Under certain circumstances it would be advantageous to select the sample from the individual stratum. That is, rather than selecting 200 students

Table 7.1
Stratification of college students by class

Class	Number of students
Freshmen	2,000
Sophomores	1,250
Juniors	1,000
Seniors	750
	5,000

from the entire population of 5,000, we might draw them from each of the strata according to their student classification and select n_i units (students) from each stratum. Thus the sample size (n) would equal $n_1 + n_2 + n_3 + n_4$, where n_1 is the number of units selected from the first stratum, n_2 is the items from the second stratum, and so on.

There are various approaches for determining the number of items to select from each stratum. Perhaps the most common is to make the number of items selected proportional to the size of the strata. Suppose in our college the student body is broken down as indicated in table 7.1.

If we were selecting units proportional to the size of the strata, we would select the sample of n_i from each stratum based on equation (7.2):

$$n_i = \left(\frac{N_i}{N}\right)n, \tag{7.2}$$

where
n_i = the stratum sample size for the ith stratum,
n = the total sample size,
N_i = the size of the ith stratum,
N = the universe size.

The last column of table 7.2 was computed using this equation.

In selecting sample units proportional to the size of the strata, we assume that the variances within each stratum are similar, or we simply choose to ignore the existence of any differences. We may wish to select the number of sample units proportional to the *size* of the stratum and to the *variance* (or *standard deviation*) within the stratum because this selection allows us to increase the proportion of the sample selected from strata that are larger and have more variability (larger variance). This approach provides for greater precision for any given sample size where strata variances are different. For example, in sampling the undergrad-

Table 7.2
Sample size for each stratum using proportional parts.

Class	Number of students (N_i)	Number selected for sample (n_i)
Freshman	2,000	80
Sophomores	1,250	50
Juniors	1,000	40
Seniors	750	30
	5,000	200

$$n_1 = \frac{2,000}{5,000}(200) = 80 \qquad n_3 = \frac{1,000}{5,000}(200) = 40$$

$$n_2 = \frac{1,250}{5,000}(200) = 50 \qquad n_4 = \frac{750}{5,000}(200) = 30$$

uate students to obtain information on grade point averages, most likely the GPA's for seniors are more homogeneous (smaller variance) than are the GPA's for freshmen. So, even if the stratum for freshmen was the same size as the stratum for seniors, we would want to take a larger size sample from the freshmen than from seniors to adjust for the greater variability of GPA's for freshmen. Equation (7.3) is a modification of equation (7.2). It is used to compute the sample size for each stratum (n_i) considering differences in strata size and variance:

$$n_i = \frac{N_i\sigma_i}{\Sigma[N_i\sigma_i]} \cdot n, \tag{7.3}$$

where
n_i = the stratum sample size for the ith stratum,
n = the total sample size,
N_i = the size of the ith stratum,
σ_i = the standard deviation for the ith stratum.

The main problem with this procedure is that often σ_i is not known or ascertainable in the real world.

A third approach used in stratified sampling is to select the number of sample units proportional to the *size* of the strata and the reciprocal of the *cost* of sampling from the strata. This procedure adjusts for cost differences of selecting sample units among different strata. It might be more (or less) expensive to select items from one stratum than another. In most situations the number of items selected from each stratum would vary directly with the size of the stratum and inversely with the cost associated with selecting items from the stratum.[1] If only cost and the relative size of each stratum are to be considered, equation (7.4) may be used to determine n_i for each stratum:

$$n_i = \frac{N_i(1/c_i)}{\Sigma[N_i(1/c_i)]} \cdot n, \qquad (7.4)$$

where

n_i = the stratum sample size for the ith stratum,
n = the total sample size,
N_i = the size of the ith stratum,
c_i = the cost of sampling from the ith stratum.

Sometimes, we may find it beneficial to include all three factors—stratum size, variance, and cost—to select the number of sample units from each stratum. Equation (7.5) would be used in this case. Here again, this formula is similar to others used for determining the size sample to select from each stratum (n_i):

$$n_i = \frac{N_i\sigma_i(1/c_i)}{\Sigma[N_i\sigma_i(1/c_i)]} \cdot n, \qquad (7.5)$$

where

n_i = the stratum sample size for the ith stratum,
n = the total sample size,
N_i = the size of the ith stratum,
σ_i = the standard deviation of the ith stratum,
c_i = the cost of sampling from the ith stratum.

The concept here is that for a given stratum, n_i is larger if (1) the stratum is large, (2) the stratum has greater variance, and (3) the sampling cost is less in the stratum.

The preceding equations are the most likely approaches one might employ in determining the number of items to be selected from each stratum. The actual selection of units from each of the strata may be conducted in the same manner as a simple random sample.

Cluster Sample

Another form of a modified random sample design is called the *cluster sample*. The cluster sample design requires that the sample units be grouped. However, unlike the stratified sample, the sample units are grouped by *clusters* in the *universe* not by homogeneous strata in the population. The *sample unit* consists of more than one *sample element*. An entire cluster or group of elements is

---

1. Depending on the cost relationship, sometimes the square root of the reciprocal of the cost, $\sqrt{1/C_i}$, is used.

randomly selected, and all (or part) of the elements in the cluster are included in the sample. Thus the sample unit is the cluster of elements.

A simple illustration of the use of a cluster sample design would be to take a sample of the 5,000 college students by selecting entire sections of courses. Rather than select individual students, we would select, at random, courses and sections. As an example, Math 134, section 2, might be randomly selected. In this case all students enrolled in this section of the course are included in the sample.

When dividing a universe into clusters, care should be taken to *avoid any overlap* of the clusters. In our example, some student might be in the sections of two different courses that have been randomly selected. As in the case of the simple random sample, duplication of individual items, as well as clusters, should be avoided. Thus, if a sample element appears in two randomly selected clusters, it should be omitted from the second cluster as well as any other clusters selected thereafter.

A more practical example of the use of a cluster sample will serve to illustrate an important advantage of this type of sample design. (See example 7.1).

Example 7.1

▷ Hy-take Supermarket, Inc., has one of its stores located near university student housing. Because it is felt that the average student purchase is lower than the average purchase among nonstudents, Hy-take has not used a trading stamp program in this store. Recently, however, the general manager questioned the assumption regarding the average purchase by students and has suggested that a sample of individual purchases be taken at the store to determine the average purchase.

The store keeps its used rolls of cash register tape for several months. Each of these rolls ordinarily contains the records of the purchases of from 60 to 90 customers.

If the rolls are treated as sample units, a simple random sample can be selected from the store's file of tapes. Each sample unit contains a *cluster* of sample elements (individual purchases in this case). That is, the roll of tape is a cluster of sample elements. In taking a random sample of tape rolls, the group at Hy-take is really taking a cluster sample. A simple random sample of purchases would have required that individual purchases (not rolls) be selected by a random method. ◁

The preceding example shows that there are circumstances when it is more expedient, or less expensive, to select a cluster of elements rather than select each individual element. However, we must be cautious about the effect the cluster sample will have on precision and the extent to which the cluster might distort the sample results. Consider a situation where abnormally high (or low) purchases are made during certain times of the day. As a result a cash register tape may have only the high (or low) purchases. In other words, *a segment or cluster may not contain anything like a representative distribution of sample elements.*

It may be interesting to note that, while the cluster sample may appear similar to the stratified sample, it is much more similar to the systematic sample. Using a systematic selection with a random start really involves selecting a cluster of elements by selecting every ith element. That is, in a systematic sample when the ith unit is selected so are all the other units in the sample. As in the systematic design, if there are i clusters in a cluster sample, then there are only i possible samples regardless of how many elements there are in each cluster. Think about it!

Multistage Sampling

The sample designs that have been mentioned are considered the basic ones among the sample selection procedures. There are others, but these four, (1) simple random sample, (2) systematic sample, (3) stratified sample, and (4) cluster sample, provide the fundamental selection techniques for many other sample systems. One such sample system is called the *multistage sample* system. It is called *multistage* because the selection procedure takes place in a hierarchy of stages. Yet these four fundamental selection techniques can be usefully applied in various combinations in a multistage sample system. Consider example 7.2 which illustrates a multistage sample system.

Example 7.2

▷ The president of Hy-take Supermarket, Inc., was so pleased with the results from sampling purchases at the university store that he has decided a sample should be taken of the purchases at 150 stores in the country. Ms. Ashley, the market researcher at the Hy-take home office, has decided that a three-stage (multistage) sample should be taken. The first stage is to select, on the basis of a simple random sample, 15 of the 150 stores. She then recommends that cash register tape rolls be randomly selected at each of the 15 stores. The final stage of the sample is to select every 20th purchase on the tape rolls using a random start. ◁

In this example, the first stage is the selection of the *primary sample unit* (the 15 stores in the example); the second is the selection of the *secondary sample unit* (tapes at each store); the third is the selection of the *tertiary sample unit* (specific purchases). Each stage of the multistage procedure may involve any of the possible sample designs mentioned earlier.

One real-world use of multistage sampling is the situation in which sampling units are dispersed over a wide geographical area. It is often less expensive to divide the geographical area into smaller geographical regions and let them be the primary sample unit, a number of which are selected. Then subregions become the secondary sample units and are selected (systematically or simple randomly). Then subregions may again be subdivided into tertiary sample units such as cities or even neighborhoods.

Final or *ultimate sample units* may be selected at any stage. If they had been selected after the primary units were selected, we would have a two-stage sample. If they were selected after the secondary units, we would have a three-stage sample; after the tertiary sample unit a four-stage sample, and so on.

Multistage sampling has a wide range of applications, but, in general, it is useful in sampling larger numbers of units, especially when cost saving is important.

Random versus Nonrandom Sampling

Usually in discussions on sampling and sampling theory, the underlying assumption is that the sample is random, and the variables and estimators are randomly distributed. In these discussions, estimations and statistical tests are based on theoretical distributions depending on random variables. In fact the formulation of the standard errors is based on these same basic distributions. In chapter 5 the formulation of the standard errors for estimation was based on the assumption of a simple random sample.

Sometimes nonrandom or *judgment samples* are used to determine characteristics of various populations. Perhaps the most common of these judgment samples is the so-called *quota sample*. Quota samples are so called because quotas of certain characteristics are established to complete the sample. For example, when the sample units are houses, the quotas may be established by the size of the house, the location of house, the structure of the house, and the design of the house. Therefore a particular survey sheet may require an interviewer to sample a 1,700 to 2,000 square foot house, located in the northern quadrant of town, with a frame structure, single-story design. The quotas are established from known characteristics of the universe. The sample is "forced" to assume these characteristics. The theory underlying quota samples is that, if the sample is deterministically designed to assume known characteristics of the universe, then it should assume the unknown characteristics. This concept is questionable; certainly there is no statistical basis for it. However, estimates from quota samples have been known to be fairly accurate.

The Sample Size

In statistical sampling the cost of the sample and the sample size are so interrelated that it is difficult to consider one factor without the other. For, other things being equal, you would expect cost to vary with the sample size. However, for presentation purposes and clarity of discussion, we consider sample size in this section and sample cost in the next.

In addition to the cost there are two other important factors affecting the size of a sample. They are

1. the level of accuracy or precision desired,
2. the variability (variance) of the items in the population.

Notice that the size of the universe (N) is *not* one of the other two important factors in determining sample size (n). The only time the universe size (N) becomes a factor is when the universe size itself is small enough to require the use of the finite population correction factor (fpc). Of course, if the universe is small in size, the sample size necessary to achieve a given level of precision may be reduced. At the same time, if the universe is quite small, a sample may be impractical anyway.

As you know, the *level of accuracy* refers to the extent to which the estimator coincides with the parameter and considers the effects of internal error, external error, and bias, while the *level of precision* refers only to internal error. In determining sample size, we are prone to consider only the level of precision because it is easily measured by the standard error. This approach may be reasonable if we always use only the simple random sample design in sampling. However, in designing a sample from among several alternative approaches and selection procedures, we cannot overlook the importance of external error and bias.

For example, it may be more accurate (less external error) to use a systematic selection of names from a list of registered voters, rather than a simple random one, if the list is compiled according to the date of registration. Similarly, using a telephone interview to gather information may evoke bias in the response. Data obtained from the use of a mail questionnaire is usually biased because only those persons with relatively strong feelings on a particular subject return the interview sheet, and there is bias from nonresponse.

Nonetheless, in using mathematical formulas to determine sample size, we usually relate sample size to levels of precision, and the standard error is one component in these equations. Generally speaking, to lower the standard error and gain a higher level of precision the sample size must be larger.

The other important factor affecting sample size is the *variability* of items in the population. In short, more variability among elements usually requires larger samples. Yet we have seen how to divide the population into strata and how to sample from these strata. If the variance *between* strata is greater than the variance of items *within*, we can get more precision with a stratified sample than a simple random sample.

If the population divides into convenient groups or segments of sample elements, then we may select sample units consisting of a number of elements. Cluster sampling usually requires a smaller sample size because the sampling units determining the sample size for selection contain more than one sample element. If the variability of the elements within the cluster is greater than the variability between clusters, a cluster sample may give better accuracy or precision for the same sample size, or dollar of sample cost, than either the stratified sample or the simple random sample.

For purposes of illustration, let us see how sample size is computed for simple random samples, considering precision (as mesaured by the standard error) and the variance (actually the standard deviation) for means and proportions.

Sample Size for Means

Using the standardized normal deviate

$$Z = \frac{\overline{X} - \mu}{\sigma_{\overline{X}}}$$

and remembering that $\sigma_{\overline{X}} = \sigma/\sqrt{n}$, then

$$Z = \frac{\overline{X} - \mu}{\sigma/\sqrt{n}},$$

$$Z = \left(\frac{\overline{X} - \mu}{\sigma}\right)\sqrt{n},$$

$$Z\sigma = (\overline{X} - \mu)\sqrt{n},$$

$$\sqrt{n} = \frac{Z\sigma}{\overline{X} - \mu},$$

$$n = \frac{Z^2\sigma^2}{(\overline{X} - \mu)^2}. \tag{7.6}$$

Now, equation (7.6) can be used to determine the sample size n necessary to provide an $\overline{X}$ *within a certain distance of* μ, $(\overline{X} - \mu)$, for a given level of confidence which will be provided by the normal distribution. The Z value will again be looked up in appendix H using the appropriate α value. The difficult feature as far as using equation (7.6) is that it assumes that the variance σ^2 is known. As we already know, this is not always the case; thus equation (7.6) for application purposes is usually rewritten as follows:

$$n = \frac{Z^2 s^2}{(\overline{X} - \mu)^2} . \tag{7.7}$$

Example 7.3

▷ Suppose the Board of Directors of a large corporation desires to know the average number of vacation days taken by its employees. Further the corporation desires that the estimate be off by no more than 2 days from the true average, and that the level of confidence be 90 percent. This project has been delegated to the Statistics Department. The department knows that it is being asked to estimate μ so that the difference, $\overline{X} - \mu$, is no more than 2 days. The department also realizes that σ is unknown and will estimate σ from a preliminary sample of 100. Suppose this sample of 100 yielded a standard deviation of 15 days. Using equation (7.7),

$$n = \frac{Z^2 s^2}{(\overline{X} - \mu)^2} = \frac{(1.64)^2(15)^2}{(2)^2} = \frac{(2.69)(225)}{4} = 151.$$

Thus, to complete its assignment, the Statistics Department will take a sample of 151 observations to obtain $\overline{X}$ which will provide an estimate of μ within ± 2 days at the 90 percent confidence level. ◁

Equations (7.6) and (7.7) do not directly utilize any information concerning the size of the population. In fact, in using these equations, you assume that the population is infinite in size. However, if the size of the population is actually finite, then equation (7.8) should be used:

$$n = \frac{NZ^2 s^2}{(N - 1)(\overline{X} - \mu)^2 + Z^2 s^2} . \tag{7.8}$$

Thus, if the corporation in example 7.3 had 1,000 employees, the sample size would be somewhat less:

$$n = \frac{NZ^2s^2}{(N-1)(\overline{X}-\mu)^2 + Z^2s^2}$$

$$= \frac{1{,}000(1.64)^2(15)^2}{(1{,}000-1)(2)^2 + (1.64)^2(15)^2}$$

$$= \frac{1{,}000(2.69)(225)}{999(4) + 2.69(225)}$$

$$= \frac{605{,}250}{3{,}996 + 605.25} = \frac{605{,}250}{4{,}601.25}$$

$$n = 131.5 \text{ or } 132.$$

Sample Size for Proportions

It is possible to determine the sample size necessary to provide a sample proportion p, that is within a certain distance of the true population proportion, π. Returning to the expression

$$Z = \frac{p - \pi}{\sigma_p},$$

where $\quad \sigma_p = \sqrt{\dfrac{\pi(1-\pi)}{n}}$

so that

$$Z = \frac{p - \pi}{\sqrt{\dfrac{\pi(1-\pi)}{n}}}$$

$$Z\sqrt{\frac{\pi(1-\pi)}{n}} = p - \pi$$

$$Z^2\left[\frac{\pi(1-\pi)}{n}\right] = (p - \pi)^2$$

$$n = \frac{\pi(1-\pi)Z^2}{(p-\pi)^2}. \tag{7.9}$$

Example 7.4

▷ A political candidate has employed a consulting firm famous for polls. He has requested the firm to determine within 3 percentage points the proportion of people who currently will vote for him. The level of confidence desired is 95 percent. Thus the firm will be using equation (7.9) to decide on the necessary sample size, so that

$$n = \frac{\pi(1-\pi)Z^2}{(p-\pi)^2}.$$

As no information is available concerning the candidate's previous popularity, π is assumed to be 0.50 which generates the largest value of n, and from appendix H, $Z = 1.96$:

$$n = \frac{(0.50)(0.50)(1.96)^2}{(0.03)^2} = \frac{(0.2500)(3.84)}{0.0009} = 1{,}067.$$

The firm must use a sample of 1,067 voters to give the candidate the information he desires. ◁

This equation (7.9) should only be used when the population is infinite in size. In situations where the size of the population is limited (finite), equation (7.10) should be applied:

$$n = \frac{NZ^2\pi(1 - \pi)}{(N - 1)(p - \pi)^2 + Z^2\pi(1 - \pi)}. \tag{7.10}$$

Suppose the candidate in example 7.4 represents an area in which there are only 5,000 voters. Then the required sample size for his poll would be determined as follows:

$$
\begin{aligned}
n &= \frac{NZ^2\pi(1 - \pi)}{(N - 1)(p - \pi)^2 + Z^2\pi(1 - \pi)} \\[2mm]
&= \frac{5{,}000(1.96)^2(0.50)(0.50)}{(5{,}000 - 1)(0.03)^2 + (1.96)^2(0.50)(0.50)} \\[2mm]
&= \frac{5{,}000(3.84)(0.50)(0.50)}{4999(0.0009) + 3.84(0.25)} \\[2mm]
&= \frac{4{,}800}{4.4991 + 0.96} = \frac{4{,}800}{5.4591}
\end{aligned}
$$

$n = 879.3$ or 880.

Cost Analysis

In developing a sample design for a given situation, cost must be considered a factor equally important to size, precision, and accuracy. Indeed, in every sampling situation we should conceptually pose the question, "Is this information worth the cost of obtaining it?"

The Cost of Sampling

Like most other situations there are both *fixed* and *variable costs* associated with taking a sample. It is usually less expensive to take a sample than to do a complete enumeration (examine 100 percent of the universe). Yet the cost of taking a sample of n items may vary considerably from one sample design to another. In considering the cost of alternative designs, we must differentiate between the fixed cost and the variable cost. The fixed cost of course will be a fixed amount whether we take a sample size 1 or one of size 10,000.

The variable cost, on the other hand, varies with the sample size. Some designs may have a high fixed cost but a low variable cost. Other designs may be just the opposite—the fixed cost is low, but the variable cost is high. However, it is usually the variable cost that is more affected by different sample designs because of the effect on sample size and operating field expenses. In any event, if you are taking a sample for yourself, boss, client, instructor, or whomever, you must eventually address the question of cost.

Example 7.5

▷ Suppose our political candidate is now considering running for the U.S. Senate from his state or for the U.S. Congress from his district, whichever he feels he'll have a better chance of winning. Much of his decision will depend upon the sample survey results.

He explains to you that he wants to take two political "polls" (sample surveys) of registered voters—one across the entire state and another across his congressional district. He wants the same level of accuracy for both. Now, he wants to know how much each poll will cost.

You examine the past political behavior of the state and district; you also compare certain characteristics of the registered voters for both geographical areas. Based on this analysis you determine that the variability of voting behavior is roughly the same when comparing the congressional district to the state as a whole.

Now, considering accuracy and variability, what sample size would you recommend, and consequently what cost estimates would you give the candidate?

If the variability is essentially the same between the state and the district, if the candidate wants the same level of accuracy or precision for each sample, and if you use the same sample design, then the sample size would be approximately the same for either the district or the state. The state contains a much larger number of registered voters than the congressional district, but this fact is of little consequence in determining the sample size. Finally, however, the state is the larger geographical area, so the cost to sample it may be higher if you expect to use face-to-face interviews or telephone calls to collect the information. ◁

The Worth of Sampling

The question of how much sample information is worth to the user often becomes the overriding consideration in determining how much to spend on a particular sample. This idea can be illustrated through a simple nonreal-world illustration, the concept of which does have real-world application.

BASIC STATISTICS

Suppose there are two poker cards which may be either kings or aces. They may be all aces or all kings or any combination of them. Further, imagine that you are engaged in some game in which you have the opportunity to wager (make an investment) on whether or not both cards are aces.

If you decide to make the wager and both cards are aces, you win $11; otherwise, you lose $3. You are limited to a single bet each time, but you are not required to play each time, and you do have the option to buy information about the cards.

A sample of one card cost $1, and a complete enumeration of both cards cost $2. In this case, is the information worth the cost? If so, do you take the sample of one card or the enumeration of both?

To use this illustration further, assume that the two cards had been drawn previously from a deck consisting only of the four kings and four aces. At this point you do not know anything about the cards except they were taken from the larger deck of eight cards as specified.

Now, suppose you decide not to purchase any information but to place the wager. Is this bet (investment) favorable or unfavorable to you?

Let's look at the probability of both cards being aces if chosen randomly from the larger deck of eight cards:

$$P(\text{Two aces}) = \frac{C_2^4 \cdot C_0^4}{C_2^8} = \frac{6}{28} = \frac{3}{14}.$$

Thus the probability of your winning is 3/14, and the probability of your losing is 11/14. The expected value (the long-run average) of a game can be calculated using the equation for mathematical expectation for a discrete random variable.

$$E(X) = \Sigma[x \cdot P(X = x)] \tag{7.11}$$

Continuing with our illustration concerning two aces:

$$E(X) = \left(\frac{3}{14}\right)(\$11) + \left(\frac{11}{14}\right)(\$ - 3)$$

$$= \frac{\$33}{14} - \frac{\$33}{14}$$

$$E(X) = 0.$$

The zero value means that, if you continue to make this wager enough times, you will break even. The "game" is said to be *fair*.

Now, let's see what happens if you decide to take an enumeration of both cards. It will cost you $2 per wager to see both cards before you decide to make the bet. Knowing both cards of course gives you all the information you need to decide whether you want to make this "investment." But does the cost outweigh the benefit? Remember, you will win 3/14 or 21.43 percent of the time; you will lose 78.57 percent of the time.

When you play knowing both cards are aces, it cost you $2 for the enumeration, and you will win, netting you $9. This event occurs with a frequency of 3/14. In this situation you will never lose because, when you determine from the enumeration that both cards are *not* aces, you simply do not make the wager. However, the information costs you $2, and this event occurs with a frequency of 11/14. Recapping this aspect of the illustration, observe then:

You net $9 with a frequency of 3/14:
$$\$9 \times 3/14 = \$27/14$$

You lose zero dollars but at a cost of $2 to obtain the information with a frequency of 11/14:
$$-\$2 \times 11/14 = \underline{-\$22/14}$$

The worth of the investment in your favor is
$$\$5/14$$

The positive $5/14 means that you will win here over the previous situation when you did not buy the information.

Now, let's go on and see what happens if you take a sample of *one* card. If you sample one card, and it is not an ace, you know not to bet because both cards cannot be aces. This sample costs you $1. If the one card had been an ace, you know the probability that the other card is an ace is 3/7. In this case, is the sample worth the cost of $1? To answer this question, we must answer some others first:

(a) How often will you not place the wager (make the investment)?

(b) How often will you wager and lose?

(c) How often will you wager and win?

Out of 8 cards in the larger deck there can be 28 combinations of two cards—king and king, king and ace (or ace and king; it makes no difference), or ace and ace:

$$C_2^8 = \frac{8!}{2!6!} = 28.$$

Of the 28 combinations, 6 are two kings, 6 are two aces, and 16 are king-ace. So, 6 times out of 28 (3/14) you will take a sample when both cards are kings. A king will show in the sample, and you will not make the wager. There are 16 out of 28 combinations of king-ace (8/14). Half the time your sample of one card will select the king, and you won't bet. The other half the time, your sample will select the ace. In this situation you will wager and lose. Then 6 times of 28 both cards are aces (3/14). Your sample indicates you should wager (make the investment), and of course you will win. Recapping these probabilities, we find the following:

(*a*) Both cards are kings, and
you do not play. Sample cost
is $1 and frequency is 3/14: $3/14 \times -\$1 = -\$\ 3/14$

(*b*) Cards are king-ace, but sample
showed king. You do not bet,
but sample cost is $1 and
frequency is 1/2 times 8/14: $4/14 \times -\$1 = -\$\ 4/14$

(*c*) Cards are king-ace, but sample
showed ace. You bet, but you
lose. Sample cost is $1; wager
is $3. Frequency is 1/2 times
8/14. $4/14 \times -\$4 = -\$16/14$

(*d*) Both cards are aces. You bet;
you win $11, but it cost $1
for the sample, netting you $10.
The frequency is 3/14: $3/14 \times \$10 = \underline{\quad \$30/14}$

(*e*) The value of this game or $\$7/14$
investment is

This positive value of 7/14 indicates that you are better off by $7/14 taking a sample of one card than taking no sample and by $2/14 ($7/14 − $5/14) from purchasing the enumeration of two cards.

While this illustration is rather fictional, there are situations in the real world that tend to parallel it. And you may use this same general approach to determine if the information or sample is really worth the cost.

Summary of Equations

7.1   Interval for systematic sample:

$$i = \frac{N}{n}$$

7.2   Number in ith strata:

$$n_i = \left(\frac{N_i}{N}\right) \cdot n$$

7.3   Number in ith strata—σ considered:

$$n_i = \frac{N_i \sigma_i}{\sum [N_i \sigma_i]} \cdot n$$

7.4   Number in ith strata—cost considered:

$$n_i = \frac{N_i(1/c_i)}{\sum [N_i(1/c_i)]} \cdot n$$

7.5   Number in ith strata—σ and cost considered:

$$n_i = \frac{N_i \sigma_i(1/c_i)}{\sum [N_i \sigma_i(1/c_i)]} \cdot n$$

7.6   Sample size for estimation of population mean—σ known:

$$n = \frac{Z^2 \sigma^2}{(\overline{X} - \mu)^2}$$

7.7   Sample size for estimation of population mean—using s:

$$n = \frac{Z^2 s^2}{(\overline{X} - \mu)^2}$$

7.8   Sample size for estimation of population mean—using finite population correction factor:

$$n = \frac{NZ^2 \sigma^2}{(N - 1)(\overline{X} - \mu)^2 + Z^2 \sigma^2}$$

7.9 Sample size for estimation of population proportion:

$$n = \frac{\pi(1 - \pi)Z^2}{(p - \pi)^2}$$

7.10 Sample size for estimation of population proportion— using finite population correction factor:

$$n = \frac{NZ^2\pi(1 - \pi)}{(N - 1)(p - \pi)^2 + Z^2\pi(1 - \pi)}$$

7.11 Expected value of the random variable:

$$E(X) = \Sigma[x \cdot P(X = x)]$$

Review Questions

1. Explain the use of each equation shown at the end of the chapter.

2. Define
(a) simple random sample
(b) systematic sample
(c) stratified sample
(d) cluster sample
(e) multistage sampling
(f) precision
(g) accuracy
(h) expected value

3. What is the difference between internal and external error?

4. What is sample bias?

Problems

1. A bank has 5,000 customers with sequential account numbers from 0000 to 4999. If a random sample of 50 customers was selected, determine the first five to be selected if appendix M was used beginning in row 18 and columns 35–38. Proceed downward in the table to select the customers.

2. Assume you have 100 households numbered 00 to 99. Which households would be in a random sample of size 10 if you use the table of random digits (appendix M) and begin in row 30 and columns 25 and 26. (Proceed downward in the table to determine the households.)

3. Assume a population of the following 15 digits: 9, 6, 5, 12, 10, 14, 4, 2, 15, 7, 1, 8, 11, 13, 3. Estimate the population mean by selecting a systematic sample of size 3 beginning with the fifth element and selecting every fifth element thereafter.

4. Use the following population of numbers to select the specified samples: 8, 5, 3, 6, 2, 3, 7, 9, 3, 5, 6, 3.

(a) Estimate the population mean by selecting a systematic sample beginning with the third element and selecting every third element thereafter.

(b) Repeat part *(a)*, but begin with the second element.

(c) Compare the results of parts *(a)* and *(b)*.

5. A large national retailer has categorized charge account customers according to the number of years the customer has maintained the account as follows:

Number of years	f
0 and under 5	8,000
5 and under 10	12,000
10 and under 15	14,000
15 and above	10,000

Determine the number to include in each stratum of a stratified sample of 500 if the method of proportional parts is used.

6. Suppose a retirement community has the following numbers in the given age groups:

Age	f
55 and under 60	2000
60 and under 65	1500
65 and under 70	1000
70 and above	700

How many should be included in each stratum of a stratified sample of 300 if the method of proportional parts is used?

7. In problem 6, if the cost of sampling is $1.00 per person for the youngest group and increases by 50¢ per person for each group, how will the sample size of 300 be distributed in each stratum?

8. Amicom specializes in "market surveys." Recently the firm received a contract from a county planning agency to conduct a sample to determine information regarding the income and spending habits of families in the area. In a conversation with an Amicom researcher, the agency head stated, "We don't know what sample technique you will use, but we would prefer a 20 percent sample rate."

(a) Discuss the problems of taking such a sample in your county.

(b) If you were with Amicom, what are some of the factors you should consider in designing this sample.

9. An organization is composed of four departments. Each department maintains a property inventory on items valued over a specified amount. The number of items on each department's inventory is as follows:

Department	Number of inventoried items
A	2,000
B	3,000
C	4,000
D	1,000

If the inventory manager desires to verify the accuracy of the inventory records by taking a stratified sample of 100 items, determine the number of items to be in each stratum using the method of proportional parts.

10. Repeat problem 9 if the sampling cost is $1.00 per item for Department A and increases 25¢ per item for B, 50¢ per item for C, and 75¢ per item for D.

11. The Washington Emergency Medical Service serves a geographical area that has 60,000 urban residents and 30,000 rural residents. From past experience the Service knows that the cost of serving the residents is normally distributed and that the standard deviation of per customer costs is $5.00 for urban residents and $20.00 for rural residents. The WEMS is preparing to take a sample of its recent records to determine current costs. They have already decided that $n = 400$, but how should the sample be divided between the strata (urban and rural)?

12. The Washington Emergency Medical Service has just discovered that it can sample the records of its rural customers using its computer so that the cost of each sample unit will be 50¢. For its urban customers, however, it must use written records so that the cost per sample unit will be $1.00. Rework problem 11 with this added information.

13. An evening newspaper plans to collect a sample from the home delivery customers. The editor of the paper is interested in learning the average amount of time a reader devoted to the editorial section of the Sunday edition. If the editor estimates the standard deviation of reading times to be 5 minutes, determine the sample size necessary to estimate to within half a minute the true average reading time. Set alpha = 0.05.

14. Rework problem 13 if the circulation of the newspaper included 5,000 home delivery customers.

15. A morning newspaper is interested in knowing more about the reading habits of the home delivery customers. If a previous survey indicated 50 percent of the readers always "look over" the editorial page, what sample size should be used to estimate within 4 points the proportion of people who always "look over" the editorial page. Set alpha = 0.10.

16. The mean average time for an aluminum fitting to be anodized is to be estimated from a simple random sample. Assume a standard deviation of 30 seconds, answer *(a)*, *(b)*, and *(c)*. What size sample should you take

(a) if the desired probability of being wrong by more than 10 seconds (under or over) is 0.01?
(b) if the desired probability is 0.10 for a tolerable error of 30 seconds?
(c) if the desired probability is 0.10 for a tolerable error of 15 seconds?
(d) Assuming a standard deviation of 15 seconds, what sample size would you take under condition *(a)*?
(e) Compare the results of *(a)* and *(d)* and then *(b)* and *(c)*.

17. Telephone, Inc., would like to estimate the proportion of families who prefer push-button phones. If 40 percent of the families were known previously to prefer push-button phones, what size simple random sample is needed to estimate the current proportion who prefer push-button phones within 4 points of the actual proportion? Set $\alpha = 0.10$.

18. New Products is interested in obtaining the average age of the customer who buys its model. What size simple random sample would be necessary to make this age estimate within 4 years if a preliminary sample of 81 customers produced a sample mean of 45 years and a standard deviation of 10 years? Set $\alpha = 0.05$.

19. In a large city a research analysis organization wants to estimate the average monthly rental expense for independent realtors for the Real Estate Association. If an estimate of the standard deviation is $75.00, what size simple random sample would be necessary to assure that the difference between the sample mean and the population mean did not exceed $5.00 at the 95 percent level of confidence.

20. A large television repair service is trying to improve the scheduling of its repair persons. The manager feels that he needs an estimate of the mean length of travel time required for a service call that is accurate to within 2 minutes. Moreover, he needs a 99 percent level of confidence for his answer. A small random sample of the travel times came up with the following times (in minutes): 10, 15, 7, 6, 24. Based upon this information, how large a sample would be necessary to meet the manager's requirements? (Assume travel times are normally distributed.)

21. If a simple random sample of registered voters is to be taken to determine how they expect to vote on a proposal to increase the property tax millage, what sample size would be necessary to keep the sample proportion within 3 points of the population proportion? Assume a 50-50 split and $\alpha = 0.05$.

22. A political candidate is interested in taking a poll of registered voters in his congressional district. There are approximately 95,000 registered voters in the district consisting of one urban county of 45,000 voters and five rural counties with a total of 50,000 voters. You have estimated that the fixed cost of the poll amounts to $500. The average variable cost per interview for personal interviews comes to $5.00 in the urban county and $6.00 in the rural counties. There are only two candidates involved in the election, and it is estimated that they are running very close to one another (roughly a 50-50 split).

(a) Suppose your candidate wants the desired risk probability of being in error by more than 4 percentage points (in either direction) to be 0.05. What size simple random sample would you take?

(b) Suppose he wanted to reduce the error to 2 percentage points at the same risk probability. What size simple random sample would you take?

(c) Compute the cost under (a) and (b), and compare.

(d) Do you think there would be any advantage to taking a stratified sample based on rural and urban classification of registered voters.

(e) In this type of sample, what other basis for stratification might you consider?

(f) Would there be any basis for using a systematic sample in this case? A cluster sample?

23. Assume a game consists of drawing two cards from a deck composed of five kings and five queens. If both cards are kings, you win $14; otherwise you lose $4. Determine the expected value of the game.

24. Rework problem 23 if you take a sample of one card at a cost of $1.

25. Use the following historical data to determine the average (expected value) monthly demand:

Monthly demand (dozens)	Probability
30	0.1
31	0.2
32	0.4
33	0.2
34	0.1

Case Study

Clark Kent announced his candidacy for governor of the state a few days ago. It appears as if it will be a difficult campaign for neither of the two candidates for the office are well known. Moreover, as of yet neither has begun to advertise, make public appearances, or really make known his position on various issues beyond the fact both are Democrats.

At the last meeting of Mr. Kent's campaign steering committee, it was decided that a public opinion survey (political poll) should be taken to determine the attitude of the state's 2.5 million voters toward various potential campaign issues and their voting intentions with respect to the two candidates.

The state has six congressional districts of approximately equal population, and it has already decided that the survey will be conducted via telephone from Mr. Kent's headquarters in the Third Congressional District. In addition, the steering committee feels that they want to be "99 percent certain" that their results are accurate within 1 percent.

Determine the sample size that should be used and how the sample should be allocated between the congressional districts considering both accuracy and cost of the survey if the only additional information available to you is that cited in the following table:

District	Percent of total number of registered voters	Estimated mean cost per telephone interview
1	19	$1.25
2	16	3.15
3	17	0.50
4	15	3.00
5	16	1.75
6	17	2.50

8

Statistical Testing:
One Sample

In chapter 6 you were introduced to the concept of confidence intervals, and you learned to construct confidence intervals for the population mean and population proportion. In this chapter we are going to investigate statistical testing with the objective of making inferences about a population mean or proportion. For the moment, however, let us limit our discussion to means in order to find out what statistical testing is all about.

Chance Difference versus a Statistically Significant Difference

We may calculate a sample mean from sample data and compare it to a population mean, and we may find that the two values differ. The question we need to answer is whether the difference between the mean of the population (μ) and the sample mean ($\overline{X}$) is real or whether the difference is due to sampling variation that we know can exist.

To illustrate, look at figure 8.1. The only $\overline{X}$'s that equal μ are the $\overline{X}$'s between the two arrows. All the other sample means, even though they are part of the sampling distribution with mean μ, do not actually equal μ. Thus, on any given occasion in

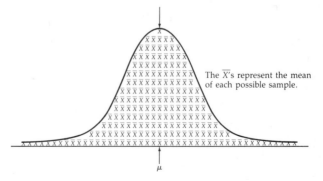

The $\overline{X}$'s represent the mean of each possible sample.

Figure 8.1   A sampling distribution of $\overline{X}$'s with mean of μ

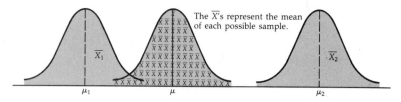

The $\overline{X}$'s represent the mean of each possible sample.

Figure 8.2 Two sample means that belong to a population with mean other than μ

which we draw a sample, any one of the sample means in the sampling distribution might be selected, and most of the sample means would not equal the true population mean. This difference that could occur between the population mean and the sample means would be due to the sampling process. Such a difference between a sample statistic (such as $\overline{X}$) and a population parameter (such as μ) is called a *chance difference* and is *not considered a statistically significant difference.*

A *real* or *statistically significant difference* **between a sample statistic and a population parameter is due to a basic difference not attributable to chance.** Such a real difference between the sample mean and the population mean indicates that the sample mean $\overline{X}$ came from a population with a true mean other than μ. Figure 8.2 illustrates this possibility. Two sample means $\overline{X}_1$ and $\overline{X}_2$ are shown which, if drawn, would indicate that a significant difference exists because $\overline{X}_1$ and $\overline{X}_2$ are from two different populations, with population means equal to μ_1 and μ_2 respectively. That is, if $\overline{X}_1$ were selected, the distance $\overline{X}_1 - \mu$ is just too great to be attributed to chance. It would be correct to state that $\overline{X}_1$ came from a population with another mean, μ_1, and μ_1 is different from μ. A similar statement can be made for $\overline{X}_2$. Let us turn to example 8.1 to demonstrate the point.

Example 8.1

▷ An ice company has a machine designed to fill bags with 10 pounds of ice. Since the machine is in continuous use for eight hours, a periodic check of bag weights is made to assure that there is not much deviation from the desired 10-pound weight. From past experience it is known that the bag weights have averaged 10 pounds with a standard deviation of 0.2 pounds. Suppose a recent sample of 64 bags produced a sample mean of 9.8 pounds. If individual bag weights are normally distributed, should management be concerned about the current performance of the machine?

An approach to answering the question would be to establish

BASIC STATISTICS

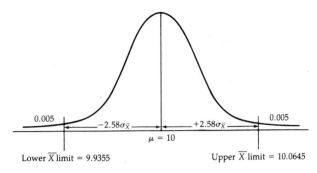

Figure 8.3 99 percent confidence limits

99 percent confidence limits about μ and see if $\overline{X}$ falls in the confidence limits. Figure 8.3 shows the idea. If $2.58\sigma_{\overline{X}}$ is added and subtracted from μ, upper and lower limits are established within which any sample mean should fall 99 times out of 100 provided the population mean remains unchanged; thus a chance error can occur 1 time in 100:

$$\text{Upper } \overline{X} \text{ limit} = \mu + 2.58\frac{\sigma}{\sqrt{n}}$$

$$= 10 + 2.58\frac{0.20}{\sqrt{64}}$$

$$= 10 + 0.0645$$

$$\text{Upper } \overline{X} \text{ limit} = 10.0645;$$

$$\text{Lower } \overline{X} \text{ limit} = \mu - 2.58\frac{\sigma}{\sqrt{n}}$$

$$= 10 - 0.0645$$

$$\text{Lower } \overline{X} \text{ limit} = 9.9355.$$

Since the limits, 9.9355 to 10.0645, do not contain $\overline{X} = 9.8$, the conclusion is drawn that the difference between $\overline{X}$ and μ is statistically significant. Thus, it is concluded that the machine is not producing at the desired average weight. The difference is real, not chance, with a probability of being correct equal to 0.99. ◁

The Test Procedure

The preceding example demonstrated the concept of a statistically significant difference. In this section a more standardized approach to statistical testing will be given, and then the previous example will be reworked.

Formulating the Null and Alternate Hypothesis

In statistical testing a null hypothesis is an assumption about a population. Sample information is used to make a decision about the null hypothesis. If the sample information supports the null hypothesis, then the null hypothesis is accepted. Acceptance of a null hypothesis implies that any difference observed between the sample information and the assumed population characteristic could be considered the result of chance variation. However, if the sample evidence does not support the null hypothesis, then the null hypothesis is rejected. Rejection of the null hypothesis implies that the sample result is significantly different from the assumption.

One additional comment on accepting the null hypothesis should be made. Some researchers prefer to say that a null hypothesis is *not rejected* rather than to accept the null hypothesis. Such a position is taken because is it difficult to prove, without any doubt, that a null hypothesis is true. However, whichever expression ("accept the null hypothesis" or "do not reject the null hypothesis") is used, just remember that such a decision is reached when the sample evidence does not justify rejecting the null hypothesis.

Symbolically, a null hypothesis about a population mean would be as follows:

$$H_0: \quad \mu = \mu_H.$$

The symbol "H_0" means null hypothesis. The preceding null hypothesis states that the population mean μ has a value of μ_H (some number). Another way to state this is that there is *no difference* between μ and μ_H. Our goal is to test this null hypothesis.

Before actually testing the null hypothesis, it is appropriate to state an alternate hypothesis. The alternate hypothesis is a statement about the situation that should exist if the null hypothesis is rejected. An alternate hypothesis may take one of three forms as shown below:

(1) $H_A: \quad \mu \neq \mu_H,$ (2) $H_A: \quad \mu < \mu_H,$ (3) $H_A: \quad \mu > \mu_H.$

The first alternate states that if the population mean is not equal to μ_H, alternatively, it may be greater than or less than μ_H. This type of alternate hypothesis is often referred to as a *two-sided test* of the null hypothesis.

Alternate hypotheses 2 and 3 are *one-sided tests* of the null hypothesis. That is, the researcher is interested only in detecting differences on one side of the hypothesized value of the mean. Hypothesis 2 states that, if the null is rejected, it will be rejected because it is believed that the true mean is less than μ_H. Hypothesis 3, on the other hand, states that the null hypothesis will be

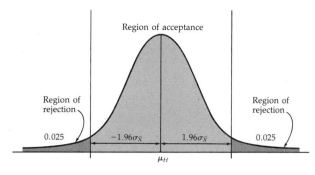

Figure 8.4   A two-sided test of the null hypothesis with $\alpha = 0.05$

rejected only if the true mean is greater than μ_H. Thus these two alternate hypotheses are designed to detect differences on only one side of the null hypothesis.

Level of Significance and Rejection Limits

The purpose of the level of significance (alpha level) is to provide a basis for deciding whether an observed difference between a sample statistic and the hypothesized value of the population parameter is a chance difference (result of sampling variation), or whether the difference should be declared a statistically significant difference. Selection of an alpha level permits the partitioning of the sampling distribution into a region of acceptance and a region of rejection.

Considering a two-sided test, figure 8.4 is partitioned into a region of acceptance and a region of rejection assuming an alpha level of 0.05 (the choice of an alpha level will be discussed near the end of this chapter). Thus the region of acceptance is represented by the area under the curve containing the 95 percent of the sample means ($\overline{X}$'s) nearest to μ_H; whereas the region of rejection is represented by the areas in the two ends of the curve containing 5 percent of the sample means ($\overline{X}$'s) farthest from the μ_H (2.5 percent at each end of the distribution). Consequently it is predetermined when H_0 is correct that 95 percent of the time any difference that is due to chance will be attributed to chance and 5 percent of the time any difference due to chance will be labeled incorrectly as a statistically significant difference. The result is, with an alpha level of 0.05, 5 percent of the time we would fail to identify that the true mean is equal to μ_H when, in fact, it is.

Figures 8.5 and 8.6 show the regions of rejection and acceptance for the second and third alternate hypotheses, respectively. Notice that the rejection region is placed on one side of the curve. This is necessary since the alternate hypotheses were one sided,

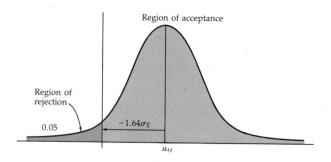

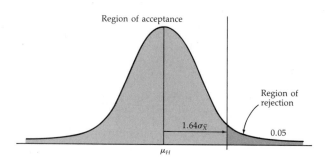

Figure 8.6   A one-sided test of the null hypothesis with $\alpha = 0.05$

and, in order to reject the null hypothesis and accept the alternate hypothesis, a sample mean $(\overline{X})$ must be less than $\mu_H - 1.64\sigma_{\overline{X}}$ (figure 8.5), or a sample mean $(\overline{X})$ must be greater than $\mu_H + 1.64\sigma_{\overline{X}}$ (figure 8.6).

Making the Decision

Once the appropriate null and alternate hypotheses are formulated and the rejection region identified, the procedure is to identify the appropriate sampling distribution (normal or t distribution) and use that distribution to decide if the difference $(\overline{X} - \mu_H)$ *is too large to be considered a chance difference.* For instance, in figure 8.4, if the value $Z_{test} = (\overline{X} - \mu_H)/\sigma_{\overline{X}}$ is greater than $+1.96$ or less than -1.96, the difference $(\overline{X} - \mu_H)$, converted to a standard normal deviate, is too large to be a chance difference. The conclusion would have to be that the difference is a statistically significant difference. For figure 8.4, the hypotheses and rejection limits would be

H_0: $\mu = \mu_H$,
H_A: $\mu \neq \mu_H$,
    $\alpha = 0.05$.

Reject H_0 if $Z_{test} < -1.96$ or $Z_{test} > 1.96$.

And for the one-sided tests, the hypotheses and rejection limits would be as follows:

BASIC STATISTICS

For figure 8.5,

H_0: $\mu = \mu_H$,
H_A: $\mu < \mu_H$,
$\alpha = 0.05$.

Reject H_0 if $Z_{test} < -1.64$.[1]

For figure 8.6,

H_0: $\mu = \mu_H$,
H_A: $\mu > \mu_H$,
$\alpha = 0.05$.

Reject H_0 if $Z_{test} > 1.64$.

The decision process is initiated given the assumption that the null hypothesis is true (this situation is analogous to a court of law presuming someone innocent unless proven guilty). Then sample information is obtained, and a decision is made whether or not to reject the null hypothesis based on the comparison of the calculated test statistic (Z_{test} in the previous discussion) and the rejection limits.

Now that we have established a test procedure, let us rework example 8.1:

1. Formulate the null hypothesis
H_0: $\mu = 10$.
2. Formulate the alternate hypothesis
H_A: $\mu \neq 10$.
3. Select the level of significance
$\alpha = 0.01$;
from appendix H the critical value of $Z = 2.58$.
4. Establish rejection limits
Reject H_0 if $Z_{test} < -2.58$ or $Z_{test} > 2.58$.
5. Calculate Z_{test}

$$Z_{test} = \frac{\overline{X} - \mu_H}{\sigma_{\overline{X}}} \quad where \quad \sigma_{\overline{X}} = \frac{\sigma}{\sqrt{n}}$$

$$= \frac{9.8 - 10}{0.2/\sqrt{64}} = \frac{9.8 - 10}{0.025}$$

$$= -\frac{0.2}{.025}$$

$$Z_{test} = -8.$$

---

1. Remember $-1.65 < -1.64$.

6. Make decision.

Since Z_{test} is less than -2.58, reject H_0 and accept H_A; conclude that the population mean is now less than 10 pounds.

Arithmetic Mean: One Sample Test

Now that we have outlined a procedure for testing hypotheses using information from a single sample, we need to investigate the appropriate test for the sampling situations we encountered in chapter 6, which were

1. normal population and known population standard deviation,
2. large samples in order to use the central limit theorem,
3. small samples from a normal population where the population standard deviation is unknown.

Each of these situations has a particular test statistic for use in step 5 shown in the previous section. For situation 1, where the sample is drawn from a normal population with known population standard deviation, the test statistic is equation (8.1):

$$Z_{test} = \frac{\overline{X} - \mu_H}{\sigma_{\overline{X}}}. \tag{8.1}$$

If situation 2 exists, where large samples are available in order to use the central limit theorem, the test statistic is equation 8.1 when the population standard deviation (σ) is known and equation (8.2) when the population standard deviation (σ) is unknown:

$$Z_{test} = \frac{\overline{X} - \mu_H}{s_{\overline{X}}}. \tag{8.2}$$

The reader is reminded that equation (8.2) is actually an approximate test. In fact, when σ is unknown, the correct test is equation (8.3) given the assumption of a normally distributed population. However, it has become rather common practice to utilize equation (8.2) for large samples ($n > 30$). And if situation 3 exists, where the t distribution is appropriate, the test statistic is equation (8.3):

$$t_{test} = \frac{\overline{X} - \mu_H}{s_{\overline{X}}}. \tag{8.3}$$

Now let us see an example for each of these situations.

Example 8.2

▷ *Test of Hypothesis about the Population Mean When Population Is Normal and Population Standard Deviation Is Known.* Suppose from the most recent census data available that the Chamber of Commerce has established that family income is normally distributed with an average of $13,000 and a standard deviation of $4,500. However, because the data are several years old, the Chamber questions that the average has remained unchanged. To investigate the possibility of change, the Chamber took a random sample of 25 families which yielded a sample mean of $14,500. At the 0.05 level of significance, does it seem likely that change has occurred?

H_0: $\mu = \$13,000,$
H_A: $\mu \neq \$13,000,$
   $\alpha = 0.05.$

Reject H_0 if $Z_{test} < -1.96$ or $Z_{test} > 1.96$:

$$Z_{test} = \frac{\overline{X} - \mu_H}{\sigma_{\overline{X}}} \quad \text{where} \quad \sigma_{\overline{X}} = \frac{\sigma}{\sqrt{n}}$$

$$= \frac{14,500 - 13,000}{4,500/\sqrt{25}}$$

$$= \frac{1,500}{900}$$

$$Z_{test} = 1.67.$$

Accept (do not reject) the null hypothesis. The difference between $\overline{X}$ and μ_H can be considered chance (sampling) variation. Thus the Chamber of Commerce concludes that average family income has not changed. ◁

Example 8.3

▷ *Test of Hypothesis about the Population Mean When Large Samples Are Selected.* Based on a study performed five years ago, a regional insurance company had determined that residential building costs averaged $34 per square foot for a certain class of house. Given the recent experience with inflation, the company believes that the cost per square foot has increased. If a random sample of 36 recently constructed homes yielded an average cost per square foot of $40 with a standard deviation of $3, at the 0.05 level of significance is the company's belief supported? (Note that this will be a one-sided test because the belief is that costs have increased.)

H_0: $\mu = 34$ or $\mu \leq 34,$
H_A: $\mu > 34,$
   $\alpha = 0.05.$

Reject H_0 if $Z_{test} > 1.64$ (see figure 8.6):

$$Z_{test} = \frac{\overline{X} - \mu_H}{s_{\overline{X}}} \quad \text{where} \quad s_{\overline{X}} = \frac{s}{\sqrt{n}}$$

$$= \frac{40 - 34}{3/\sqrt{36}}$$

$$= \frac{6}{3/6}$$

$$= \frac{6}{0.5}$$

$$Z_{test} = 12.$$

Reject the null hypothesis; the difference between $\overline{X}$ and μ_H is a statistically significant difference. The company's belief is supported. ◁

Example 8.4

▷ *Test of Hypothesis about the Population Mean When Small Samples Are Selected from a Normal Population with Unknown Population Standard Deviation.* A chicken grower claims that the average weight of a particular group of 500 chickens is at least 30 ounces. Food Stores, Inc. is interested in purchasing the chickens on the basis of the claim. Before agreeing to a purchase, however, the Food Stores' management selected a random sample of 25 chickens which yielded a sample mean of 29 ounces and a standard deviation of 1.5 ounces. If the weights can be considered to be normally distributed, should the claim be rejected at the 0.05 level of significance?

$H_0: \quad \mu = 30 \quad \text{or} \quad \mu \geq 30,$
$H_A: \quad \mu < 30,$
$\qquad \alpha = 0.05.$

Reject H_0 if $t_{test} < -1.711$ (obtain 1.711 from appendix I using $n - 1$ or $25 - 1 = 24$ degrees of freedom):

$$t_{test} = \frac{\overline{X} - \mu_H}{s_{\overline{X}}} \quad \text{where} \quad s_{\overline{X}} = \frac{s}{\sqrt{n}} \sqrt{\frac{N - n}{N - 1}}$$

$$= \frac{29 - 30}{(1.5/\sqrt{25})\sqrt{(500 - 25)/500 - 1}}$$

$$= \frac{-1}{(0.3)(0.976)} = \frac{-1}{0.293}$$

$$t_{test} = -3.413.$$

Reject the null hypothesis. Based on the sample evidence, management should not purchase the chickens. ◁

Proportions: One Sample Test

Just as we can construct confidence intervals for proportions, we can test hypotheses about the population proportion. The null hypothesis would be of the form,

$$H_0: \quad \pi = \pi_H,$$

and the three possible alternatives would be of the form

(1) $H_A: \quad \pi \neq \pi_H,$ (2) $H_A: \quad \pi < \pi_H,$ (3) $H_A: \quad \pi > \pi_H,$

just as existed for the mean. The test procedure would also be the same as outlined for population means if a large sample is taken in order to use the central limit theorem. The appropriate test statistic is shown in equation (8.4):

$$Z_{test} = \frac{p - \pi_H}{\sigma_p}, \qquad\qquad (8.4)$$

$$where \ \sigma_p = \sqrt{\frac{\pi_H(1 - \pi_H)}{n}}.$$

Notice that the standard error of the proportion σ_p is calculated using π_H, the hypothesized value of the population proportion, rather than p, the sample proportion. The reasoning is that, since we have hypothesized that $\pi = \pi_H$, and perform the test under the assumption the null hypothesis is true, it is appropriate to use π_H when calculating σ_p. This approach differs, however, from the estimation method discussed in chapter 6 where, under the assumption of no knowledge about π, p was used as the approximation for π.

Example 8.5

▷ Mr. Candidate believes he is going to carry a particular city by a large margin. In fact he believes he should get more than 60 percent of the vote. His campaign manager, who is not so optimistic, has employed the firm of Election Forecasters to check out Mr. Candidate's belief. If, in a sample of 400 voters, 252 indicated they would vote for Mr. Candidate, at the 0.05 level of significance do you think his optimism is justified?[2]

---

2. Remember that percentages are proportions multiplied by 100.

$$p = \frac{X}{n} = \frac{252}{400} = 0.63,$$

$$\sigma_p = \sqrt{\frac{\pi_H(1 - \pi_H)}{n}}$$

$$= \sqrt{\frac{(0.60)(0.40)}{400}}$$

$$= \sqrt{0.0006}$$

$$\sigma_p = 0.0245,$$

H_0: $\pi = 0.60$ or $\pi \leq 0.60$,

H_A: $\pi > 0.60$,

$\alpha = 0.05$.

Reject H_0 if $Z_{test} > 1.64$:

$$Z_{test} = \frac{p - \pi_H}{\sigma_p}$$

$$= \frac{0.63 - 0.60}{0.0245}$$

$$Z_{test} = 1.22.$$

Accept (do not reject) H_0, the difference between the sample proportion (0.63) and the hypothesized population proportion (0.60) can be considered a chance difference. Therefore Mr. Candidate's belief is not supported. ◁

Type I and Type II Errors

In our discussion and examples in this chapter there has not been any indication as to what possible errors the researcher might make when he is testing hypotheses. Table 8.1 shows the possible situations that exist when a hypothesis is tested. From the table you can see that there are four possibilities, two of which are the correct decision and two of which lead to erroneous decisions.

Table 8.1
Possible outcomes when testing hypotheses

	H_0 is accepted	H_A is accepted
H_0 is correct	Correct decision $1 - \alpha$	Type I error α
H_A is correct	Type II error β	Correct decision $1 - \beta$

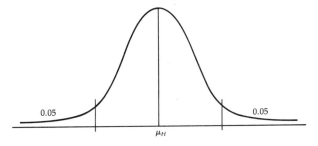

Figure 8.7   Type I error set at 0.10 for two-sided test

The first error in table 8.1, denoted as *Type I error*, occurs when the correct null hypothesis is rejected and the alternate, which is incorrect, is accepted. As you have probably noticed, the probability of Type I error is α and results from the level of significance selected by the researcher. It is the probability that a researcher will fail to identify a *true* null hypothesis.

Figure 8.7 shows a two-sided situation where the α level is 0.10. This is a situation where the Type I error has been set at 0.10. Figure 8.8 shows a one-sided situation where the α level is 0.10. Notice that in the one-sided case the Type I error occurs at only one end of the sampling distribution.

The choice of a value for α is a managment decision rather than a statistical decision. The decision has to be made based on a judgment concerning the seriousness of committing Type I error. The general guideline is to set α at a relatively low value (0.01 or 0.05) if rejecting a correct null hypothesis would be very costly and/or hazardous. Conversely, if accepting an incorrect null hypothesis (Type II error) is costly or hazardous, the α value should be high (0.25 or higher).

Looking again at table 8.1, the other error that appears is called *Type II error*, and the probability of Type II error is designated β. Type II error is committed when the researcher accepts an incor-

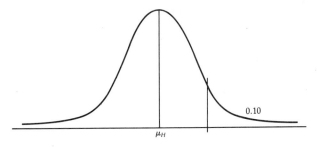

Figure 8.8   Type I error set at 0.10 for one-sided test

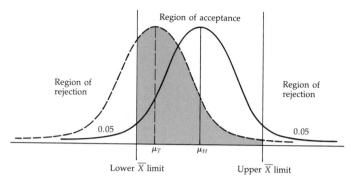

Figure 8.9   Type II error

rect null hypothesis. From table 8.1 we see this indicates that the alternate hypothesis was the correct one, but the null hypothesis was accepted.

A Type II error has a prerequisite in that this particular error can occur only if the hypothesized mean is incorrect (H_0 is false). Or it is often said that Type II error occurs when an incorrect H_0 is accepted (not rejected). Figure 8.9 shows how this happens.

Figure 8.9 shows a situation where the null and alternate hypotheses would have been

H_0: $\mu = \mu_H$,
H_A: $\mu \neq \mu_H$,

where μ_H is the hypothesized or assumed value of the population mean. Using an α level of 0.10, we know that, if a sample $\overline{X}$ were obtained that fell between the upper and lower $\overline{X}$ limits (region of acceptance), we would accept H_0. However, if the true mean is not μ_H but is really μ_T, then the real sampling distribution is the dotted distribution. In this situation the correct decision is to reject H_0 and accept H_A. Failure to do this results in Type II error.

The probabilty of Type II error is the shaded area of figure 8.9. Type II error can be calculated by determining the area under the dotted sampling distribution curve between the upper and lower $\overline{X}$ limits. Example 8.6 demonstrates the calculation of Type II error.

Example 8.6

▷ Suppose a random sample of 100 items yielded a sample mean of 50 and a standard deviation of 10. If the hypothesized mean is 52, but the true mean is really 51, what is the probability of Type II error if $\alpha = 0.05$?

BASIC STATISTICS

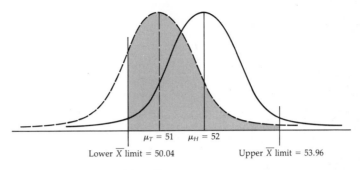

$\mu_T = 51$ $\mu_H = 52$

Lower $\overline{X}$ limit = 50.04   Upper $\overline{X}$ limit = 53.96

Figure 8.10   Location of Type II error

Figure 8.10 shows the problem as it is perceived. The curve represented by the solid line is the sampling distribution that exists if $\mu_H = 52$. The upper $\overline{X}$ limit and the lower $\overline{X}$ limit are calculated using μ_H, and the region between them is the region of acceptance. The dotted curve represents the sampling distribution associated with the actual mean of 51. The shaded area is the portion of the dotted curve within the region of acceptance, and this shaded area is the likelihood of committing Type II error.

$$LL_{\overline{X}} = \mu_H - Z_\alpha s_{\overline{X}} \qquad\qquad UL_{\overline{X}} = \mu_H + Z_\alpha s_{\overline{X}}$$
$$= 52 - 1.96(s/\sqrt{n}) \qquad\qquad = 52 + 1.96(s/\sqrt{n})$$
$$= 52 - 1.96(10/\sqrt{100}) \qquad = 52 + 1.96(10/\sqrt{100})$$
$$= 52 - 1.96(1) \qquad\qquad = 52 + 1.96(1)$$
$$= 52 - 1.96 \qquad\qquad = 52 + 1.96$$
$$LL_{\overline{X}} = 50.04, \qquad\qquad UL_{\overline{X}} = 53.96.$$

Figure 8.11 demonstrates how to calculate Type II error. The shaded area is divided into two regions. The areas of these two

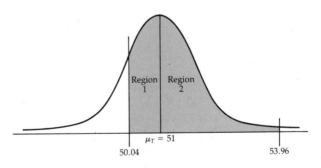

Region 1   Region 2

$\mu_T = 51$

50.04   53.96

Figure 8.11   Computing Type II error

regions added together make up the probability of Type II error, or β:

$$Z_1 = \frac{LL_{\bar{X}} - \mu_T}{s_{\bar{X}}} = \frac{50.04 - 51}{1}$$

$$Z_1 = \frac{-0.96}{1} = -0.96,$$

Area 1 = 0.3315;

$$Z_2 = \frac{UL_{\bar{X}} - \mu_T}{s_{\bar{X}}} = \frac{53.96 - 51}{1}$$

$$= \frac{2.96}{1}$$

$$Z_2 = 2.96,$$
Area 2 = 0.4985.

Thus the probability of committing Type II error is 0.8300 (0.3315 + 0.4985). ◁

Operating Characteristic Curve and Power of the Test

An operating characteristic curve presents the *probability of accepting a null hypothesis* for various values of the population parameter at a given α level using a particular sample size. To construct an operating characteristic curve, different values of the population parameter (μ or π) are assumed, and the probability of accepting H_0 is determined for each assumed value. Then the assumed values of the parameter are plotted on the abscissa, while the probability of accepting H_0 is plotted on the ordinate. Thus any point on the graph represents the probability of accepting the null hypothesis for a given value of the parameter.

Figure 8.12 is the operating characteristic curve for example

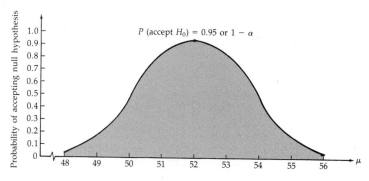

Figure 8.12   Operating characteristic curve, H_0: $\mu = 52$, $\alpha = 0.05$, $n = 100$

8.6. Notice in figure 8.12, when the population mean is 52, the probability of acceptance is 0.95, which is $1 - \alpha$. This occurrence is consistent with the original design of the test. That is, if $\mu = 52$, the null hypothesis should be accepted 95 percent of the time and rejected 5 percent of the time. Also notice in figure 8.12 that, when $\mu = 51$, the probability of accepting the null hypothesis is 0.8300. This value, 0.8300, is the Type II error calculated in example 8.6. Thus all points on the operating characteristic curve are obtained by calculating β as demonstrated in example 8.6. (All probabilities are β except at the peak of the curve which is $1 - \alpha$.)

Looking at figure 8.12 again, if the actual value of the mean is either 51 or 53 (one unit from the hypothesized value of the mean), the probability of β error is 0.8300. Further, if the actual value of the mean is either 50 or 54 (two units from the hypothesized value of the mean), then the probability of β error is 0.4840. Consequently, when the actual value of the mean is two units above or below the hypothesized value of the mean, the test will fail to detect this fact 48.4 percent of the time.

Thus use of the operating characteristic (OC) curve allows an evaluation of the decision (accept H_0) that might be reached for a given sample size and a given level of significance. The OC curve reveals that the test constructed for the situation described in example 8.6 is prone toward the β error if the actual value of the population mean is close (one or two units) to the hypothesized value of the mean.

If these β errors are not acceptable risks to the decision maker, there are two courses of action that will reduce them. One action would be to increase the value of α, which would increase the probability of rejecting H_0, thereby reducing the β error. The other action would be to increase the sample size. This action has the effect of reducing the region of acceptance by reducing $s_{\bar{x}} = s/\sqrt{n}$. Either of these actions would improve what is known as the *power of the test* $(1 - \beta)$.

The power of the test is *the inverse function of the operating characteristic curve*. That is, the power of the test is the probability of rejecting the null hypothesis for various possible values of the population parameter. Table 8.2 presents the power of the test values for the situation described in example 8.6.

As you see from table 8.2, the power of the test increases as the values of the mean get farther away from the hypothesized value of 52. For the values of μ, except 52, the power of the test $(1 - \beta)$

Table 8.2
Operating characteristic curve values and power of the test,
$H_0: \mu = 52$, $\alpha = 0.05$, $n = 100$

Possible values of μ	Probability of accepting H_0 (operating characteristic curve)	Probability of rejecting H_0 (power of the test)
48	0.0207 ⎫	0.9793 ⎫
49	0.1492 ⎪ β	0.8508 ⎪ $1 - \beta$
50	0.4840 ⎬	0.5160 ⎬
51	0.8300 ⎭	0.1700 ⎭
52	0.9500 $1 - \alpha$	0.0500 α
53	0.8300 ⎫	0.1700 ⎫
54	0.4840 ⎪ β	0.5160 ⎪ $1 - \beta$
55	0.1492 ⎬	0.8508 ⎬
56	0.0207 ⎭	0.9793 ⎭

is the probability of making the correct decision. And for this reason it is often used instead of the operating characteristic curve.

Figure 8.13 is a graphic presentation of the power of the test and is usually referred to as the *power function*. As μ changes from 52, the probability values can be thought of as the probability of rejecting H_0 when H_0 should be rejected. And from figure 8.13 it can be seen that the probability of rejection does increase as the value of μ moves farther and farther away from 52.

Sample Size Determination: Controlling Alpha and Beta

In chapter 7 sample size determination was discussed in conjunction with the estimation of population parameters, μ and π. You may recall that such sample size determination considered only the level of confidence, or $1 - \alpha$. In fact, then, it considered only

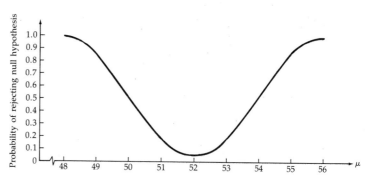

Figure 8.13   Power function, H_0: $\mu = 52$, $\alpha = 0.05$, $n = 100$

Type I error. However, a similar sample size determination for the purpose of hypothesis testing should consider both Type I and Type II errors.

Procedurally, it is necessary to establish two criteria:

1. An acceptable probability of Type I error.

2. An acceptable probability of Type II error for a specified possible true value of the population parameter (different from the hypothesized value).

After specifying the acceptable alpha level and the possible true value of the parameter associated with a designated probability (β) of Type II error, an appropriate sample size for a hypothesis test on μ can be determined using equation (8.5) if the population standard deviation is known:

$$n = \frac{(Z_\alpha + Z_\beta)^2 \sigma^2}{(\mu_T - \mu_H)^2},$$ (8.5)

where

$Z_\alpha = $ the Z-value determined by the desired level of alpha (if H_A is two-sided, use $\alpha/2$),

$Z_\beta = $ the Z-value determined by the desired β error (if $\beta > 0.50$, the Z-value is negative and determined from appendix H using $\beta - 0.50$),

$(\mu_T - \mu_H) = $ the maximum desired difference between a possible true value of the population mean (μ_T) and μ_H.

And, as was the case in chapter 7, if the population standard deviation is not known, equation (8.6) can be used:

$$n = \frac{(Z_\alpha + Z_\beta)^2 s^2}{(\mu_T - \mu_H)^2}.$$ (8.6)

If the parameter under investigation is the population proportion (π), the appropriate sample size for a fixed alpha and beta error can be determined using equation (8.7):

$$n = \frac{\left[Z_\alpha \sqrt{\pi_H(1 - \pi_H)} + Z_\beta \sqrt{\pi_T(1 - \pi_T)}\right]^2}{(\pi_T - \pi_H)^2}.$$ (8.7)

Examples 8.7 and 8.8 are presented to demonstrate the calculation of the appropriate sample size for a fixed alpha and beta error.

Example 8.7

▷ Let us return to the situation in example 8.6 where the hypothesized mean (μ_H) was 52 and alpha was 0.05. Suppose that it is desired that the difference $(\mu_T - \mu_H)$ be no greater than 4 and that the probability of beta error should be 0.02. Using equation (8.6) and $s = 10$, the appropriate sample size can be determined:

$$n = \frac{(Z_\alpha + Z_\beta)s^2}{(\mu_T - \mu_H)^2},$$

$Z_\alpha = Z_{0.05/2} = Z_{0.025} = 1.96$ (from appendix I),
$Z_\beta = Z_{0.02} = 2.05$ (from appendix H),

$$n = \frac{(1.96 + 2.05)^2(10)^2}{(4)^2}$$

$$= \frac{(4.01)^2(10)^2}{(4)^2}$$

$n = 100.5$ or $100.$ ◁

Notice in example 8.7 that the sample size (n) was determined to be 100. Now, examine table 8.2, and note that β error associated with $\mu = 48$ and $\mu = 56$. As you can see, the β error is 0.0207 or 0.02 as used in example 8.7. Remembering that example 8.6 used $n = 100$, it is apparent that the sample size determined in example 8.7 was calculated for the two entries ($\mu = 48$ and $\mu = 56$) of table 8.2. Thus for the required maximum spread of 4 (56 − 52 or 46 − 52), the required β error of 0.02, and the required alpha error of 0.05, the sample size is 100.

Example 8.8

▷ A political office holder recently has taken a strong position on a very controversial legislative proposal. In the past the office holder has been supported by at least 65 percent of his constituents. Given the controversial nature of the proposed legislation, the office holder plans to take a sample to determine if there has been a decrease in support. If it is desired that the probability of Type I error should not exceed 0.05 and that the probability of Type II error should not exceed 0.10 when $\pi_T = 0.60$, what is the appropriate sample size?

$$n = \frac{\left[Z_\alpha \sqrt{\pi_H(1 - \pi_H)} + Z_\beta \sqrt{\pi_T(1 - \pi_T)}\right]^2}{(\pi_T - \pi_H)^2},$$

$Z_\alpha = Z_{0.05} = 1.645$ (from appendix I),
$Z_\beta = Z_{0.10} = 1.282$ (from appendix I),

$$n = \frac{\left[1.645\sqrt{0.65(1 - 0.65)} + 1.282\sqrt{0.60(1 - 0.60)}\right]^2}{(0.60 - 0.65)^2}$$

$$= \frac{\left[1.645\sqrt{0.65(0.35)} + 1.282\sqrt{0.60(0.40)}\right]^2}{(-0.05)^2}$$

$$= \frac{\left[1.645\sqrt{0.2275} + 1.282\sqrt{0.2400}\right]^2}{(-0.05)^2}$$

$$= \frac{(0.7846 + 0.6280)^2}{0.0025}$$

$$= \frac{(1.4126)^2}{0.0025}$$

$$= \frac{1.995}{0.0025}$$

$$n = 798. \triangleleft$$

A Final Comment

As you reflect on the material presented in this chapter, a pattern for hypothesis testing is probably developing. In general, the difference between a sample result and a null hypothesis is compared, and the difference is declared a chance difference or a statistically significant difference. If the difference is a statistically significant difference, then the researcher is able to reject the null hypothesis and accept some alternate hypothesis. However, if the difference is considered to be a chance difference, then the researcher cannot reject the null hypothesis.

The decision to reject the null hypothesis indicates that the sample statistic is in the rejection region which is established by the chosen value of alpha. That is, for an alpha value of 0.10, a decision to reject the null hypothesis implies that the sample statistic differed from the hypothesized parameter—such as $(\overline{X} - \mu)$ or $(p - \pi)$—by an amount large enough that it could have occurred by chance only 10 percent of the time.

The question that might still be raised, however, is, "Was the difference strong or weak at the 0.10 level of significance?" That

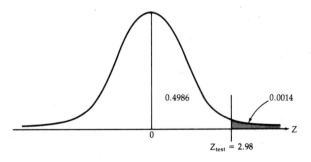

Figure 8.14   The specific probability

is, could we also have rejected at the 0.05, 0.01, or 0.001 levels of significance? Such a question could be answered by reporting the *specific probability* associated with the difference. For instance, if a hypothesis test yielded $Z_{test} = 2.98$, from appendix H we could determine that the probability of obtaining $Z > 2.98$ is 0.0014 (see figure 8.14).

Using this type of reporting, the researcher is identifying the smallest alpha level at which the null hypothesis is rejected. You will find that computer packages use this *smallest alpha* method of reporting.

Summary of Equations

8.1 Hypothesis test for population mean—σ known and normal population:

$$Z_{\text{test}} = \frac{\overline{X} - \mu_H}{\sigma_{\overline{X}}}$$

8.2 Hypothesis test for population mean—using s and large sample:

$$Z_{\text{test}} = \frac{\overline{X} - \mu_H}{s_{\overline{X}}}$$

8.3 Hypothesis test for population mean—using s and small sample selected from a normal population:

$$t_{\text{test}} = \frac{\overline{X} - \mu_H}{s_{\overline{X}}}$$

8.4 Hypothesis test for population proportion:

$$Z_{\text{test}} = \frac{p - \pi_H}{\sigma_p}$$

8.5 Sample size for hypothesis test of population mean—σ known:

$$n = \frac{(Z_\alpha + Z_\beta)^2 \sigma^2}{(\mu_T - \mu_H)^2}$$

8.6 Sample size for hypothesis test of population mean—using s:

$$n = \frac{(Z_\alpha + Z_\beta)^2 s^2}{(\mu_T - \mu_H)^2}$$

8.7 Sample size for hypothesis test on population proportion:

$$n = \frac{\left[Z_\alpha \sqrt{\pi_H(1 - \pi_H)} + Z_\beta \sqrt{\pi_T(1 - \pi_T)}\right]^2}{(\pi_T - \pi_H)^2}$$

Review Questions

1. Identify the use of each equation shown at the end of the chapter.

2. Define

(a) null hypothesis
(b) alternate hypothesis
(c) Type I error
(d) Type II error
(e) operating characteristic curve
(f) power of the test

3. Distinguish between a chance difference and a statistically significant difference.

4. List the steps in the statistical testing of a hypothesis.

5. Explain when Type II error can occur.

6. Discuss level of significance.

Problems

1. Use the following to test the null hypothesis that $\mu = 50$ at the 0.05 level of significance. Assume the population is normally distributed.

$$\sigma = 10$$
$$\overline{X} = 52$$
$$n = 100$$
$$H_A: \quad \mu \neq 50$$

2. Test the null hypothesis $\mu = 190$ against the alternate hypothesis $\mu \neq 190$ if a random sample of 49 items yielded an $\overline{X} = 182$. If the distribution of the population in unknown, but the sample standard deviation is known to be 21, perform the test at an α level of 0.10.

3. A soft drink filling machine is designed to fill cups with 6 ounces of liquid. Each machine is tested before it is shipped. If a random sample of 100 cups yielded a sample mean of 6.10 ounces with a sample standard deviation of 0.2 ounces, is the machine properly adjusted? Set alpha = 0.01.

4. Use the following sample results to test the null hypothesis $\mu = 75$ against the alternate hypothesis $\mu > 75$. Set $\alpha = 0.01$.

$$n = 64$$
$$s = 24$$
$$\overline{X} = 80$$

5. Work problem 3 if $N = 500$. Does your conclusion change?

6. Use the following sample results to test the null hypothesis $\mu = 120$ against the alternate hypothesis $\mu < 120$. Set $\alpha = 0.01$.

$n = 64$
$s = 32$
$\overline{X} = 100$

7. The manager of a large hotel believes that the average length of stay by a hotel guest is at least 5 days. If a random sample of 25 guest records indicates a sample mean of 4.6 days and a sample standard deviation of one day, do the results support the manager's claim? Set $\alpha = 0.05$. Assume a normal distribution.

8. The manufacturer of a battery-operated toy car has made modifications on the motor of the car. One result of the modifications seems to be an extended lifetime for the battery. From past data it is known that the battery in the car had an average operating life of 30 hours. To investigate if the modifications did improve battery life, a random sample of 100 cars with the new motor was obtained yielding a mean operating lifetime of 31 hours and a standard deviation of 0.75 hours. At the 0.05 level of significance can we infer that the battery lifetime was extended?

9. A chicken grower claims that the average weight of a particular group of chickens exceeds 30 ounces. Before agreeing to a purchase, a particular customer selected a sample of 25 chickens, which yielded a sample mean of 30.75 ounces and a standard deviation of 1.5 ounces. If the weights can be considered to be normally distributed, should the claim be rejected at the 0.05 level of significance?

10. A bottle-filling process has a setting of 16 ounces. A random sample of 49 bottles produced a sample mean of 15.95 ounces and a standard deviation of 0.35 ounces. At the 0.05 level of significance, what can you conclude about the machine setting?

11. Suppose the situation in problem 10 had included only 16 bottles in the sample:
(a) Would a t-test be appropriate?
(b) If the t-test is appropriate, would your conclusion drawn in problem 11 differ from problem 10?

12. An outboard motor has been designed to have a trouble-free life of 2,000 hours. A random sample of 25 engines experienced malfunction at an average of 1,600 hours with a standard deviation of 400 hours. If $\alpha = 0.10$, what can you conclude about the trouble-free life? Assume a normal distribution.

13. A credit department has records indicating the average balance for customers with charge accounts is $65. If a recent sample of 64 customers yielded a sample mean of $70 and a sample standard deviation of $6, is the average balance increasing? Set $\alpha = 0.05$.

14. A certain developing solution is advertised to last at least 25 days under normal use. If a random sample of 36 customers yielded a sample mean of 23 with a standard deviation of 4 days, is the advertising claim supported? Set $\alpha = 0.01$.

15. A company maintains the time required for a service call does not exceed 25 minutes. If a random sample of 36 calls yields a sample mean of 26.6 minutes with a standard deviation of 3 minutes, at a 0.05 level of significance, what can be concluded?

16. In the preceding problem, if the sample size had been 9, would your conclusion change? What assumption would be necessary?

17. A credit department has records that indicate the average number of days (measured from the first of the month) for a payment is 8. If a recent sample of 200 customers revealed a sample mean of 10 days and sample standard deviation of 3 days, has the average number of days increased? Set $\alpha = 0.02$.

18. Test the null hypothesis $\pi = 0.70$ against the alternate hypothesis $\pi \neq 0.70$ if a sample proportion of 0.76 was obtained from a sample size of 300. Set $\alpha = 0.05$.

19. A company plans to market nationally a new product only if more than 70 percent of the test market responds favorably to the product. If a random sample of 1,000 persons from the test market yielded 730 favorable responses, at the 0.05 level of significance what can you conclude?

20. A union composed of several thousand employees is preparing to vote on a new contract. If a random sample of 400 members yielded 212 who planned to vote yes, should you conclude that the new contract will receive more than 50 percent yes votes? Set $\alpha = 0.05$.

21. An insurance company has maintained records for the past ten years regarding the proportion of claims for a particular type of insurance program. The ten-year proportion is 0.30. If a random sample of 200 policies contains 70 claims, is there indication that the proportion of claims is increasing? Set $\alpha = 0.01$.

22. A sample of 100 items from a large shipment of Part WK produced 8 defective parts. If the manufacturer guaranteed 97 percent acceptable parts, should the shipment be accepted? Set $\alpha = 0.01$.

23. Rework problem 22 if the shipment contained 500 parts.

24. A store manager believes more than 70 percent of the purchases exceed $5.00. If a random sample of 60 shoppers yielded 48 persons with purchases in excess of $5.00, is the manager's claim substantiated? Set $\alpha = 0.05$.

25. New Products, Inc., has a standing rule for test marketing a new product. The rule necessitates a test market acceptance of greater than 80 percent before actual mass production of an item. If the latest product received 1,950 favorable responses from the 2,500 interviews, should mass production begin? Set $\alpha = 0.05$.

26. A switchboard operator believes that at least 40 percent of the calls received are long distance. If a random sample of 200 calls produced 70 long distance calls, do you agree with the operator? Set $\alpha = 0.10$.

27. A sample of 150 registered voters was taken to determine if more than 50 percent of the registered voters were in favor of a particular issue. If the sample has 80 yes responses, at the 0.05 level of significance what can you conclude?

28. A sample of 100 items from a normal population produced a sample mean of 30 and a standard deviation of 5. If the intent was to test the null hypothesis $\mu = 29$ against the alternate hypothesis $\mu \neq 29$, what is the probability of committing Type II error when the true mean is 28.5? Set $\alpha = 0.05$.

29. A processing plant has a machine that places lids on all filled jars. The machine supposedly is set at a 30-pound rating with a known standard deviation of 8 pounds. At the 0.05 level of significance, what is the probability of committing Type II error if a sample of 64 jars is taken and the true mean setting is 28 pounds?

30. For the preceding problem, construct an operating characteristic curve and a power curve.

31. A study was done on the daily cash balances of a bank to investigate the hypothesis that the average cash balance was $250,000. The sample of 100 yielded a mean of $248,000 and a standard deviation of $30,000. If the true mean was actually $246,000, what is the probability of Type II error? Set $\alpha = 0.05$.

32. For the preceding problem, construct an operating characteristic curve and a power curve.

33. A filling machine is designed to fill containers with 4 pounds of product. If a hypothesis test (H_0: $\mu = 4$) is to be conducted at the 0.01 alpha level, what sample size would be necessary to assure that the β error did not exceed 0.20 if the machine was actually filling the containers to a mean weight of 3.96 pounds? Assume σ is known to be 0.11 pounds.

34. A battery calculator is designed to have an average lifetime of 600 hours. Determine the sample size necessary to test the hypothesis (H_0: $\mu \geq 600$) at an alpha level of 0.025 with the β error not to exceed 0.1 if the actual mean lifetime is 596 hours. Assume σ is known to equal 8 hours.

35. A legislator is interested in knowing if the registered voters are still strongly favoring a tax proposal that was passed a few years earlier and supported by 70 percent of the people. If it is desired that the probability of Type I error should not exceed 0.10 and that the probability of Type II error should not exceed 0.10 when the support has dropped to 60 percent, what is the appropriate sample size?

36. A monthly magazine claims that at least 80 percent of their subscribers are college graduates. Determine the sample size necessary to investigate the claim if it is desired that the probability of Type I error should not exceed 0.05 and that the probability of Type II error should not exceed 0.20 if the actual percentage of college graduates is only 70.

Case Study

Although the sample of 196 high-risk drivers in Florida (see the case at the end of chapter 6) determined that 27 percent of those surveyed had received one or more citations for driving while intoxicated, the management team of Lock Underwriters decided to expand their business into Florida. When one of the stockholders asked about that decision, Mr. Jackson responded by saying that he was 95 percent confident that Florida was not significantly different from the rest of the country.

A few days later, the stockholder filed a legal action against Mr. Jackson alleging mismanagement. The stockholder cited as part of the facts in the case a recent publication containing data provided by the American Association of Casualty Underwriters showing that in the United States, 24.3 percent of all high-risk insured drivers have received one or more citations for driving while intoxicated. Moreover the stockholder has an expert in insurance underwriting that will testify to the effect that when 26 percent of the high-risk insured drivers in a state have one or more citations for driving while intoxicated, an insurance underwriter will lose money.

What do you believe the facts are in this case?

9

Confidence Intervals and Hypothesis Testing: Two Samples

In chapter 5, you learned how to calculate the mean and standard deviation of the distribution of sample means ($\overline{X}$'s). You also learned several things about the shape of the distribution of sample means. There are situations, however, where we may select two samples and perform analysis based on the results of both. To be able to do this, we need to investigate the sampling distribution of sample statistics obtained from two random samples.

The Sampling Distribution of the Difference between Two Sample Means ($\overline{X}_1 - \overline{X}_2$)

Analogous to what we did in chapter 5, if you take all possible samples of n size, from a population of N, and compare the mean of each sample with the mean of all the other possible samples, you will find that the arithmetic mean of the *difference* in sample means is zero (0). That is, the sampling distribution of differences in sample means will have a mean of 0 if the samples all came from the same population. Moreover the differences will be normally distributed if the population is normally distributed or if the sample size is large (central limit theorem). If the sample size is small, the differences in sample means selected from a normally distributed population will be Student t distributed.

But what if we have samples from two different populations? As a matter of fact, the mean of the distribution of differences in sample means will be equal to the difference in the two population means, or $\mu_{\overline{X}_1 - \overline{X}_2} = \mu_1 - \mu_2$. The sampling distribution of the differences will be normally distributed if *both* populations are normal distributions with known variances or if the sample sizes are large. The distribution will be Student t shaped if *the samples are small, the standard deviations of both populations are equal, and the populations are normal.*

These conditions should sound somewhat familiar to you

Table 9.1
Comparisons of sampling distributions

Sampling distribution	Number of samples	
	One	Two
Normal	Normal population, σ is known	Two normal populations, σ_1 and σ_2 are known
Normal, via central limit theorem	Large sample	Two large samples
Student t	Small sample, normal population, σ is unknown	Two small samples, Two normal populations, $\sigma_1 = \sigma_2$ (unknown)

since the sampling distribution of differences in two sample means is similar to the sampling distribution of sample means. These similarities are summarized in Table 9.1.

So far, we have mentioned the standard deviation of the population, or populations, from which the samples are taken, but we have not considered the standard deviation of the sampling distribution of differences. This standard deviation is called the *standard error of the difference between two sample means* and can be calculated using equation (9.1) if the standard deviations of the two populations are known:

$$\sigma_{\bar{X}_1 - \bar{X}_2} = \sqrt{\frac{\sigma_1^2}{n_1} + \frac{\sigma_2^2}{n_2}}, \qquad (9.1)$$

where
σ_1^2 = variance of the first population,
σ_2^2 = variance of the second population,
n_1 = the size of the first sample,
n_2 = the size of the second sample.

If the standard deviations of the populations are unknown, but both samples are large (more than 30), then equation (9.2) can be used to estimate the standard error of the difference between two sample means:

$$s_{\bar{X}_1 - \bar{X}_2} = \sqrt{\frac{s_1^2}{n_1} + \frac{s_2^2}{n_2}}, \qquad (9.2)$$

where

$$s_1^2 = \frac{\Sigma(X - \overline{X}_1)^2}{n_1 - 1} = \text{variance of the first sample,}$$

$$s_2^2 = \frac{\Sigma(X - \overline{X}_2)^2}{n_2 - 1} = \text{variance of the second sample.}$$

When the sample sizes are small (30 or less) *and* the standard deviations of the two populations are unknown but equal, equation (9.3) can be used to calculate an esitmate of the standard error:

$$s_{\overline{X}_1 - \overline{X}_2} = \sqrt{\frac{s_1^2(n_1 - 1) + s_2^2(n_2 - 1)}{n_1 + n_2 - 2}} \sqrt{\frac{1}{n_1} + \frac{1}{n_2}}. \qquad (9.3)$$

Constructing Confidence Interval Estimate of Difference between Two Population Means ($\mu_1 - \mu_2$)

In order to construct a confidence interval estimate of the difference between two population means we must first determine whether

1. the populations are normally distributed,
2. σ_1 and σ_2 are known,
3. $\sigma_1 = \sigma_2$,
4. whether the sample sizes (n_1 and n_2) are greater than 30.

If both populations are normally distributed, and σ_1 and σ_2 are known, then equation (9.4) is appropriate to use:

$(1 - \alpha)\%$ confidence interval for
$$\mu_1 - \mu_2 = (\overline{X}_1 - \overline{X}_2) \pm Z\sigma_{\overline{X}_1 - \overline{X}_2}, \qquad (9.4)$$

where $\sigma_{\overline{X}_1 - \overline{X}_2} = \sqrt{\dfrac{\sigma_1^2}{n_1} + \dfrac{\sigma_2^2}{n_2}}.$

If, however, the population variances are unknown, and the shapes of the population distributions are not necessarily normal, then equation (9.5) can be used provided both sample sizes are large (n_1 and $n_2 > 30$) so that the central limit theorem is effective:

$(1 - \alpha)\%$ confidence interval for
$$\mu_1 - \mu_2 = (\overline{X}_1 - \overline{X}_2) \pm Zs_{\overline{X}_1 - \overline{X}_2}, \qquad (9.5)$$

where $s_{\overline{X}_1 - \overline{X}_2} = \sqrt{\dfrac{s_1^2}{n_1} + \dfrac{s_2^2}{n_2}}.$

And, if both populations are normally distributed, the population variances are unknown but equal ($\sigma_1 = \sigma_2$), and the sample sizes are 30 or less, a confidence interval estimate for $\mu_1 - \mu_2$ can be made using equation (9.6):

$(1 - \alpha)\%$ confidence interval for
$$\mu_1 - \mu_2 = (\overline{X}_1 - \overline{X}_2) \pm ts_{\overline{X}_1 - \overline{X}_2}, \qquad (9.6)$$

where

$$s_{\overline{X}_1 - \overline{X}_2} = \sqrt{\frac{s_1^2(n_1 - 1) + s_2^2(n_2 - 1)}{n_1 + n_2 - 2}} \sqrt{\frac{1}{n_1} + \frac{1}{n_2}},$$

$n_1 + n_2 - 2 =$ degrees of freedom.

Now let us examine some example problems using each of the preceding confidence interval equations.

Example 9.1

▷ Two union representatives in a large company wonder if their respective groups are paid the same hourly wage. In order to investigate this situation further, the first representative takes a sample of 25 employees and determines that their average hourly wage is $4.65. At the same time the second representative has determined from a sample of size 20 that the average hourly wage of the second group is $4.52. It is known that the two populations from which the samples were drawn are normally distributed and that $\sigma_1 = \$0.25$ and $\sigma_2 = \$0.20$. Use this information to make a 90 percent confidence interval estimate for the difference between the two population means. Since $\alpha = 0.10$, then $Z = 1.64$, and

$(1 - \alpha)\%$ confidence interval for
$$\mu_1 - \mu_2 = (\overline{X}_1 - \overline{X}_2) \pm Z\sigma_{\overline{X}_1 - \overline{X}_2},$$

$(1 - 0.10)$ confidence interval for

$$\mu_1 - \mu_2 = (\$4.65 - \$4.52) \pm 1.64 \sqrt{\frac{(0.25)^2}{25} + \frac{(0.20)^2}{20}},$$

$$= \$0.13 \pm 1.64 \sqrt{\frac{0.0625}{25} + \frac{0.0400}{20}}$$

$$= \$0.13 \pm 1.64 \sqrt{0.0025 + 0.0020}$$

$$= \$0.13 \pm 1.64 \sqrt{0.0045}$$
$$= \$0.13 \pm 1.64(0.067)$$
$$= \$0.13 \pm \$0.11$$
$$\$0.02 \le (\mu_1 - \mu_2) \le \$0.24. \ ◁$$

Example 9.2

▷ A manufacturing company assembles its product on production lines known as Line A and Line B. Line A is composed of employees with at least two years experience whereas Line B is composed of employees with less than six months experience. A recent 35-day study of Line A revealed an average number of assemblies each day of 70 with a standard deviation of 5 units. A similar study of Line B conducted for 40 days revealed an average of 64 units assembled per day with a standard deviation of 8 units. Construct a 95 percent confidence interval estimate for the difference in the daily production of the two lines. Since $\alpha = 0.05$, then $Z = 1.96$, and

$(1 - \alpha)\%$ confidence interval for
$\mu_1 - \mu_2 = (\bar{X}_1 - \bar{X}_2) \pm Zs_{\bar{X}_1 - \bar{X}_2}$,
$(1 - 0.05)$ confidence interval for

$$\mu_A - \mu_B = (70 - 64) \pm 1.96 \sqrt{\frac{5^2}{35} + \frac{8^2}{40}}$$

$$= 6 \pm 1.96 \sqrt{\frac{25}{35} + \frac{64}{40}}$$

$$= 6 \pm 1.96 \sqrt{0.714 + 1.6}$$

$$= 6 \pm 1.96 \sqrt{2.314}$$
$$= 6 \pm 1.96(1.52)$$
$$= 6 \pm 2.98$$
$$3.02 \leq (\mu_A - \mu_B) \leq 8.98 \triangleleft$$

Example 9.3

▷ The manufacturer of two types of boots (A and B) is interested in the prices that are being charged for the boots. A random sample of 12 retailers who stock Type A yielded an average price of $50.00 with a standard deviation of $5.00, and a random sample of 10 retailers who stock Type B yielded an average price of $55.00 with a $6.00 standard deviation. Under the assumption that the populations of prices are normally distributed and that the population standard deviations, though unknown, are equal, construct a 98 percent confidence interval for the difference between the two average prices.[1] (Remember that the degrees of freedom (df) are $n_1 + n_2 - 2$.) Since $\alpha = 0.02$ and $df = 12 + 10 - 2$, then $t = 2.528$, and

---

1. Chapter 10 contains a statistical test to determine if two unknown variances are equal.

$(1 - \alpha)\%$ confidence interval for $\mu_1 - \mu_2 = (\overline{X}_1 - \overline{X}_2) \pm ts_{\overline{X}_1 - \overline{X}_2}$

$(1 - 0.02)\%$ confidence interval for

$$\mu_A - \mu_B = (\$50 - \$55)$$

$$\pm 2.528 \sqrt{\frac{5^2(12 - 1) + 6^2(10 - 1)}{12 + 10 - 2}} \sqrt{\frac{1}{12} + \frac{1}{10}}$$

$$= -\$5 \pm 2.528 \sqrt{\frac{25(11) + 36(9)}{20}} \sqrt{0.183}$$

$$= -\$5 \pm 2.528 \sqrt{29.95} \sqrt{0.183}$$

$$= -\$5 \pm 2.528(5.47)(0.428)$$
$$= -\$5 \pm \$5.92$$
$$-\$10.92 \le (\mu_A - \mu_B) \le \$0.92 \lhd$$

Testing Hypotheses about the Difference between Two Population Means ($\mu_1 - \mu_2$)

When we test a hypothesis about the difference between two population means, we are testing the following null hypothesis:

H_0: $\mu_1 - \mu_2 = 0$ or $\mu_1 = \mu_2$

against one of the three alternate hypotheses,

(1) H_A: $\mu_1 - \mu_2 \ne 0$,   (2) H_A: $\mu_1 - \mu_2 < 0$,
(3) H_A: $\mu_1 - \mu_2 > 0$.

As was the case in chapter 8, the first alternate is the two-sided test of the null hypothesis that there is no difference between the two population means, and the second and third alternates are the one-sided tests. Figure 9.1 shows the two-sided test, and figures 9.2 and 9.3 show the one-sided tests for the second and third alternate hypotheses, respectively. Notice again that on the one-

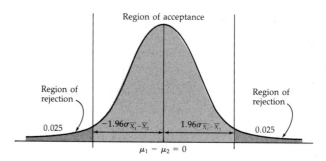

Figure 9.1   A two-sided test of the null hypothesis with $\alpha = 0.05$

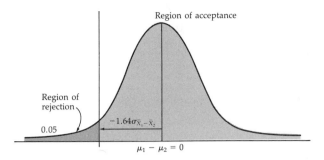

Figure 9.2   A one-sided test of the null hypotheses with $\alpha = 0.05$

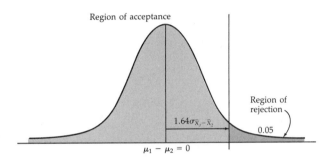

Figure 9.3   A one-sided test of the null hypothesis with $\alpha = 0.05$

sided tests all of the Type 1 error (α error) goes in a single tail of the sampling distribution.

In order to test the hypothesis, one of three following test equations will be used. Equation (9.7) is used when both populations are normally distributed and σ_1 and σ_2 are known. Equation (9.8) is used when n_1 and n_2 are large samples, which enables the central limit theorem to be used, and equation (9.9) is used when n_1 and n_2 are small samples selected from two normal populations that have the same standard deviation.

$$Z_{\text{test}} = \frac{(\overline{X}_1 - \overline{X}_2) - (\mu_1 - \mu_2)}{\sigma_{\overline{X}_1 - \overline{X}_2}}, \tag{9.7}$$

$$Z_{\text{test}} = \frac{(\overline{X}_1 - \overline{X}_2) - (\mu_1 - \mu_2)}{s_{\overline{X}_1 - \overline{X}_2}}, \tag{9.8}$$

$$t_{\text{test}} = \frac{(\overline{X}_1 - \overline{X}_2) - (\mu_1 - \mu_2)}{s_{\overline{X}_1 - \overline{X}_2}}. \tag{9.9}$$

Example 9.4

▷ An insurance company has agents in two different states. A random sample of 200 claims in state one produced an average claim of $750.00, and a random sample of 250 claims in state two produced an average claim of $720.00. If it is known that the two populations from which the samples were drawn are normally distributed and that $\sigma_1 = \$100$ and $\sigma_2 = \$80$, at the 0.10 level of significance are the population means different?

H_0: $\mu_1 - \mu_2 = 0$,
H_A: $\mu_1 - \mu_2 \neq 0$,
  $\alpha = 0.10$.

Reject H_0 if $Z_{test} < -1.64$ or $Z_{test} > 1.64$:

$$Z_{test} = \frac{(\overline{X}_1 - \overline{X}_2) - (\mu_1 - \mu_2)}{\sigma_{\overline{X}_1 - \overline{X}_2}}$$

$$= \frac{(750 - 720) - 0}{\sqrt{\dfrac{(100)^2}{200} + \dfrac{(80)^2}{250}}}$$

$$= \frac{30}{\sqrt{50 + 25.6}}$$

$$= \frac{30}{8.69}$$

$Z_{test} = 3.45$.

Reject H_0; the two states do not have the same average claim. ◁

Example 9.5

▷ Stores Inc. is investigating the average dollar sale of two stores in cities with trade areas that are about the same size. From store one, a random sample of 100 sales produced an average sale of $20.00 with a standard deviation of $4.00; from the second store a random sample of 125 sales produced an average sale of $18.00 with a standard deviation of $5.00. At the 0.05 level of significance, does store one have a larger average dollar sale?

H_0: $\mu_1 - \mu_2 = 0$,
H_A: $\mu_1 - \mu_2 > 0$,
  $\alpha = 0.05$.

Reject H_0 if $Z_{test} > 1.64$ using equation (9.8):

$$Z_{test} = \frac{(\overline{X}_1 - \overline{X}_2) - (\mu_1 - \mu_2)}{s_{\overline{X}_1 - \overline{X}_2}}$$

$$= \frac{(20 - 18) - 0}{\sqrt{\dfrac{4^2}{100} + \dfrac{5^2}{125}}}$$

$$Z_{test} = \frac{20 - 18}{\sqrt{0.16 + 0.20}} = \frac{2}{0.6} = 3.33.$$

Reject H_0; store one has a higher average sale. ◁

Example 9.6

▷ The data in example 9.3 can be used to test the hypothesis that the average price of Type A is less than the average price of Type B:

H_0: $\mu_A - \mu_B = 0$,
H_A: $\mu_A - \mu_B < 0$,
       $\alpha = 0.05$.

Reject H_0 if $t_{test} < -1.725$ using equation (9.9):

$$t_{test} = \frac{(\overline{X}_A - \overline{X}_B) - (\mu_A - \mu_B)}{s_{\overline{X}_A - \overline{X}_B}}$$

$$= \frac{(50 - 55) - 0}{\sqrt{\frac{5^2(12 - 1) + 6^2(10 - 1)}{12 + 10 - 2}} \sqrt{\frac{1}{12} + \frac{1}{10}}}$$

$$= \frac{-5}{2.34}$$

$t_{test} = -2.137$.

Reject H_0, Type A price is less than Type B price. ◁

Dependent Samples

An unstated requirement for all of the previous discussion in this chapter is that the two random samples must be *independent*. For our purposes, independence can be taken to imply that *the individual observations of sample one have no known relationship to the individual observations in sample two*. There are times, however, when this assumption of independence is not valid; whenever this is the case, the preceding equations are not to be used. Instead, we must proceed with an analysis designed for *dependent samples*.

Dependent samples exist when the individual observations of one sample can be matched with individual observations of the other sample. Thus the samples form what can be referred to as *matched pairs*. This matching of pairs can be accomplished based on characteristics such as age, sex, professional skill, and other physical characteristics; or by having an observation on the same individual in both samples such as a pre- and post-test. When we do have these matched pairs, what we actually analyze is the difference in the two observations and treat this difference the same as the single samples discussed in chapters 6 and 8. Instead of working with sample means, we work with the mean of the differences ($\overline{d}$):

$$\overline{d} = \frac{\Sigma d}{n},\qquad\qquad(9.10)$$

where

$\bar{d}$ = average of the differences (d) in all matched pairs from the two samples,

d = difference between each matched pair of observations,

n = number of matched pairs;

and

$$s_d = \sqrt{\frac{\Sigma(d - \bar{d})^2}{n - 1}},$$ (9.11)

where s_d = estimate of the standard deviation for all the matched pairs.

Figure 9.4 shows pictorially the type of sampling distribution we would be working with when we have matched pairs. The mean of the sampling distribution, $\mu_{\bar{d}}$, is actually $\mu_1 - \mu_2$. The measure of dispersion for this sampling distribution, standard error of the difference of the matched pairs, is

$$\sigma_{\bar{d}} = \frac{\sigma_d}{\sqrt{n}},$$ (9.12)

and, when σ_d is unknown, we will estimate σ_d with equation (9.11). If the samples are large so that the number of pairs is large, it is reasonable to assume that the $\bar{d}$'s will be normally distributed as shown in figure 9.4. In this situation the test statistic is shown in equation (9.13):

$$Z_{\text{test}} = \frac{\bar{d} - \mu_{\bar{d}}}{s_{\bar{d}}}, \quad \textit{where} \quad s_{\bar{d}} = \frac{s_d}{\sqrt{n}}$$ (9.13)

If the number of pairs is small, however, we will consider the $\bar{d}$'s to be t distributed provided it is reasonable to assume that the original populations were both normally distributed. For this situation the test statistic is shown in equation (9.14):

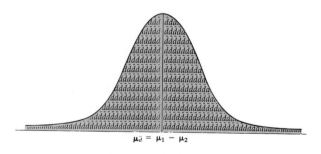

$$\mu_{\bar{d}} = \mu_1 - \mu_2$$

Figure 9.4   Sampling distribution for $\bar{d}$

$$t_{\text{test}} = \frac{\bar{d} - \mu_{\bar{d}}}{s_{\bar{d}}}. \tag{9.14}$$

Example 9.7

▷ Suppose we select 25 employees who have not completed high school and give them a reading test. After this first test, we give them some formal vocabulary training and then retest their reading skill. Table 9.2 shows the results of the two reading tests. Can we conclude that the vocabulary training was beneficial?

H_0: $\mu_{\bar{d}} = 0$ or $\mu_1 = \mu_2$,
H_A: $\mu_{\bar{d}} \neq 0$ or $\mu_1 \neq \mu_2$,
$\alpha = 0.05$,
$df = 25 - 1 = 24$.

Table 9.2
Data for example 9.7

First test	Second test	Difference (d)	$(d - \bar{d})$	$(d - \bar{d})^2$
65	67	2	−1	1
72	70	−2	−5	25
64	72	8	5	25
43	50	7	4	16
55	54	−1	−4	16
84	86	2	−1	1
72	80	8	5	25
52	50	−2	−5	25
49	62	13	10	100
80	81	1	−2	4
38	56	18	15	225
93	90	−3	−6	36
77	78	1	−2	4
62	64	2	−1	1
69	72	3	0	0
58	57	−1	−4	16
45	55	10	−7	49
90	88	−2	−5	25
60	62	2	−1	1
54	52	−2	−5	25
72	70	−2	−5	25
49	53	4	1	1
53	56	3	0	0
82	84	2	−1	1
66	70	4	1	1
		$\overline{75}$		$\overline{648}$

$$\bar{d} = \frac{\sum d}{n} = \frac{75}{25} = 3 \qquad s_d = \sqrt{\frac{\sum(d - \bar{d})^2}{n - 1}} = \sqrt{\frac{648}{25 - 1}}$$

$$s_d = \sqrt{27} = 5.20$$

Reject H_0 if $t_{test} < -2.064$ or $t_{test} > 2.064$:

$$t_{test} = \frac{\bar{d} - \mu_{\bar{d}}}{s_{\bar{d}}} \quad where \quad s_{\bar{d}} = \frac{s_d}{\sqrt{n}}$$

$$= \frac{3.0}{1.04} \qquad\qquad = \frac{5.20}{\sqrt{25}}$$

$$t_{test} = 2.88, \qquad\qquad s_{\bar{d}} = 1.04.$$

Reject H_0; conclude that the vocabulary test was beneficial to the employees. ◁

The Difference between Two Proportions

Just as we can construct confidence intervals and test hypotheses about the difference between two population means, we can do the same for the difference between two population proportions $(\pi_1 - \pi_2)$. In order to do this, we will select large samples so that we can use the central limit theorem to assert that the sampling distribution of differences between the sample proportions $(p_1 - p_2)$ is normally distributed. The mean of the sampling distribution of differences between sample proportions $(p_1 - p_2)$ is the difference between the population proportions $(\pi_1 - \pi_2)$, and the standard deviation (*standard error of the difference between sample proportions*) is

$$\sigma_{p_1 - p_2} = \sqrt{\frac{\pi_1(1 - \pi_1)}{n_1} + \frac{\pi_2(1 - \pi_2)}{n_2}}. \tag{9.15}$$

Confidence Interval Estimate for the Difference between Two Population Proportions

When n_1 and n_2 are both large samples ($n_1 p_1$, $n_1(1 - p_1)$, $n_2 p_2$, $n_2(1 - p_2)$ should all be ≥ 5), the sampling distribution of the difference between two sample proportions $(p_1 - p_2)$ is normally distributed. Consequently it is possible to estimate the difference between two population proportions $(\pi_1 - \pi_2)$ by the construction of a confidence interval as shown in equation (9.16):

$(1 - \alpha)\%$ confidence interval for
$$\pi_1 - \pi_2 = (p_1 - p_2) \pm Z\sigma_{p_1 - p_2}. \tag{9.16}$$

Notice in equation (9.16) the standard error term is $\sigma_{p_1 - p_2}$ which is given in equation (9.15). Inspection of equation (9.15) indicates that π_1 and π_2 are assumed to be known. However, if π_1 and π_2 were known, there would be no need for the interval estimate. Thus, when we are sampling, we will have to work with sample proportions, p_1 and p_2, and we can use equation (9.17) to estimate the standard error of the difference between two sample proportions:

$$s_{p_1-p_2} = \sqrt{\frac{p_1(1 - p_1)}{n_1} + \frac{p_2(1 - p_2)}{n_2}}. \qquad (9.17)$$

Using $s_{p_1-p_2}$, we can then construct a confidence interval estimate for the difference between two population proportions as shown in equation (9.18):

$(1 - \alpha)\%$ confidence interval for

$$\pi_1 - \pi_2 = (p_1 - p_2) \pm Zs_{p_1-p_2}. \qquad (9.18)$$

Example 9.8

▷ Two organizations are known to favor a proposed tax cut. There is some concern, however, as to the difference in the proportion of support in the two groups. A sample of 200 from Organization 1 yielded 120 who favored the tax cut, and a sample of 250 from Organization 2 yielded 125 who favored the tax cut. Based on the sample results, a 95 percent confidence interval estimate of the difference between the two population proportions can be constructed:

$$p_1 = \frac{120}{200} = 0.60,$$

$$p_2 = \frac{125}{250} = 0.50,$$

$(1 - \alpha)\%$ confidence interval for $\pi_1 - \pi_2 = (p_1 - p_2) \pm Zs_{p_1-p_2}$, 95% confidence interval for

$$\pi_1 - \pi_2 = (0.60 - 0.50) \pm 1.96 \sqrt{\frac{0.6(1 - 0.6)}{200} + \frac{0.5(1 - 0.5)}{250}}$$

$$= (0.10) \pm 1.96 \sqrt{\frac{(0.60)(0.40)}{200} + \frac{(0.50)(0.50)}{250}}$$

$$= 0.10 \pm 1.96 \sqrt{0.0012 + 0.001}$$

$$= 0.10 \pm 1.96 \sqrt{0.0022}$$
$$= 0.10 \pm 1.96(0.0469)$$
$$= 0.10 \pm 0.092$$
$$0.008 \leq \pi_1 - \pi_2 \leq 0.192.$$

Thus the difference between the two population proportions may be as great as 0.192 or as small as 0.008. ◁

Hypothesis Test for Two Population Proportions

At times it is desirable to test the null hypothesis that two population proportions are equal:

H_0: $\pi_1 = \pi_2$.

In order to test the preceding null hypothesis, it is necessary to estimate the standard error of the difference between two sample proportions. As stated previously, equation (9.15) assumes that the population proportions are known and if π_1 and π_2 were known, there would be no reason to be sampling. Thus, as in the case of confidence interval construction, we will have to work with sample proportions, p_1 and p_2. Assuming that $\pi_1 = \pi_2$, we will estimate the common population proportion as

$$\hat{p} = \frac{X_1 + X_2}{n_1 + n_2}. \tag{9.19}$$

Using $\hat{p}$, you can estimate the standard error $s_{p_1-p_2}$ as follows:

$$s_{p_1-p_2} = \sqrt{\frac{\hat{p}(1-\hat{p})}{n_1} + \frac{\hat{p}(1-\hat{p})}{n_2}}. \tag{9.20}$$

The appropriate test statistic is shown in equation (9.21):

$$Z_{\text{test}} = \frac{(p_1 - p_2) - (\pi_1 - \pi_2)}{s_{p_1-p_2}}. \tag{9.21}$$

Example 9.9

▷ The manufacturer of Kornles believes that her munchy product being introduced in Region Y will be more popular than in Region X, where it is currently produced and distributed. In order to check this hypothesis, a random sample from each region was taken. The sample in Region X contained 700 people, 560 of whom claimed to prefer the taste of Kornles. In Region Y, 525 of the 750 people sampled responded favorably. Based on these results, does it appear that Kornles will be more popular in Region Y than they have been in Region X? (Set $\alpha = 0.05$.)

H_0: $\pi_Y - \pi_X = 0$ or $\pi_Y = \pi_X$,
H_A: $\pi_Y - \pi_X > 0$ or $\pi_Y > \pi_X$,
   $\alpha = 0.05$.

Reject H_0 if $Z_{\text{test}} > 1.64$;

$$p_X = \frac{X_X}{n_X}$$

$$= \frac{560}{700}$$

$$p_X = 0.80,$$

$$p_Y = \frac{X_Y}{n_Y}$$

$$= \frac{525}{750}$$

$$p_Y = 0.70,$$

$$\hat{p} = \frac{X_X + X_Y}{n_X + n_Y}$$

$$= \frac{560 + 525}{700 + 750}$$

$$\hat{p} = 0.748.$$

$$Z_{test} = \frac{(p_Y - p_X) - (\pi_Y - \pi_X)}{s_{p_Y - p_X}}$$

$$= \frac{(0.70 - 0.80) - 0}{\sqrt{\frac{\hat{p}(1 - \hat{p})}{n_Y} + \frac{\hat{p}(1 - \hat{p})}{n_X}}}$$

$$= \frac{-0.10}{\sqrt{\frac{0.748(0.252)}{750} + \frac{0.748(0.252)}{700}}}$$

$$= \frac{-0.10}{\sqrt{0.000251 + 0.000269}}$$

$$= \frac{-0.10}{0.0228}$$

$$Z_{test} = -4.39.$$

Accept (do not reject) H_0; the product is not doing better in the new market. In fact, based on the sample results, the product may be less popular in Region Y. ◁

Summary of Equations

9.1 Standard error of the difference between two sample means—known σ:

$$\sigma_{\overline{X}_1 - \overline{X}_2} = \sqrt{\frac{\sigma_1^2}{n_1} + \frac{\sigma_2^2}{n_2}}$$

9.2 Standard error of the difference between two sample means—using s_1 and s_2 from large samples:

$$s_{\overline{X}_1 - \overline{X}_2} = \sqrt{\frac{s_1^2}{n_1} + \frac{s_2^2}{n_2}}$$

9.3 Standard error of the difference between two sample means—using small samples:

$$s_{\overline{X}_1 - \overline{X}_2} = \sqrt{\frac{s_1^2(n_1 - 1) + s_2^2(n_2 - 1)}{n_1 + n_2 - 2}} \sqrt{\frac{1}{n_1} + \frac{1}{n_2}}$$

9.4 Confidence interval for difference between two population means—known σ and normal populations:

$(1 - \alpha)\%$ confidence interval for
$\mu_1 - \mu_2 = (\overline{X}_1 - \overline{X}_2) \pm Z\sigma_{\overline{X}_1 - \overline{X}_2}$

9.5 Confidence interval for difference between two population means—using s_1 and s_2 from large samples:

$(1 - \alpha)\%$ confidence interval for
$\mu_1 - \mu_2 = (\overline{X}_1 - \overline{X}_2) \pm Zs_{\overline{X}_1 - \overline{X}_2}$

9.6 Confidence interval for difference between two population means—using t distribution:

$(1 - \alpha)\%$ confidence interval for
$\mu_1 - \mu_2 = (\overline{X}_1 - \overline{X}_2) \pm ts_{\overline{X}_1 - \overline{X}_2}$

9.7 Hypothesis test for difference between two population means—known σ and normal populations.

$$Z_{\text{test}} = \frac{(\overline{X}_1 - \overline{X}_2) - (\mu_1 - \mu_2)}{\sigma_{\overline{X}_1 - \overline{X}_2}}$$

9.8 Hypothesis test for difference between two population means—using s_1 and s_2 from large samples:

$$Z_{\text{test}} = \frac{(\overline{X}_1 - \overline{X}_2) - (\mu_1 - \mu_2)}{s_{\overline{X}_1 - \overline{X}_2}}$$

9.9 Hypothesis test for difference between two population means—using t distribution:

$$t_{test} = \frac{(\overline{X}_1 - \overline{X}_2) - (\mu_1 - \mu_2)}{s_{\overline{X}_1 - \overline{X}_2}}$$

9.10 Average of the differences in matched pairs:

$$\overline{d} = \frac{\Sigma d}{n}$$

9.11 Standard deviation for the differences of matched pairs:

$$s_d = \sqrt{\frac{\Sigma(d - \overline{d})^2}{n - 1}}$$

9.12 Standard error of the difference of matched pairs:

$$\sigma_{\overline{d}} = \frac{\sigma_d}{\sqrt{n}}$$

9.13 Hypothesis test for difference between two population means—using dependent samples:

$$Z_{test} = \frac{\overline{d} - \mu_{\overline{d}}}{s_{\overline{d}}}$$

9.14 Hypothesis test for difference between two population means—using t distribution and dependent samples:

$$t_{test} = \frac{\overline{d} - \mu_{\overline{d}}}{s_{\overline{d}}}$$

9.15 Standard error of the difference between two sample proportions—using π_1 and π_2:

$$\sigma_{p_1 - p_2} = \sqrt{\frac{\pi_1(1 - \pi_1)}{n_1} + \frac{\pi_2(1 - \pi_2)}{n_2}} \, ,$$

9.16 Confidence interval for difference between two population proportions:

$(1 - \alpha)\%$ confidence interval for
$$\pi_1 - \pi_2 = (p_1 - p_2) \pm Z\sigma_{p_1 - p_2}$$

9.17 Standard error of the difference between two sample proportions—using p_1 and p_2:

$$s_{p_1-p_2} = \sqrt{\frac{p_1(1-p_1)}{n_1} + \frac{p_2(1-p_2)}{n_2}}$$

9.18 Confidence interval for difference between two population proportions—using $s_{p_1-p_2}$:

$(1-\alpha)\%$ confidence interval for

$\pi_1 - \pi_2 = (p_1 - p_2) \pm Zs_{p_1-p_2}$

9.19 Estimate of the common population proportion:

$$\hat{p} = \frac{X_1 + X_2}{n_1 + n_2}$$

9.20 Standard error of the difference between two population proportions—using $\hat{p}$:

$$s_{p_1-p_2} = \sqrt{\frac{\hat{p}(1-\hat{p})}{n_1} + \frac{\hat{p}(1-\hat{p})}{n_2}}$$

9.21 Hypothesis test for difference between two population proportions:

$$Z_{test} = \frac{(p_1 - p_2) - (\pi_1 - \pi_2)}{s_{p_1-p_2}}$$

Review Questions

1. Identify the use of each equation shown at the end of the chapter.

2. Under what conditions will the differences between sample means be normally distributed?

3. Under what conditions will the differences between sample means be t distributed?

4. What are independent samples?

5. What are dependent samples?

6. Under what conditions will the sampling distribution of the difference between two sample proportions ($p_1 - p_2$) be normally distributed?

Problems

1. From two populations, N_1 with numbers 4, 5, and 6, and N_2 with numbers 1, 2, 3, determine all possible pairs of sample means and their differences. Use a sample size of two from both populations.

2. Use the results of problem 1 to
(a) plot a histogram of the differences of pairs of sample means.
(b) calculate the mean and standard deviation of the differences.

3. Use the data in problem 1 to
(a) show that the average of the differences between all pairs of sample means equals $\mu_1 - \mu_2$.
(b) calculate the standard deviation of the difference between the two sample means using equation (9.1). Is this the same answer as the standard deviation determined in problem 2(b).

4. Two random samples are selected from two normal populations with standard deviations equal to 4 and 6. If the first sample (from population with $\sigma_1 = 4$) is size 24 and yields a mean of 41 and the second sample (from population with $\sigma_2 = 6$) is size 36 and yields a mean of 40, construct a 95 percent confidence interval estimate for the difference between the two population means.

5. Use the data in problem 4 to test the null hypothesis $\mu_1 = \mu_2$ against the alternate hypothesis $\mu_1 \neq \mu_2$. Set $\alpha = 0.05$.

6. Use the data in problem 4 to test the null hypothesis $\mu_1 = \mu_2$ against the alternate hypothesis $\mu_1 - \mu_2 < 0$. Set $\alpha = 0.05$.

7. The corporation that operates a chain of restaurants is considering the purchase of the Eatery chain. A random sample of 10 restaurants already in the chain showed an average volume of 1,400 meals per day with a standard deviation of 120, whereas a survey of 15 of the Eatery shops showed a mean volume of 1,675 meals with a standard deviation of 250. At the 0.05 level of significance, would the corporation be safe in acquiring the Eatery shops if it desires comparable operations? Assume normal populations.

8. Two different production runs on the same brand of golf ball produced the following results concerning the diameter of the balls.

	First run	Second run
Mean	1.65 inches	1.70 inches
Standard deviation	0.04 inches	0.06 inches
Sample size	100	125

At the 0.01 level of significance, did the runs produce the same size golf ball?

9. The manager of a full-service station is interested in comparing activity in the morning and afternoon. A random sample of 12 mornings produced an average number of customers of 18 with a standard deviation of 3, while a random sample of 14 afternoons yielded an average number of customers of 22 with a standard deviation of 4. At the 0.05 level of significance, are the morning and afternoon activities different? Assume the samples were obtained from normal populations.

10. A department of a large store has conducted a study regarding the average dollar purchase of male and female customers. Using the following information, test to see if the average purchase is different. Set alpha = 0.01.

Female	Male
$n_1 = 49$	$n_2 = 64$
$\overline{X} = \$25$	$\overline{X} = \$22$
$\sigma = \$5$	$\sigma = \$4$

11. A manufacturer of paper produces two types of stock. A random sample of 5 rolls from each type yielded the following tear strengths:

A	B
154	149
143	162
135	160
140	154
128	175

At the 0.05 level of significance, are the tear strengths different? State any assumptions necessary.

12. Two random samples are selected. The first sample of size 70 yields a mean of 135 and a standard deviation of 14. The second sample of size 60 yields a mean of 110 and a standard deviation of 10. Construct a 99% confidence interval estimate for the difference between the two population means.

13. Use the data in problem 10 to test the null hypothesis $\mu_1 = \mu_2$ against the alternate hypothesis $\mu_1 = \mu_2 < 0$. Set $\alpha = 0.10$.

14. Use the data in problem 10 to test the null hypothesis $\mu_1 = \mu_2$ against the alternate hypothesis $\mu_1 - \mu_2 > 0$. Set $\alpha = 0.10$.

15. Two random samples were taken of members of two different nationwide professional organizations to determine their annual incomes. The following results were determined from the samples.

		Organization	
I		II	
Mean income	$27,700	Mean income	$28,300
Standard deviation	600	Standard deviation	700
Sample size	226	Sample size	226

At the 0.05 level of significance, does a significant difference in the average incomes of the two groups exist?

16. The management of a very large plant wished to investigate the effect of the 4-day work week on absenteeism. Two random samples of size 40 were selected; group I of employees worked 10-hour days (4-day week), and group II of employees worked 8-hour days (5-day week). If group I averaged 4 hours of absenteeism per week with a standard deviation of 1.2 and group II averaged 4.4 hours of absenteeism per week with a standard deviation of 1.5, should we conclude that the shorter work week reduces absenteeism? Set $\alpha = 0.05$.

17. Two organizations are meeting at the same convention hotel. A sample of 10 members of The Cranes revealed an average daily expenditure on food of $32.00 with a standard deviation of $4.00; and a sample of 15 members of The Penguins revealed an average daily expenditure on food of $34.00 with a standard deviation of $6.00.

(a) If $\alpha = 0.01$, are the average expenditures of the two organizations different?
(b) What assumptions were necessary to perform a test of hypothesis in part *(a)*?

18. Use the data in problem 17 to see if the members of organization II are bigger spenders than the members of organization I.

19. A paint manufacturer is experimenting with a new, less expensive paint. A random sample of 144 gallons of the old paint produced an average coverage of 480 square feet with a standard deviation of 25 square feet, and a random sample of 138 gallons of the new paint produced an average coverage of 500 square feet with a standard deviation of 30 square feet. At the 0.01 level of significance, do you think the new paint provides greater coverage than the old paint?

20. The bank has two major locations. A recent survey of loan applications indicated that location B issued loans to 60 of 100 applicants and location D issued loans to 80 of 125 applicants. Construct a 95% confidence interval estimate for the difference between the proportions.

21. A cereal manufacturer has two production plants. If a sample of 50 boxes from plant I had 5 boxes improperly filled or packaged, and a sample of 48 boxes had 4 boxes improperly filled or packaged, construct a 90% confidence interval estimate for the difference between the population proportions.

22. Two production processes are designed to manufacture a particular bottle. If a random sample of 100 bottles produced by the first process yielded 8 defective bottles, and a random sample of 200 bottles from the second process yielded 40 defective bottles, are the population proportions different? Set $\alpha = 0.10$.

23. If a random sample of 100 registered voters yielded 70 yes respondents, and another random sample of 150 voters yielded 112 yes respondents, can you conclude at the 0.05 level of significance that the proportions of yes respondents are different?

24. A sample of 150 automobiles in the southern part of a state revealed 30 cars that could not pass the safety inspection. In the northern part of the state, a sample of 250 automobiles revealed 70 cars that could not pass the safety inspection. At the 0.01 level of significance determine if the population proportions are different.

25. Two small business owners met at a convention. After some discussion, it was determined that Owner A believed she experienced a larger proportion of telephone sales than Owner B. To decide the issue, each owner collected data on incoming calls. The results are recorded as follows. Using these results, do you think Owner A is correct? Set $\alpha = 0.05$.

	Owner *A*	Owner *B*
Number of sales	400	300
Number of telephone sales	100	60

26. To encourage participation the proprietor of Gameplace gave free instruction on how to improve scores to all who desired it. From a sample of 20 of the participants, scores were obtained before and after the instruction period. Based on the following results, would you regard the instruction as beneficial? Set $\alpha = 0.01$.

Participant	Before score	After score
1	90	95
2	105	112
3	78	90
4	125	126
5	150	148
6	85	90
7	100	102
8	102	115
9	94	99
10	129	125
11	152	153
12	80	90
13	76	80
14	140	145
15	95	110
16	104	112
17	89	100
18	110	115
19	115	122
20	99	108

Case Study

A shoe manufacturer has developed two versions (A and B) of a new jogging shoe and is attempting to decide which one to introduce to the market. The firm decided to subject the two versions to a test by having joggers try both versions and then, considering various attributes, rate each version on a scale of 0 to 100. Their sample consisted of 40 joggers randomly selected in the Tulsa area, but they believe the sample is representative of joggers across the country.

After seeing the results of the test given in the table that follows, the president of the firm concluded that since 67.5 percent of the sample preferred version B, the company should introduce that version in sizes for both men and women and in a variety of colors. As an after thought, however, he handed the data to you with the request that you "take a look and see if there is anything there I've missed."

Jogger	Sex	Rating score Version A	Version B	Jogger	Sex	Rating score Version A	Version B
1	F	50	71	21	F	80	84
2	F	33	48	22	F	78	79
3	M	76	82	23	F	70	73
4	F	90	61	24	F	71	75
5	F	75	49	25	F	62	66
6	F	46	47	26	M	76	75
7	M	58	57	27	M	60	65
8	M	82	89	28	F	49	50
9	F	52	53	29	M	52	70
10	M	79	80	30	M	98	95
11	F	61	64	31	M	83	92
12	F	60	62	32	F	65	52
13	M	90	88	33	M	78	81
14	F	10	13	34	F	73	74
15	M	90	91	35	M	95	93
16	M	65	64	36	M	90	92
17	M	73	80	37	F	38	41
18	F	37	24	38	F	25	20
19	M	74	75	39	F	42	43
20	F	83	82	40	M	89	88

10

The F Distribution and an Introduction to Analysis of Variance

In chapter 9 you were introduced to the concept of testing hypotheses concerning two population means. There are situations where more than two means may be under investigation. In order to test the hypotheses concerning the equality of more than two means, the approach known as *analysis of variance* is appropriate. However, before studying analysis of variance, it is necessary to become familiar with the F distribution.

The F Distribution

Suppose we select all possible samples of size n_1 and size n_2 from a normal population. Then for each of these samples let us calculate $\overline{X}_1$ and $\overline{X}_2$ and use the statistics to calculate s_1^2 and s_2^2. Now suppose we form all possible ratios of s_1^2/s_2^2. The result is known as the F *distribution*. A graph of the F distribution is shown in figure 10.1.

Actually, figure 10.1 is only *one of many* F distributions since the F distribution is *a function of the degrees of freedom of the numerator and denominator of the ratio*. Thus use of this distribution is similar to the use of the t distribution. Appendix J contains the F values

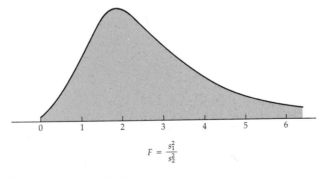

$$F = \frac{s_1^2}{s_2^2}$$

Figure 10.1   The F distribution

associated with an 0.05 and the 0.01 probabilities in the right-hand end of the distribution.

Notice in appendix J that the smallest F ratio is 1.00 (lower right corner of appendix). This is because this appendix contains only the right side values of F associated with the probabilities. Use of this distribution can best be demonstrated by an example.

Example 10.1

▷ A manufacturer believes he has two types of tires (assume tire wear to be normally distributed) whose variability in wear is the same. If a sample of 25 tires of the first type produced a variance of 2,000 miles and a sample of 20 tires of the second type produced a variance of 1,600 miles, is the manufacturer correct?

$$H_0: \quad \sigma_1^2 = \sigma_2^2,$$
$$H_A: \quad \sigma_1^2 \neq \sigma_2^2,$$
$$\alpha = 0.05.$$

Reject H_0 if $F_{test} > 2.11$.

The F value is a function of the degrees of freedom of the numerator and denominator. Since appendix J contains only F values of 1.00 or greater, it is necessary to use

$$F_{test} = \frac{s_{larger}^2}{s_{smaller}^2}, \tag{10.1}$$

which means that the numerator degrees of freedom is associated with the larger sample variance and the denominator degrees of freedom is associated with the smaller sample variance. In this example, the larger sample variance came from the sample of 25 tires, so the degrees of freedom for the numerator is $25 - 1 = 24$. The smaller sample variance came from the sample of 20 tires, so the degrees of freedom for the denominator is $20 - 1 = 19$.

Thus the F value is found by locating 24 across the top of the appendix and 19 down the side of the appendix. Now within appendix J, when $\alpha = 0.05$, the F associated with this set of degrees of freedom is 2.11.

To complete the example, let us calculate the F_{test}:

$$F_{test} = \frac{s_{larger}^2}{s_{smaller}^2}$$

$$= \frac{2,000}{1,600}$$

$$F_{test} = 1.25.$$

Accept (do not reject) H_0; conclude that the difference in the variances is a chance difference. ◁

Analysis of Variance

In chapter 9 you were shown how to test the null hypothesis that $\mu_1 = \mu_2$ using either the normal or t distributions. Now suppose you wanted to test a null hypothesis concerning the equality of three or more population means. An approach to testing more than two sample means is known as *analysis of variance* (ANOVA). As the name implies, analysis of variance will employ variances to examine the possible equality of several population means. Analysis of variance is based on the realization that there are two ways to estimate the population variance from sample data. Since there are two ways to estimate the same variance, *the value of the F ratio formed by the two estimates will vary from 1 by chance alone if the data came from one normally distributed population with a mean and variance of μ and σ^2. If, however, the sample data came from two or more normal populations that have different means and a common variance, the F ratio formed by the two variance estimates will tend to be greater than 1, indicating more than one population.*

Before we demonstrate the two methods of estimating the population variance it is helpful to introduce a common way that collected data are organized. Notice in table 10.1 that the individual data items are double subscripted. That is, each item has an ij subscript such that X_{ij} represents the row i and column j element.

To demonstrate the computations, let us return to the population of four numbers, 1, 2, 3, and 4, which was used in chapter 5. This particular population has a mean of 2.5 and a variance of 1.25. Table 10.2 shows all possible samples of size two. Using the sample data, we can demonstrate two methods of determining that $\sigma^2 = 1.25$.

The first method is derived from our knowledge that $\sigma_{\bar{X}} = \sigma/\sqrt{n}$, which, when rewritten, shows that $\sigma^2 = n\sigma_{\bar{X}}^2$.

Table 10.1
General presentation of ANOVA data

| Observations (i) | Samples (j) | | | | |
	1	2	3	. . .	n'
1	X_{11}	X_{12}	X_{13}	. . .	$X_{1n'}$
2	X_{21}	X_{22}	X_{23}	. . .	$X_{2n'}$
3	X_{31}	X_{32}	X_{33}	. . .	$X_{3n'}$
.	.	.	.		.
.	.	.	.		.
.	.	.	.		.
n	X_{n1}	X_{n2}	X_{n3}	. . .	$X_{nn'}$
Sample mean ($\bar{X}_j$)	$\bar{X}_1$	$\bar{X}_2$	$\bar{X}_3$	. . .	$\bar{X}_{n'}$

Table 10.2
All possible samples of size two

Sample number (j): $n' = 16$																
Observation (i)	1	2	3	4	5	6	7	8	9	10	11	12	13	14	15	16
1	1	1	1	1	2	2	2	2	3	3	3	3	4	4	4	4
2	1	2	3	4	1	2	3	4	1	2	3	4	1	2	3	4
Sample mean ($\overline{X}_j$)	1.0	1.5	2.0	2.5	1.5	2.0	2.5	3.0	2.0	2.5	3.0	3.5	2.5	3.0	3.5	4.0

Table 10.3
Determining the population variance from $\sigma_{\overline{X}}^2$

Sample number	Observed values	Sample mean ($\overline{X}_j$)	$\overline{X}_j - \mu$	$(\overline{X}_j - \mu)^2$
1	1, 1	1.0	−1.5	2.25
2	1, 2	1.5	−1.0	1.00
3	1, 3	2.0	−0.5	0.25
4	1, 4	2.5	0	0
5	2, 1	1.5	−1.0	1.00
6	2, 2	2.0	−0.5	0.25
7	2, 3	2.5	0	0
8	2, 4	3.0	0.5	0.25
9	3, 1	2.0	−0.5	0.25
10	3, 2	2.5	0	0
11	3, 3	3.0	0.5	0.25
12	3, 4	3.5	1.0	1.00
13	4, 1	2.5	0	0
14	4, 2	3.0	0.5	0.25
15	4, 3	3.5	1.0	1.00
16	4, 4	4.0	1.5	2.25
		40.0	0.0	10.00

$$\mu = \frac{\sum \overline{X}_j}{n'} = \frac{40}{16} = 2.5$$

$$\sigma_{\overline{X}} = \sqrt{\frac{\sum_{j=1}^{n'} (\overline{X}_j - \mu)^2}{n'}} = \sqrt{\frac{10}{16}} = \sqrt{0.625} = 0.791$$

$$\sigma_{\overline{X}}^2 = (0.791)^2 = 0.625$$

$$\sigma^2 = n\sigma_{\overline{X}}^2 = 2\sigma_{\overline{X}}^2 = 2(0.625) = 1.25$$

Thus, if we calculate $\sigma_{\overline{X}}^2$ and multiply it by 2, we should have $\sigma^2 = 1.25$. Table 10.3 shows how it is done. First we calculate μ, which is the average of the sample means. Then we use μ to calculate the measure of dispersion for the sampling distribution,

$\sigma_{\bar{X}}^2$. And to obtain the variance of the population, we multiply $\sigma_{\bar{X}}^2$ by 2 to obtain 1.25 as shown in table 10.3. The standard error of the mean, $\sigma_{\bar{X}}$, is calculated using equation (10.2):

$$\sigma_{\bar{X}} = \sqrt{\frac{\sum_{j=1}^{n'} (\bar{X}_j - \mu)^2}{n'}}. \qquad (10.2)$$

where

n' = number of samples.

The second method is demonstrated in table 10.4. In this method the concept that the population variance is the average of all sample variances of a given sample size, n, will be employed:

$$\sigma^2 = \frac{\sum_{j=1}^{n'} s_j^2}{n'} \qquad (10.3)$$

Table 10.4 shows the calculation which produces $\sigma^2 = 1.25$.

Table 10.4
Determination of the population variance using all the sample variances

Sample number	Observed values	Sample mean	Sample variance $= \dfrac{\sum_{i=1}^{n} (X_{ij} - \bar{X}_j)^2}{n - 1}$
1	1, 1	1.0	0
2	1, 2	1.5	0.50
3	1, 3	2.0	2.00
4	1, 4	2.5	4.50
5	2, 1	1.5	0.50
6	2, 2	2.0	0
7	2, 3	2.5	0.50
8	2, 4	3.0	2.00
9	3, 1	2.0	2.00
10	3, 2	2.5	0.50
11	3, 3	3.0	0
12	3, 4	3.5	0.50
13	4, 1	2.5	4.50
14	4, 2	3.0	2.00
15	4, 3	3.5	0.50
16	4, 4	4.0	0
			20.00

$$\sigma^2 = \frac{\sum_{j=1}^{n'} s_j^2}{n'} = \frac{20}{16} = 1.25$$

Thus it is easy to see that the ratio

$$\frac{n\sigma_{\bar{X}}^2}{\displaystyle\sum_{j=1}^{n'} s_j^2/n'} = \frac{\sigma^2}{\sigma^2} = 1.00.$$

Now what happens in analysis of variance is that the numerator, $n\sigma_{\bar{X}}^2$, does not contain the *actual* value of $\sigma_{\bar{X}}^2$ because we don't select all possible samples of a given size. What is used in analysis of variance is an *estimate of $\sigma_{\bar{X}}^2$ denoted $\hat{\sigma}_{\bar{X}}^2$*, which is obtained from the actual samples collected. Similarly, in the denominator, the sum, Σs^2, contains only a few of the possible s^2's rather than all sixteen as in our demonstration. Still, however, the F ratio, which is formed by the two estimates of the population variance using a few samples, should vary from 1 only by chance if there is just one population involved. Thus the F ratio for analysis of variance, when we do not use all possible samples of a given size, is given in equation (10.4):

$$F = \frac{\dfrac{n\left[\displaystyle\sum_{j=1}^{n'} (\bar{X}_j - \hat{\mu})^2\right]}{n' - 1}}{\dfrac{\displaystyle\sum_{j=1}^{n'}\left[\dfrac{\displaystyle\sum_{i=1}^{n}(X_{ij} - \bar{X}_j)^2}{n - 1}\right]}{n'}} \tag{10.4}$$

where

$$\hat{\mu} = \frac{\displaystyle\sum_{j=1}^{n'} \bar{X}_j}{n'},$$

$n' = $ number of samples.

In the numerator of this F ratio, the divisor is $n' - 1$ because $\hat{\mu}$ is an estimate of μ and not μ calculated from all possible sample means. This divisor, $n' - 1$, is the degrees of freedom for the numerator, and the degrees of freedom for the denominator is $n'(n - 1)$. Now let us examine example 10.2 to see how the two variance estimates are determined from sample data.

Example 10.2

▷ Suppose data on the daily productivity of three machine operators have been obtained. Operator A has produced in five days 100, 110, 92, 95, and 108 parts. Operator B has produced in five days, 94, 97, 90, 101, and 98 parts. Operator C has produced 98, 104, 113, 97, and 103 parts in five days. Assuming normal populations with equal variances, determine, at the 0.05 level of significance, if the three operators are not producing at the same average daily rate.

H_0: $\mu_A = \mu_B = \mu_C$,
H_A: at least two means are different,
   $\alpha = 0.05$.

Reject H_0 if $F_{test} > 3.88$:

df for numerator $= n' - 1$
$\qquad\qquad\qquad = 3 - 1 = 2;$
df for denominator $= n'(n - 1)$
$\qquad\qquad\qquad 3(5 - 1) = 12.$

Observation (i)	Operator (j):n' = 3		
	1	2	3
1	100	94	98
2	110	97	104
3	92	90	113
4	95	101	97
5	108	98	103
Sample mean ($\overline{X}_j$)	101	96	103

Sample mean calculations:

$$\overline{X}_j = \frac{\sum\limits_{i=1}^{n} X_{ij}}{n},$$

$$\overline{X}_1 = \frac{100 + 110 + 92 + 95 + 108}{5}$$

$$\overline{X}_1 = \frac{505}{5} = 101,$$

$$\overline{X}_2 = \frac{94 + 97 + 90 + 101 + 98}{5}$$

$$\overline{X}_2 = \frac{480}{5} = 96,$$

$$\overline{X}_3 = \frac{98 + 104 + 113 + 97 + 103}{5}$$

$$\overline{X}_3 = \frac{515}{5} = 103.$$

Sample standard deviation calculations:

$$s_j^2 = \frac{\sum_{i=1}^{n} (X_{ij} - \overline{X}_j)^2}{n - 1},$$

$$s_1^2 = [(100 - 101)^2 + (110 - 101)^2 + (92 - 101)^2 + (95 - 101)^2 + (108 - 101)^2]/5 - 1$$

$$= \frac{(-1)^2 + 9^2 + (-9)^2 + (-6)^2 + 7^2}{4}$$

$$= \frac{1 + 81 + 81 + 36 + 49}{4}$$

$$s_1^2 = \frac{284}{4} = 62,$$

$$s_2^2 = [(94 - 96)^2 + (97 - 96)^2 + (90 - 96)^2 + (101 - 96)^2 + (98 - 96)^2]/5 - 1$$

$$= \frac{(-2)^2 + 1^2 + (-6)^2 + 5^2 + 2^2}{4}$$

$$= \frac{4 + 1 + 36 + 25 + 4}{4}$$

$$s_2^2 = \frac{70}{4} = 17.5,$$

$$s_3^2 = [(98 - 103)^2 + (104 - 103)^2 + (113 - 103)^2 + (97 - 103)^2 + (103 - 103)^2]/5 - 1$$

$$= \frac{(-5)^2 + 1^2 + 10^2 + (-6)^2 + 0}{4}$$

$$= \frac{25 + 1 + 100 + 36 + 0}{4}$$

$$s_3^2 = \frac{162}{4} = 40.5;$$

$$\hat{\mu} = \frac{\sum_{j=1}^{n'} \overline{X}_j}{n'}$$

$$= \frac{101 + 96 + 103}{3}$$

$$\hat{\mu} = \frac{300}{3} = 100;$$

$$F_{\text{test}} = \cfrac{\cfrac{n\left[\displaystyle\sum_{j=1}^{n'}(\overline{X}_j - \hat{\mu})^2\right]}{n' - 1}}{\cfrac{\displaystyle\sum_{j=1}^{n'}\left[\displaystyle\sum_{i=1}^{n}(X_{ij} - \overline{X}_j)^2\right]}{n - 1}}{n'}$$

$$= \cfrac{5\left[\cfrac{(101 - 100)^2 + (96 - 100)^2 + (103 - 100)^2}{3 - 1}\right]}{\cfrac{62 + 17.5 + 40.5}{3}}$$

$$= \cfrac{5\left[\cfrac{1 + 16 + 9}{2}\right]}{\cfrac{62 + 17.5 + 40.5}{3}}$$

$$= \cfrac{65}{40}$$

$$F_{\text{test}} = 1.625.$$

Accept (do not reject) H_0; conclude that the operators are equally as good. ◁

The preceding discussion and example are designed to provide insight into how analysis of variance works. The computational effort associated with equation (10.4), however, is laborious and cumbersome for routine use. Fortunately, there is a much more satisfactory solution approach.

Reexamining equation (10.4), notice that both the numerator and denominator are estimates of the population variance, and these estimators are ratios of sum of squares divided by the appropriate degrees of freedom. Thus equation (10.4) can be described as the ratio of the numerator sum of squares divided by the numerator degrees of freedom and the denominator sum of squares divided by the denominator degrees of freedom.

Determination of the appropriate sum of squares can be accomplished by utilizing the fact that there is a total sum of squares (total sample variation of all items obtained from all samples about the overall mean) which can be calculated as follows:

Sum of squares for total $(SST) = \displaystyle\sum_{i=1}^{n}\sum_{j=1}^{n'}(X_{ij} - \hat{\mu})^2,$ \hfill (10.5) and the working equation is

$$\text{Sum of squares for total } (SST) = \sum_{i=1}^{n} \sum_{j=1}^{n'} X_{ij}^2 - \frac{\left(\sum_{i=1}^{n} \sum_{j=1}^{n'} X_{ij} \right)^2}{nn'}.$$

(10.6)

This total sum of squares can be partitioned into two components called the *between-sample* sum of squares and the *within-sample* sum of squares:

Sum of squares total = sum of squares between + sum of squares within. (10.7)

The between-sample sum of squares refers to the sum of squares obtained when the individual sample means differ from the overall mean (μ). Looking at equation (10.4), the numerator has the term $\sum_{j=1}^{n'} (\overline{X}_j - \hat{\mu})^2$, which is the between-sample sum of squares. The sum of squares between can be calculated using the following equation:

$$\text{Sum of squares between } SSB = \frac{\sum_{j=1}^{n'} \left(\sum_{i=1}^{n} X_{ij} \right)^2}{n} - \frac{\left(\sum_{i=1}^{n} \sum_{j=1}^{n'} X_{ij} \right)^2}{nn'}.$$

(10.8)

The within-sample sum of squares refers to the sum of squares obtained from the individual samples. That is, when each sample variance (s_j^2) is determined, the numerator, $\sum_{i=1}^{n} (X_{ij} - \overline{X}_j)^2$, of each sample is a sum of squares employing a sample mean. If, as in example 10.2, there are three samples, then there are three sum of squares, one using each sample mean. If we add these sum of squares (those obtained from each sample) together, we have determined the within sample sum of squares. Equation (10.9) can be used to calculate the within sample sum of squares:

$$\text{Sum of squares within } (SSW) = \sum_{i=1}^{n} \sum_{j=1}^{n'} X_{ij}^2 - \frac{\sum_{j=1}^{n'} \left(\sum_{i=1}^{n} X_{ij} \right)^2}{n}.$$

(10.9)

Once the between-sample sum of squares and within-sample sum of squares have been determined, the F ratio becomes

$$F_{\text{test}} = \frac{SSB/n' - 1}{SSW/n'(n - 1)}.$$

(10.10)

Table 10.5
ANOVA summary table format

Source of variation	Sum of square	Degrees of freedom	Mean squares	F
Between	SSB	$n' - 1$	$SSB/n' - 1$	$\dfrac{SSB/n' - 1}{SSW/n'(n - 1)}$
Within	SSW	$n'(n - 1)$	$SSW/n'(n - 1)$	
Total	SST	$n'n - 1$		

When the calculations follow the sum of squares approach, the results are often reported in an ANOVA summary table such as table 10.5.

In table 10.5 you should note that the mean squares are variances. That is, the ratio of the sums of squares divided by degrees of freedom is a variance. In the ANOVA summary table the variances are called mean squares, and the ratio of the mean squares forms the F-test. To demonstrate how to use the sum of squares approach, let us rework example 10.2:

Example 10.3 ▷

Observation (i)	Operator (j):$n' = 3$		
	1	2	3
1	100	94	98
2	110	97	104
3	92	90	113
4	95	101	97
5	108	98	103
Sample mean ($\overline{X}_j$)	101	96	103

Sum of squares for total

$$SST = \sum_{i=1}^{n} \sum_{j=1}^{n'} X_{ij}^2 - \frac{\left(\sum_{i=1}^{n} \sum_{j=1}^{n'} X_{ij} \right)^2}{nn'}$$

$$\begin{aligned}
= &(100^2 + 110^2 + 92^2 + 95^2 + 108^2 + 94^2 \\
&+ 97^2 + 90^2 + 101^2 + 98^2 + 98^2 \\
&+ 104^2 + 113^2 + 97^2 + 103^2) \\
&- (100 + 110 + 92 + 95 + 108 \\
&+ 94 + 97 + 90 + 101 + 98 \\
&+ 98 + 104 + 113 + 97 + 103)^2/(5)(3)
\end{aligned}$$

$$= (10,000 + 12,100 + 8,464 + 9,025 + 11,664$$
$$+ 8,836 + 9,409 + 8,100 + 10,201 + 9,604$$
$$+ 9,604 + 10,816 + 12,769 + 9,409 + 10,609)$$
$$- \frac{(1,500)^2}{15}$$

$$= 150,610 - \frac{2,250,000}{15}$$

$$= 150,610 - 150,000$$

$$SST = 610.$$

Sum of squares between

$$SSB = \frac{\sum\limits_{j=1}^{n'} \left(\sum\limits_{i=1}^{n} X_{ij} \right)^2}{n} - \frac{\left(\sum\limits_{i=1}^{n} \sum\limits_{j=1}^{n'} X_{ij} \right)^2}{nn'}$$

$$= \frac{(100 + 110 + 92 + 95 + 108)^2}{5}$$

$$+ \frac{(94 + 97 + 90 + 101 + 98)^2}{5}$$

$$+ \frac{(98 + 104 + 113 + 97 + 103)^2}{5} - 150,000$$

$$= \frac{(505)^2 + (480)^2 + (515)^2}{5} - 150,000$$

$$= 150,130 - 150,000$$

$$SSB = 130.$$

Sum of squares within

$$SSW = \sum\limits_{i=1}^{n} \sum\limits_{j=1}^{n'} X_{ij}^2 - \frac{\sum\limits_{j=1}^{n'} \left(\sum\limits_{i=1}^{n} X_{ij} \right)^2}{n}$$

$$= 150,610 - 150,130$$

$$SSW = 480.$$

Table 10.6
ANOVA summary table

Source of variation	Sum of squares	Degrees of freedom	Mean squares	F
Between	130	2	65	1.625
Within	480	12	40	
Total	610	14		

General Comments

The use of the F ratio test is predicated on sampling from *normal populations*. If the samples are large, the central limit theorem can be used as the basis for normality.

The preceding section was a brief introduction to analysis of variance for samples of equal size. An example of what is called *one-way analysis* of variance was provided. More advanced texts should be consulted for expanded discussions or applications of analysis of variance.

Summary of Equations

10.1 Hypothesis test for two population variances:

$$F_{\text{test}} = \frac{s_{\text{larger}}^2}{s_{\text{smaller}}^2}$$

10.2 Standard error of the mean—using all possible $\overline{X}$'s:

$$\sigma_{\overline{X}} = \sqrt{\frac{\sum_{j=1}^{n'} (\overline{X}_j - \mu)^2}{n'}}$$

10.3 Population variance—using all possible sample variances:

$$\sigma^2 = \frac{\sum_{j=1}^{n'} s_j^2}{n'}$$

10.4 F ratio for analysis of variance:

$$F = \frac{n \left[\dfrac{\sum_{j=1}^{n'} (\overline{X}_j - \hat{\mu})^2}{n' - 1} \right]}{\dfrac{\sum_{j=1}^{n'} \left[\dfrac{\sum_{i=1}^{n} (X_{ij} - \overline{X}_j)^2}{n - 1} \right]}{n'}}$$

10.5 Total variation from overall mean:

$$\text{Sum of squares total } (SST) = \sum_{i=1}^{n} \sum_{j=1}^{n'} (X_{ij} - \mu)^2$$

10.6 Total sum of squares—working equation:

Sum of squares for total

$$SST = \sum_{i=1}^{n} \sum_{j=1}^{n'} X_{ij}^2 - \frac{\left(\sum_{i=1}^{n} \sum_{j=1}^{n'} X_{ij} \right)^2}{nn'}$$

10.7 Partioning of total sum of squares:

Sum of squares total (SST) = sum of squares between (SSB) + sum of squares within (SSW)

10.8 Between-sample sum of squares:

Sum of squares between

$$SSB = \frac{\sum_{j=1}^{n'}\left(\sum_{i=1}^{n} X_{ij}\right)^2}{n} - \frac{\left(\sum_{i=1}^{n}\sum_{j=1}^{n'} X_{ij}\right)^2}{nn'}$$

10.9 Within-sample sum of squares:

Sum of squares within

$$SSW = \sum_{i=1}^{n}\sum_{j=1}^{n'} X_{ij}^2 - \frac{\sum_{j=1}^{n'}\left(\sum_{i=1}^{n} X_{ij}\right)^2}{n}$$

10.10 F_{test}—using sum of squares:

$$F_{test} = \frac{SSB/n' - 1}{SSW/n'(n - 1)}$$

Review Question

1. Discuss analysis of variance.

Problems

1. Two random samples are selected. The first sample of size 76 yields a mean of 135 and a standard deviation of 14. The second sample of size 61 yields a mean of 110 and a standard deviation of 10. At the 0.05 level of significance, determine if it is reasonable to assume that the variances are equal.

2. Two organizations are meeting at the same convention hotel. A sample of 35 members of The Cranes revealed an average daily expenditure on food of $12.00 with a standard deviation of $1.00; and a sample of 25 members of The Penguins revealed an average daily expenditure on food of $14.00 with a standard deviation of $2.00. At the 0.01 level of significance, determine if it is reasonable to assume that the variances are equal.

3. Two random samples are selected from two normal populations. If the first sample of size 31 yielded a standard deviation of 6 and the second sample of size 30 yielded a standard deviation of 4, at the 0.05 level of significance are the variances different?

4. The following data were obtained from the members of three organizations at their last annual convention. Determine if the average daily food expenditures are different. Set $\alpha = 0.01$.

Daily expenditure of five members of each organization

The Cranes	The Penguins	The Robins
$25	$17	$30
18	18	28
22	16	25
21	20	21
24	14	26

5. A milk company has three routes. The times (in hours) to complete the routes have been recorded for the last five days. Use the data to determine if the average route times are equal. Set alpha = 0.05.

Route 1	Route 2	Route 3
8.5	8.2	8.4
8	7.8	8.3
7.9	7.7	8.5
8.6	7.6	8.7
7.8	8.1	9.0

6. A distributor of electronic games is considering three types of batteries for the games. Each manufacturer has submitted a sample of batteries to be tested. The distributor collected the following data on the lifetime (measured in hours) of the batteries.

Manufacturer I	II	III
60	65	48
62	70	52
70	68	56
58	72	50
64	74	54

Use the data to test if the means of battery lifetimes are equal. Set alpha = 0.05.

7. A manufacturer of portable calculators is considering three batteries to use in the new calculator. The following data on the life between charges (in hours) have been collected on each battery:

Battery		
I	II	III
12	17	21
19	25	15
19	13	16
16	31	29
20	36	32
24	15	14
30	26	18
15	35	26

At the 0.05 level of significance, is there a difference in the average life between charges?

8. A paper manufacturer wishes to determine if there is any significant difference in the moisture prevention of four methods of storage. The results of testing the four methods of storage are given as follows:

Methods			
A	B	C	D
5.2	4.3	6.0	5.6
5.3	3.7	5.0	8.0
6.5	3.8	5.6	5.4
5.4	4.6	4.9	6.5
7.6	4.1	4.5	8.5

At the 0.01 level of significance, is there a difference in average moisture content?

9. A retailer has classified the charge customers into three categories based on the frequency of purchases. Recently, the credit manager has prepared data on the number days required for payment (measured from the first of the month) for each class of customer. Use the data that follows to determine if the three categories have different average payment times. Set alpha = 0.01.

Customer category		
1	2	3
8	5	9
6	6	10
7	6	8
7	5	10
8	4	9

10. The following are hypothetical data for three random samples:

I	II	III
25	27	30
32	36	34
27	28	31
33	32	32
18	22	28

Under the assumption of normal populations and equal population variances, test the null hypothesis $\mu_I = \mu_{II} = \mu_{III}$. Set $\alpha = 0.01$.

Case Study

Mr. Dwayne Thomason, Director of Data Processing at the Vance Corporation, recently read an article that discussed possible variables affecting productivity. The article mentioned that the temperature of the work place might have an effect upon the output of the workers. Mr. Thomason decided to test to find out whether or not room temperature had a significant effect upon the output of his data entry operators.

In designing the test procedure, Mr. Thomason knew that there were several "outside" factors affecting output that needed to be considered and possibly avoided during the testing process. Thus he decided to not run the tests on Monday so as to avoid the "Monday Blues" and to not run the tests on Friday to avoid the "TGIF Syndrome." Also he decided to not monitor output during breaktimes when one or more of the operators might be away from their stations. Hoping to measure "normal" output, Mr. Thomason decided to be very discreet about monitoring output.

The data in the table show the output of the four operators in characters per hour for selected one-hour intervals. At the 0.05 level of significance, does there appear to be a significant difference in output at the three temperature levels?

	Temperature	Characters/hour for the four operators (in 000)			
		8:30–9:30	10:00–11:00	1:15–2:15	3:00–4:00
Tuesday	72°	580	550	550	490
Wednesday	68°	450	575	560	475
Thursday	70°	480	515	520	460

Contributed by Vicky L. Crittenden, Florida State University/Tallahassee Community College and William F. Crittenden, Florida State University.

11

Nonparametric Statistical Tests

In earlier chapters we presented rather extensive discussions on hypothesis testing. In applying those testing procedures, you may recall that we either assumed some *population parameter was known* (or could be estimated) or that *we knew something about the form of the population distribution* from which the sample was drawn.

Nonparametric and Distribution-Free Techniques

As you might suspect, situations arise where we can make *no statement* about the value of a parameter, but some statistical test of a hypothesis would appear to be useful. In such cases we would use a *nonparametric test*. In other instances we may want to test a hypothesis, but we may not know or would not want to assume the precise form of the population distribution. To apply a statistical test in this situation, we would use some *distribution-free technique*.

Of the two terms, nonparametric and distribution-free, the term, nonparametric, seems to be the most widely used even though the term, distribution-free, may be more descriptive. That is, nonparametric statistics is not meant to imply that no interest exists regarding the population parameters. On the contrary, it may still be desirable to learn more about a population parameter, but no classical parametric test (Z-test, t-test, F-test, and so on) is available because there is no prior knowledge about the population distribution (normal, binomial, and so on).

Given such situations it is still possible to test hypotheses using distribution-free methods that do away with the need to specify the population distribution. Consequently the hypotheses are tested on the premise that knowledge of the sampling distribution does not depend on prior knowledge of the population distribution. Thus let us define nonparametric statistics as the branch of statistics where hypotheses testing is performed without making specific assumptions about the population distribution and without having to specify certain parameter values.

Probably the most widely used nonparametric test is the *chi square (χ^2) test*. While there are several applications of the chi square test, we will examine only three of them. But, first, let us consider the nature of the *chi square distribution*.

The chi square distribution is a continuous probability distribution. In fact there is not simply one chi square distribution. Like the Student *t* distribution, there is a different chi square (χ^2) distribution for every number of *degrees of freedom*. The distributions shown in figure 11.1 are representative of three different numbers of degrees of freedom. Notice that with fewer degrees of freedom the distributions are more positively skewed. With a greater number of degrees of freedom the distribution becomes approximately normal.

The chi square distribution has several properties that make it easy to use:

1. The mean of the chi square distribution is equal to the number of degrees of freedom (*df*).

2. The variance is two times the number of degrees of freedom, 2(*df*).

3. The sum of two or more independent chi square variables yields a chi square variable.

4. The square of a standard normal variable is a chi square variable.

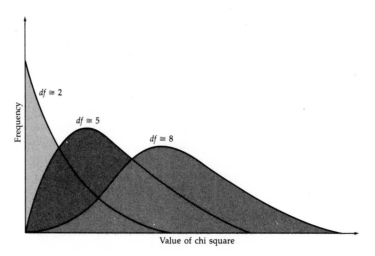

Figure 11.1   Chi square distributions

Appendix K presents values from the chi square distribution for selected values of alpha and degrees of freedom. This appendix will be needed when the chi square distribution is used in hypothesis testing.

Degrees of Freedom

When you learned how to compute the sample standard deviation that can be used to estimate the population standard deviation, you encountered the term *degrees of freedom*. At that point in chapter 2 it was explained that one degree of freedom is lost when you compute s as an estimate of σ. Similarly, when you are using chi square distributions to test hypotheses, degrees of freedom are lost. In other words, $df < n$. The value of df is determined in different ways depending upon the way in which the chi square distribution is being used. Specifically, if a chi square technique is being applied to test the shape (goodness-of-fit) of a distribution:

$$df = g - 1 - m,$$

where

g = the number of groups, or classes, in the frequency distribution,

m = the number of population parameters that must be estimated from sample statistics in order to test the hypothesis.

If chi square is utilized to test for homogeneity or contingency (these are discussed in detail in the following pages), then

$$df = (r - 1)(c - 1),$$

where

r = the number of rows in the table,

c = the number of columns in the table.

Recalling that there is a different χ^2 distribution for each number of degrees of freedom, it is important to determine the correct value of df.

Goodness-of-Fit Test

Chi square is frequently used to test *goodness-of-fit*. That is, it is used to ascertain whether the distribution of values in a sample supports a hypothesized population distribution. In order to apply the chi square distribution in this manner, the critical value of chi square is expressed as[1]

---

1. The expression $(O - E)/\sqrt{E}$ is normally distributed. Utilizing properties 3 and 4 of the chi square distribution, $\sum_{i=1}^{n} [(O_i - E_i)^2/E_i]$ is chi square distributed.

$$\chi^2_{\text{test}} = \sum_{i=1}^{g} \frac{(O_i - E_i)^2}{E_i},$$

(11.1)

where

O_i = observed frequency of the variable (sample),
E_i = expected frequency (based on the hypothesized population distribution).

The goodness-of-fit test is performed by establishing a null hypothesis (such as the data representing a uniform distribution or a normal distribution or some other distribution of interest). Then the expected frequencies (E) are determined based on the probability distribution stated in the null hypothesis. These calculated expected frequencies (E) are compared with the observed frequencies (O). If the expected and observed frequencies appear to agree, then the null hypothesis is accepted. If the observed and expected frequencies do not seem to agree, the null hypothesis is rejected. Examples 11.1 and 11.2 demonstrate the goodness-of-fit test.

Example 11.1

▷ A firm markets its product using door-to-door sales. The firm employs two hundred sales people nationwide. Recently, the sales manager has become interested in the number of sales call made by each of the employees. The sales manager has reasoned that if all of the employees are working hard that they should make the same number of calls during a set period of time. In order to investigate this hypothesis, the sales manager uses a sample of five employees to provide the observed frequencies shown in table 11.1. At the 0.01 level of significance, is the manager's idea supported? The test procedure is

H_0: sales calls are uniformly distributed,
H_A: sales calls are not uniformly distributed,
$\alpha = 0.01$

$df = g - 1 - m = 5 - 1 - 0 = 4$.

Reject H_0 if $\chi^2_{\text{test}} > 13.2767$ (obtain 13.2767 from appendix K).

The appropriate test statistic is

$$\chi^2_{\text{test}} = \sum_{i=1}^{g} \frac{(O_i - E_i)^2}{E_i}.$$

The expected frequencies (E) are determined based on the null hypothesis. In this example, the null hypothesis of a uniform

Table 11.1
The observed frequencies

Employee	Number of calls
I	31
II	62
III	59
IV	40
V	58
Total	250

distribution implies that the total (250) should be distributed equally to each employee. Thus observed = 250/5 = 50, and

$$\chi^2_{test} = \sum_{i=1}^{g} \frac{(O_i - E_i)^2}{E_i}$$

$$= \frac{(31 - 50)^2}{50} + \frac{(62 - 50)^2}{50} + \frac{(59 - 50)^2}{50}$$

$$+ \frac{(40 - 50)^2}{50} + \frac{(58 - 50)^2}{50}$$

$$= 7.22 + 2.88 + 1.62 + 2 + 1.28$$

$$\chi^2_{test} = 15.$$

Reject H_0; conclude that the employees are not making the same number of calls. ◁

In the preceding example, the number of degrees of freedom was determined using the expression, $df = g - 1 - m$, discussed earlier. Notice that $m = 0$ because, when testing for a uniform distribution, no population parameters are estimated. In the next example, however, parameters will have to be estimated as the goodness-of-fit test is applied to the per capita income data given in table 11.2.

Table 11.2
Per capita income in a sample of 103 counties

Class interval	Frequency
$2,500 and under $3,100	12
3,100 and under 3,700	15
3,700 and under 4,300	31
4,300 and under 4,900	20
4,900 and under 5,500	15
5,500 and under 6,100	10
	103

Example 11.2

▷ The data in table 11.2 are to be tested to determine if they are normally distributed. Set alpha = 0.05:

H_0: per capita incomes are normally distributed,
H_A: per capita incomes are not normally distributed,
$\alpha = 0.05$,
$df = g - 1 - m$, where $m = 2$ (use of the normal distribution requires the estimation of μ using $\overline{X}$ and the estimation of σ using s); then

$df = 6 - 1 - 2 = 3$.

Reject H_0 if $\chi^2 > 7.81473$ (obtain 7.81473 from appendix K).

The expected frequencies (E) are calculated by determining the probability of an observation being in an interval and multiplying the probability times the total number of observations. Table 11.3 presents the expected frequencies.

Given that $\overline{X} = \$4{,}239$ and that $s = \$875$, an illustration of calculating the expected frequency for the third class interval is presented (an example of grouped data computations of $\overline{X}$ and s is given at the end of chapter 2):

$$Z_1 = \frac{3{,}700 - 4{,}239}{875} = 0.62,$$

Area 1 = 0.2324.

$$Z_2 = \frac{4{,}300 - 4{,}239}{875} = \frac{61}{875} = 0.07,$$

Area 2 = 0.0279.

Thus

$$P(3{,}700 < X < 4{,}300) = 0.2324 + 0.0279$$
$$= 0.2603,$$

and

Expected frequency $(E) = (0.2603)(103)$
$$= 26.81,$$

$$\chi^2_{test} = \frac{(12 - 9.97)^2}{9.97} + \frac{(15 - 17.59)^2}{17.59}$$
$$+ \frac{(31 - 26.81)^2}{26.81} + \frac{(20 - 25.56)^2}{25.56}$$
$$+ \frac{(15 - 15.32)^2}{15.32} + \frac{(10 - 7.71)^2}{7.71}$$
$$= 0.413 + 0.381 + 0.655 + 1.209 + 0.067 + 0.680$$
$$\chi^2_{test} = 3.405.$$

Table 11.3
Expected frequencies

Class interval	Probability	Expected frequency
Under $3,100	0.0968	9.97
3,100 and under 3,700	0.1708	17.59
3,700 and under 4,300	0.2603	26.81
4,300 and under 4,900	0.2485	25.56
4,900 and under 5,500	0.1487	15.32
Above $5,500	0.0749	7.71

Accept (do not reject) H_0; conclude that the distribution is normal. ◁

Test of Independence

At times, data obtained from a random sample may be classified by two or more attributes. When we have two or more attributes so that we cross classify the data, the resulting presentation is called a *contingency table*. Consider the following generalized presentation showing two attributes (R and C) having m categories of R and n categories of C. In each RC cross-classified category, there are O_{ij} observed units obtained from the sample. Thus O_{11} is the number of observations classified in the R_1C_1 category.

		Columns							Row Total
		C_1	C_2	C_3	.	. .	.	C_n	
	R_1	O_{11}	O_{12}	O_{13}	.	. .	.	O_{1n}	$O_1.$
	R_2	O_{21}	O_{22}	O_{23}	.	. .	.	O_{2n}	$O_2.$
Rows	R_3	O_{31}	O_{32}	O_{33}	.	. .	.	O_{3n}	$O_3.$
	.	.	.	.	.	. .	.	.	.
	.	.	.	.	.	. .	.	.	.
	.	.	.	.	.	. .	.	.	.
	R_m	O_{m1}	O_{m2}	O_{m3}	.	. .	.	O_{mn}	$O_m.$
Column Total		$O._1$	$O._2$	$O._3$	.	. .	.	$O._n$	$O..$

Figure 11.2   A contingency table

Such cross classification is usually done in order to investigate the relationship between the row attribute and the column attribute. An investigation of this possible relationship is a statistical test for independence. The general procedure is to determine an expected frequency number (E_{ij}) for each row-column classification and to compare the expected frequencies with the observed frequencies (each O_{ij}). If the expected frequencies seem compatible with the observed frequencies, the independence of the two attributes is presumed. If, however, the expected and observed frequencies seem incompatible, then it is presumed that a dependency relationship between the row and column attributes exists.

In order to compare the observed and expected frequencies, it is necessary to calculate the expected frequencies (E_{ij}). Expected frequencies are calculated under the assumption of independence between the row and column attributes. Under this assumption, the probability of any observation being located in any row or column is equal to the probability of being in that particular row or column. Thus, using figure 11.2, the probability of being in row i could be written:

$$P(R_i) = \frac{\text{row } i \text{ total}}{\text{overall total}} = \frac{O_{i.}}{O_{..}}. \tag{11.2}$$

Likewise, the probability of any observation being located in column j could be written;

$$P(C_j) = \frac{\text{column } j \text{ total}}{\text{overall total}} = \frac{O_{.j}}{O_{..}}. \tag{11.3}$$

Continuing with this line of reasoning, then the probability of any observation being located in row i and column j should be equal to the joint probability: $P(R_i)P(C_j)$. Thus

$$P(R_iC_j) = \frac{(O_{i.})(O_{.j})}{(O_{..})(O_{..})}. \tag{11.4}$$

Then, if we want to estimate the number of observations (E_{ij}) for any location, the following expression would apply:

$$E_{ij} = [P(R_iC_j)]n \tag{11.5}$$

Rewriting in terms of the observed frequencies, equation (11.5) becomes

$$E_{ij} = \left(\frac{O_{i.}O_{.j}}{O_{..}O_{..}}\right)(O_{..}) = \frac{(O_{i.})(O_{.j})}{O_{..}}. \tag{11.6}$$

Expressed as a word equation, equation (11.6) is

$$E_{ij} = \frac{(\text{row total } i)(\text{column total } j)}{\text{overall total } n}.$$

These expected frequencies (E_{ij}) are then used in equation (11.7) to perform a test of statistical independence. Consider Example 11.3 for further explanation.

$$\chi^2_{\text{test}} = \sum_{i=1}^{m} \sum_{j=1}^{n} \frac{(O_{ij} - E_{ij})^2}{E_{ij}}. \tag{11.7}$$

BASIC STATISTICS

Example 11.3

▷ A survey was conducted by a large mail order supplier. In a recent mailout of "specials" to a current list of customers, a random sample of 1,200 customers received three types of response envelopes. Two of the envelopes were conventional, one with prepaid postage, and one without postage. The third envelope was a fold-up type where the order form actually folded into an envelope with postage required. The responses were cross classified as shown in table 11.4.

The purpose of the investigation is to test the relationship between response and type of envelope. The expected frequencies (E_{ij}) are calculated using equation (11.6). The results are shown in table 11.5.

The hypothesis test would be performed in the following manner, using alpha = 0.01:

H_0: response is independent of type of envelope,
H_A: response is not independent of type of envelope,
$\alpha = 0.01$,
$df = (r - 1)(c - 1) = (2 - 1)(3 - 1) = 2$.
Reject H_0 if $\chi^2_{test} > 9.21034$:

$$\chi^2_{test} = \sum_{i=1}^{m} \sum_{j=1}^{n} \frac{(O_{ij} - E_{ij})^2}{E_{ij}}$$

$$= \frac{(300 - 260.42)^2}{260.42} + \frac{(150 - 208.33)^2}{208.33}$$

$$+ \frac{(175 - 156.25)^2}{156.25} + \frac{(200 - 239.58)^2}{239.58}$$

$$+ \frac{(250 - 191.67)^2}{191.67} + \frac{(125 - 143.75)^2}{143.75}$$

$$= 6.016 + 16.332 + 2.25 + 6.539 + 17.751 + 2.446$$

$$\chi^2_{test} = 51.334.$$

Reject the null hypothesis; conclude that response rate and type of envelope are not independent. There is a relationship between response and envelope enclosed.

Table 11.4
Results of the mail survey

	Prepaid postage	No postage	Fold-up return	Total
Response	300	150	175	625
No response	200	250	125	575
Total	500	400	300	1,200

Table 11.5
Computation of expected frequencies

Prepaid postage		No postage		Fold-up return	
Response	$\dfrac{(625)(500)}{1,200} = \dfrac{260.42}{(E_{11})}$		$\dfrac{(625)(400)}{1,200} = \dfrac{208.33}{(E_{12})}$		$\dfrac{(625)(300)}{1,200} = \dfrac{156.25}{(E_{13})}$
No response	$\dfrac{(575)(500)}{1,200} = \dfrac{239.58}{(E_{21})}$		$\dfrac{(575)(400)}{1,200} = \dfrac{191.67}{(E_{22})}$		$\dfrac{(575)(300)}{1,200} = \dfrac{143.75}{(E_{23})}$

Before you continue, an additional issue needs to be considered. First, notice how the number of degrees of freedom was determined. As was pointed out earlier in this chapter, $df = (r - 1)(c - 1)$ when testing for contingency. Theoretically, when $df = 1$, a correction factor should be applied when computing χ^2_{test} so that the equation (11.7) is changed to the following:

$$\chi^2_{\text{test}} = \sum_{i=1}^{m} \sum_{j=1}^{n} \left[\frac{(|O_{ij} - E_{ij}| - 1/2)^2}{E_{ij}} \right] \qquad (11.8)$$

In using equation (11.8), if $(|O_{ij} - E_{ij}| - 1/2) < 0$, treat it as if it is 0.

Test of Homogeneity

Whereas the test for independence assumes one random sample, the test for homogeneity is a test to determine if two or more independent random samples were drawn from the same population. The mathematical approach (calculation of E_{ij}) is the same as for the contingency table. Thus the chi square test used in example 11.2 is appropriate for the homogeneity test. However, the reasoning by which the E_{ij}'s are calculated is quite different.

Referring to figure 11.3, in the homogeneity situation both the rows and columns are not considered *active* classifications. That is, in the test for independence, the intent is to investigate a rela-

		Columns						Row Total
		C_1	C_2	C_3	. . . .	C_n		
	R_1	O_{11}	O_{12}	O_{13}	. . . .	O_{1n}		$O_{1\cdot}$
	R_2	O_{21}	O_{22}	O_{23}	. . . .	O_{2n}		$O_{2\cdot}$
	R_3	O_{31}	O_{32}	O_{33}	. . . .	O_{3n}		$O_{3\cdot}$
Rows	.	.	.	.	. . . .	.		.
	.				. . . .			
	.				. . . .			
	R_m	O_{m1}	O_{m2}	O_{m3}	. . . .	O_{mn}		$O_{m\cdot}$
Column Total		$O_{\cdot 1} = n_1$	$O_{\cdot 2} = n_2$	$O_{\cdot 3} = n_3$	. . . .	$O_{\cdot n} = n_n$		$O_{\cdot\cdot}$

Figure 11.3 A cross-classification table

tionship between the row and column attributes; thus both influence location of an observation. In the homogeneity layout, however, only one attribute (columns in our discussion) is to influence location. The other attribute (rows in our discussion) represents a classification that contains categories common to the active columns.

Notice also in figure 11.3 that the column totals are fixed. That is, the column totals are the sample sizes for each sample taken. This feature of the column totals being fixed differs from the contingency table situation depicted in figure 11.2. Referring back to figure 11.2, the only fixed total is the overall total $(O_{..})$ which is the sample size (remember contingency tables use only one random sample). Thus the homogeneity table has column totals and the overall total fixed, whereas the contingency table only has the overall total fixed.

Returning our attention to figure 11.3, we can estimate the expected frequencies (E_{ij}) under the null hypothesis of homogeneity which assumes that the samples came from one population. To estimate the expected frequencies, we must examine the probabilities associated with being in any row. Thus the probability of an observation being located in row one is

$$P(R_1) = \frac{\text{row one total}}{\text{overall total}} = \frac{O_{1.}}{O_{..}}.$$

And more generally, the probability of an observation being located in a particular row can be determined using equation (11.9):

$$P(R_i) = \frac{\text{row } i \text{ total}}{\text{overall total}} = \frac{O_{i.}}{O_{..}}. \tag{11.9}$$

You will probably recall that this approach to calculating row probabilities is identical to the approach used in contingency tables discussed earlier. The difference between the homogeneity table computation of expected frequencies and the contingency table computation of expected frequencies is that the homogeneity computation does not involve the joint probability approach shown in equation (11.4). That is, the homogeneity approach focuses only on the row probabilities since the column probabilities are fixed because the sample sizes are fixed. Thus the computation of the E_{ij} for the homogeneity table is more direct, as shown in equations (11.10) and (11.11). You should note that equation (11.11) is identical to equation (11.6) even though the reasoning is quite different.

$$E_{ij} = P(R_i)n_j,$$ (11.10)

or

$$E_{ij} = \frac{O_{i.}O_{.j}}{O_{..}}.$$ (11.11)

Examine example 11.4 to see how the homogeneity test is performed.

Example 11.4

▷ A major insurance company is planning a nationwide promotional campaign via TV. Of interest to the company is whether the same campaign should be used nationwide, or whether the campaigns should be "regionalized." As a method to investigate the need for regionalization, the company selected samples from each region and cross-classified the responses according to current type-of-insurance interest and geographic region as shown in table 11.6. The investigation used alpha = 0.05.

H_0: no differences in regional preferences,
H_A: differences in regional preferences,
$\alpha = 0.05$,
$df = (r - 1)(c - 1) = (3 - 1)(4 - 1) = 6$.

Reject H_0 if $\chi^2_{test} > 12.5916$ (obtain 12.5916 from appendix K).

Table 11.6
Data for homogeneity test

	North	South	East	West	
Whole life	60	40	45	55	200
Level term	35	30	40	45	150
Decreasing term	25	30	20	25	100
	120	100	105	125	450

Table 11.7
Calculation of the expected frequencies for homogeneity test

	North	South	East	West
Whole life	$\frac{(200)(120)}{450} = 53.33$	$\frac{(200)(100)}{450} = 44.44$	$\frac{(200)(105)}{450} = 46.67$	$\frac{(200)(125)}{450} = 55.56$
Level term	$\frac{(150)(120)}{450} = 40$	$\frac{(150)(100)}{450} = 33.33$	$\frac{(150)(105)}{450} = 35$	$\frac{(150)(125)}{450} = 41.67$
Decreasing term	$\frac{(100)(120)}{450} = 26.67$	$\frac{(100)(100)}{450} = 22.22$	$\frac{(100)(105)}{450} = 23.33$	$\frac{(100)(125)}{450} = 27.78$

BASIC STATISTICS

The expected frequencies are determined in table 11.7 using equation (11.11):

$$\chi^2_{test} = \sum_{i=1}^{m} \sum_{j=1}^{n} \frac{(O_{ij} - E_{ij})^2}{E_{ij}}$$

$$= \frac{(60 - 53.33)^2}{53.33} + \frac{(40 - 44.44)^2}{44.44} + \frac{(45 - 46.67)^2}{46.67}$$

$$+ \frac{(55 - 55.56)^2}{55.56} + \frac{(35 - 40)^2}{40} + \frac{(30 - 33.33)^2}{33.33}$$

$$+ \frac{(40 - 35)^2}{35} + \frac{(45 - 41.67)^2}{41.67} + \frac{(25 - 26.67)^2}{26.67}$$

$$+ \frac{(30 - 22.22)^2}{22.22} + \frac{(20 - 23.33)^2}{23.33} + \frac{(25 - 27.78)^2}{27.78}$$

$$= 0.834 + 0.444 + 0.060 + 0.006 + 0.625 + 0.333 + 0.714$$
$$+ 0.266 + 0.105 + 2.724 + 0.475 + 0.278$$

$$\chi^2_{test} = 6.864.$$

Accept (do not reject) H_0; conclude that there are no differences in regional preferences. ◁

Your introduction to chi square analysis has been brief. Even with this limited knowledge, however, you will find that the chi square test is a useful technique if you will follow these simple rules in using the test:

1. No classification should have an expected frequency of less than 5. This constraint may sometimes necessitate combining classifications in order to raise the number of expected frequencies, but be careful, for in some cases the combination may not be meaningful.

2. Chi square distributions are continuous. If a situation involves a discrete distribution, chi square analysis may still be used, but a continuity correction factor may be necessary. If you desire further clarification or explanation of its use, consult an advanced text dealing with statistical methods.

Runs Test

We have frequently referred to *random occurrences*. In fact you have learned that, in order to use a sample statistic properly to make a probability statement concerning a population parameter, the sample must be random.

The importance of some assurance of randomness may seem rather trivial to you, for randomness may seem obvious. Yet experiments have shown that subjects' concept of randomness differ greatly from what is actually random. For instance, have you ever noticed what is customarily called random tile flooring? Usually such flooring is not random but has been laid purposely to look random.

One approach to investigating randomness is the *runs test*. A run is composed of repetitive sequences of identical occurrences. Considering the toss of a coin ten times, the following results might be obtained:

$$\underset{1}{\underline{TT}} \quad \underset{2}{\underline{HH}} \quad \underset{3}{\underline{T}} \quad \underset{4}{\underline{HH}} \quad \underset{5}{\underline{TT}} \quad \underset{6}{\underline{H}}.$$

By grouping the identical occurrences, we can establish that there are six runs. This total number of runs can be used to investigate randomness. The idea is that too few runs (such as one or two: HHHHHHHHHH or TTTTTHHHHH) or too many runs (such as ten: THTHTHTHTH) do not support the concept of randomness. If the sample is large (number of heads or tails exceeds 20), the sampling distribution of the number of runs is approximately normal with mean, $\overline{R}$, and standard deviation, s_R, determined from equations (11.12) and (11.13), respectively. Equation (11.14) shows the test statistic:[2]

$$\overline{R} = \frac{2n_1 n_2}{n_1 + n_2} + 1,$$ (11.12)

$$s_R = \sqrt{\frac{2n_1 n_2(2n_1 n_2 - n_1 - n_2)}{(n_1 + n_2)^2(n_1 + n_2 - 1)}},$$ (11.13)

where
$\overline{R}$ = mean number of runs,
n_1 = number of outcomes of one type,
n_2 = number of outcomes of the other type,
s_R = standard deviation of the distribution of the number of runs;

$$Z_{test} = \frac{R \pm \frac{1}{2} - \overline{R}}{s_R},$$ (11.14)

where R = the number of runs.

Example 11.5

▷ The manager of a department that investigates complaints has classified the complaints into two categories, customer service

---

2. The $\pm 1/2$ is used in equation (11.14) to convert the discrete data (number of runs) to a continuous distribution. Use $+1/2$ if $R < \overline{R}$, and use $-1/2$ if $R > \overline{R}$.

Table 11.8
Data and computations for runs test

$\overline{CC}$ $\overline{P}$ $\overline{C}$ $\overline{P}$ $\overline{C}$ $\overline{PPP}$ $\overline{C}$ $\overline{PP}$ $\overline{CC}$ $\overline{P}$ $\overline{C}$ $\overline{P}$ $\overline{CCC}$ $\overline{PP}$ $\overline{C}$ $\overline{P}$
$\overline{CCCC}$ $\overline{PP}$ $\overline{C}$ $\overline{PP}$ $\overline{CC}$ $\overline{P}$ $\overline{CC}$ $\overline{P}$ $\overline{C}$ $\overline{PC}$ $\overline{P}$ $\overline{C}$ $\overline{PP}$ $\overline{C}$ $\overline{PPP}$
$\overline{CCC}$ $\overline{P}$ $\overline{CC}$ $\overline{P}$ $\overline{C}$ $\overline{P}$ $\overline{C}$ $\overline{P}$ $\overline{CCC}$

Number of customer related complaints = n_1 = 35

Number of product related complaints = n_2 = 29

Number of runs = R = 41

$$\overline{R} = \frac{2n_1 n_2}{n_1 + n_2} + 1$$

$$= \frac{2(35)(29)}{35 + 29} + 1$$

$$\overline{R} = 31.72 + 1 = 32.72$$

$$s_R = \sqrt{\frac{2n_1 n_2 (2n_1 n_2 - n_1 - n_2)}{(n_1 + n_2)^2 (n_1 + n_2 - 1)}}$$

$$= \sqrt{\frac{2(35)(29)[2(35)(29) - 35 - 29]}{(35 + 29)^2 (35 + 29 - 1)}}$$

$$s_R = \sqrt{\frac{3,990,980}{258,048}} = \sqrt{15.466} = 3.93$$

related and product related. Even though no records of this nature have been maintained, the manager believes that a pattern exists. To investigate his concern, the next 64 complaints were classified into one of the two categories. The results were used to test the following null hypothesis at alpha = 0.10:

CCPCPCPPPCPPCCPCPCCCPPCP
CCCCPPCPPCCPCCPCPCPCPPCPPP
CCCPCCPCPCPCCC

H_0: nature of the complaints is random,
H_A: nature of the complaints is not random,

$\alpha = 0.10$.

Reject H_0 if $Z_{test} < -1.64$ or $Z_{test} > 1.64$:

$$Z_{test} = \frac{R \pm \frac{1}{2} - \overline{R}}{s_R}$$

$$Z_{test} = \frac{41 - \frac{1}{2} - 32.72}{3.93} = 1.98.$$

Reject H_0. It would appear that there is a pattern to the complaints. We could not conclude at the 0.05 level of significance that the complaints were random. ◁

Sign Test—One Sample

In chapter 8 we used the Z-test and t-test to make inferences about a population mean, μ. In both cases the difference between a sample mean, $\overline{X}$, and a hypothesized value of the population mean, μ_H, was examined in order to decide whether the difference should be called a chance difference or a statistically significant difference. In either case the test depended on a hypothesized value of μ and the assumption of the normal distribution or the central limit theorem. As discussed earlier, such tests are parametric tests.

Suppose, however, that a situation exists where it is not realistic to assume a normally distributed population, and the sample size may or may not be large enough to assume the central limit theorem. Suppose, further, that it is desirable to draw some conclusion about the middle of the data. One approach would be to employ the sign test to investigate the location of the median. The hypothesis to be tested is associated with the value of the population median, Md, and is analogous to the parametric test of the population mean.

In general, the procedure is to hypothesize a value for the population median and then observe whether each value of the random sample is greater than (plus sign) or less than (minus sign) the hypothesized value of the median (if there is no difference, the observation is ignored, and the sample reduced accordingly). If there are a disproportionate number of pluses or minuses, then the null hypothesis is rejected.

Symbolically, the null hypothesis might be stated as follows:

H_0: $Md = Md_H$,

and the alternate hypothesis might be stated in one of three ways:

H_A: $Md \neq Md_H$, or H_A: $Md < Md_H$, or H_A: $Md > Md_H$.

As we think about analyzing the null hypothesis by noting the plus and minus signs, it should be understood that, given the assumption that the null hypothesis is true, the number of pluses and number of minuses should be equal. Thus the null hypothesis can be restated as

H_0: $\pi = 0.5$ [probability of plus (or minus)],

and the alternate hypothesis restated in the same manner:

1. H_A: $\pi \neq 0.5$ [probability of plus (or minus) not equal 0.5],
2. H_A: $\pi < 0.5$ [probability of plus (or minus) less than 0.5],
3. H_A: $\pi > 0.5$ [probability of plus (or minus) greater than 0.5].

The null hypothesis can then be tested using the binomial distribution. The decision to reject H_0 or not to reject H_0 will involve calculating the probability that the particular sample result and any more extreme sample results could have occurred. This calculated probability is then compared to the chosen alpha level, and, if the calculated probability is less than the alpha level, the null hypothesis is rejected. Symbolically, the rejection region is

Reject H_0 if

$$\text{Two tail}\ \left[\ \text{or}\ \begin{cases} P(X \geq x \text{ for a given } n \text{ and } \pi = 0.50) < \alpha/2, \\ P(X \leq x \text{ for a given } n \text{ and } \pi = 0.50) < \alpha/2; \end{cases}\right.$$

One tail $P(X \geq x \text{ for a given } n \text{ and } \pi = 0.50) < \alpha$.

Consider example 11.6 to see how the median test is performed.

Example 11.6

▷ A local retailer has a rather large charge account business. When the business was small, the retailer knew that the average time before a delinquent charge account was brought current was 30 days (that is, about one month later). The retailer is interested to know if the median time of delinquency has increased. If a random sample of 10 delinquent accounts revealed the information given in table 11.9, at the 0.05 level of significance should the retailer conclude that the median for days of delinquency has increased?

H_0: $\pi = 0.5$ (probability of plus $= 0.5$),
H_A: $\pi > 0.5$ (probability of plus > 0.5),
     $\alpha = 0.05$.
Reject H_0 if $P(X \geq x) < 0.05$.

Table 11.9
Time of delinquent accounts

Account	Days delinquent, X	$X - Md_H$	Sign
A	27	− 3	−
B	32	+ 2	+
C	41	+11	+
D	31	+ 1	+
E	25	− 5	−
F	33	+ 3	+
G	34	+ 4	+
H	32	+ 2	+
I	29	− 1	−
J	28	− 2	−

$Md_H = 30$; hypothesized median value is 30

From the cumulative binomial distribution, appendix E, we obtain

$P(X \geq 6$, when $\pi = 0.5$ and $n = 10) = 0.377$.

Since $0.377 > 0.05$, accept (do not reject) H_0; the sample result does not suggest that the median has increased. ◁

If the sample size is large ($n \geq 30$), the sign test can be approximated using the Z-test. In order to use the Z-test, the population mean and standard deviation are estimated using equations (11.15) and (11.16):

$$E(k) = n\pi, \tag{11.15}$$

$$s_k = \sqrt{n\pi(1 - \pi)}, \tag{11.16}$$

where
$E(k)$ = the expected (average) number of pluses or minuses,
$\quad s_k$ = the standard deviation of the random variable (number of pluses or minuses),
$\quad n$ = the sample size,
$\quad \pi$ = the population proportion (0.50 when the number of pluses is expected to equal the number of minuses).

These estimators $E(k)$ and s_k, are used in the test statistic presented in equation (11.17):

$$Z_{\text{test}} = \frac{k \pm 1/2 - E(k)}{s_k}, \tag{11.17}$$

where
k = the number of pluses or minuses.

Example 11.7

▷ Suppose that in example 11.6 the retailer had examined 50 accounts and discovered 33 accounts had delinquent balances exceeding 30 days. At the 0.05 level of significance, should the retailer reject the null hypothesis?

H_0: $\quad \pi = 0.5$ (probability of plus = 0.5),
H_A: $\quad \pi > 0.5$ (probability of plus > 0.5),
$\quad \alpha = 0.05$.

Reject H_0 if $Z_{\text{test}} > 1.64$:

$E(k) = n\pi$
$\quad\quad = 50(0.5)$
$E(k) = 25$,

$s_k = \sqrt{n\pi(1 - \pi)}$
$\quad = \sqrt{50(0.5)(1 - 0.5)}$

$$= \sqrt{12.5}$$
$$s_k = 3.54,$$
$$Z_{test} = \frac{k \pm 1/2 - E(k)}{s_k}$$
$$= \frac{(33 - 0.5) - 25}{3.536}$$
$$= \frac{7.5}{3.536}$$
$$Z_{test} = 2.12.$$

Reject H_0; the sample result does suggest that the median has increased. ◁

Sign Test–Related Samples

You may encounter business problems where it is necessary to compare two sample series consisting of related samples (matched pairs of observations) in order to determine which series is the larger of the two, or at least if there is any difference between them. If the magnitude of the differences is not important then the two series may be compared on the basis of the signs ($+$ or $-$) of the differences. Thus, under the assumption that magnitudes are not to be considered, a hypothesis that the values in two series are not different in size can be tested using the sign test.

The sign test for related samples (dependent samples) is used in the same manner that it is used for the median test. If the samples are small, the binomial distribution is appropriate, and if the samples are large ($n \geq 30$) the Z-test given in equation (11.17) is a reasonably good approximation. Examples 11.8 and 11.9 illustrate the use of the sign test for dependent samples.

Example 11.8

▷ An insurance company is considering a new plan for determining the cash value of each policy. The company knows that the new plan will be beneficial to all new policyholders. However, it is not known whether the new plan would be beneficial to current policyholders. Consequently the company has decided to take a sample of 40 current policyholders and compute the future cash values under both the old and new method. If more than half of the current policyholders might benefit from the new plan, the option of converting to the new plan will be made available to all current policyholders. The sample results are given in table 11.10. At alpha $= 0.01$, should the option be made available to old policyholders?

Table 11.10
Data for sign test: related samples

| Policyholder | Projected cash value (in $10,000) | | |
	Cash value, old plan	Cash value, new plan	Sign (new–old)
1	9.0	10.0	+
2	9.5	10.1	+
3	10.2	10.1	−
4	13.0	13.8	+
5	8.1	8.0	−
6	11.9	12.2	+
7	9.6	9.8	+
8	11.2	11.5	+
9	6.3	6.4	+
10	12.7	13.6	+
11	12.5	13.7	+
12	9.2	9.2	(omit)
13	11.0	11.9	+
14	11.3	12.1	+
15	6.8	6.8	(omit)
16	7.0	7.5	+
17	7.0	6.9	−
18	14.2	14.3	+
19	10.0	10.8	+
20	13.1	13.0	−
21	14.0	14.8	+
22	12.7	12.8	+
23	11.9	11.5	−
24	12.0	12.0	(omit)
25	7.2	7.1	−
26	10.3	10.5	+
27	6.8	7.3	+
28	11.7	11.8	+
29	9.2	10.0	+
30	13.7	13.5	−
31	12.8	13.2	+
32	10.7	11.0	+
33	11.2	11.1	−
34	12.3	12.9	+
35	9.1	9.0	−
36	8.7	8.7	(omit)
37	9.9	10.2	+
38	13.1	12.5	−
39	12.4	13.3	+
40	11.7	12.0	+

$n = 36$

$E(k) = n\pi = 36(0.5) = 18$

$s_k = \sqrt{n\pi(1 - \pi)}$

$s_k = \sqrt{36(0.5)0.5} = \sqrt{9} = 3$

H_0: $\pi = 0.5$ (probability of plus = 0.5),
H_A: $\pi > 0.5$ (probability of plus > 0.5),
    $\alpha = 0.01$.
Reject H_0 if $Z_{test} > 2.33$.

Using equation (11.17) and $k = 26$ pluses,

$$Z_{test} = \frac{k \pm 1/2 - E(k)}{s_k}$$

$$= \frac{26 + 1/2 - 18}{3}$$

$Z_{test} = 2.83$.

Reject H_0. Apparently, the new plan does benefit the majority of the current policyholders. ◁

As we stated earlier, the sample distribution of pluses and minuses is actually a binomial distribution, but the sign test we have illustrated assumes that a normal distribution can be used to carry out the test with a reasonable degree of accuracy. As a general rule this assumption can be made when the sample size n is at least 30. In the calculation of Z, the addition of 1/2 to k is necessary because the binomial distribution represents discrete events, while the normal distribution is continuous. (From a practical standpoint, when n becomes large failing to subtract 1/2 will not alter your solutions.) Carrying out a sign test using the binomial distribution (when n is less than 30) is a relatively easy, but sometimes lengthy, task. It requires that the binomial $[\pi + (1 - \pi)]$ be expanded to the n power so that the probability of obtaining a certain number or greater pluses can be determined. If this probability is less than the specified value of alpha for the test, the null hypothesis is rejected. Example 11.9 illustrates this type of test.

Example 11.9

▷ Suppose that in our insurance situation we had obtained only every fourth observation in table 11.10. This would have resulted in 7 usable data points (numbers 4, 8, 16, 20, 28, 32, and 40), 6 pluses and 1 minus. We then test the following null hypothesis at the 0.10 α level:

H_0: $\pi = 0.5$ (probability of plus = 0.5),
H_A: $\pi > 0.5$ (probability of plus > 0.5),
    $\alpha = 0.10$.
Reject H_0 if $P(X \geq x) < 0.10$.

From the cumulative binomial distribution, appendix E, we obtain:

$$P(X \geq 6) = 1.0 - P(X \leq 5)$$
$$= 1.0 - 0.9375$$
$$P(X \geq 6) = 0.0625.$$

Since $0.0625 < 0.10$, we conclude the outcome of 6 pluses from 7 occurrences could not be due to chance, and we reject the null hypothesis. ◁

Mann-Whitney U-test

In chapter 9 we learned to use the t-test to determine if two independent random samples were drawn from the same population or from different populations. But the Student t distribution requires the assumptions that the populations from which the samples are drawn be normally distributed and have equal variances. An alternative to using the t-test is the Mann-Whitney U-test which can be used to decide if two distributions have the same shape.

The U-test is based upon a comparison of the rank of the values of one sample in an array of the combined values of the two samples. It is fairly obvious that if the two samples come from the same population, or two populations with equal means, the average ranks for each sample should be equal. The form of the null and alternate hypotheses is as follows:

H_0: population distributions are identical (this is equivalent to testing $\mu_I = \mu_{II}$),

1. H_A: the population distributions are not identical,

2. H_A: population distribution I is to the right of population distribution II,

3. H_A: population distribution I is to the left of population distribution II.

When n_1 and $n_2 \geq 10$, a normal distribution having mean $(\overline{U})$ given in equation (11.18) and standard deviation s_u given in equation (11.19) can be employed:[3]

$$\overline{U} = \frac{n_1 n_2}{2}, \tag{11.18}$$

$$s_u = \sqrt{\frac{n_1 n_2 (n_1 + n_2 + 1)}{12}}, \tag{11.19}$$

---

3. Smaller samples may be tested using special tables of the Mann-Whitney U-statistic.

where

n_1 = the size of the first sample,
n_2 = the size of the second sample.

The appropriate test statistic is given in equation (11.20):

$$Z_{test} = \frac{U - \overline{U}}{s_u} \qquad (11.20)$$

The value of U required in equation (11.20) is obtained from either equation (11.21) or equation (11.22) based on the formulation of the alternate hypotheses:

$$U_1 = n_1 n_2 + \frac{n_1(n_1 + 1)}{2} - R_1, \qquad (11.21)$$

$$U_2 = n_1 n_2 + \frac{n_2(n_2 + 1)}{2} - R_2, \qquad (11.22)$$

where

R_1 = the sum of the ranks of the first sample,
R_2 = the sum of the ranks of the second sample.

If the alternate hypothesis is two-tailed, the U can be either U_1 or U_2. If the alternate hypothesis is one-tailed, then U must be chosen from the sample obtained from the population designated as population distribution I, as specified in alternative hypotheses 2 and 3.

Example 11.10

▷ Daily sales data for two sales people have been accumulated as shown in table 11.11. At the 0.05 level of significance, are the sample results different?

Table 11.11
Daily sales

Smith	Jones
$100	$ 82
125	98
60	115
137	143
82	65
99	123
150	128
98	93
143	135
122	120
95	
110	

Table 11.12
Mann-Whitney U-test

Array of daily sales	Salesman	Rank	Smith rank	Jones rank
60	Smith	1	1	
65	Jones	2		2
82	Smith	3.5	3.5	
82	Jones	3.5		3.5
93	Jones	5		5
95	Smith	6	6	
98	Smith	7.5	7.5	
98	Jones	7.5		7.5
99	Smith	9	9	
100	Smith	10	10	
110	Smith	11	11	
115	Jones	12		12
120	Jones	13		13
122	Smith	14	14	
123	Jones	15		15
125	Smith	16	16	
128	Jones	17		17
135	Jones	18		18
137	Smith	19	19	
143	Smith	20.5	20.5	
143	Jones	20.5		20.5
150	Smith	22	22	
			139.5	113.5

$$U_1 = n_1 n_2 + \frac{n_1(n_1 + 1)}{2} - R_1 \qquad U_2 = n_1 n_2 + \frac{n_2(n_2 + 1)}{2} - R_2$$

$$= (12)(10) + \frac{12(12 + 1)}{2} - 139.5 \qquad = (12)(10) + \frac{10(10 + 1)}{2} - 113.5$$

$$U_1 = 58.5 \qquad\qquad U_2 = 61.5$$

Thus $U = 58.5$ or 61.5

$$\overline{U} = \frac{n_1 n_2}{2}$$

$$\overline{U} = 60$$

$$s_u = \sqrt{\frac{n_1 n_2(n_1 + n_2 + 1)}{12}}$$

$$= \sqrt{\frac{(12)(10)(12 + 10 + 1)}{12}}$$

$$= \sqrt{230}$$

$$s_u = 15.17$$

H_0: population distributions are identical,
H_A: population distributions are not identical,
$\alpha = 0.05$.

Reject H_0 if $Z_{test} < -1.96$ or $Z_{test} > 1.96$:

$$Z_{test} = \frac{U - \overline{U}}{s_u}$$

$$= \frac{58.5 - 60}{15.17}$$

$$= \frac{-1.5}{15.17}$$

$$Z_{test} = -0.099.$$

Accept (do not reject) H_0. Based on the results of the U-test, both Smith and Jones appear to sell equal amounts. ◁

The U-test not only has the advantage of not requiring assumptions regarding the population distribution, but it is also generally easier to apply. However, like most of the nonparametric techniques, it is a somewhat weaker test than comparable parametric tests as a general rule. Still the nonparametric methods can be a useful addition to your bag of statistical tools.

Kruskal-Wallis Test

There may be instances when you have three or more *independent random samples* and you need a test to determine if they come from populations with identical distributions. This can be accomplished with the Kruskal-Wallis test. Using this method, a statistic called H (equation 11.23) is computed that is approximately chi square distributed with $df = n' - 1$, where n' is the number of independent samples and no sample is smaller than 5. This statistic (H) is based upon the sum of the ranks of the values in each sample with the ranks being determined by first combining all samples and constructing an ordered array. The formula for computing H is

$$H = \frac{12}{n(n + 1)} \Sigma \frac{R_i^2}{n_i} - 3(n + 1), \qquad (11.23)$$

where
   n = the size of all samples combined,
   R_i = the sum of the ranks of the ith sample,
   n_i = the size of the ith sample.

Example 11.11

▷ Smokey Motors Corporation has developed three new diesel engines and needs to select one that might be suitable for use in a high-performance car it hopes to market. All three engines are equally suitable with respect to most performance and design criteria, and the company has just completed testing the fuel consumption of each of the three engines. In order to determine the fuel consumption, they built a limited number of each of the engine types and tested them under "normal" road conditions. After looking at the results, shown in table 11.13, the Vice-President for Marketing remarked that it appeared that there might be differences in fuel consumption among the three engines. At the 0.01 level of significance, are the sample results different?

H_0: distributions of the three populations are identical,
H_A: distributions are not identical,
$\alpha = 0.01$,
df = number of samples −1.

In this instance, $df = 3 - 1 = 2$. Reject H_0 if $H > 9.21034$ (obtained from appendix K).

Since $H = 10.632 > \chi_\alpha^2 = 9.21034$, the null hypothesis is rejected. There may be a difference in the consistency of the fuel consumption of the three engines. ◁

In our example, there were no "ties" in the ranks. If there had been, we would have assigned their mean ranks to the tied values. To illustrate, had the 23rd and 24th items both been 32, we would have assigned each of them the rank 23.5. Then it would have been necessary to adjust the statistic H to one called H';

$$H' = \frac{H}{C},$$ (11.24)

Table 11.13
Fuel consumption of three engine types (kilometers per liter)

Type 1	Type 2	Type 3
31	12	17
30	13	20
29	26	23
22	24	9
32	11	15
25	16	18
	27	19
	10	14
		8
		5

Table 11.14
Computation of H for the Kruskal-Wallis test

KPL	Type 1	Type 2	Type 3
5			1
8			2
9			3
10		4	
11		5	
12		6	
13		7	
14			8
15			9
16		10	
17			11
18			12
19			13
20			14
22	15		
23			16
24		17	
25	18		
26		19	
27		20	
29	21		
30	22		
31	23		
32	24		
R_i	$\overline{123}$	$\overline{88}$	$\overline{89}$

$$\frac{R_1^2}{n_1} = \frac{(123)^2}{6} = 2{,}521.5 \qquad \frac{R_2^2}{n_2} = \frac{(88)^2}{8} = 968 \qquad \frac{R_3^2}{n_3} = \frac{(89)^2}{10} = 792.1$$

$$H = \frac{12}{n(n+1)} \, \Sigma \, \frac{R_i^2}{n_i} - 3(n+1)$$

$$= \frac{12}{24(25)}(2{,}521.5 + 968 + 792.1) - 3(25)$$

$$H = 0.02(4{,}281.6) - 75 = 85.632 - 75 = 10.632$$

$$C = 1 - \frac{\Sigma(t_i^3 - t_i)}{n^3 - n}, \tag{11.25}$$

where

t_i = the number of values in each set of tied values,
n = the combined total of all samples.

Unless there are several tied values in the samples, the values of H and H' will be very close to one another. In any case, H' is compared to χ_α^2 in testing of the hypothesis instead of H.

Summary of Equations

11.1   Hypothesis test for goodness-of-fit:

$$\chi^2_{test} = \sum_{i=1}^{n} \frac{(O_i - E_i)^2}{E_i}$$

11.2   Probability of occurrence located in row i:

$$P(R_i) = \frac{O_{i.}}{O_{..}}$$

11.3   Probability of occurrence located in column j:

$$P(C_j) = \frac{O_{.j}}{O_{..}}$$

11.4   Probability of occurrence located in row i and column j:

$$P(R_iC_j) = \frac{(O_{i.})(O_{.j})}{(O_{..})(O_{..})}$$

11.5   Estimate of number of observations in the row i-column j location:

$$E_{ij} = [P(R_iC_j)]n$$

11.6   Estimate of number of observations—using row, column, and overall totals:

$$E_{ij} = \frac{O_{i.}O_{.j}}{O_{..}}$$

11.7   Hypothesis test for independence and for homogeneity:

$$\chi^2_{test} = \sum_{i=1}^{m} \sum_{j=1}^{n} \left[\frac{(O_{ij} - E_{ij})^2}{E_{ij}} \right]$$

11.8   Chi-square test when $df = 1$:

$$\chi^2_{test} = \sum_{i=1}^{m} \sum_{j=1}^{n} \left[\frac{(|O_{ij} - E_{ij}| - \frac{1}{2})^2}{E_{ij}} \right]$$

11.9   Probability of occurrence located in row i—same as equation (11.2):

$$P(R_i) = \frac{O_{i.}}{O_{..}}$$

11.10   Estimate of number of observations in the row i-column j location:

$$E_{ij} = P(R_i)n_j$$

11.11   Estimate of number of observations—using row, column, and overall totals; same as equation (11.6):

$$E_{ij} = \frac{O_{i.}O_{.j}}{O_{..}}$$

11.12   Average number of runs:

$$\overline{R} = \frac{2n_1n_2}{n_1 + n_2} + 1$$

11.13   Standard deviation of the sampling distribution of the number of runs:

$$s_R = \sqrt{\frac{2n_1n_2(2n_1n_2 - n_1 - n_2)}{(n_1 + n_2)^2(n_1 + n_2 - 1)}}$$

11.14   Runs test:

$$Z_{\text{test}} = \frac{R \pm \frac{1}{2} - \overline{R}}{s_R}$$

11.15   Expected number of pluses or minuses:

$$E(k) = n\pi$$

11.16   Standard deviation of the number of pluses or minuses:

$$s_k = \sqrt{n\pi(1 - \pi)}$$

11.17   Sign test:

$$Z_{\text{test}} = \frac{k \pm \frac{1}{2} - E(k)}{s_k}$$

11.18   Mean value for Mann-Whitney U-test:

$$\overline{U} = \frac{n_1 n_2}{2}$$

11.19   Standard deviation for Mann-Whitney U-test:

$$s_U = \sqrt{\frac{n_1 n_2 (n_1 + n_2 + 1)}{12}}$$

11.20   Mann-Whitney U-test:

$$Z_{\text{test}} = \frac{U - \overline{U}}{s_U}$$

11.21   Used to determine U value for Mann-Whitney U-test:

$$U_1 = n_1 n_2 + \frac{n_1(n_1 + 1)}{2} - R_1$$

11.22   Used to determine U value for Mann-Whitney U-test:

$$U_2 = n_1 n_2 + \frac{n_2(n_2 + 1)}{2} - R_2$$

11.23   Statistic for Kruskal-Wallis test:

$$H = \frac{12}{n(n + 1)} \Sigma \frac{R_i^2}{n_i} - 3(n + 1)$$

11.24   Equation to adjust for ties when using Kruskal-Wallis test:

$$H' = \frac{H}{C}$$

11.25   Equation to adjust for ties when using Kruskal-Wallis test:

$$C = 1 - \frac{\Sigma(t_i^3 - t_i)}{n^3 - n}$$

Review Questions

1. Define nonparametric statistics.

2. Identify the use of each equation given at the end of the chapter.

3. Why is there a need for nonparametric statistics?

4. Distinguish between the test for independence and the test for homogeneity.

1. A statistics instructor just completed preparing his grades for the semester, and they are as follows:

Grade	Number
A	2
B	40
C	25
D	13
F	20

At the outset of the semester he told his classes that, over time, his grades had been uniformly distributed. If the above grades are considered a random sample, test the instructor's statement. Set $\alpha = 0.01$.

2. A single die was thrown 120 times with the following results. Based on the observed frequencies, is the die fair? Set $\alpha = 0.05$.

Number of spots	Observed frequency
1	18
2	21
3	16
4	22
5	23
6	20

3. A restaurant manager believes that the arrivals every half hour follow a uniform distribution. The records from a typical Friday are given as follows:

Time of arrival	Number of tables served
7:00–7:30	13
7:30–8:00	15
8:00–8:30	16
8:30–9:00	15
9:00–9:30	14
9:30–10:00	11

Test the manager's belief. Set alpha $= 0.01$.

4. A student has maintained the following information on typing errors:

Number of errors	Number of assignments
0	11
1	14
2	15
3 or more[a]	12

[a] The total number of errors is 47.

Is this a uniform distribution? Set $\alpha = 0.01$.

5. The manager of Stop-and-Wait, a local fast-service food chain, has kept records of the number of people waiting in line to be served at his restaurant. The other day, a customer complained about the number of people waiting, and the manager replied. "I know from experience that the number of people waiting is normally distributed." If the following information is considered to be a random sample, was the manager's statement correct? Set $\alpha = 0.05$.

Customers in line	Frequency
0 but less than 2	15
2 but less than 4	20
4 but less than 6	35
6 but less than 8	25
8 but less than 10	35
10 but less than 12	40
12 but less than 14	20
14 but less than 16	15

6. A random sample of the records of an automobile insurance company yielded the following information on the accident experience of its policy holders. After examining the distribution, the company statistician concluded the number of accidents is normally distributed. Using $\alpha = 0.01$, is he correct?

Number of accidents	Frequency
0	10
1	40
2	100
3	150
4	200
5	125
6	75
7	50
8	30
9	20

7. A local retail store has experimented with enclosing a return envelope in some of the monthly statements. The results are given as follows:

State of payment	With envelope	Without envelope
Paid when due	150	170
Not paid when due	50	30

At the 0.01 level of significance is there any indication that the enclosure of an envelope promotes prompt payment?

8. The owner of the Specialty Shop has recorded information on 500 customers:

	Female	Male
Purchase	110	60
No purchase	150	180

At the 0.05 level of significance, do you believe that whether a purchase is made or not is independent of the sex of the customer?

9. A retailer with stores in three cities has been stocking three brands of oversized rackets. Sales data for the past month are presented in the following table. At the 0.01 alpha level, are racket sales related to store location?

	Racket size		
Store	Large	Big	Hugh
A	50	30	45
B	100	60	40
C	40	80	55

10. The Robb Insurance Agency has recently completed a study of 300 salesmen relating the price of automobiles driven and the salesmen's home office ratings.

	Salesmen's rating		
Auto price	Successful	Fair	Poor
High	80	40	15
Medium	55	30	15
Low	15	30	20

At the 0.01 level of significance, would you conclude that price of the car and salesmen's ratings are independent?

11. The manager of a retail outlet in a large city prepared two advertising campaigns to be tested on a sample of customers with charge accounts. Six hundred customers were selected and randomly assigned to campaign I or II. The advertising literature was mailed with the monthly statements, and records were maintained as to whether a sale resulted from the mail-out. (The advertising literature was limited to one product, and any purchaser of this product during the following month was asked about any literature received.) The results of the study are given in the following table. At the 0.05 alpha level, was the type of campaign related to whether or not a purchase was made?

	Campaign	
	I	II
Purchase	200	110
No purchase	120	170

12. Premium Product Promotions, a market research firm, is trying to determine if there are differences in the brand of beer preferred by various customer groups. Formulate and test a hypothesis for them using $\alpha = 0.025$ if the following data are the preferences of a sample of 800.

Customer group	Brand of beer		
	Pale	Golden	Heavy
Housewives	75	20	5
Businessmen	50	130	20
Factory worker	5	25	170
College students	100	100	100

13. The following distribution represents even and odd digits from a table of numbers believed to be random. Test at the 0.05 level of significance to see if the digits are random.

E E O E O E E O E E O E O O E O E E O O O E E
O E E E O O E E O E E O E O E O O O E E O O

14. In the Student Union the red and green tiles are supposedly laid on the floor as random tiles. Perform a runs test to see, at the 0.05 level of significance, if the red and green tiles are actually laid in a random manner. (We have used only one row of tiles in the sample.)

R R G R G R G G R G R R G R R G G R R G G G
G R G G G R G R G G R G R G G R R R G R G R
G R R G R R

15. An airport manager has classified aircraft departures as private or commercial. Based on the following observations, are the departures random? Set alpha = 0.01.

P C C P P C C C C P C C C C P P P C C C C P P P C
C C P C C C C P P P C C C P C C C C C P C C C C

16. Ten students, randomly selected, participated in a memory course. After the course, 6 students scored higher on a reading comprehension test, 2 had the same score, and the second score of 2 was lower than the first. At the 0.05 level of significance, do you think the memory course improved the reading comprehension of the students?

17. The vice-president of sales would like to initiate a new formula for calculating sales commissions. He believes that the new formula will increase the dollar amount of the commissions of the majority of the employees. In order to investigate his belief, he selects a random sample from past sales and calculates the commission using both the old and new formulas. Set alpha = 0.05, and test to see if there is a difference in commissions.

Employee	Commissions	
	Old plan	New plan
1	219	240
2	150	148
3	180	190
4	200	210
5	210	225
6	426	450
7	330	315
8	325	340
9	208	215
10	105	100
11	400	420
12	360	375
13	219	200
14	210	190
15	460	480
16	135	145
17	215	210
18	80	75
19	120	140
20	518	500
21	600	620
22	585	595
23	75	60
24	50	45

	Commissions	
Employee	Old plan	New plan
25	175	190
26	280	270
27	455	475
28	425	400
29	600	650
30	700	695
31	150	200
32	375	410
33	400	390
34	65	75
35	110	115
36	700	720
37	45	50
38	810	800
39	292	301
40	40	50

18. During the past thirteen weeks, weekly sales records for a particular item have been maintained at two different stores.

Store A	Store B
$10	$18
29	13
12	7
19	4
22	36
14	6
50	25
46	15
5	28
39	11
42	17
15	20
24	16

At the 0.05 level of significance, would you conclude that there are differences in sales at the two locations?

19. The Specialty specializes in two types of dinners, a steak dinner or a seafood dinner. Two chefs are employed, one for each type of dinner. The management, in order to determine if one chef is better than the other, has maintained a record of complaints for the past 10 weeks. Determine, at the 0.05 level, if the chefs are equally popular.

Number of complaints Steak chef	Seafood chef
8	3
9	0
12	13
2	16
11	6
7	8
1	5
15	10
19	14
4	1

20. A publisher of several magazines distributed nationwide relies on telephone contacts to sell subscriptions. The publisher has divided the country into two sections and employs salespeople to phone for subscriptions. A question of interest to the publisher is whether the salespeople in both sections make the same number of daily calls. A random sample from each section produced the information given in the following table. Use the given information to test for a difference in the number of daily calls. Set alpha = 0.01.

Telephone contacts Section I	Section II
62	75
49	80
83	65
59	60
68	82
71	89
56	72
50	92
70	88
58	90

21. The Dean has developed some data on the academic performance of students from three high schools. The following is a sample of that data:

Sleepy high	Big time high	Grubby high
1.6	2.5	0.5
2.0	2.8	1.4
1.8	3.0	3.2
2.7	2.9	1.3
1.0	3.5	2.6
2.1		2.4
2.3		

Test the Dean's hypothesis that there is no difference in the distribution of the academic performance between students from the various high schools. Set $\alpha = 0.05$.

22. Mr. S. Lagree is the office manager for a company that provides part-time office help. He insists that the four secretaries he employs be highly accurate typists, but lately he has begun to suspect that the distribution of numbers of errors they make is different. To test this, he has taken a sample of their work for several days and recorded the errors per page. Test Mr. Lagree's notion. Set $\alpha = 0.01$.

Errors per page

Ann	Bill	Cindy	Donnie
2.1	0.0	1.2	4.1
3.3	1.9	1.3	2.8
3.8	1.5	0.1	2.4
2.7	2.2	0.7	4.2
5.1	2.3	2.6	3.5
4.0	0.8	1.1	3.4

Case Study

The President of the shoe-manufacturing firm has become somewhat more cautious in his evaluation of the potential for the two versions of the firm's new jogging shoe. (See the case study at the end of chapter 9 for background information and data.) More specifically he has recognized that you may have made some assumptions concerning normal distributions and $\sigma_A = \sigma_B$ in your previous analysis of the test data on the proposed new jogging shoe. At this point, he gives you a new set of instructions. "If you know some ways to analyze the data without making those assumptions, do so."

Prepare a report containing the results of your latest analysis, contrast the new results with those obtained when you use assumptions of normality and $\sigma_A = \sigma_B$, and explain why your results differ if, in fact, they do.

12

Correlation and Regression Analysis: The Simple Linear Case

Until now we have focused our attention on describing and drawing inferences relating to a single variable. This chapter addresses itself to describing, in a linear fashion, the relationship between two variables. The appendix to this chapter is devoted to one technique that can be used to describe the relationship between two variables in a nonlinear fashion. Finally, chapter 13 will be devoted to describing the relationship between more than two variables.

Nature of Correlation and Regression Analysis

Correlation and regression analysis involves determining and measuring the relationship (covariation or going-together) between two or more variables. Regression deals with determining a quantitative expression to describe the relationship between two or more variables, while the purpose of correlation is to measure the degree of the relationship.

Frequently, statistical analysts attempt to differentiate further between correlation and regression analysis. We have made a distinction between them only for the purpose of defining how each contributes to the general analysis and interpretation of the data. But further distinctions will not be attempted because, in an analytical sense, they are intertwined, and it serves no useful purpose here to make them.

This chapter deals with simple linear correlation and regression analysis. The word *simple* in this case does not mean *easy*. Rather, it refers to the notion that we are concerned with the relationship between only two variables—a *dependent* variable and an *independent* variable. The idea of *linear* correlation and regression analysis relates to the assumption that the best representation of perfect correlation is a *straight* (linear) regression line fitted to the observed data. That is, in a linear analysis we describe the mathematical relationship by a straight line using the general form given in equation (12.1):

$$Y = \alpha + \beta X. \tag{12.1}$$

So, simple means only *two variables,* and linear means a *straight-line (arithmetic) relationship* between them.

Before we get too involved with correlation and regression analysis per se, let us consider the variables that are the subject of the analysis. First, what is the nature of the data that comprise the two variables? Second, how do we determine which is the *dependent* and which is the *independent* variable? The answer to the first question may seem very simple. Yet there are some serious implications related to your response. Specifically, will the data be treated as an *enumeration* of a population, or shall they be treated as a *sample?* If treated as an enumeration, then measures computed from the data (α and β) will be treated as *parameters.* But if the data represent a sample, the measures (a and b) computed will be treated as *estimators.* In most cases (perhaps in all cases) data used for correlation and regression analysis should be considered sample data. Most, if not all, of the time, data represent a slice (sample), either out of time or space, and the correlation and regression analysis projects to a much broader base—a population. Thus, using sample data, we will determine estimates of α and β, denoted a and b, and form the *regression equation:*

$$Y_c = a + bX. \tag{12.2}$$

Example 12.1

▷ The faculty at State U. has been concerned about the number of hours students spend in the library. In fact the Dean of Women has decided to do a study relating the number of hours spent in the library with the student grade point average. There are 1,000 students at State. While it might be expensive, a census of all 1,000 students could be taken. An estimate of the average number of hours spent per week in the library and the student's cumulative grade point averages could be obtained for all 1,000 students, and a correlation and regression analysis could then be made. In this case, can the Dean assume that she is dealing with a population and not a sample? After all, she has 1,000 values for average hours in the library and 1,000 values for grade point averages. In other words, she has information on all 1,000 students. Is she really dealing with a complete universe? Maybe not! Let us see what she might do with correlation and regression analysis and these data. ◁

If the Dean's analysis deals only with the present 1,000 pairs of values and nothing more, she is concerned with a population, and any measures computed may be considered parameter values. But, if she plans to make any projection or any inferences regarding the State U. students during other semesters, the Dean is dealing with a sample. Thus any measures computed are sample estimators, not parameters. For instance, if the Dean generalizes (say, students who spend more time in the library make higher GPA's) in any way, she is treating the 1,000 observations as a sample. Generalizations might extend across time—next semester or next year—or to other students or other campuses. If the extension is to other semesters or other years (except for the time the actual study was performed), the data represent a sample out of time (a temporal sample). If the results are extended to other campuses or other student bodies, the sample is a spatial one. Moreover any time a projection of the relationship is extended beyond the limits of the observed data, the data must be considered a sample and treated accordingly.

The second point to consider regarding the variables is how to determine which one is the independent variable and which one is the dependent variable. In correlation and regression analysis one variable is expressed as a function of the other in mathematical terms. But in terms of actual data the analysis itself cannot determine which variable is a function of the other. This determination is made on the basis of knowledge of the subject matter itself. Consider example 12.1. Is the grade point average a function of the number of hours spent in the library, or is the number of hours spent in the library a function of the grade point average? Which variable is the independent one, and which is the dependent one? Much depends on the relationship that is under consideration. In this illustration the faculty of State U. is probably trying to show that the more hours spent in the library, the higher the grade point average. Thus the number of hours spent in the library is treated as the independent variable, and the GPA is treated as the dependent variable. Thus hours are plotted on the X axis, and grade points are plotted on the Y axis.

Let us now examine the steps involved in conducting simple linear correlation and regression analysis. These steps are:

1. Plot the scatter diagram.
2. Compute the regression equation.
3. Compute the standard error of estimate.
4. Test hypothesis concerning β.

5. Compute correlation coefficient.

6. Construct confidence intervals.

Each of these steps will be developed in the sections that follow in this chapter.

The Scatter Diagram

A scatter diagram consists of coordinate points of the two variables plotted on a graph. The points are not connected together. Rather they look like a scattering of dots, not too unlike a pattern of bird shot. In simple regression analysis there are only two variables—the independent variable is plotted on the X axis (the abscissa) and is sometimes referred to as the X *variable,* while the dependent, or Y *variable,* is plotted on the Y axis (the ordinate).

The purpose of the scatter diagram is to provide a picture of any relationship between the independent and dependent variables. To this extent, the scatter diagram may assist the analyst in three ways:

1. It indicates generally whether or not there is an apparent relationship between the two variables.

2. If there is a relationship, it may suggest whether it is *linear* or *nonlinear.*

3. If the relationship is linear, the scatter diagram will show whether the relationship is *positive* or *negative.*

Figure 12.1 shows four scatter diagrams. The first one (figure 12.1a) shows two variables that have no apparent correlation. Figures 12.1b and c show linear correlation between the two variables plotted. In figure 12.1b the linear correlation is positive— shows a positive relationship; the scatter diagram of figure 12.1c indicates a negative relationship. The fourth illustration, figure 12.1d, does not suggest a linear relationship at all. In fact the scatter strongly indicates that there is a nonlinear relationship (curvilinear) between the X and Y variables. Since this chapter deals with simple linear correlation and regression analysis, we will wait until the appendix for further discussion of this type of relationship.

The Regression Equation

The *regression equation* expresses the relationship between the X and Y variables more precisely. It is one thing to say, "Y is a function of X." It is another thing (more explicit) to express this relationship in a mathematical equation. Given the assumption of linearity, the equation showing the relationship between the X and Y variables is just a mathematical function for a straight line:

$Y_c = a + bX.$

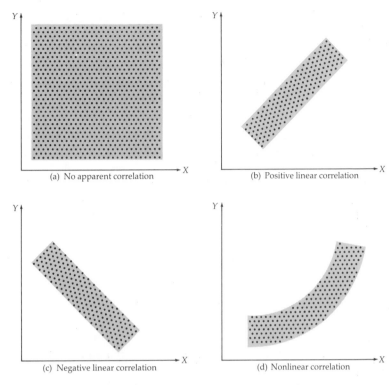

Figure 12.1   Four scatter diagrams

In this equation we must distinguish between the actual dependent variable represented by Y, and the computed or regression value of the dependent variable, represented by Y_c. When the actual dependent variable Y is plotted with X, the result is the scatter diagram. However, when the Y_c value is plotted with X and the points are connected together, the result is called the *regression line*.

Mathematical Definition

One of the important functions of the regression equation, or the resulting regression line, is to provide a mathematical expression defining the relationship between the X and Y variables. The accuracy of this definition depends on the extent to which the equation or line actually represents the data. If the scatter diagram falls in a linear pattern, as in figures 12.1b and c, the line may be reasonably representative of the actual data. On the other hand, a straight line fitted to the data in figure 12.1d is not very representative, because the coordinated points of the scatter diagram do not form a linear pattern. Thus the more closely the scatter diagram conforms to a straight line (positive or negative), the more accurately the line describes the relationship.

Standard of Perfect
Correlation

Another way to view the relationship between the dots in the scatter diagram and the regression line is that the regression line is a standard of perfect correlation. Thus the closer the dots conform to the line, the higher the correlation between X and Y. The regression line provides the base to which we compare the actual observations (the coordinate points) to determine the extent of the correlation. If the regression line has any slope to it (positive or negative), and all dots fall directly on it, there is perfect correlation. Of course we must remember that the occurrence of perfect correlation is highly improbable. Rather it is a defined state of perfection. As such, then, the regression line is a standard of perfect correlation.

Determination of the
Regression Equation

Let us look at the equation again:

$$Y_c = a + bX$$

where

Y_c = the estimate of the average value of Y for a specified value of X,

X = a value of the independent variable,

a = the estimate of the Y intercept (α) of the regression line (a constant),

b = the estimate of the slope (β) of the regression line (a constant).

There are several ways to determine the regression equation. One can simply draw "freehand" a line that appears to represent the relationship between the two variables as displayed in a scatter diagram. The regression equation can then be written for the freehand line. There are, however, some rather obvious problems with determining the regression line in this fashion. Specifically, the placement of the line depends upon individual judgment, and no two people could be expected to draw the line in exactly the same place.

A frequently used technique for determining the regression equation is the *least-squares method*. Since the values of a, the ordinate intercept, and b, the slope, are mathematically determined, the results are reproducible. That is, two or more people using the same data set and the least-squares method will arrive at identical regression equations, barring arithmetic errors. The line determined using this technique is unique in that *the sum of the squared deviations* of the values of Y from the regression line are a *minimum*. Thus the least-squares regression line and the arithmetic mean are similar in this respect. Moreover the sampling distribution associated with a least-squares regression line has similar characteristics to the sampling distribution of the arithmetic mean.

Before you apply the least-squares method of regression to a set of data, you should consider the following assumptions associated with its use:

1. There is a linear relationship between X and Y.

2. Y is a random variable, and the values of X are known constants.

3. For every value of X there is a subset of values of the dependent variable Y, and each of these subsets of Y values has the same standard deviation σ. The condition described by this assumption, equal standard deviations, is called *homoscedasticity*.

4. The deviations $Y - Y_c$ between the actual values of Y and the estimated average value using regression are independent.

5. The subsets of Y values for each X value are normally distributed. (This assumption is necessary only if the researcher desires to make inferences about the population parameters.)

We sometimes apply the least-squares method to a data set that does not have all the characteristics described by these assumptions. When we do this, we are simply adding another assumption. That is, there is not enough difference to be important. However, the assumptions still exist, and you should recognize that your results, and any decisions based upon them, may be incorrect if the data set does not meet the assumptions.

To compute the values of a and b in the linear regression equation, we use two *normal equations*. These basic normal equations are

$$\Sigma Y = na + b\Sigma X, \tag{12.3}$$
$$\Sigma XY = a\Sigma X + b\Sigma X^2, \tag{12.4}$$

where X and Y refer to the actual values of the independent and dependent variables respectively, and n is the number of observations.

To use the normal equations in their present forms, with two unknowns in each equation, we must employ some method (such as the elimination procedure) to solve for a and b. Consider the following normal equations. By the elimination process we compute the values of a and b as follows:

$$40 = 5a + 20b,$$
$$100 = 20a + 140b.$$

1. Multiply this first equation by -4 so the coefficients of a are the same in both equations:

$$-4(40) = -4(5a + 20b).$$

Thus

$$-160 = -20a - 80b,$$
$$100 = 20a + 140b.$$

2. Sum the two equations. That is, fuse them into one equation with one unknown:

$$-60 = 60b.$$

3. Solve for the remaining unknown:

$$b = -1.$$

4. Substitute and solve for the other unknown:

$$-160 = -20a - 80(-1),$$
$$a = 12.$$

5. Check results:

$$100 = 20(12) + 140(-1),$$
$$100 = 240 - 140.$$

Another way to solve for two unknowns in two equations is to solve for one unknown in terms of the other in one equation and then substitute in the second equation. The results of such a procedure would be the following equations:

$$b = \frac{n\Sigma XY - (\Sigma X)(\Sigma Y)}{n(\Sigma X^2) - (\Sigma X)^2}, \qquad (12.5)$$

$$a = \frac{\Sigma Y}{n} - b\left(\frac{\Sigma X}{n}\right) = \overline{Y} - b\overline{X}. \qquad (12.6)$$

Example 12.2

▷ Chromie Motors, a manufacturer of large automobiles, has recently hired a consultant to help boost its sales and thereby the morale of its management (and hopefully its employees). The consultant, a marketing specialist, has reasoned that sales may be a function of advertising expenditures. Thus he has decided to analyze the relationship between sales and advertising expenditures for the company. The sales and advertising expenditure data are given in table 12.1. With these data we can now compute the regression equation for the consultant.

Table 12.1
Chromie Motors sales and advertising expenditures

Year	Advertising expenditures	Sales
1967	$100,000	$9,000,000
1968	105,000	8,000,000
1969	90,000	5,000,000
1970	80,000	2,000,000
1971	80,000	4,000,000
1972	85,000	6,000,000
1973	87,000	4,000,000
1974	92,000	7,000,000
1975	90,000	6,000,000
1976	95,000	7,000,000
1977	93,000	5,000,000
1978	85,000	5,000,000
1979	85,000	4,000,000
1980	70,000	3,000,000
1981	85,000	3,000,000

X ($1,000)	Y ($1,000,000)	XY	X^2	Y^2
100	9	900	10,000	81
105	8	840	11,025	64
90	5	450	8,100	25
80	2	160	6,400	4
80	4	320	6,400	16
85	6	510	7,225	36
87	4	348	7,569	16
92	7	644	8,464	49
90	6	540	8,100	36
95	7	665	9,025	49
93	5	465	8,649	25
85	5	425	7,225	25
85	4	340	7,225	16
70	3	210	4,900	9
85	3	255	7,225	9
1,322	78	7,072	117,532	460

$$b = \frac{(15)(7,072) - (1,322)(78)}{(15)(117,532) - (1,322)^2} = 0.1938$$

$$a = \frac{(78)}{(15)} - (0.194)\frac{(1,322)}{(15)} = -11.8802$$

$$Y_c = -11.8802 + 0.1938X$$

$$Y_c = a + bx$$

Table 12.2
Chromie Motors advertising expenditures and estimated sales

X ($1,000)	Y_C ($1,000,000)
70	1.6858
80	3.6238
85	4.5928
87	4.9807
90	5.5618
92	5.9494
93	6.1432
100	7.4998
105	8.4688

Predictions Based on the Regression Equation

One of the purposes of regression analysis is to make predictions of the values of the dependent variable based on certain values of the independent variable. If we look at the equation

$$Y_c = -11.8802 + 0.1938X,$$

we see that once we have computed values for a and b, we can assign values to X and solve for Y_c. Through this procedure Y_c becomes an estimate of the average value of Y given a value of X. Using the equation developed for the data in example 12.2, the Y_c values may be computed as shown in table 12.2.

By plotting Y_c and X, the regression line is determined. We need only plot any two points here, because the regression line is a straight line. However, it may be a good idea to plot at least three points to avoid mistakes. If there has been no error in computing Y_c, given the assigned values of X, the regression line should pass through all three points. Once the regression line is plotted (see figure 12.2), it becomes fairly obvious that, in this example, there is reasonably high correlation.

Standard Error of Estimate

The *standard error of estimate is the standard deviation for the observed values of Y from the regression line*. It's a computed value that shows the average amount by which the individual Y values vary from the regression line. Graphically, the standard error of estimate shows, on the average, how much the individual dots (observations) in a scatter diagram vary from the regression line. Since the regression line represents a standard of perfect correlation, the larger the standard error of estimate relative to the magnitude of the Y_c values, the lower the coefficient of correlation. If the stand-

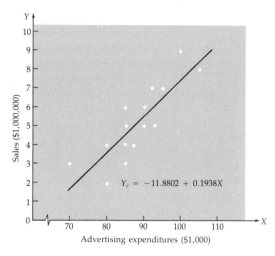

Figure 12.2   Scatter diagram and regression line for Chromie Motors data

ard error of estimate were 0, all the dots in the scatter diagram would fall precisely on the regression line, and therefore there would be perfect correlation.

In this section we will present two formulas for computing the standard error of estimate. The first formula, equation (12.7), is sometimes referred to as the long method (definition equation), and it is the one that most graphically represents the relationship between the individual Y values and the regression line values, Y_c:

$$s_{yx} = \sqrt{\frac{\Sigma(Y - Y_c)^2}{n - 2}}, \tag{12.7}$$

where

s_{yx} = the standard error of estimate,
Y = actual (observed) values of the dependent variable,
Y_c = regression values computed from the regression equation,
n = number of observations.

As is shown in this equation, the standard error represents the sum of the squares of the differences between the actual values of the dependent variable Y and the estimates of values of the dependent variable Y_c computed from the regression equation. Thus the standard error of estimate relates the actual dependent variable Y to the computed dependent variable Y_c. *The degrees of freedom in this case are represented by n − 2 because two parameters (α and β) have been estimated.* Using them in this equation assumes that we are involved in a sampling situation and recognizes the idea that we are computing an estimator of a parameter for the standard error of estimate.

The short method for computing the standard error of estimate requires a longer formula, equation (12.8), than equation (12.7):

$$S_{yx} = \sqrt{\frac{\Sigma Y^2 - a\Sigma Y - b\Sigma XY}{n - 2}}.$$ (12.8)

Equation (12.7) requires that each value of Y_c be computed from the regression equation, then subtracted from Y, and then squared; whereas the formula for the short method requires only one additional element, ΣY^2, to be computed. The other information in the formula has already been obtained in computing the regression equation. That is, we have already determined ΣY and ΣXY. Obviously, we know the constants a and b from the regression equation. The value of ΣY^2 can be easily computed by squaring each value of Y and summing the squares. Both the long and short calculation methods are illustrated in example 12.3.

Example 12.3

▷ The consultant for Chromie Motors has decided he needs to measure the dispersion about the regression line.

X	Y	Y_c	$(Y - Y_c)$	$(Y - Y_c)^2$
100	9	7.4998	1.5002	2.2506
105	8	8.4688	-0.4688	0.2198
90	5	5.5618	-0.5618	0.3156
80	2	3.6238	-1.6238	2.6367
80	4	3.6238	0.3762	0.1415
85	6	4.5928	1.4072	1.9802
87	4	4.9804	-0.9804	0.9612
92	7	5.9494	1.0506	1.1038
90	6	5.5618	0.4382	0.1920
95	7	6.5308	0.4692	0.2201
93	5	6.1432	-1.1432	1.3069
85	5	4.5928	0.4072	0.1658
85	4	4.5928	-0.5928	0.3514
70	3	1.6858	1.3142	1.7271
85	3	4.5928	-1.5928	2.5370
				16.1097

Long method

$$S_{yx} = \sqrt{\frac{\Sigma(Y - Y_c)^2}{n - 2}}$$

$$S_{yx} = \sqrt{\frac{16.1097}{15 - 2}}$$

$$S_{yx} = \sqrt{1.2392} = 1.113$$

Short method

$$S_{yx} = \sqrt{\frac{\Sigma Y^2 - a\Sigma Y - b\Sigma XY}{n - 2}}$$

$$= \sqrt{\frac{460 - (-11.8802)(78) - (0.1938)(7072)}{15 - 2}}$$

$$S_{yx} = \sqrt{1.2386} = 1.113$$

◁

Variation Analysis

The main objective of correlation analysis is to try to relate the changes in the values of the dependent variable Y with the movements or changes in the values of the independent variable X. That is, the dependent variable is varying (otherwise it would not be considered a variable), and we are concerned with the extent of the variation that can be related to the independent variable.

To begin with, notice that in example 12.3 the standard error is the square root of the mean of the squared variations of Y and Y_c, adjusted for lost degrees of freedom. *The unexplained variation is that portion of the total variation of Y which is not associated with, or which cannot be related to, X.* Thus the more unexplained variation there is in analyzing a two-variable situation, the less correlation there is between the two variables X and Y under consideration. The concepts of unexplained variation, total variation, and explained variation will be presented in the next few pages.

Figure 12.3 shows a graphic analysis representing the three variations involved in a simple linear correlation and regression analysis. This figure depicts a scatter diagram with a horizontal line drawn through it at the value of the mean of Y (figure 12.3a). The mean line for the Y variable is of course a horizontal line because the mean of Y is the same for every value of X. On the other hand, the regression line, which is shown in the figure

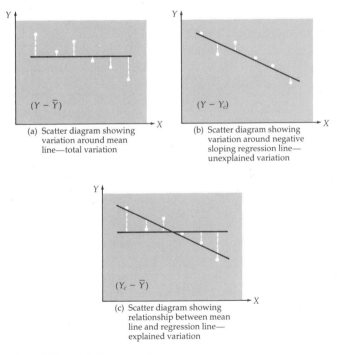

(a) Scatter diagram showing variation around mean line—total variation

(b) Scatter diagram showing variation around negative sloping regression line— unexplained variation

(c) Scatter diagram showing relationship between mean line and regression line— explained variation

Figure 12.3   Variation concepts

12.3b diagram, would be a straight line with either a negative or positive slope. Since the variation around the mean of Y represents the total variation of Y, the total variation may be graphically depicted by the scatter about the horizontal (mean) line. We can depict the unexplained variation graphically by the scatter about the regression line. So total variation is depicted by the scatter about the mean line, and unexplained variation is depicted by the scatter about the regression line.

Now let us consider the graphic representation of the explained variation. Perhaps it is not quite so obvious as the other two, but it is the variation in Y that is related to changes in the independent variable. This portion of the variation of the dependent variable can be viewed as the variation of Y_c about the mean and represented by the difference between the regression line and the mean line. In fact the explained variation is graphically depicted by the distance between the mean line and the regression line at each value of X as shown in figure 12.3c.

Another way to view the explained, unexplained, and total variation is in terms of *squared variation*, commonly referred to as *sum of squares*. This is shown in equations (12.9) through (12.11).

Total sum of squares:

$$SST = \Sigma(Y - \overline{Y})^2. \tag{12.9}$$

Regression (explained) sum of squares:

$$SSR = \Sigma(Y_c - \overline{Y})^2. \tag{12.10}$$

Error (unexplained) sum of squares:

$$SSE = \Sigma(Y - Y_c)^2. \tag{12.11}$$

As a matter of fact the total sum of squares can be partitioned into two components: regression sum of squares and error sum of squares; that is,

$$SST = SSR + SSE \tag{12.12}$$

or

$$\Sigma(Y - \overline{Y})^2 = \Sigma(Y_c - \overline{Y})^2 + \Sigma(Y - Y_c)^2. \tag{12.13}$$

The calculations of the sums of squares for Chromie Motors are shown in table 12.3.

Inference about Regression

Now that we have determined a, b, and s_{yx}, it is necessary to investigate whether the regression really exists. That may seem to be an unnecessary step since you might be reasonably sure that your calculations of a and b are accurate. However, if the

Table 12.3
Total, regression, and error sum of squares

Y	Y_c	$(Y - \overline{Y})$	$(Y - \overline{Y})^2$	$(Y - Y_c)$	$(Y - Y_c)^2$	$(Y_c - \overline{Y})$	$(Y_c - \overline{Y})^2$
9	7.4998	3.8	14.44	1.5002	2.2506	2.2998	5.2891
8	8.4688	2.8	7.84	−0.4688	0.2198	3.2688	10.6851
5	5.5618	−0.2	0.04	−0.5618	0.3156	0.3618	0.1309
2	3.6238	−3.2	10.24	−1.6238	2.6367	−1.5762	2.4844
4	3.6238	−1.2	1.44	0.3762	0.1415	−1.5762	2.4844
6	4.5928	0.8	0.64	1.4072	1.9802	−0.6072	0.3687
4	4.9804	−1.2	1.44	−0.9804	0.9612	−0.2196	0.0482
7	5.9494	1.8	3.24	1.0506	1.1038	0.7494	0.5616
6	5.5618	0.8	0.64	0.4382	0.1920	0.3618	0.1309
7	6.5308	1.8	3.24	0.4692	0.2201	1.3308	1.7710
5	6.1432	−0.2	0.04	−1.1432	1.3069	0.9432	0.8896
5	4.5928	−0.2	0.04	0.4072	0.1658	−0.6072	0.3687
4	4.5928	−1.2	1.44	−0.5928	0.3514	−0.6072	0.3687
3	1.6858	−2.2	4.84	1.3142	1.7271	−3.5142	12.3496
3	4.5928	−2.2	4.84	−1.5928	2.5370	−0.6072	0.3687
			54.40 (SST)		16.1097 (SSE)[a]		38.2996 (SSR)[a]

$SST = SSR + SSE$

a. Rounding errors cause the sum of SSR and SSE to differ slightly from SST.

data with which you are working are sample data, the values of a and b are only estimates of α and β, respectively. Since the values in your data set are assumed to be randomly selected, the values you calculate for a and b may not accurately reflect α and β. Put another way, we must ask whether b, the estimate of β, is really different from 0, *because, if $\beta = 0$, there is no slope to the regression line and thus no relationship between X and Y.*

In order to determine if the relationship really exists, it is necessary to test the null hypothesis that $\beta = 0$. Using the test statistic shown in equation (12.14) is one way do to this:

$$t_{test} = \frac{b - \beta_H}{s_b},$$
(12.14)

where
b = estimate of β from normal equations,
β_H = hypothesized value of β,
s_b = standard error (standard deviation) of b.

Also either equation (12.15) or (12.16) has to be used to calculate s_b, the standard error of the sample b's:

$$s_b = \frac{s_{yx}}{\sqrt{\Sigma(X - \overline{X})^2}},$$
(12.15)

$$s_b = \frac{s_{yx}}{\sqrt{\Sigma X^2 - \dfrac{(\Sigma X)^2}{n}}} . \qquad (12.16)$$

The degrees of freedom for the test are $n - 2$. And finally, to use equation (12.14) legitimately, we must assume that the Y values will be normally distributed for any Y_c or have a large n in order to rely on the central limit theorem.

Example 12.4

▷ The consultant for Chromie Motors now must determine if the regression line he has developed really has a significant slope:

H_0: $\beta = 0$,
H_A: $\beta \neq 0$,
     $\alpha = 0.05$,
     $df = n - 2 = 15 - 2 = 13$.

Reject H_0 if $t_{test} < -2.160$ or $t_{test} > +2.160$:

$$t_{test} = \frac{b - \beta_H}{s_b}$$

$$t_{test} = \frac{b - \beta_H}{\dfrac{s_{yx}}{\sqrt{\Sigma X^2 - \dfrac{(\Sigma X)^2}{n}}}}$$

$$= \frac{0.1938 - 0}{\dfrac{1.113}{\sqrt{117{,}532 - \dfrac{(1{,}322)^2}{15}}}}$$

$$t_{test} = \frac{0.1938 - 0}{\dfrac{1.113}{\sqrt{117{,}532 - 116{,}521.26}}}$$

$$= \frac{0.1938 - 0}{\dfrac{1.113}{\sqrt{1{,}019.74}}}$$

$$= \frac{0.1938 - 0}{\dfrac{1.113}{31.933}}$$

$$= \frac{0.1938 - 0}{0.0349}$$

$t_{test} = 5.559$.

Reject H_0; conclude that $\beta \neq 0$ and that the regression relationship does exist. ◁

In chapter 10, there is a discussion of analysis of variance and the
F distribution. This procedure provides an alternative for testing
the null hypothesis of $\beta = 0$. Analysis of variance yields the
same results as the t-test in equation (12.14) when you are testing
the slope of a simple linear regression equation. However, analysis of variance has broader possibilities for use in that it can be
applied to multiple regression equations to test several independent variables at one time, while the t-test can be applied
only to each independent variable separately.

When testing regression equations, analysis of variance utilizes the regression sum of squares and the error sum of squares.
Each sum of squares is divided by the appropriate degrees of
freedom to calculate what are known as mean squares. The two
mean squares are called *regression mean square (MSR)* and *error
mean square (MSE)* and are shown in equations (12.17) and
(12.18):

$$MSR = \frac{SSR}{m}, \tag{12.17}$$

where $SSR = \Sigma(Y_c = \overline{Y})^2$; \hfill (12.10)

$$MSE = \frac{SSE}{n - m - 1}, \tag{12.18}$$

where
$SSE = \Sigma(Y - Y_c)^2,$ \hfill (12.11)
  m = number of independent variables,
  n = sample size.

These mean squares are sample variances and can be used in the
F ratio to test for regression. The test statistic is

$$F_{\text{test}} = \frac{MSR}{MSE}, \tag{12.19}$$

and the results are frequently presented in an ANOVA summary
table as shown in table 12.4.

For table 12.4 the following computational equations can be
used to determine the sum of squares:

$$SST = \Sigma Y^2 - \frac{(\Sigma Y)^2}{n}, \tag{12.20}$$

$$SSE = \Sigma Y^2 - a\Sigma Y - b\Sigma XY, \tag{12.21}$$

$$SSR = SST - SSE. \tag{12.22}$$

Table 12.4
Anova summary table: general presentation

Source of variation	Sum of squares	Degrees of freedom	Mean squares	F_{test}
Regression	SSR	m	$MSR = \dfrac{SSR}{m}$	$\dfrac{MSR}{MSE}$
Error	SSE	$n - m - 1$	$MSE = \dfrac{SSE}{n - m - 1}$	
Total	SST	$n - 1$		

Thus, using the data in table 12.1, we can test β in the following manner:

H_0: $\beta = 0$,
H_A: $\beta \neq 0$,
  $\alpha = 0.05$.

Reject H_0 if $F_{test} > 4.67$:

$$SST = \Sigma Y^2 - \frac{(\Sigma Y)^2}{n} = 460 - \frac{(78)^2}{15}$$

$SST = 54.40$,
$SSE = \Sigma Y^2 - a\Sigma Y - b\Sigma XY$
  $= 460 - (-11.8802)(78) - 0.1938(7072)$
  $= 460 + 926.6556 - 1370.5536$
$SSE = 16.102$,
$SSR = 54.40 - 16.102 = 38.298$,

$$F_{test} = \frac{\dfrac{38.298}{1}}{\dfrac{16.102}{13}} = 30.92.$$

Since $F_{test} > 4.67$, reject H_0. The above calculations are summarized in table 12.5. From these results we can conclude $\beta \neq 0$ and regression does exist.

In this example the critical value of $F(4.67)$ was found by using the table in appendix J. The degrees of freedom are m and $(n - m - 1)$, with $df = m$ across the top and $df = (n - m - 1)$ down the side.

If you look back at example 12.4, you can see that the results using analysis of variance are consistent with that achieved using the t-test.

Table 12.5
Anova summary table

Source of variation	Sum of squares	Degrees of freedom	Mean squares	F_{test}
Regression	38.298	1	38.298	30.92
Error	16.102	13	1.2386	
Total	54.40	14		

The Coefficients— Measuring the Relationship

Earlier in this chapter we said that correlation analysis was primarily concerned with measuring the relationship between the variables. For instance, we may ask, "How will income levels and education levels correlate?" In correlation analysis such relationships between variables are measured by computing certain coefficients. There are four such coefficients associated with simple linear correlation and regression analysis:[1]

1. The coefficient of determination, r^2.

2. The coefficient of correlation, r.

3. The coefficient of nondetermination, k^2.

4. The coefficient of alienation, k.

While the *coefficient of correlation, r,* is perhaps the best known of the four, we shall soon learn that the *coefficient of determination* is more meaningful. Also note that the coefficient of correlation, r, is the square root of the coefficient of determination, r^2. The same relationship exists between the *coefficient of alienation* and the *coefficient of nondetermination*.

Computation and Meaning of the Coefficients

The coefficient of determination, r^2, is defined as the proportion (or percent) of the variation in the values of the dependent variable that is "explained" by variations in the values of the independent variable. Thus, if the total variation squared (SST) is 54.40 and the explained variation squared (SSR) is 38.2996, then the coefficient of determination equals 0.704 or 70.4 percent. That is,

$$r^2 = \frac{\Sigma(Y_c - \overline{Y})^2}{\Sigma(Y - \overline{Y})^2} \qquad (12.23)$$

$$= \frac{38.2996}{54.40}$$

$$r^2 = 0.704 \text{ or } 70.4\%.$$

1. In this text, r, r^2, k, and k^2 are all sample statistics. To be more precise estimates of the population parameters, the sample statistics should be adjusted for degrees of freedom.

Or using computational formulas for determining the numerator and denominator, equation (12.23) becomes

$$r^2 = \frac{SSR}{SST} = \frac{SST - SSE}{SST}$$

$$= \frac{\left[\Sigma Y^2 - \frac{(\Sigma Y)^2}{n}\right] - [\Sigma Y^2 - a\Sigma Y - b\Sigma XY]}{\Sigma Y^2 - \frac{(\Sigma Y)^2}{n}} \qquad (12.24)$$

$$= \frac{\left[460 - \frac{(78)^2}{15}\right] - \left[460 - (-11.8802)(78) - (0.1938)(7072)\right]}{460 - \frac{(78)^2}{15}}$$

$$= \frac{[460 - 405.6] - [460 + 926.6556 - 1370.5536]}{460 - 405.60}$$

$$r^2 = \frac{54.40 - 16.102}{54.40} = \frac{38.298}{54.40} = 0.704.$$

And if the actual sums of squares are not needed, the following equation (called the product-moment formula) is a more direct method for determining r and r^2. This equation, however, does not distinguish between the dependent and independent variables.

$$r = \frac{n\Sigma XY - \Sigma X\Sigma Y}{\sqrt{[n\Sigma X^2 - (\Sigma X)^2][n\Sigma Y^2 - (\Sigma Y)^2]}} \qquad (12.25)$$

$$r^2 = (r)^2;$$

$$r = \frac{15(7072) - (1322)(78)}{\sqrt{[15(117532) - (1322)^2][15(460) - 78^2]}}$$

$$= \frac{106080 - 103116}{\sqrt{(1762980 - 1747684)(6900 - 6084)}}$$

$$= \frac{2964}{\sqrt{(15296)(816)}} = \frac{2964}{\sqrt{12481536}} = \frac{2964}{3533}$$

$$r = 0.839,$$
$$r^2 = 0.704.$$

The coefficient of correlation is an abstract measure of the degree of relationship between variables. The measure is based on a scale value between 0 and 1 in the case of positive correlation and 0 and −1 in the case of negative correlation. A value of 0 means there is no correlation, and a value of 1 or −1 means perfect correlation. The

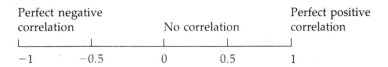

Figure 12.4 The meaning of correlation values

sign of $r(+$ or $-)$ depends on the sign of b in the regression equation. If b is positive, the correlation is positive; if b is negative, the correlation is inverse or negative (figure 12.4).

Like all scale values, the coefficient of correlation is difficult to interpret. Unlike the coefficient of determination, it cannot be given a direct interpretation. Yet, we can say that, in general, the higher the absolute value (to allow for negative correlation) of the coefficient, the higher the correlation.

The coefficient of nondetermination is the proportion of the variation in the dependent variable that is not explained by changes in the values of the independent variable. It is equal to the unexplained variation divided by the total variation. Using example 12.3,

$$k^2 = \frac{\Sigma(Y - Y_c)^2}{\Sigma(Y - \overline{Y})^2} \qquad (12.26)$$

$$= \frac{16.09}{54.40}$$

$$k^2 = 0.296 \text{ or } 29.6\%.$$

From our previous discussion of variations, it can be shown that there is another way to calculate the coefficient of nondetermination. The two coefficients (determination and nondetermination) must sum to one (1) if all variation is explained or unexplained. Thus, if r^2 is already known, then equation (12.27) may be used:

$$k^2 = 1 - r^2. \qquad (12.27)$$

The coefficient of alienation is an abstract measure of the lack of correlation and is equal to the square root of the coefficient of nondetermination. That is, $k = \sqrt{k^2}$. Like the coefficient of correlation, k is a rating on a scale. But unlike r, k cannot be positive or negative. It is an unsigned (absolute value) rating. The sign attached to r means direction of a relationship. The coefficient of alienation k means the absence of a relationship, no direction, so it can take on no sign.

Earlier we discussed the relationship between correlation and regression analysis and sampling. If we consider that we are dealing with a sample situation in the analysis, then the coefficients that we computed are really not parameter values; they are estimators based on the observed (sample) data. However, r is a biased estimator of the population coefficient of correlation in that the sample coefficient of correlation tends to be greater than the population value. This is particularly true when you are working with small samples of population values. While we will not present the method here, in more advanced texts you will find a technique of adjusting r so that it becomes an unbiased estimator of the population coefficient of correlation.

Types of Relationships

One frequent error we are prone to make in correlation and regression analysis is to interpret a high correlation between the variables as a cause-and-effect relationship. Correlation analysis measures a relationship; it does not define the basis for it. Thus a high coefficient of determination is interpreted to mean a high proportion of explained variation, but the term "explained" as used in this context does not mean "causal." It means that the variables are closely related in some undefined way. The relationship between the variables may be purely coincidental (a casual relationship), or it may be due to seriality where some third phenomenon or variable is providing the basis for the relationship.

Suppose it can be shown that there is high correlation between the salaries of professional marriage counselors and the divorce rate in America. It is highly unlikely that one variable here is causing the other. They may show high, positive correlation because each variable is being affected by a changing cultural pattern in the country—a serial effect. Yet, we might find that the high negative correlation between popcorn sales and the incidence of television viewing is causal since popcorn sales occur at movie theaters.

In no case does correlation analysis itself, then, define the type of relationship. At best, one might say there must be correlation for a meaningful relationship to exist. Thus, if two variables show little or no correlation, it would be difficult to conclude that change in one is possibly causing change in the other.

Confidence Intervals

Earlier in the chapter we utilized the expression, $Y_c = a + bX$, to estimate values of Y for given values of X. Actually, these estimated values of Y, designated Y_c, are estimates of the *average value of Y for a given X*. If we are using sample data, we can use these averages to construct confidence intervals for the Y values for a given X. In particular, we will be able to construct two sets

BASIC STATISTICS

of confidence intervals. The first interval is for the average value of Y (denoted Y_c) that might have been obtained from the sample information, and the second interval is for an individual value of Y, which is sometimes referred to as the forecasted or predicted value of Y.

Confidence Interval for Y_c

Construction of a confidence interval for Y_c is actually a construction of *a range for the possible average value for Y for a given X*. Such an interval exists because, from regression analysis, only *one* of many possible regression lines is obtained from a single sample. Since a and b are subject to sampling variation, any estimates that use a and b are subject to sampling error. Y_c reflects this variation, and so we need a measure of dispersion to account for this sampling error. This measure will be called the *standard error of Y_c*, and we will use equation (12.28) to calculate this measure of dispersion. Then, to construct confidence intervals for Y_c, use equation (12.29):

$$s_{yc} = s_{yx} \sqrt{\frac{1}{n} + \frac{(X - \bar{X})^2}{\Sigma X^2 - \frac{(\Sigma X)^2}{n}}}, \tag{12.28}$$

$(1 - \alpha)$ confidence interval for $Y_c = (a + bX) \pm ts_{yc}.$ (12.29)

Example 12.5

▷ Continuing with Chromie Motors, let us construct a 95 percent confidence interval for Y_c if X is 80 thousand dollars.

95% confidence interval for Y_c

$$= (a + bX) \pm ts_{yx} \sqrt{\frac{1}{n} + \frac{(X - \bar{X})^2}{\Sigma X^2 - \frac{(\Sigma X)^2}{n}}},$$

$$= [-11.8802 + (0.1938)(80)]$$

$$\pm (2.16)1.113 \sqrt{\frac{1}{15} + \frac{(80 - 88.13)^2}{117532 - 116512.27}}$$

$$= (-11.8802 + 15.504) \pm 2.4041 \sqrt{\frac{1}{15} + \frac{66.0969}{1019.73}}$$

$$= 3.6238 \pm 2.4041 \sqrt{0.0667 + 0.0648}$$

$$= 3.6238 \pm 2.4041(0.3626)$$

$$= 3.6238 \pm 0.8717,$$

$$2.7521 \leq Y_c \leq 4.4955.$$

Thus the average value of sales when the advertising expenditure is 80 thousand dollars has a lower limit of 2.7521 million dollars and an upper limit of 4.4955 million dollars. ◁

Confidence Interval for Y

The confidence interval previously constructed is for the average value of sales (Y_c) when X is a particular value. At times, however, it is more desirable to construct a confidence interval for the single *predicted* value (Y) rather than an average value. In order to do this, it is necessary to have the proper measure of dispersion. Such a measure includes the sampling error present in s_{yc}, but it also must include the variation due to the individual observations being scattered about the regression line. We will call this measure of dispersion the *standard error of a predicted value of Y* and use equation (12.30) to calculate s_{yp}:

$$s_{yp} = s_{yx} \sqrt{1 + \frac{1}{n} + \frac{(X - \bar{X})^2}{\Sigma X^2 - \frac{(\Sigma X)^2}{n}}} . \tag{12.30}$$

Then, to construct confidence intervals for a predicted value of Y for a given X, use equation (12.31):

$(1 - \alpha)$ confidence interval for $Y = (a + bX) \pm ts_{yp}$. (12.31)

Example 12.6

▷ Let us now construct a 95 percent confidence interval for the predicted sales of Chromie Motors when the advertising expenditure is 80 thousand dollars.

95% confidence interval for Y_c

$$= [a + bX] \pm ts_{yx} \sqrt{1 + \frac{1}{n} + \frac{(X - \bar{\bar{X}})^2}{\Sigma X^2 - \frac{(\Sigma X)^2}{n}}} ,$$

$= [-11.8802 + 0.1938(80)]$
$\quad \pm (2.16)1.113\sqrt{1 + 0.0667 + 0.0648}$
$= 3.6238 \pm 2.4041\sqrt{1.1315}$
$= 3.6238 \pm 2.4041(1.0637)$
$= 3.6238 \pm 2.5572,$
$1.0666 \le Y \le 6.1810.$ ◁

You should notice that the interval for the predicted value of an individual Y is larger than the interval for the average value of Y. This is true because predicting individual values is subject to more sampling variation, thus more variability, than predicting average values.

Figure 12.5 shows the situation. The dotted limits are for the predicted values of Y for the possible X's. Also the interval becomes larger as the value of the independent variable X moves

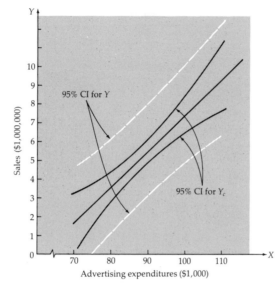

Figure 12.5 Chromie Motors regression line and confidence intervals for sales

away from $\overline{X}$. Thus the predictors become more vulnerable or, suspect, as we let X vary more and more from $\overline{X}$.

A Complete Example

A student organization at State U. is interested in the relationship between student grade point averages (GPA) and the score made on the school's entrance examination. Moreover the organization leaders are also wondering whether any real meaningful predictions of students' GPA's can be made from these entrance test scores.

To answer the questions, an investigation of the relationship between GPA and entrance test scores was undertaken. A sample of 30 records has been randomly selected from the entire senior class (all fourth-year students). The GPA's and entrance test scores for the 30 students are shown in table 12.6.

1. Figure 12.6 shows a scatter diagram between test scores (X) and the dependent variable, GPA (Y).

Also the following values are determined from the sample data:

$$
\begin{array}{ll}
n = 30 & \Sigma Y^2 = 213.12 \\
\overline{X} = 129.7 & \Sigma X = 3{,}891 \\
\overline{Y} = 2.573 & \Sigma XY = 10{,}677.68 \\
\Sigma Y = 77.19 & \Sigma X^2 = 542{,}455
\end{array}
$$

Table 12.6
Admission test scores and grade point averages

Student number	Admission test score (200 pts. possible), X	GPA (4.0 possible), Y
128	92	1.75
934	145	2.81
156	180	3.50
408	142	2.75
389	102	1.95
515	77	1.65
423	155	3.20
056	135	2.80
109	70	1.40
310	170	3.49
915	72	1.54
405	105	2.00
985	168	3.46
361	87	1.65
020	112	2.35
490	108	2.05
581	96	2.02
095	100	1.91
430	175	3.60
756	132	2.75
615	165	2.41
324	148	3.05
049	170	3.29
519	101	1.93
804	102	3.15
439	119	2.28
650	130	2.50
522	168	3.45
888	185	3.75
158	180	2.75

2. The regression equation is computed as follows:

$$b = \frac{30(10677.68) - (3891)(77.19)}{30(542455) - (3891)^2}$$

$$b = \frac{19984.11}{1133769} = 0.0176.$$

$$a = \overline{Y} - b\overline{X}$$

$$a = 2.573 - 129.7(0.0176)$$

$$a = 0.2903,$$

$$Y_c = 0.2903 + 0.0176X.$$

The regression line is shown in figure 12.6.

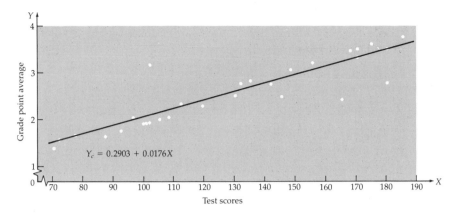

$Y_c = 0.2903 + 0.0176X$

Figure 12.6   Scatter diagram and regression line showing pattern of relationship between grade point average and entrance test scores at State U. for 30
randomly selected seniors

3. The standard error of estimate is

$$s_{yx} = \sqrt{\frac{213.12 - (77.19)(0.2903) - (10677.68)(0.0176)}{30 - 2}}$$

$$s_{yx} = \sqrt{\frac{2.7845}{28}} = \sqrt{0.0994} = 0.3153.$$

4. If we want to test the value of b, then

H_0: $\beta = 0$,
H_A: $\beta \neq 0$,
    $\alpha = 0.05$,
    $df = n - 2 = 30 - 2 = 28$.

Reject H_0 if $t_{test} < -2.048$ or $t_{test} > 2.048$.

$$t_{test} = \frac{b - \beta_H}{\dfrac{s_{yx}}{\sqrt{\sum X^2 - \dfrac{(\sum X)^2}{n}}}}$$

$$t_{test} = \frac{0.0176 - 0}{\dfrac{0.3153}{\sqrt{542{,}455 - \dfrac{(3{,}891)^2}{30}}}}$$

$$= \frac{0.0176}{\dfrac{0.3153}{\sqrt{37{,}792.3}}}$$

$$t_{test} = \frac{0.0176}{0.00162} = 10.86.$$

Reject H_0, and conclude the regression relationship does exist.

Alternatively, the hypotheses can be tested using analysis of variance. In that case

H_0: There is no regression
H_A: There is regression
$\alpha = 0.05$

Reject H_0 if $F_{test} > 4.20$,

$$SST = \Sigma Y^2 - \frac{(\Sigma Y)^2}{n}$$

$$= 213.12 - \frac{(77.19)^2}{30}$$

$$= 213.12 - 198.6099$$

$$SST = 14.5101,$$

$$SSE = \Sigma Y^2 - a\Sigma Y - b\Sigma XY$$
$$= 213.12 - 0.2903(77.19) - 0.0176(10,677.68)$$
$$= 213.12 - 22.4083 - 187.9272$$

$$SSE = 2.7845,$$

$$SSR = SST - SSE$$
$$= 14.5101 - 2.7845$$

$$SSR = 11.7256,$$

$$MSR = \frac{SSR}{m}$$

$$= \frac{11.7256}{1}$$

$$MSR = 11.7256,$$

$$MSE = \frac{SSE}{n - m - 1}$$

$$= \frac{2.7845}{30 - 1 - 1}$$

$$MSE = 0.0994,$$

$$F_{test} = \frac{MSR}{MSE}$$

$$= \frac{11.7256}{0.0994}$$

$$F_{test} = 117.96.$$

Reject H_0, and conclude the regression relationship does exist.

5. The correlation coefficients are computed as follows:

$$r = \frac{n\Sigma XY - \Sigma X\Sigma Y}{\sqrt{[n\Sigma X^2 - (\Sigma X)^2][n\Sigma Y^2 - (\Sigma Y)^2]}}$$

$$= \frac{30(10,677.68) - (3,891)(77.19)}{\sqrt{[30(542,455) - (3,891)^2][30(213.12) - (77.19)^2]}}$$

$$= \frac{320,330.4 - 300,346.29}{\sqrt{(16,273,650 - 15,139,881)(6,393.6 - 5,958.3)}}$$

$$= \frac{19,984.11}{\sqrt{(1,133,769)(435.3)}} = \frac{19,984.11}{\sqrt{493,529,645.7}}$$

$$r = \frac{19,984.11}{22,215.52} = 0.90,$$

$$r^2 = (r)^2$$
$$r^2 = 0.81,$$
$$k^2 = 1 - r^2$$
$$= 1 - 0.81$$
$$k^2 = 0.19,$$
$$k = \sqrt{k^2}$$
$$k = 0.4359.$$

6. To make a 95 percent confidence interval estimate of the mean of Y_c at $X = 150$:

$(1 - \alpha)$ confidence interval for Y_c

$$= [a + bX] \pm ts_{yx} \sqrt{\frac{1}{n} + \frac{(X - \bar{X})^2}{\Sigma X^2 - \dfrac{(\Sigma X)^2}{n}}},$$

$$= [0.2903 + 0.0176(150)]$$

$$\pm 2.048(0.3153) \sqrt{\frac{1}{30} + \frac{(150 - 129.7)^2}{542,455 - \dfrac{(3,891)^2}{30}}}$$

$$= [0.2903 + 2.64] \pm 0.646 \sqrt{\frac{1}{30} + \frac{412.09}{37792.3}}$$

$$= 2.93 \pm 0.646\sqrt{0.033 + 0.011}$$
$$= 2.93 \pm 0.646\sqrt{0.044}$$
$$= 2.93 \pm 0.646(0.210)$$
$$= 2.93 \pm 0.136,$$
$$2.794 \le Y_c \le 3.066.$$

7. To determine the 95 percent confidence interval estimate for a single value of Y when $X = 150$,

$(1 - \alpha)$ confidence interval for Y

$$= [a + bX] \pm ts_{yx} \sqrt{1 + \frac{1}{n} + \frac{(X - \overline{X})^2}{\Sigma X^2 - \frac{(\Sigma X)^2}{n}}},$$

$$= [0.2903 + 0.0176(150)]$$

$$\pm 2.048(0.3153) \sqrt{1 + \frac{1}{30} + \frac{(150 - 129.7)^2}{542,455 - \frac{(3,891)^2}{30}}}$$

$$= [0.2903 + 2.64] \pm 0.646 \sqrt{\frac{31}{30} + \frac{412.09}{37,792.3}}$$

$$= 2.93 \pm 0.646\sqrt{1.033 + 0.011}$$
$$= 2.93 \pm 0.646\sqrt{1.044}$$
$$= 2.93 \pm 0.646(1.022)$$
$$= 2.93 \pm 0.660,$$
$$2.27 \leq Y \leq 3.59.$$

This example can also serve to illustrate another point. The highest possible score on the entrance test is 200 points, and the highest grade point average is 4.0. Thus X cannot exceed 200 and Y cannot be greater than 4.0. In other words, there are limits to this particular regression equation.

Appendix
Data Transformation

There may be times when the two variables you are attempting to correlate do not appear to have a linear relationship. For example, a scatter diagram, such as that depicted by figure 12.7, might show there is a relationship between the dependent and inde-

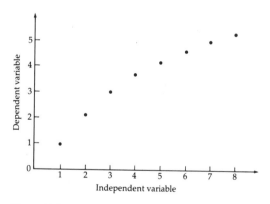

Figure 12.7   An illustration of a nonlinear relationship

pendent variables but not a linear one. However, it may still be possible to utilize linear correlation and regression analysis to analyze the data if one, or both, of the variables are *transformed* prior to the analysis.

Data transformation is a systematic conversion, or restatement, of all the numeric values of some variable. More specifically, data are transformed when the values are raised to some power, reciprocals are taken, logarithmic conversions are made, and so on.

Table 12.7 contains simplified illustrations of two transformation schemes. The most common of the two is the use of logarithms. That will be illustrated in example 12.7. But, before we look at the example, there are some things you should be aware of when transforming data.

When a given variable can take on both positive and negative values, as well as zero, data transformation by itself cannot be

Table 12.7
Examples of data transformation

Independent variable, X	Transformation by converting to logarithms	
	Dependent variable, Y	Logarithmic transformed dependent variable, $\log Y$
1	1.0000	0.00000
2	1.5000	0.17609
3	2.2500	0.35218
4	3.3750	0.52827
5	5.0625	0.70432
6	7.5938	0.88047
7	11.3906	1.05652
8	17.0859	1.23274

Independent variable, X	Transformation by squaring	
	Dependent variable, Y	Squared dependent variable, Y^2
1	1.0000	1.000
2	2.2361	5.000
3	3.0000	9.000
4	3.6056	13.000
5	4.1231	17.000
6	4.5826	21.000
7	5.0000	25.000
8	5.3852	29.000

used to convert nonlinear data to a linear form. For example, there is no logarithm for zero. Next, since the basic objective of data transformation is to allow you to use linear regression in analyzing data that are behaving in some nonlinear fashion, any equation that you develop will not provide for a change in direction. That is, the dependent variable will always be treated as if it has a positive or negative slope over its entire range of values. Finally, a linear equation fitted to data that have been transformed using logarithms implies a constant rate of change. If the slope of the dependent variable is positive, a constant rate of change is, in effect, a compound rate of growth, so that a constant amount of change in the independent variable yields accelerating amounts of change in the dependent variable.

Exponential Regression and Correlation Analysis

Let us consider the first of the two nonlinear cases mentioned in the previous section, where Y is changing by a certain rate while X is changing by a certain amount. If we were to plot a scatter diagram of such a situation on regular arithmetic graph paper, the scatter might appear something like the pattern shown in figure 12.8a. But if the data were plotted on semilog graph paper, then the scatter would appear as shown in figure 12.8b. The linear relationship shown in this figure, 12.8b, indicates that the regression equation should be an exponential one. That is, the dependent variable is changing by a constant rate rather than by a constant amount.

The Exponential Regression Equation

There are two ways in which this type of relationship can be expressed as a regression equation. It can be expressed in terms of Y_c itself, or it can be expressed in terms of the log Y_c. In the first case the equation reads as follows:

$$Y_c = ab^X \tag{12.32}$$

In terms of log Y_c the equation is

$$\log Y_c = \log a + X \log b \tag{12.33}$$

A close examination of the second equation reveals that it is a linear equation in its logarithmic form.[2] This bears out what the scatter diagram in figure 12.8b indicates. In the log form of Y, the relationship is linear.

---

2. If you do not understand logarithms nor remember how to determine antilogs from logs, look at the section at the beginning of appendix A.

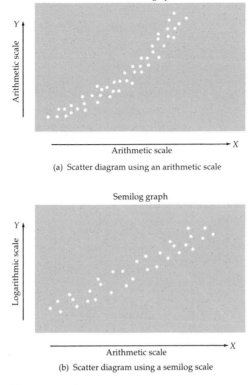

Figure 12.8   Scatter diagrams

Given this linear relationship for log Y and X, the procedure for finding Y_c requires that you first transform all Y values into log Y. Then you compute log b and log a using equations (12.34) and (12.35). After finding log b and log a, you can use equation (12.33) to compute log Y_c. Next, you find Y_c by determining the antilog of log Y_c.

$$\log b = \frac{[n\Sigma(X \log Y)] - [(\Sigma X)(\Sigma \log Y)]}{n(\Sigma X^2) - (\Sigma X)^2}, \tag{12.34}$$

$$\log a = \frac{\Sigma \log Y}{n} - \log b \frac{\Sigma X}{n}. \tag{12.35}$$

Example 12.7

▷ The Vice-President for Sales of Our Favorite Publisher, Inc., has been collecting data from the expense reports of the company's book salesmen. The salesmen frequently entertain prospective customers by taking them to dinner. In casually looking at the data, the Vice-President has decided that the entertainment cost appears to increase faster than the size of the party being entertained. The Vice-President decided to pursue his idea a little further and came up with the data and calculations in table 12.8.

The equation shown in table 12.8 can also be written in antilog form:

$$Y_c = 6.073(1.289)^X.$$

Looking at the equation written in this fashion, we can interpret the b coefficient. Specifically, b is one plus the *rate* of increase in the dependent variable as the independent variable increases one unit. In this case the value of b tells us that entertainment

Table 12.8
Entertainment cost of Our Favorite Publisher salesmen (logarithmic regression equation)

Size of party, X	Entertainment cost, Y	log Y	X log Y	X²	(log Y)²
1	$ 5.00	0.69897	0.69897	1	0.488559
2	10.20	1.00860	2.01720	4	1.017274
3	15.35	1.18611	3.55833	9	1.406857
4	20.50	1.31175	5.24700	16	1.720688
5	25.95	1.41414	7.07070	25	1.999792
6	32.20	1.50786	9.04716	36	2.273642
7	38.50	1.58546	11.09822	49	2.513683
8	46.00	1.66276	13.30208	64	2.764771
9	53.80	1.73078	15.57702	81	2.995599
10	62.00	1.79239	17.92390	100	3.212662
55	$309.50	13.89882	85.54058	385	20.393527

$$\log b = \frac{n\Sigma\,(X \log Y) - (\Sigma X)(\Sigma \log Y)}{n(\Sigma X^2) - (\Sigma X)^2}$$

$$= \frac{(10)(85.54058) - (55)(13.89882)}{(10)(385) - (55)^2}$$

$$= \frac{855.4058 - 764.4351}{3850 - 3025}$$

$$= \frac{90.9707}{825}$$

$$\log b = 0.11027$$

$$\log a = \frac{\Sigma \log Y}{n} - \log b\left(\frac{\Sigma X}{n}\right)$$

$$= \frac{13.89882}{10} - (0.11027)\left(\frac{55}{10}\right)$$

$$= 1.389882 - 0.606485$$

$$\log a = 0.783397$$

$$\log Y_c = 0.78340 + 0.11027\,X$$

costs increases 28.90 percent each time the party size increases by 1. ◁

Just as with other regression equations, it is possible to substitute a value for X and calculate Y_c. Suppose we use the equation from table 12.8 to calculate the value of Y_c when the size of the party (X) is 20:

$$\log Y_c = 0.78340 + 0.11027X$$
$$= 0.78340 + 0.11027(20)$$
$$= 0.78340 + 2.20540$$
$$\log Y_c = 2.98880,$$
$$Y_c = \text{antilog} (2.98880)$$
$$Y_c = \$974.50.$$

Perhaps you are surprised at the magnitude of Y_c in this example. You should remember that, in using exponential equations, we are assuming a *constant rate of change,* and this can cause our estimate of the dependent variables (Y_c) to *explode* with relatively small increases in the magnitude of the independent variable.

Coefficients of Correlation and Determination

The quality of the relationship between the two variables can also be measured using the exponential equation as the standard of perfect correlation. In the same manner as shown earlier in the chapter, we can calculate the coefficient of correlation using equation (12.36). This calculation, using the data from example 12.7,

$$r = \frac{n\Sigma(X \log Y) - \Sigma X(\Sigma \log Y)}{\sqrt{[n\Sigma X^2 - (\Sigma X)^2][n\Sigma(\log Y)^2 - (\Sigma \log Y)^2]}} \qquad (12.36)$$

$$= \frac{10(85.54058) - 55(13.89882)}{\sqrt{[10(385) - (55)^2][10(20.393527) - (13.89882)^2]}}$$

$$= \frac{855.4058 - 764.4351}{\sqrt{(3,850 - 3,025)[10(20.393527) - 193.17719]}}$$

$$r = \frac{90.9707}{\sqrt{(825)(203.93527 - 193.17719)}}$$

$$= \frac{90.9707}{\sqrt{(825)(10.75808)}}$$

$$= \frac{90.9707}{\sqrt{8,875.416}}$$

$$r = \frac{90.9707}{94.2094} = 0.9656.$$

Thus, with an $r = 0.9656$, the coefficient of determination (r^2) is

0.9324. In other words, using the exponential regression equation, variations in the independent variable explain 93.2 percent of the variation in the dependent variable.

Other Comments on Nonlinear Regression and Correlation Analysis

The previous discussion has been only a brief introduction to nonlinear analysis through data transformation. However, you should recognize that the various statistical measures you studied in linear regression can also be applied here.

Summary of Equations

12.1   General form of a straight-line equation:

$$Y = \alpha + \beta X$$

12.2   Regression equation—sample data:

$$Y_c = a + bX$$

12.3   Normal equation used to solve for a and b:

$$\Sigma Y = na + b\Sigma X$$

12.4   Normal equation used to solve for a and b:

$$\Sigma XY = a\Sigma X + b\Sigma X^2$$

12.5   Equation to solve for b:

$$b = \frac{n\Sigma XY - (\Sigma X)(\Sigma Y)}{n(\Sigma X^2) - (\Sigma X)^2}$$

12.6   Equation to solve for a:

$$a = \frac{\Sigma Y}{n} - b\left(\frac{\Sigma X}{n}\right) = \overline{Y} - b\overline{X}$$

12.7   Standard error of estimate:

$$s_{yx} = \sqrt{\frac{\Sigma(Y - Y_c)^2}{n - 2}}$$

12.8   Standard error of estimate:

$$s_{yx} = \sqrt{\frac{\Sigma Y^2 - a\Sigma Y - b\Sigma XY}{n - 2}}$$

12.9   Total sum of squares:

$$SST = \Sigma(Y - \overline{Y})^2$$

12.10   Regression (explained) sum of squares:

$$SSR = \Sigma(Y_c - \overline{Y})^2$$

12.11   Error (unexplained) sum of squares:

$$SSE = \Sigma(Y - Y_c)^2$$

12.12   Partitioned total sum of squares:

$$SST = SSR + SSE$$

12.13   Partitioned total sum of squares:

$$\Sigma(Y - \overline{Y})^2 = \Sigma(Y_c - \overline{Y})^2 + \Sigma(Y - Y_c)^2$$

12.14   Test statistic for the estimate of β:

$$t_{\text{test}} = \frac{b - \beta_H}{s_b}$$

12.15   Standard error of b:

$$s_b = \frac{s_{yx}}{\sqrt{\Sigma(X - \overline{X})^2}}$$

12.16   Standard error of b:

$$s_b = \frac{s_{yx}}{\sqrt{\Sigma X^2 - \dfrac{(\Sigma X)^2}{n}}}$$

12.17   Regression mean square:

$$MSR = \frac{SSR}{m}$$

12.18   Error mean square:

$$MSE = \frac{SSE}{n - m - 1}$$

12.19   Analysis of variance test statistic for the estimate of β:

$$F_{\text{test}} = \frac{MSR}{MSE}$$

12.20   Total sum of squares:

$$SST = \Sigma Y^2 - \frac{(\Sigma Y)^2}{n}$$

12.21   Error sum of squares:

$$SSE = \Sigma Y^2 - a\Sigma Y - b\Sigma XY$$

12.22 Regression sum of squares:

$$SSR = SST - SSE$$

12.23 Coefficient of determination:

$$r^2 = \frac{\Sigma(Y_c - \overline{Y})^2}{\Sigma(Y - \overline{Y})^2}$$

12.24 Coefficient of determination:

$$r^2 = \frac{SSR}{SST} = \frac{SST - SSE}{SST}$$

$$= \frac{\Sigma Y^2 - \dfrac{(\Sigma Y)^2}{n} - [\Sigma Y^2 - a\Sigma Y - b\Sigma XY]}{\Sigma Y^2 - \dfrac{(\Sigma Y)^2}{n}}$$

12.25 Coefficient of correlation:

$$r = \frac{n\Sigma XY - \Sigma X\Sigma Y}{\sqrt{[n\Sigma X^2 - (\Sigma X)^2][n\Sigma Y^2 - (\Sigma Y)^2]}}$$

12.26 Coefficient of nondetermination:

$$k^2 = \frac{\Sigma(Y - Y_c)^2}{\Sigma(Y - \overline{Y})^2}$$

12.27 Coefficient of nondetermination:

$$k^2 = 1 - r^2$$

12.28 Standard error of Y_c:

$$s_{yc} = s_{yx}\sqrt{\frac{1}{n} + \frac{(X - \overline{X})^2}{\Sigma X^2 - \dfrac{(\Sigma X)^2}{n}}}$$

12.29 Confidence interval estimate of Y_c:

$(1 - \alpha)$ confidence interval for $Y_c = (a + bX) \pm ts_{yc}$

12.30 Standard error of a predicted value of Y:

$$s_{yp} = s_{yx} \sqrt{1 + \frac{1}{n} + \frac{(X - \bar{X})^2}{\Sigma X^2 - \frac{(\Sigma X)^2}{n}}}$$

12.31 Confidence interval for a predicted value of Y:

$(1 - \alpha)$ confidence interval for $Y = (a + bX) \pm t s_{yp}$

12.32 Exponential regression equation:

$Y_c = ab^X$

12.33 Logarithmic form of an exponential regression equation:

$\log Y_c = \log a + X \log b$

12.34 Equation to solve for $\log b$:

$$\log b = \frac{[n\Sigma(X \log Y)] - [(\Sigma X)(\Sigma \log Y)]}{n(\Sigma X^2) - (\Sigma X)^2}$$

12.35 Equation to solve for $\log a$:

$$\log a = \frac{\Sigma \log Y}{n} - \log b \frac{\Sigma X}{n}$$

12.36 Coefficient of correlation using logarithmic form of an exponential regression equation:

$$r = \frac{n\Sigma(X \log Y) - \Sigma X(\Sigma \log Y)}{\sqrt{[n\Sigma X^2 - (\Sigma X)^2][n\Sigma \log Y)^2 - (\Sigma \log Y)^2]}}$$

Review Questions

1. Define
(a) regression
(b) correlation
(c) scatter diagram
(d) standard error of the estimate
(e) coefficient of determination
(f) coefficient of nondetermination

2. What is the explained variation?

3. What is the unexplained variation?

4. What is the purpose of testing the null hypothesis $\beta = 0$?

5. In the regression equation $Y_c = a + bX$, what are the a and b?

Problems

1. The management of an insurance company is interested in learning more about the variation in the number of automobile accidents. Among other things, they would like to determine what, if any, relationship there is between temperature and the frequency of accidents. They have collected the data presented as follows:

Daily low temperature (X)	Number of auto accidents (Y)
31	4
46	5
75	10
61	8
68	9
53	6
58	7
22	3

(a) Plot a scatter diagram for the data.
(b) Determine the regression equation.
(c) Test b using $\alpha = 0.05$.
(d) Compute the coefficient of correlation.
(e) Estimate the number of accidents (Y_c), with a 99 percent confidence interval, when $X = 50$.
(f) Construct a 99 percent confidence interval for the predicted value of Y when $X = 72$.

2. Over a period of several years, the Administration has required that students register bicycles. In an effort to plan future facilities, the following data have been assembled:

Number of bicycles registered	Mean price of gasoline (per gallon)
60	$0.50
65	0.50
80	0.75
92	0.90
120	0.95
150	1.00
200	1.20

(a) Determine the linear, least-squares regression equation to estimate the number of bicycles, and test the slope of that equation using $\alpha = 0.05$.

(b) Make a quantitative statement with respect to the quality of the regression equation determined in part (a).

(c) If the price of gasoline increases to $1.50 per gallon, estimate the number of bicycles (Y_c) using a level of confidence of 90 percent.

3. A company has recorded the following data on performance rating and employment test scores:

Performance (Y)	Test scores (X)
10	3
15	5
25	6
40	8
30	11
50	15

(a) Plot a scatter diagram for the data.

(b) Determine the regression equation.

(c) Is the regression significant at $\alpha = 0.05$?

(d) Estimate the performance rating for an employee with a test score of 12.

(e) Construct a 95 percent confidence interval for Y_c when $X = 12$.

(f) Construct a 95 percent confidence interval for a predicted value of Y when $X = 12$.

(g) Determine the coefficient of correlation and the coefficient of determination.

4. A series of tests to investigate the relationship between highway speeds and gasoline mileage yielded the following results.

Speed (mph) X	Gasoline mileage (mpg) Y
50	16
90	8
70	13
60	15
80	10
55	15
65	14
45	18
75	12
85	10

(a) Plot a scatter diagram for the data.
(b) Determine the regression equation.
(c) Test the null hypothesis $\beta = 0$ at the 0.05 level of significance.
(d) Construct a 99 percent confidence interval for Y_c when $X = 60$.
(e) Construct a 99 percent confidence interval for an individual value of gasoline mileage when speed $= 60$.
(f) Determine the coefficients of correlation and determination for the data.

5. It is sometimes said that the incidence of urban violent crime is directly related to the size of the city. The following is from a random sample of crime statistics classified by city size:

City population (in 10,000's)	Crime per 1000 residents
10	6
2	1
7	4
5	3
4	2
9	5

(a) Determine the regression equation.
(b) Test (b) with $\alpha = 0.1$.
(c) Estimate Y_c if $X = 8$.
(d) Compute the correlation coefficient.

6. In the following data, median family income is stated to the nearest $10,000. New car expenditures are given to the nearest $100,000. These data are for a group of seven counties.

Median family income (X)	New car expenditures (Y)
1	2
2	2
3	3
4	4
5	5
6	6
7	6

(a) Determine the regression equation for the given data.
(b) Compute the coefficient of correlation for the data.
(c) Test $H_0: \beta = 0$, using $\alpha = 0.01$.
(d) Estimate the mean new car expenditures (Y_c) when $X = 3$.

7. Assume a sample was taken to collect data on family income and food expenditures for home consumption. The results are given as follows. (Save your work to use in problem 8.)

X ($1,000) Family income	Y ($1,000) Expenditure
24.0	3.6
15.0	2.9
18.2	2.9
12.6	2.6
8.0	2.0
9.5	2.4
21.0	3.0
11.4	2.5
6.4	1.8
13.2	2.4

(a) Plot a scatter diagram for the data.
(b) Determine the regression equation.
(c) Test for regression with $\alpha = 0.05$.
(d) Determine the coefficient of determination.

8. Use the data and results of problem 7 to
(a) estimate the average expenditure for home consumption if the family income is $20,000.
(b) construct a 95 percent confidence interval for the value estimated in part (a).
(c) construct a 95 percent confidence interval for the forecasted home consumption expenditure when family income is $20,000.

9. Data have been collected on family income and away-from-home food expenditures. (Save your results for use in problem 10.)
(a) Plot a scatter diagram for the data.
(b) Determine the regression equation.
(c) Test the null hypothesis, $\beta = 0$, at the 0.10 level of significance.
(d) Determine the coefficient of determination.

Y ($1,000) Expenditures	X ($1,000) Family income
1.8	24.0
0.9	15.0
1.1	18.2
0.6	12.6
0.3	8.0
0.4	9.5
1.2	21.0
0.8	11.4
0.1	6.4
0.7	13.2

10. Use the data and results of problem 9 to

(a) estimate the average away-from-home expenditure for a family with annual income of $20,000.

(b) construct a 90 percent confidence interval for the value estimated in part (a).

(c) construct a 90 percent interval for the predicted away-from-home expenditure when family income is $20,000.

11. Delivery. Ltd. has recorded data on delivery time and weight load on deliveries that are about the same distance.

Y Delivery time (hours)	X Weight (hundreds of pounds)
2	20
3	22
1.5	15
6	45
4	32
2.5	29
3	25
5	40

(a) Determine the regression equation.

(b) Calculate the standard error of the estimate.

(c) Test for regression at the 0.05 level of significance.

(d) Determine the coefficient of determination.

12. $n = 12$ $\Sigma XY = 2000$

$\overline{Y} = 19$ $\Sigma Y^2 = 4612$

$\overline{X} = 8$ $\Sigma X^2 = 1502$

(a) Determine the regression equation.

(b) Test for regression at the 0.05 level of significance.

13. The owner of the Freshaire Hotel has been concerned with the number of complaints received from guests about poor room service. The hotel has 400 rooms on floors 2 through 7. For the past two months records have been maintained according to the

X Floor number	Y Number of complaints
2	50
3	40
4	45
5	36
6	32
7	25

number of room service complaints by floor. These data are shown on page 359. Is there a relationship between floor number and number of complaints? Set $\alpha = 0.05$.

14. If $n = 25$, $\Sigma(Y - \overline{Y})^2 = 500$, $\Sigma(Y - Y_c)^2 = 125$, and $\Sigma(Y_c - \overline{Y})^2 = 375$,
(a) determine the standard error of the estimate.
(b) determine the coefficient of determination.
(c) determine the coefficient of nondetermination two ways.

15. If $n = 100$, $\Sigma(Y - \overline{Y})^2 = 8,000$, and $\Sigma(Y_c - \overline{Y})^2 = 7,000$,
(a) determine the coefficient of correlation.
(b) determine the standard error of the estimate.

16. A local sailboat manufacturer has been gathering data for some time on the level of expenditures for advertising and the sales of boats in various market areas. The data are presented below.

Advertising expenditures	Boat sales
$2,000	$20,000
1,500	18,000
3,000	25,000
3,200	30,000
2,800	24,000
2,500	22,000
2,700	23,000

(Note: If you restate advertising in hundreds of dollars and sales in thousands of dollars, it will simplify the arithmetic.)
(a) Determine the regression equation for sales as a function of advertising.
(b) Test the slope of the regression equation using $\alpha = 0.05$.
(c) Compute the coefficient of correlation
(d) Estimate actual sales when $3,100 is spent on advertising. Use a 95 percent level of confidence.
(e) Answer the following: What percentage of the variation in sales is explained by variations in advertising? How many dollars in sales would you expect $100 in advertising to yield?

17. Management of the Postal Inspection Service is interested in predicting the salary level of employees based on years of education. A random sample of 10 employees yielded the following results:

Years of education	Salary ($1,000)
12.5	20.2
14.0	23.7
12.0	19.8
16.0	21.5
11.0	21.8
16.0	35.7
18.0	39.2
11.2	22.6
15.0	25.9
12.7	17.3

(a) Plot a scatter diagram for the data.

(b) Determine the regression equation.

(c) Test the null hypothesis, $\beta = 0$, at the 0.05 level of significance.

(d) Determine the coefficient of determination.

(e) Based on the results from parts *(c)* and *(d)*, what can you tell management about using years of education to predict salaries?

18. Henning and Kazmier recently conducted a study to compare IQ scores made by learning disabled students on both the Peabody Picture Vocabulary Test Form and the Wechsler Intelligence Scale for Children-Revised Full-Scale IQ test. Nineteen verified learning disabled students, ages 7 through 12 from three Hamilton County, Tennessee, public schools, were included in the study. The following data were obtained:

PVVT	WISC-R Full Scale
125	89
107	71
121	90
105	80
108	88
107	88
104	87
88	76
93	82
90	81
93	84
93	88
86	84
109	110
100	102
78	82
78	84
77	85
85	96

(a) Plot a scatter diagram for the data.

(b) Determine the regression equation using WISC-R Full-Scale as the dependent variable.

(c) Test for regression at the 0.05 level of significance.

(d) Determine the coefficient of correlation.

(e) Based on the results of the previous four parts of this problem, do you think it is necessary to give students both examinations? Why?

19. A local law enforcement officer has noticed what he believes to be a direct relationship between the maximum temperature on a given day and the incidence of major crimes in his city. He has selected a random sample of 10 days and determined both the high temperature and the number of major crimes.

High temperatures (degrees)	Number of major crimes
12	2
89	15
40	6
52	8
75	14
60	12
50	7
20	3
32	5
90	16

Using linear least-squares regression,

(a) at $\alpha = 0.05$, is the officer correct?

(b) as a predictor of major crimes, what is the quality of daily high temperatures?

(c) with 95 percent confidence, what is the average number of crimes the officer should expect when the high temperature is 80 degrees?

(d) If the officer wants to have enough manpower on hand to handle the number of major crimes 95 percent of the time, how many crimes should he have officers for when he expects the temperature to be 60 degrees?

20. There has been considerable discussion of the tax burden of citizens in recent months. As a result you have been asked to answer the following questions, assuming the data given you are a random sample:

(a) What equation expresses the linear, least-squares relationship between per capita income and state and local tax burden?

(b) Is the regression coefficient (b) significant with $\alpha = 0.05$?

(c) What percent of the variation in state and local tax burden is explained by changes in per capita income?

(d) Over what interval would you expect state and local tax burden to fall, with 95 percent confidence, when per capita income is $6,625?

(e) What mean state and local tax burden would you expect, with 95 percent confidence, when per capita income is $6,625?

(f) Transform state and local tax burden data using logs and then determine the regression equation.

State	Per capita income ($100)	State and local tax burden ($100)
Alabama	56	5
Arizona	65	8
Arkansas	55	5
Florida	67	6
Georgia	60	6
Illinois	80	9
Indiana	69	7
Maine	58	7
Michigan	76	7
Ohio	71	6
Oregon	70	8
Texas	68	6

21. A service company has kept records on the number of employees it needs for various levels of business activity. A portion of those records is given below.

Number of service calls	Number of employees required
10	1
15	2
20	3
25	5
30	8
35	12
40	15

(a) Transform the data on numbers of employees using logs and fit a regression equation to the data.

(b) Using the regression equation, estimate the number of employees required if the level of business reaches 45 calls.

(c) Compute the coefficient of correlation.

(d) What is the percentage increase in personnel required for each additional service call?

22. A certain stretch of highway has become infamous for the number of traffic accidents. Records kept by the highway department for several years are presented in the following table:

Year	July 4 traffic count	Number of accidents
1973	452	2
1974	425	1
1975	500	6
1976	517	5
1977	350	1
1978	480	4
1979	603	7
1980	611	10
1981	650	15
1982	718	18

(a) Use logs to transform the data on number of accidents, and fit a regression equation to the transformed data.
(b) In 1983 the traffic count on July 4 was 700. Estimate the number of accidents in that year.
(c) Compute the coefficient of correlation.

Case Study

Risk, Inc., a firm that specializes in providing venture capital for new businesses, is evaluating the short-term outlook for the automobile industry. Part of their data base was extracted from various issues of the *Survey of Current Business* and is presented in the following table.

Prepare a report that includes scatter diagrams and the regression equations that show the relationship between the sales of automotive dealers and newspaper advertising, magazine advertising, and total personal income, respectively. Include in your report 95 percent confidence interval estimates of automotive dealer sales if

Newspaper advertising = $230 million,
Magazine advertising = $300 million,
Total personal income = $2,600 billion.

Which of your regression equations do you believe is the most statistically reliable for explaining changes in the sales of automotive dealers? Why?

Year	Retail sales automotive dealers ($ millions)	Automotive advertising		Total personal income ($ billions)
		Newspapers ($ millions)	Magazines ($ millions)	
1970	64,966	92.8	95.3	803.6
1971	78,916	100.8	111.3	863.5
1972	88,612	98.0	102.1	944.9
1973	100,661	99.8	120.4	1,055.0
1974	93,089	108.8	104.7	1,154.7
1975	102,105	93.3	101.5	1,249.7
1976	125,685	127.0	142.3	1,382.7
1977	148,444	144.5	177.1	1,531.6
1978	168,086	150.6	220.8	1,717.4
1979	177,251	196.0	212.6	1,943.8
1980	162,309	182.4	231.1	2,160.4
1981	180,722	225.6	290.1	2,415.8

13

Correlation and Regression Analysis: The Multivariate Case

In the previous chapter the subject of correlation and regression analysis was discussed under the following assumptions:

1. The regression equation was assumed to be linear—a straight line.
2. There were two variables—a dependent variable and an independent one.

Relationships in the real world are frequently neither linear nor confined to two variables. Thus in this chapter we relax one of the assumptions made in the preceding chapter and look at correlation and regression analysis where the relationship is *linear,* but there are more *than two variables*—one dependent variable and several independent ones.

Multivariate Regression and Correlation Analysis

Multivariate linear correlation and regression analysis is just an extension of the simple linear case, so that there is more than one independent variable. Thus the assumption regarding only two variables (one dependent and one independent) is relaxed. There may be 2, 3, 4, or more independent variables. Each independent variable, however, is linearly related to the single dependent variable.[1]

In simple linear regression we attempt to associate the variation of the dependent variable (Y) with a single independent variable (X) and measure the proportion of explained (associated)

---

1. Note: This statement is technically correct though nonlinearity can be introduced to multivariate regression analysis by transforming the values of one or more of the variables. (The values may be transformed by squaring them, converting to logarithms, and so on.) The transformed values are then used in the regression equation.

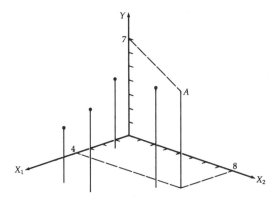

Figure 13.1   Facsimile of a three-dimensional scatter diagram

variation. However, theory, logic, or simply intuition may tell us that the value of the dependent variable is determined by more than a single independent variable. For instance, the number of hours you study is obviously not the only factor that will determine your grade in a course. Other factors such as reading speed, reading comprehension, and prerequisite preparation may be of equal, or greater, importance with the number of study hours so that an effort to analyze variations in grades might require you to incorporate all of these factors, and perhaps more. Multiple regression analysis is one statistical method that may be utilized to accomplish this.

Scatter Diagram

Figure 13.1 is a facsimile of a scatter diagram with two independent variables. The points in the scatter diagram represent the coordinate points of Y, X_1, and X_2. For instance, point A in figure 13.1 shows the point at which $Y = 7$, $X_1 = 4$, and $X_2 = 8$ intersect. Obviously, it would not be possible to show the same sort of relationship when there were four such variables—a dependent and three independent variables. However, a series of two-variable scatter diagrams might be constructed in which each independent variable in the multivariate analysis would be related to the dependent variable. Thus there would be a separate scatter diagram for each independent variable. In each case the scatter diagram would indicate the linear relationship between the dependent variable and the particular variable being used.

Multivariate Regression Equation

A multivariate equation for describing the relationship between a dependent variable (Y) and m number of independent variables (X_i) takes the following form:

$$Y = \alpha + \beta_1 X_1 + \beta_2 X_2 + \cdots + \beta_m X_m + \varepsilon. \tag{13.1}$$

BASIC STATISTICS

Just as in simple linear regressions, α is the Y intercept of the line. Moreover each independent variable (X_i) has its own slope (β_i) but in the case of multivariate regression analysis β_i is the slope of the line *if the remaining independent variables are held constant*. In other words, the value of β_1 is the average change in Y, assuming other independent variables are held constant, and X_1 varies by one unit. The β_i values in a multiple regression equation are called *coefficients of partial regression*, while the β value in a simple linear regression equation is called the *coefficient of simple regression*.

The symbol ε (epsilon) is one that has not been introduced to you before. As it is used in equation (13.1), ε represents *random error*. We must recognize random error (ε) in the equation because we are dealing with a *probabilistic model* rather than a *deterministic model*. Put another way, our equation may not fully explain all variations in the dependent variable, and therefore the estimates of Y and the actual values of Y may differ. Actually, a probabilistic model provides a mean, or *expected*, value of Y so that *the difference between the expected and actual values of Y is random error* (ε). Because in a deterministic model X has a specific value, Y can have only one value, and so there is no error. The following discussion will cover only probabilistic models for they allow us to evaluate *errors in prediction* whereas deterministic models do not.

Fundamental Assumptions

When you attempt to describe the relationship between two or more variables with a probabilistic model, certain fundamental conditions must be assumed to exist. The first assumption is that for every value of the independent variable (X), the values of dependent variable (Y) are normally distributed with a variance equal to σ^2. The second assumption is that each value of Y is independent of all other values of Y.

Given these assumptions, the behavior of the random error term (ε) in equation (13.1) can be described in the following way. First, the values of ε will be normally distributed with a mean of zero and a variance of σ^2. Second, the values of ε will be independent of one another. It is because of these two conditions that we can evaluate the prediction errors of a probabilistic model.

Estimating the Multivariate Regression Equation

As we did with simple linear regression, we *estimate multivariate regression equations* using sample data. Therefore the estimated multivariate regression equation takes the following form:

$$Y_c = a + b_1 X_1 + b_2 X_2 + \cdots + b_m X_m. \tag{13.2}$$

Here again, Y_c is the computed regression value for the dependent variable. The value a is still the Y intercept, but it is the value of Y_c when all independent variables (X's) equal zero. There is a separate b value (or slope) for each of the independent variables.

The constants a, b_1, b_2, . . . b_m in the equation are computed from normal equations. Thus the multivariate equation assumes the characteristics of a least-squares equation. The general forms of the normal equations are

$$\Sigma Y = na + b_1 \Sigma X_1 + b_2 \Sigma X_2 + \cdots + b_m \Sigma X_m, \tag{13.3}$$

$$\Sigma X_i Y = a\Sigma X_1 + b_1 \Sigma X_1^2 + b_2 \Sigma X_1 X_2 + \cdots + b_m \Sigma X_1 X_m, \tag{13.4}$$

$$\Sigma X_2 Y = a\Sigma X_2 + b_1 \Sigma X_1 X_2 + b_2 \Sigma X_2^2 + \cdots + b_m \Sigma X_2 X_m, \tag{13.5}$$

$$\Sigma X_m Y = a\Sigma X_m + b_1 \Sigma X_1 X_m + b_2 \Sigma X_2 X_m + \cdots + b_m \Sigma X_m^2. \tag{13.6}$$

Consider a regression equation with three variables—a dependent variable and two independent ones. The equation would appear as

$$Y_c = a + b_1 X_1 + b_2 X_2, \tag{13.7}$$

and the normal equations would be

$$\Sigma Y = na + b_1 \Sigma X_1 + b_2 \Sigma X_2, \tag{13.8}$$

$$\Sigma X_1 Y = a\Sigma X_1 + b_1 \Sigma X_1^2 + b_2 \Sigma X_1 X_2, \tag{13.9}$$

$$\Sigma X_2 Y = a\Sigma X_2 + b_1 \Sigma X_1 X_2 + b_2 \Sigma X_2^2. \tag{13.10}$$

When we presented simple linear regression in chapter 12, we discussed testing the b coefficient of the equation to determine if it was significantly different from zero. Multiple regression requires that we do the same thing except now we must test the b value for each of the variables in the equation. Just as in simple linear regression, if the value of b for a particular variable is not significantly different from zero, we cannot assume that there is a relationship between the dependent variable and that particular independent variable.

In order to test each b value in the multiple regression equation, we must first calculate an estimate of the *standard error of multiple regression* (this is sometimes called *error variance*). This measure is similar to the standard error of estimate of a simple linear regression equation in that it is a measure of the average amount that the Y values differ from the estimated (Y_c) values. In terms of our assumption this is a measure of random error (ε).

Thus the standard error of multiple regression (s_m) can be viewed as a measure of closeness-of-fit for the regression equation.

Now using the estimate of the standard error of multiple regression, it is possible to estimate the standard error of each coefficient of partial regression (s_{b_i}). Then the null hypothesis, $\beta_i = 0$, can be tested, using a t-test for each independent variable to determine if in the multiple regression equation there is any relationship between that independent variable X_i and the dependent variable. We will look at a detailed example later.

You should be aware that an independent variable may be accepted as having a significant regression relationship with a given dependent variable in a simple linear regression equation, but may be rejected, using the same significance level, when one or more additional independent variables are added to the equation. If this happens, it is because *multicollinearity* exists between the independent variables. That is, two or more of the independent variables have relatively high correlation among themselves and they are said to contribute like information to the model.

If the sample data you are using is a time series, you may also encounter another type of problem in attempting to estimate a multiple regression model. One of the conditions we assume to exist is that the random error values are independent. When this is not true, *serial correlation* is said to exist. Serial correlation frequently occurs when you fail to include some important independent variable in the model, and as a result those independent variables you have elected to include will appear to be more statistically significant (higher t values) than they really are.

There are statistical procedures available to test for both multicollinearity and serial correlation. However, having called your attention to these two possible problem areas, we will leave any discussion of the techniques for testing to a more advanced text.

After testing each b_i value to determine if it is significant, it is now time to consider testing the total regression effect of the independent variables on the dependent variable. In other words, there is a way to test the entire equation. This can be accomplished by the use of analysis of variance, in the form initially presented in chapter 12, to test the null hypothesis $\beta_1 = \beta_2 = \cdots = \beta_m = 0$, where m is the number of independent variables. Just as in other applications of analysis of variance, the test statistic, a value of F, is found using the following formula:

$$F_{\text{test}} = \frac{MSR}{MSE},$$ (13.11)

where

$$MSR = SSR/m, \tag{13.12}$$

$$MSE = SSE/(n - m - 1). \tag{13.13}$$

Multiple Correlation

Like simple linear correlation we can compute a measure of the quality of the relationship between the dependent variable and the independent variables. The measure is called the *coefficient of multiple correlation (R), and R^2 is the coefficient of multiple determination.* R^2 can be interpreted to mean the proportion of the total variation in the dependent variable that is explained by changes in all the independent variables. Equation (13.14) is used to compute R^2:

$$R^2 = \frac{SSR}{SST}, \tag{13.14}$$

where

SSR = regression sum of squares,
SST = total sum of squares.

Multiple Regression and the Computer

Our discussion of multiple correlation and regression has been in general terms up until now. In addition you may have noticed that you were given very few equations. As a matter of fact the computations required for multiple regression analysis are rather time-consuming. Consequently these computations are usually carried out by using either computers or, for problems with relatively few independent variables, programmable electronic calculators, and multiple regression programs are readily available for both of them.

In table 13.1 there is a set of data which has been used to develop an illustration of the output that might be provided from a

Table 13.1
Data for a multiple regression example

State	Dependent variable	Independent variables		
	Motor vehicle thefts per 100,000 population in 1978	Police protection, number of employees per 1,000 population in 1978	High school graduates as a percent of population in 1976	Metropolitan area population as a percent of state in 1978
ME	253	2.1	67.8	22.8
NH	306	2.3	70.3	36.5
VT	214	2.1	69.7	0.0
MASS	1,096	2.9	72.3	86.4
RI	807	2.8	61.7	92.2
CT	655	2.5	70.3	88.0
NY	672	4.0	66.2	88.4

| State | Dependent variable

Motor vehicle thefts per 100,000 population in 1978 | Independent variables | | |
		Police protection, number of employees per 1,000 population in 1978	High school graduates as a percent of population in 1976	Metropolitan area population as a percent of state in 1978
NJ	561	3.6	66.4	91.8
PA	341	2.4	64.8	80.4
OHIO	402	2.3	67.7	79.7
IND	411	2.0	67.0	70.4
ILL	511	3.1	66.1	81.4
MICH	526	2.5	68.6	81.3
WIS	230	2.4	70.3	62.9
MINN	325	1.8	72.4	64.3
IOWA	235	1.9	72.3	37.2
MO	369	2.8	64.1	63.6
N DAK	148	1.8	67.6	34.5
S DAK	160	1.9	68.9	28.1
NEBR	217	2.1	74.3	44.6
KANS	248	2.2	73.1	46.3
DEL	497	2.8	69.5	68.1
MD	425	3.0	69.3	84.6
VA	229	2.2	65.7	65.5
W VA	176	1.8	64.2	36.1
NC	201	2.1	55.3	45.2
SC	276	2.4	57.1	48.3
GA	357	2.5	58.7	56.9
FLA	389	3.0	64.8	85.9
KY	251	2.0	53.3	45.1
TENN	328	2.3	54.9	62.7
ALA	303	2.3	55.5	62.0
MISS	152	2.0	52.3	26.8
ARK	200	1.9	56.2	38.5
LA	362	3.3	58.3	63.1
OKLA	351	2.3	65.6	55.7
TEX	444	2.3	64.5	79.6
MONT	307	2.3	72.5	24.3
IDAHO	244	2.6	71.5	17.4
WYO	343	3.1	75.3	0.0
COLO	488	2.7	78.1	80.8
N MEX	314	2.9	65.7	33.7
ARIZ	464	3.1	72.5	74.6
UTAH	334	2.4	80.2	78.6
NEV	602	3.6	75.7	81.4
WASH	395	2.2	76.3	71.1
OREG	395	2.4	75.5	59.4
CALIF	691	2.7	74.0	92.5
ALASKA	662	3.5	79.6	44.6
HAWAII	571	2.9	73.0	80.3

Source: *Statistical Abstract of the United States, 1980,* Bureau of the Census, U.S. Department of Commerce, 1980.

VARIABLE	REGRESSION COEFFICIENT	STANDARD ERROR	T STATISTIC	PROBABILITY OF A LARGER ABSOLUTE T	MEAN
POLICE PROTECTION	142.7140	36.74	3.8841	0.000327	2.522
HIGH SCHOOL GRAD	4.5317	2.464	1.8391	0.072368	67.540
METRO AREA POP	3.6871	7.626	4.8351	0.000015	58.872

MEAN FOR DEPENDENT VARIABLE (MOTOR VEHICLE THEFTS) IS 388.760
CONSTANT FOR SIMPLIFIED MODEL = −494.304

SOURCE	SUM OF SQUARES	DF	MEAN SQUARE	F VALUE
REGRESSION	1084551.699	3	361517.233	26.18
ERROR	635101.421	46	13806.553	
TOTAL	1719653.120	49		

THE PROBABILITY OF OBTAINING F THIS LARGE OR LARGER BY CHANCE WHEN THE HYPOTHESIS OF NO CORRELATION IS TRUE = 0.00000006

MULTIPLE R**2 = 0.63068050

Figure 13.2   Sample computer output from a multiple regression program: three independent variables

multiple regression computer program. The example of this computer output is given in figure 13.2.

In looking at the illustration, you should note that the number of motor vehicle thefts is the dependent variable and there are three independent variables. The form of the equation in the illustration is

$$Y_c = -494.304 + 142.7140X_1 + 4.5317X_2 + 3.6871X_3.$$

If you wish to test the values of b_i, the output also contains information useful for that. Given the following three sets of hypotheses,

H_0: $\beta_1 = 0$, H_0: $\beta_2 = 0$, H_0: $\beta_3 = 0$,
H_A: $\beta_1 \neq 0$, H_A: $\beta_2 \neq 0$, H_A: $\beta_3 \neq 0$,

they can be tested at any level of α with the data in the column headed "Probability of a larger absolute T." If $\alpha = 0.05$, you would reject H_0: $\beta_1 = 0$ and H_0: $\beta_3 = 0$, but you would not reject H_0: $\beta_2 = 0$ because the probability (0.072368) of a larger t value is greater than 0.05. In other words, there does not appear to be a relationship between the number of high school graduates as percent of population and motor vehicle thefts per 100,000 population, but there does appear to be one between police protection and motor vehicle thefts as well as metropolitan area population as a percent of state population and motor vehicle thefts.

Now that you have discovered this, you may wish to recompute the multiple regression equation using only two independent variables—police protection and metropolitan area population as a percent of state population. Figure 13.3 contains an illustration of the output your computer might yield. First of all, notice that the partial regression coefficients are not the same as those shown in figure 13.2 but they are still significant at a level of $\alpha = 0.05$. The values of the coefficients of partial regression become changed because there is some degree of multicollinearity between the three independent variables, and now we no longer have the effects of the third independent variable. The printout also contains the coefficient of multiple determination (R^2) as well as an F statistic. This latter is the same F_{test} for either the regression equation or R^2. Finally, at the end of the sample output is a statement concerning the probability of an F value this large or larger. In this instance, an $F_{test} = 35.77$ with $df = 2$ and 46 can be interpreted to mean that the null hypothesis $\beta_1 = \beta_3 = 0$ would be rejected at any $\alpha > 0.00000006$ and the alternate hypothesis (at least one $\beta_i \neq 0$) would be accepted.

Actually, this is a very simplified example of multiple regression analysis using a computer. The output resulting from many computer programs contains somewhat more information than that presented here. Moreover the equations presented in figures 13.2 and 13.3 are not all the ones that should be considered if your objective is to analyze the data thoroughly. For example, we

VARIABLE	REGRESSION COEFFICIENT	STANDARD ERROR	T STATISTIC	PROBABILITY OF A LARGER ABSOLUTE T	MEAN
POLICE PROTECTION	156.5655	36.863	4.2473	0.000101	2.522
METRO AREA POP	3.6703	0.782	4.6958	0.000023	58.872

MEAN FOR DEPENDENT VARIABLE (MOTOR VEHICLE THEFTS) IS 388.760
CONSTANT FOR SIMPLIFIED MODEL = −222.175

SOURCE	SUM OF SQUARES	DF	MEAN SQUARE	F VALUE
REGRESSION	1037856.388	2	518928.194	35.77
ERROR	681796.732	47	14506.313	
TOTAL	1719653.120	49		

THE PROBABILITY OF OBTAINING F THIS LARGE OR LARGER BY CHANCE WHEN THE HYPOTHESIS OF NO CORRELATION IS TRUE = 0.00000006

MULTIPLE R**2 = 0.60352659

Figure 13.3   Sample computer output from a multiple regression program: two independent variables

have not shown you an equation using just police protection and the number of high school graduates as independent variables. As a matter of fact, you might find it interesting to make police protection the dependent variable and the number of motor vehicle thefts one of the independent variables.

Appendix
Adjusted Sums of Squares Method of Determining Multiple Regression Equations

Solving the normal equations required to carry out multiple regression analysis can be rather laborious. Using the adjusted sums of squares method reduces the number of normal equations by one and therefore simplifies the task somewhat. Using this procedure for a problem with two independent variables, we can rewrite equations (13.8), (13.9), and (13.10), as equations (13.15) and (13.16). Note that instead of X_1 and X_2 we now have x_1 and x_2.

$$\Sigma x_1 y = b_1 \Sigma x_1^2 + b_2 \Sigma x_1 x_2, \tag{13.15}$$

$$\Sigma x_2 y = b_1 \Sigma x_1 x_2 + b_2 \Sigma x_2^2, \tag{13.16}$$

where

$$\Sigma y^2 = \Sigma Y^2 - n\overline{Y}^2, \tag{13.17}$$

$$\Sigma x_1^2 = \Sigma X_1^2 - n\overline{X}_1^2, \tag{13.18}$$

$$\Sigma x_2^2 = \Sigma X_2^2 - n\overline{X}_2^2, \tag{13.19}$$

$$\Sigma x_1 y = \Sigma X_1 Y - n\overline{X}_1\overline{Y}, \tag{13.20}$$

$$\Sigma x_2 y = \Sigma X_2 Y - n\overline{X}_2\overline{Y}, \tag{13.21}$$

$$\Sigma x_1 x_2 = \Sigma X_1 X_2 - n\overline{X}_1\overline{X}_2. \tag{13.22}$$

If the adjusted sums of squares method is used, the value of a in the multiple regression equation is found using equation (13.23):

$$a = \overline{Y} - b_1 \overline{X}_1 - b_2 \overline{X}_2. \tag{13.23}$$

Similarly, in order to test each b value, we can write the equations for the standard error of b_1 and b_2 using adjusted sums of squares. In order to do so, though, we first calculate the standard error of multiple regression using equation (13.24):

$$s_m = \sqrt{\frac{\Sigma y^2 - b_1 \Sigma x_1 y - b_2 \Sigma x_2 y}{n - 3}}. \tag{13.24}$$

Then we can use equations (13.25) and (13.26) to compute the *standard error of b_1 and b_2:*

$$s_{b_1} = \frac{s_m}{\sqrt{\Sigma x_1^2 - \frac{(\Sigma x_1 x_2)^2}{\Sigma x_2^2}}} \tag{13.25}$$

$$s_{b_2} = \frac{s_m}{\sqrt{\Sigma x_2^2 - \frac{(\Sigma x_1 x_2)^2}{\Sigma x_1^2}}} \qquad (13.26)$$

Now we are ready to establish the hypotheses and test the b values using equation (13.27):

$$t_{\text{test}} = \frac{b_i - \beta_H}{s_{b_i}}. \qquad (13.27)$$

It is also possible to write an equation for R^2 using this method:

$$R^2 = \frac{b_1 \Sigma x_1 y + b_2 \Sigma x_2 y}{\Sigma y^2}. \qquad (13.28)$$

Now that you have at your disposal a set of equations that can be used to do multiple regression analysis, let's look at an example that illustrates their application.

Example 13.1

▷ Dr. Mac Michael, a professor of statistics, has been a regular Friday afternoon customer of the Library (an establishment dedicated to quenching the thirst of the university community) for many years. During his visits to the Library, Dr. Mac has noticed a great variation in the number of customers. Consequently he has also observed that the more customers the Library has, the longer it takes for him to be served. Having stimulated his intellectual curiosity, as well as his thirst, Dr. Mac has suggested to the Library management that they employ him as a consultant, and they did.

Dr. Mac decided that his first task should be to find some way to make a reasonably accurate estimate, using multiple regression analysis, of how many customers the Library might expect so that it could improve planning and hope to provide faster service. Since most of the establishment's customers are university students, it seemed logical to the professor that the number might be related to the size of the student body. He also recognized that students came to the Library only if they had money to spend and, while he does not know the income of students, he does know how much they contribute to the Student Mascot Fund. Dr. Mac decided that contributions to the fund might be a good indicator of student income and therefore might also be related to the number of customers at the Library.

The data in table 13.2 are what Dr. Mac had to work with. The table also includes his calculations for a multiple regression equation.

Table 13.2
Dr. Mac's multiple regression problem

Customers, Y	Student enrollment, X_1	Student contribution to mascot fund, X_2	Y^2	X_1^2	X_2^2	YX_1	YX_2	X_1X_2
10	700	$ 50	100	490,000	2,500	7,000	500	35,000
15	750	65	225	562,500	4,225	11,250	975	48,750
20	760	80	400	577,600	6,400	15,200	1,600	60,800
15	800	100	225	640,000	10,000	12,000	1,500	80,000
20	842	105	400	708,964	11,025	16,840	2,100	80,000
16	910	85	256	828,100	7,225	14,560	2,100	88,410
18	965	90	324	931,225	8,100	17,370	1,360	77,350
22	1010	100	484	1,020,100	10,000	22,220	1,620	86,850
24	1070	110	576	1,144,900	12,100	25,680	2,200	101,000
30	1100	100	900	1,210,000	10,000	33,000	2,640	117,700
190	8907	$885	3,890	8,113,389	81,575	175,120	3,000	110,000

$$\Sigma y^2 = \Sigma Y^2 - n\bar{Y}^2$$
$$= 3,890 - 10(19)^2$$
$$= 3,890 - 3,610$$
$$\Sigma y^2 = 280$$
$$\Sigma x_1^2 = \Sigma X_1^2 - n\bar{X}_1^2$$
$$= 8,113,389 - 10(890.7)^2$$
$$= 8,113,389 - 7,933,464.9$$
$$\Sigma x_1^2 = 179,924.1$$
$$\Sigma x_2^2 = \Sigma X_2^2 - n\bar{X}_2^2$$
$$= 81,575 - 10(88.5)^2$$
$$= 81,575 - 78,322.5$$
$$\Sigma x_2^2 = 3,252.5$$

$$\Sigma x_1 y = \Sigma X_1 Y - n\bar{X}_1\bar{Y}$$
$$= 175,120 - 10(890.7)(19)$$
$$= 175,120 - 169,233$$
$$\Sigma x_1 y = 5,887$$
$$\Sigma x_2 y = \Sigma X_2 Y - n\bar{X}_2\bar{Y}$$
$$= 17,495 - 10(88.5)(19)$$
$$= 175,120 - 169,233$$
$$\Sigma x_2 y = 680$$
$$\Sigma x_1 x_2 = \Sigma X_1 X_2 - n\bar{X}_1\bar{X}_2$$
$$= 805,860 - 10(890.7)(88.5)$$
$$= 805,860 - 788,269.5$$
$$\Sigma x_1 x_2 = 17,590.5$$

I $\Sigma x_1 y = b_1\Sigma x_1^2 + b_2\Sigma x_1 x_2$
II $\Sigma x_2 y = b_1\Sigma x_1 x_2 + b_2\Sigma x_2^2$
I $5,887 = 179,924.1b_1 + 17,590.5b_2$
II $680 = 17,590.5b_1 + 3,252.5b_2$

I $5,887 = 179,924.1b_1 + 17,590.5b_2$
II (-10.2285) $- 6,955.4 = -179,924.1b_1 - 33,268.2b_2$
                $-1,068.4 = -15,677.7b_2$
                $0.06815 = b_2$

$680 = 17,590.5b_1 + 3,252.5(0.06815)$
$680 = 17,590.5b_1 + 221.65788$
$458.34212 = 17590.5b_1$
$0.02606 = b_1$
$a = \bar{Y} - b_1\bar{X}_1 - b_2\bar{X}_2$
$a = 19.0 - 0.02606(890.7) - (0.06815)(88.5)$
$= 19.0 - 23.21164 - 6.03128$
$a = -10.24292$
$Y_c = -10.24292 + 0.02606X_1 + 0.06815X_2$

Now Dr. Mac is ready to establish the hypothesis and test the b values. First, however, he must determine an appropriate value for the level of significance. While he frequently uses relatively small alpha values, he finally decided that $\alpha = 0.50$ would be good enough for the problem at hand. After all, that would mean that Dr. Mac would be rejecting a correct H_0 only 50 percent of the time:

$$H_0: \quad \beta_1 = 0, \qquad H_0: \quad \beta_2 = 0,$$
$$H_A: \quad \beta_1 \neq 0, \qquad H_A: \quad \beta_2 \neq 0.$$

Using the values from table 13.2, the calculations for the standard error of the estimate of each of b values are

$$s_m = \sqrt{\frac{\Sigma y^2 - b_1 \Sigma x_1 y - b_2 \Sigma x_2 y}{n - 3}}$$

$$= \sqrt{\frac{280 - (0.02606)(5,887) - (0.06815)(680)}{10 - 3}}$$

$$= \sqrt{\frac{280 - 153.41522 - 46.34200}{7}} = \sqrt{\frac{80.24278}{7}} = \sqrt{11.46325}$$

$$s_m = 3.38574,$$

$$s_{b_1} = \frac{s_m}{\sqrt{\Sigma x_1^2 - \dfrac{(\Sigma x_1 x_2)^2}{\Sigma x_2^2}}} = \frac{3.38574}{\sqrt{179,924.1 - \dfrac{(17,590.5)^2}{3,252.5}}}$$

$$= \frac{3.38574}{\sqrt{179,924.1 - \dfrac{309,425,690.25}{3,252.5}}} = \frac{3.38574}{\sqrt{179,924.1 - 95,134.7}}$$

$$= \frac{3.38574}{\sqrt{84,789.4}} = \frac{3.38574}{291.19}$$

$$s_{b_1} = 0.01163,$$

$$s_{b_2} = \frac{s_m}{\sqrt{\Sigma x_2^2 - \dfrac{(\Sigma x_1 x_2)^2}{\Sigma x_1^2}}} = \frac{3.38574}{\sqrt{3,252.5 - \dfrac{(17,590.5)^2}{179,924.1}}}$$

$$= \frac{3.38574}{\sqrt{3,252.5 - \dfrac{309,425,690.25}{179,924.1}}} = \frac{3.38574}{\sqrt{3,252.5 - 1,719.76}}$$

$$= \frac{3.38574}{\sqrt{1,532.74}} = \frac{3.38574}{39.15}$$

$$s_{b_2} = 0.08648.$$

Continuing with the calculations on Dr. Mac's regression equation,

H_0: $\beta_1 = 0$, H_0: $\beta_2 = 0$,
H_A: $\beta_1 \neq 0$, H_A: $\beta_2 \neq 0$,
$\alpha = 0.50$, $\alpha = 0.50$.

Reject H_0 if $t_{test} < -0.711$ or $t_{test} > 0.711$:

$$t_{test} = \frac{b_1 - \beta_H}{s_{b_1}},\qquad t_{test} = \frac{b_2 - \beta_H}{s_{b_2}},$$

$$t_{test} = \frac{0.2606 - 0}{0.01163},\qquad t_{test} = \frac{0.06815 - 0}{0.08648},$$

$t_{test} = 2.241$. $t_{test} = 0.788$.
Reject H_0. Reject H_0.

Thus at the 0.50 level of significance the professor can accept both independent variables as having a regression relationship with the number of customers. However, it should be noted that b_1 is significant at a much lower α level than is b_2.

Once again using the values in table 13.2, we can compute R and R^2 to check the quality of the multiple regression equation:

$$R = \sqrt{\frac{b_1\Sigma x_1 y + b_2\Sigma x_2 y}{\Sigma y^2}}$$

$$= \sqrt{\frac{(0.02606)5{,}887 + (0.06815)(680)}{280}}$$

$$= \sqrt{\frac{153.41522 + 46.34200}{280}}$$

$$R = \sqrt{\frac{199.75722}{280}} = \sqrt{0.71342} = 0.84464,$$

$$R^2 = 0.71342.$$

Changes in the two independent variables explain 71.34 percent of the changes in the dependent variable. The two independent variables, however, do have some overlap in this example. In other words, they are not fully independent of one another. If the R^2 is not large enough, then we may decide to try another independent variable instead of the present X_2 or we can add a third, or fourth, or more independent variables. We can't have more than eight in our example because we would exhaust our degrees of freedom since $df = n - (1 + \text{number of independent variables})$.

Finally, Dr. Mac has just learned that students have contributed $100.00 to the Student Mascot Fund. Because he already knows that enrollment is 1,200 students, he can now make an estimate as to how many customers the Library should expect:

$$Y_c = -10.24292 + 0.02606X_1 + 0.06815X_2$$
$$= -10.24292 + 0.02602(1,200) + 0.06815(100)$$
$$= -10.24292 + 31.27200 + 6.81500$$
$$Y_c = 27.84408.$$

If Dr. Mac is correct, the Library can make plans to serve approximately 28 customers. ◁ ·

Summary of Equations

13.1   Multivariate regression equation—probabilistic model:

$$Y = \alpha + \beta_1 X_1 + \beta_2 X_2 + \cdots + \beta_m X_m + \varepsilon$$

13.2   Multivariate regression equation—sample data:

$$Y_c = a + b_1 X_1 + b_2 X_2 + \cdots + b_m X_m$$

13.3   Normal equation to solve for regression coefficients, m number of independent variables:

$$\Sigma Y = na + b_1 \Sigma X_1 + b_2 \Sigma X_2 + \cdots + b_m \Sigma X_m$$

13.4   Normal equation to solve for regression coefficients, m number of independent variables:.

$$\Sigma X_1 Y = a\Sigma X_1 + b_1 \Sigma X_1^2 + b_2 \Sigma X_1 X_2$$
$$+ \cdots + b_m \Sigma X_1 X_m$$

13.5   Normal equation to solve for regression coefficients, m number of independent variables:

$$\Sigma X_2 Y = a\Sigma X_2 + b_1 \Sigma X_1 X_2 + b_2 \Sigma X_2^2$$
$$+ \cdots + b_m \Sigma X_2 X_m$$

13.6   Normal equation to solve for regression coefficients, m number of independent variables:

$$\Sigma X_m Y = a\Sigma X_m + b_1 \Sigma X_1 X_m + b_2 \Sigma X_2 X_m$$
$$+ \cdots + b_m \Sigma X_m^2$$

13.7   Regression equation, two independent variables:

$$Y_c = a + b_1 X_1 + b_2 X_2$$

13.8   Normal equation to solve for regression coefficients, two independent variables:

$$\Sigma Y = na + b_1 \Sigma X_1 + b_2 \Sigma X_2$$

13.9   Normal equation to solve for regression coefficients, two independent variables:

$$\Sigma X_1 Y = a\Sigma X_1 + b_1 \Sigma X_1^2 + b_2 \Sigma X_1 X_2$$

13.10 Normal equation to solve for regression coefficients, two independent variables:

$$\Sigma X_2 Y = a\Sigma X_2 + b_1\Sigma X_1 X_2 + b_2\Sigma X_2^2$$

13.11 Test statistic for regression of m number of independent variables:

$$F_{\text{test}} = \frac{MSR}{MSE}$$

13.12 Regression mean square:

$$MSR = \frac{SSR}{m}$$

13.13 Error mean square:

$$MSE = \frac{SSE}{n - m - 1}$$

13.14 Coefficient of multiple determination:

$$R^2 = \frac{SSR}{SST}$$

13.15 Normal equation to solve for b_1 and b_2, adjusted sums of squares method:

$$\Sigma x_1 y_1 = b_1\Sigma x_1^2 + b_2\Sigma x_1 x_2$$

13.16 Normal equation to solve for b_1 and b_2, adjusted sums of squares method:

$$\Sigma x_2 y = b_1\Sigma x_1 x_2 + b_2\Sigma x_2^2$$

13.17 Identity required for equations (13.24) and (13.28):

$$\Sigma y^2 = \Sigma Y^2 - n\overline{Y}^2$$

13.18 Identity required for equations (13.15), (13.25), and (13.26):

$$\Sigma x_1^2 = \Sigma X_1^2 - n\overline{X}_1^2$$

13.19 Identity required for equations (13.16), (13.25), and (13.26):

$$\Sigma x_2^2 = \Sigma X_2^2 - n\overline{X}_2^2$$

13.20 Identity required for equations (13.15), (13.24), and (13.28):

$$\Sigma x_1 y = \Sigma X_1 Y - n\overline{X}_1\overline{Y}$$

13.21 Identity required for equations (13.16), (13.24), and (13.28):

$$\Sigma x_2 y = \Sigma X_2 Y - n\overline{X}_2\overline{Y}$$

13.22 Identity required for equations (13.15), (13.16), (13.25), and (13.26):

$$\Sigma x_1 x_2 = \Sigma x_1 X_2 - n\overline{X}_1\overline{X}_2$$

13.23 Value of a in multiple regression equation, two independent variables:

$$a = \overline{Y} - b_1\overline{X}_1 - b_2\overline{X}_2$$

13.24 Standard error of multiple regression, two independent variables, adjusted sums of squares method:

$$s_m = \sqrt{\frac{\Sigma y^2 - b_1\Sigma x_1 y - b_2\Sigma x_2 y}{n-3}}$$

13.25 Standard error of b_1 two independent variables, adjusted sums of squares method:

$$s_{b_1} = \frac{s_m}{\sqrt{\Sigma x_1^2 - \dfrac{(\Sigma x_1 x_2)^2}{\Sigma x_2^2}}}$$

13.26 Standard error of b_2, two independent variables, adjusted sums of squares method:

$$s_{b_2} = \frac{s_m}{\sqrt{\Sigma x_2^2 - \dfrac{(\Sigma x_1 x_2)^2}{\Sigma x_1^2}}}$$

13.27   Test statistic for β_i values:

$$t_{\text{test}} = \frac{b_i - \beta_H}{s_{b_i}}$$

13.28   Coefficient of multiple determination, adjusted sums of squares method:

$$R^2 = \frac{b_1 \Sigma x_1 y + b_2 \Sigma x_2 y}{\Sigma y^2}$$

Review Questions

1. Explain random error.

2. What is the difference between deterministic and probabilistic models?

3. What are the fundamental assumptions associated with a probabilistic model?

4. Explain multicolinearity.

5. Explain serial correlation.

Problems

1. Economic Analysis Associates has been retained to study the market for flea powder. On behalf of their client, they are seeking a way of predicting sales when entering a new market and wish to investigate the feasibility of doing this on the basis of a regression equation derived from the information given in the following table:

Area	Company sales ($1,000's)	Number of dog licenses (100's)	Number of supermarkets carrying product
1	76	15	25
2	38	10	29
3	25	7	15
4	98	21	25
5	93	14	11
6	54	8	13
7	78	14	22
8	85	24	14
9	65	9	13
10	88	23	11

(a) Calculate the regression equation, test the partial regression coefficients and compute R^2.

(b) Should the sales manager use this equation to predict sales in new markets?

2. In the table that follows are some data for Burpee sales at a convenience store. The sales for a random selection of days and the high temperature, as well as the traffic count on the street, are given for each day.

Temperature	Traffic count	Number of 6-paks of Burpee sold
78	720	61
80	623	83
40	301	15
70	850	90
68	352	45
45	424	38
92	605	103
42	457	58
75	783	79
50	311	27

(a) Compute a multiple regression equation to predict Burpee sales using temperature and traffic count as the independent variables.
(b) Test the b coefficients at $\alpha = 0.2$.
(c) Compute the coefficients of multiple correlation for the equation in part *(a)*.

3. The strawberry crop seems to vary with the amount of rainfall and the amount of fertilizer used. Records of one farmer are given in the table that follows:

Amount of rainfall (inches)	Fertilizer used (tons)	Strawberry crop (crates)
16	510	1000
22	450	950
23	500	1200
13	425	700
17	450	800
25	475	1100
18	515	1050
20	500	1150
21	490	1000
19	510	950
22	525	1300

(a) Construct a multiple regression equation using rainfall as X_1 and fertilizer as X_2.
(b) Test the b coefficients with $\alpha = 0.01$.
(c) Compute the coefficient of multiple determination.

4. The following matrix contains the correlation coefficients for five variables. Based on these data, what conclusions would you reach concerning the use of variables 3, 4, and 5 as independent variables in the same multiple regression equation?

Variable	1	2	3	4	5
1	1.0000	0.9869	0.9473	0.1591	0.9941
2	0.9869	1.0000	0.9329	0.2106	0.9810
3	0.9473	0.9329	1.0000	−0.1286	0.9652
4	0.1591	0.2106	−0.1286	1.0000	0.1006
5	0.9941	0.9810	0.9652	0.1006	1.0000

5. The State Department of Higher Education has been concerned about the recently declining number of graduates of colleges and universities, and its staff has been asked to analyze the situation. In doing so, they have gathered the data presented in the following table:

Year	Number of bachelor's degrees granted	Median family income (1982 dollars)	Population aged 18 to 24 (1,000's)	Unemployment rate
1966	3,683	11,251	16.1	5.5
1967	3,689	11,358	17.0	6.7
1968	3,878	11,457	17.7	5.5
1969	4,164	11,710	18.3	5.7
1970	4,665	12,065	18.8	5.2
1971	5,012	12,444	20.3	4.5
1972	5,202	12,874	21.4	3.8
1973	5,581	13,445	22.3	3.8
1974	6,339	13,717	22.9	3.6
1975	7,282	14,236	23.7	3.5
1976	7,915	14,689	24.7	4.9
1977	8,397	14,531	25.8	5.9
1978	8,873	14,531	25.9	5.6
1979	9,221	15,103	26.4	4.9
1980	9,459	15,373	26.9	5.6
1981	9,401	15,155	27.6	8.5
1982	9,333	15,060	27.1	9.1

(a) Using population aged 18 to 24, median family income, and the unemployment rate as independent variables, calculate the multiple regression equation to predict the number of bachelor's degree granted.
(b) Test the partial regression coefficients.
(c) Compute R^2.

6. Using the data in problem 5,
(a) calculate the multiple regression equation if population aged 18 to 24 and median family income are used as independent variables and bachelor's degrees granted is the dependent variable.
(b) test the partial regression coefficients.
(c) compute R^2.

7. Someone on the staff of the State Department of Higher Education looked at the data given in problem 5 and raised questions concerning seriality and multicolinearity. Analyze the data set, and determine if there is any basis for their concern.

8. The owner of a small insurance underwriting firm asked her ten employees to rate their job satisfaction. That information, along with their scores on an aptitude test and their number of days absent, is given in the table that follows. Determine the multiple regression equation to estimate job satisfaction using the test scores and days absent, test the partial regression coefficients using $\alpha = 0.05$ and compute R^2.

Job satisfaction	Aptitude test	Days absent
32	16	2
36	18	1
35	17	1
38	19	2
43	22	0
28	16	4
25	15	4
40	20	1
30	17	3
35	19	2

9. Fat-Is-Fun has twelve retail outlets located in shopping centers along the east coast. Executives of the company want to determine if advertising expenditures and population density, when considered together, can be used to predict sales. Do the necessary analysis. Use $\alpha = 0.10$.

Population density per square mile	Annual advertising expenditures ($1,000's)	Annual sales ($1,000's)
70	4	40
50	3	30
80	6	50
70	4	40
75	5	50
30	1	20
40	2	25
50	2	24
60	3	32
85	6	55
30	1	18
90	7	60

10. The State Property Tax Assessment Coordination Division would like to determine if the age, number of rooms, and total square footage can be used to determine the market value of houses. They have gathered the information presented in the table that follows. Determine the multiple regression equation using all the variables shown, and comment on its use to predict market value.

Total rooms	Total square footage	Age of structure (years)	Market value of house ($1,000's)
6	1,900	7	$ 73
11	4,500	20	170
5	800	4	31
9	3,300	21	118
6	900	3	33
7	1,800	4	67
8	2,800	10	113
6	1,800	5	65
7	1,100	5	48
7	1,200	5	52
7	1,800	12	62
8	2,600	15	112
8	2,500	12	100
7	2,400	13	90
7	2,100	10	71

Case Study

At the end of chapter 12 there is a case study requiring the use of simple linear regression. Now that you have covered multiple regression, rework that problem

1. using all three independent variables in one equation,

2. combining newspaper and magazine advertising to form one independent variable and using that combined advertising series along with total personal income as the independent variables.

Which of your new equations do you believe is the most statistically reliable for explaining changes in the sales of automotive dealers? Why?

14 Time-Series Analysis: Trend

A *time series* is a set of data classified chronologically. That is, there are two variables (bivariate) in a time-series data set. The independent variable is always a unit of time—days, weeks, months, years, and so on. The dependent variable is often some economic or business-type variable like sales, employment, interest rates, or gross national product, but it may be anything measured through time such as mean ACT scores of high school graduates, electronic impulses, and rainfall.

Time-series analysis is just what the name implies; it is the analysis of the dependent variable in relation to the independent variable (time). The primary assumption is that the values of the dependent variable are a function of time. Thus time-series analysis involves determining and measuring the bases for changes in values of time-series (dependent) variables over some specified period of time.

Time-Series Variables and Components

In time-series analysis the values of the independent variable may be expressed in any interval or unit of time. Time, itself, is infinitely divisible so, theoretically, we can conceivably measure the dependent variable relative to any time interval regardless of how brief it may be—minutes, seconds, or even microseconds. In most instances the periods (units of time) used to express the independent variables and for measurement of the dependent variable are days, weeks, months, quarters, or years. Some examples of frequently analyzed time series are

1. weekly department store sales,
2. monthly interest rates,
3. quarterly employment,
4. annual college enrollment.

Moreover the relationship between the two variables can be evaluated over different time intervals for the same dependent variable. That is, we can analyze the behavior of the dependent variable when it is measured weekly, monthly, quarterly, or annually. Using our example, we might analyze weekly department store sales, monthly department store sales, quarterly department store sales, or annual department store sales. Obviously, the time interval used depends on the purpose of the analysis. Yet the significance of the choice of time period rests with the sources or causes of the behavior of the dependent variable. In general, these sources may be categorized as

1. *seasonal* variation (*S*),
2. *cyclical* variation (*C*),
3. *irregular* variation (*I*),
4. *secular trend* factors (*T*).

The effects of seasonal variation (*S*) in a dependent variable can only occur in data expressed in time units of less than a year because seasonal variation is a *within-year* type of time variation. Also this type fluctuation tends to repeat itself from year to year. Factors that contribute to seasonal variation are weather, custom, and social mores. June weddings, swimming in the summer, snow skiing in the winter, and school starting in the fall and ending in the spring are all seasonal events.

Seasonality usually affects monthly or quarterly data because they are in time units of less than a year. Thus you might expect seasonal variation in the rental of tuxedos, sales of swimwear, vacation expenditures, and sales of school clothes. Figure 14.1 is a graphic example of seasonal variation.

Cyclical variation (*C*) in a dependent variable occurs over a longer time period than seasonal variation. Cyclical variation is characterized by a period of general expansion (growth) followed

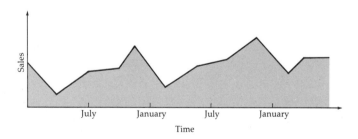

Figure 14.1   Seasonal variation

BASIC STATISTICS

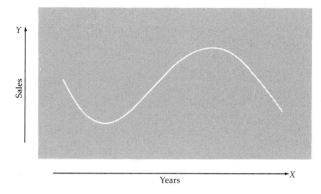

Figure 14.2   Cyclical variation

by a period of contraction. The periods of expansion and contraction vary in length and usually it requires several years to complete a cycle. The wavelike nature of cyclical variation is illustrated in figure 14.2.

Irregular variation (*I*) might best be called *random variation*. Unlike the other three types of variation, there is no pattern to its behavior. Random variations are unpredictable, both as to when they will occur and as to their duration. Periods of extreme fashions (fads), wars, and unseasonable variations in the weather are examples of sporadic events that contribute to this form of variation in economic data. The graph in figure 14.3 illustrates the irregular variation in the sales of a company.

Secular trend (*T*), often referred to simply as *trend*, shows the *long-term behavior* of a time-series variable. It is a generalized expression of the growth, positive (increasing) or negative (decreasing), in the value of the variable. Thus the *trend* is usually represented graphically by a smooth curve

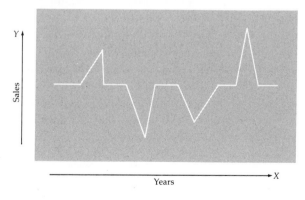

Figure 14.3   Irregular variation

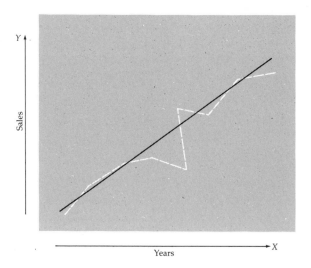

Figure 14.4　Trend

In figure 14.4 the trend is represented by the straight line superimposed over the dashed lines. The dashed lines represent the actual sales over time, and the trend is a representation of the long-term directional movement (*the trend*) of the data. In this case trend is presented as a straight line because it seems appropriate here. However, you will learn later in this chapter that there may be other shaped curves to describe the general trend behavior of the variable.

For economic, industrial, and agricultural variables, trend is usually associated with population changes (growth or decline) and changes in technology. However, trend may also be the result of changes in energy sources, capital formation, consumer taste and acceptance, or other factors that may affect the values of variables over a sustained period of time.

The General Model

At any given point in time, then, the value of a variable (economic data) is determined by the four types of variations (seasonal, cyclical, irregular, and trend). A specific value of the variable may be considered to be the product of these types of variation or their sum. In equation form the relationships are expressed as either a product model, equation (14.1) or an additive model, equation (14.2).

$$Y = T \times S \times C \times I, \tag{14.1}$$

$$Y = T + S + C + I, \tag{14.2}$$

BASIC STATISTICS

where

Y = the value of dependent variable at some point in time,

T = that part of the value of the dependent variable attributable to trend—stated as an absolute value in both models,

S = that part of the value of the dependent variable attributable to seasonal variation—stated as an index in the product model and as an absolute value in the additive model,

C = that part of the value of the dependent variable attributable to cyclical variation—stated as an index in the product model and as an absolute value in the additive model,

I = that part of the value of the dependent variable attributable to random variation—stated as an index in the product model and as an absolute value in the additive model.

Annual data exhibit no seasonal variation; so the models for annual data are

$$Y = T \times C \times I, \tag{14.3}$$

$$Y = T + C + I. \tag{14.4}$$

The additive model assumes that each component of the model is independent of the other components. For example, such an assumption implies that the seasonal variation for each month is the same amount every year regardless of the general trend or the cyclical position of the data. The product model, however, assumes that the components are related even though their causes are different. Thus in the product model one of the components (usually trend) is expressed in absolute units, and the other components are expressed as a percentage of the first component. The product model is more widely used and will receive most of the discussion in this text.

Justification for time-series analysis, and the models, stems from two principal activities. One is the study of the effect on data of each of the types of variation. The other is the utilization of the knowledge of the behavior of the data to forecast future performance. If the latter use can be carried out with some degree of accuracy, and if you understand its limitations, you will have an extremely useful statistical tool at your disposal.

Editing Data for Time-Series Analysis

Since the data to be analyzed have been gathered over a broad span of time, there are several problems that must be dealt with before we can begin a time-series analysis. The following is not intended as an all-inclusive list but, rather, to introduce you to the type of potential difficulties to be aware of when you are analyzing time-series data:

1. Data gathered over a long time span, or from many sources, may have a particular problem of consistency or comparability. Definitions may change, the geographic area covered may vary, or the method of calculation may be altered. As some examples of this, unemployment has not always been defined as it now is; accountants determine costs in several ways; and the United States has not always included fifty states. An investigation of the data should be made so that, if necessary, alternative sources of data can be sought or adjustments in the data can be made. Even when no alternative data sources exist that will improve the consistency of the data, the investigation will produce an awareness by the analyst of any weaknesses in the data.

2. Months, except February, always contain the same number of days each year but they do not always have the same number of "working days." There are also other calendar variations, such as the precise location of holidays and the beginning of the college year. Data may reflect this calendar variation so that, for example, data for June of one year may not be strictly comparable to that for June of another year. The possibility of calendar variations is often ignored in time-series analysis, but in some types of monthly data, or data for shorter time periods, it may be very significant.

3. Time-series analysis is bivariate analysis. That is, there is only one independent and one dependent variable. In measuring the relationship between two variables, you should exclude, as far as possible, the influence of other independent variables on the dependent variable. One such independent variable, when working with data measured in dollars, that will influence the dependent variable is price changes. For example, if you do time-series analysis on gross national product data expressed in given year dollars, both real output changes and price changes will be included in the dependent variable. Generally, then, data should be converted to constant dollars (you will learn how to do this in chapter 16) before the analysis begins.

Population changes can present a problem similar to that created by price changes. Consider the situation where consumption of a product increases by 20 percent, but, at the same time, population increases by 50 percent. An analysis of the data without considering the population change would certainly lead to conclusions quite different from an analysis of the data adjusted for population changes. If the dependent variable is one that population changes can influence, converting the data to a per capita basis will remove the potential distortion.

Trend

Initially in trend analysis, two decisions must be made (1) what shape line, or curve, will be used to represent the relationship between the two variables, and (2) what procedure will be used to determine the coefficients of an equation, corresponding to the shape selected, that will describe the relationship. Both decisions are critical, but at your present level in statistics, the first is the most significant. If you go on to study time-series analysis in greater depth, many questions will be raised concerning methods of determining coefficients, or as it is sometimes called, "fitting curves to data." The procedures presented in this text are relatively easy to understand, and they are probably the most commonly used techniques when time-series analysis is applied to business and economic data. They are, however, only a limited introduction to the techniques available for your use at more sophisticated levels of analysis.

Selecting the Shape of the Trend Curve

Early in the text (chapter 1) you learned about the use of graphs in statistical analysis. Scatter diagrams (a graph type) are basic to selecting the shape of the curve to describe the trend. Figure 14.5 illustrates some of the results you might obtain when you make a scatter diagram using your own data.

Notice in the graphs (figure 14.5) that we have plotted only *annual data.* For trend analysis, the use of data for shorter time periods (such as monthly or quarterly data) is unnecessary. Furthermore the use of something less than annual data would involve a greater amount of work (arithmetic) and would probably lead to a greater number of errors. If you find that you need to know how to express trend in some form other than annual data, look in the appendix to this chapter, but only after you have learned what is presented in the pages between here and there.

Graph (a) in figure 14.5 shows a set of data in which the dependent variable appears to have a positive, *linear* relationship with the independent variable. The relationship is positive because the dependent variable is generally increasing as time elapses. *Linear* describes the relationship because the dependent variable is changing by a reasonably uniform amount as time elapses. A relationship of this nature can be represented by a straight line, for which the general equation is $Y_c = a + bX$. In this and subsequent equations, Y_c is the estimated value of the

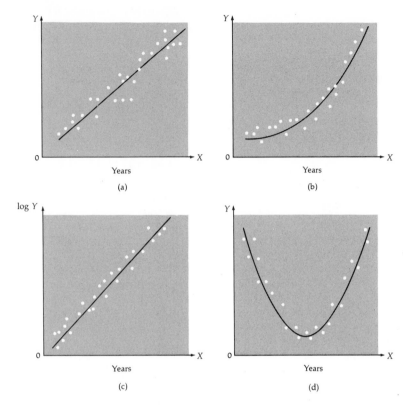

Figure 14.5   Examples of scatter diagrams

dependent variable and X represents time, the independent variable. The values (coefficients) a and b are constants that are determined for each set of data.

The data scatter in graph (b) requires that the relationship be expressed in a different manner. The trend is positive, but a straight line does not accurately express the general behavior of the dependent variable. Rather than increasing by a constant amount (as a straight line implies), it appears to be increasing at a constant rate. This type of behavior can be represented by a *power curve.*

Fortunately, there is another method of displaying the type of relationship shown in graph (b). Graph (c) shows the same data plotted with a *logarithmic* rather than arithmetic scale on the ordinate. Look at how the pattern of the scatter has changed. Data that are described by a power curve on an arithmetic graph appear as a straight line on a semilog graph. Thus $Y_c = ab^X =$ antilog of (log a + X log b).

Graph (d) in figure 14.5 presents the scatter of still another possible set of data. This relationship between the variables is distinctly different from that presented in the first two graphs. In the earlier years the relationship appears negative, but in the most recent time periods it appears positive. Neither a straight line nor a power curve will adequately describe such behavior. If the trend changes its direction, a second-degree curve is needed to portray the relationship between the variables. The general equation for a second degree curve is $Y_c = a + bX + cX^2$. The additional coefficient c has been added to the straight-line equation. At another point we will discuss the limitations of using this type of curve in time-series analysis.

Selecting the appropriate equation to describe the relationships in a set of data is not always so obvious as our illustrations might lead you to believe. If you encounter data that appear to have no pattern in their scatter, or you cannot visualize how well a given shape will describe the pattern, do what we have done in figure 14.6. Simply "rough in" a straight line on the scatter dia-

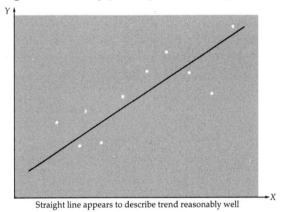

Straight line appears to describe trend reasonably well

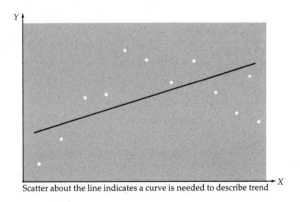

Scatter about the line indicates a curve is needed to describe trend

Figure 14.6   Selecting the shape of the trend line

gram. If the scatter alternates both above and below the line, but there is no substantial interval where all the points are either above or below, then a straight line is probably the best choice to generalize on the behavior. Should you find that there is a pattern to the arrangement of the points about the line you have drawn, a curve may be needed. The existence of change in direction in the relationship can justify selecting a second-degree curve. Otherwise, the power curve would probably be the most suitable. From a practical standpoint, as your experience with trend analysis increases, you will generally be able to select the appropriate shape from a visual inspection of the scatter diagram.

Determining the Coefficients for a Linear Trend Equation

When studying chapter 12, you learned how to "fit a line (or curve) to the data." The same concepts you learned in that chapter are applicable now, since trend analysis is simply regression analysis with time as the independent variable. In the general equation for a straight line, $Y_c = a + bX$, it is necessary to determine a and b. In this chapter two methods of determining a and b will be examined: (1) freehand and (2) least squares.

Freehand Method

The freehand method can be used to *approximate* a linear trend equation. The procedure is to plot the data and then draw a straight line through the data, trying to keep the sums of the areas above and below the trend line about the same. Figure 14.7 illustrates the procedure.

Notice in figure 14.7, the values of a and b are determined by observation. Thus, from the figure, a is observed to be 82. To determine b, the value of Y is observed for the first and last point on the regression line. That is, $Y = 82$ in 1975 and $Y = 152$ in 1981. Then

$$b = \frac{\text{last } Y \text{ value} - \text{first } Y \text{ value}}{\text{number of periods}}$$

$$b = \frac{152 - 82}{6} = \frac{70}{6} = 11.67.$$

And the regression equation is $Y_c = 82 + 11.67X$, which is shown in figure 14.7. Also given in figure 14.7 are the origin and the units of the variables. Notice that the origin is July 1, 1975. The middle of the year is used because the total sales for a year are accumulated over the year, and thus the annual sales value does not represent any point in time. Consequently convention has been to plot data of this nature in the middle of the period from which they were collected.

BASIC STATISTICS

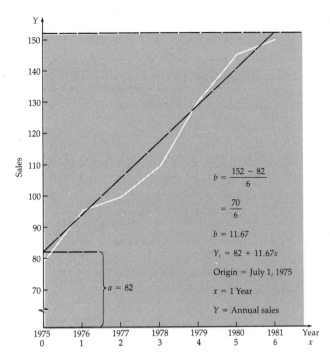

Figure 14.7 Freehand method for determining trend

Notice also that X and Y are defined so that anyone can identify easily the dependent variable and the period for the independent variable. In figure 14.7 the time period (X) is one year, and it is shown in *coded* form as the x value under each year. The coding is necessary because the value of x has to be 0 at the origin. Thus, with the origin set as 1975, then $x = 0$ in 1975, $x = 1$ in 1976, and so forth. This coding of the X variable will prove very useful throughout this chapter.

The advantage of the freehand method is that it is very easy to understand and apply. The disadvantage is that it is a method by which different people can get different answers for the same problem. Thus it is not a consistently reproducible approach.

Method of Least Squares

The least-squares method has been discussed previously in chapter 12. Tables 14.1 and 14.2 illustrate the least-squares method for determining the trend (regression) equation for an odd number of years and an even number of years, respectively. In both tables the coded form of the independent variable is used to reduce the arithmetic and the level of algebra required to solve for the constants, a and b.

In table 14.1 the origin is established as the middle year, and all of the other values of x are obtained by subtracting 1978 from

Table 14.1
Fitting a straight line, least-squares method (odd number of years)

Year X	x	x^2	Sales Y	xY
1975	−3	9	$ 80	−240
1976	−2	4	95	−190
1977	−1	1	100	−100
1978	0	0	110	0
1979	+1	1	130	130
1980	+2	4	145	290
1981	+3	9	150	450
	0	28	810	+340

$Y_c = a + bx$

$$a = \frac{\Sigma Y}{n} \qquad\qquad b = \frac{\Sigma xY}{\Sigma x^2}$$

$$a = \frac{810}{7} = 115.71 \qquad\qquad b = \frac{340}{28} = 12.14$$

$Y_c = 115.71 + 12.14x$
$x = 1$ year, $Y =$ annual sales
Origin = July 1, 1978
$Y_{c\ 1982} = 115.71 = 12.14(+4)$
$\qquad\quad = 115.71 + 48.56$
$Y_{c\ 1982} = 164.27$

each of the other years. That is, $x = +1 = (1979–1978)$, and $x = -1 = (1977–1978)$, and so on. Actually, there is a quicker way to code the data than by subtracting the middle year from the other values of the independent variable. Simply designate $x = 0$ as the middle year, $x = +1$ for the following year, $x = -1$ for the preceding year, and so forth, until all years have been assigned a coded x value. Now study table 14.1 observing the coding, the solution procedure, and the designation of the origin and the x and Y units.

In table 14.2 the computations are the same as in table 14.1. However, there are an even number of years, and the coding has to be handled differently to assure that $\Sigma x = 0$. Since it is an even number of years, there is no middle time period. That is, the middle of the data does not occur at the middle of a year; it occurs at the end of one year or at the beginning of another year. Thus in table 14.2, $x = 0$ occurs at the end of 1977, or, if you prefer, at the beginning of 1978. As discussed earlier, however, data that are associated with a period of time are usually plotted in the center of the period. Thus, to get the data centered opposite the year they came from, it is necessary to move six months ($x = 0.5$) for-

Table 14.2
Fitting a straight line, least-squares method (even number of years)

Year X	x	x^2	Output (1,000) Y	xY
1974	−3.5	12.25	12	−42.0
1975	−2.5	6.25	13	−32.5
1976	−1.5	2.25	15	−22.5
1977	−0.5	0.25	17	−8.5
1978	+0.5	0.25	18	9.0
1979	+1.5	2.25	19	28.5
1980	+2.5	6.25	21	52.5
1981	+3.5	12.25	23	80.5
	0	42.00	138	+65.0

$Y_c = a + bx$

$$a = \frac{\Sigma Y}{n} \qquad\qquad b = \frac{\Sigma xY}{\Sigma x^2}$$

$$a = \frac{138}{8} = 17.25 \qquad\qquad b = \frac{65}{42} = 1.5476$$

$Y_c = 17.25 + 1.55x$
$x = 1$ year, $Y =$ annual output
Origin = January 1, 1978
$Y_{c\ 1982} = 17.25 + 1.55(4.5)$
$Y_{c\ 1982} = 24.225$ (in thousands)

ward or six months ($x = -0.5$) backward. So, for 1977, $x = -0.5$, and for 1978, $x = 0.5$. From then on, the movement is twelve months so that $x = 1.5$ for 1979, $x = 2.5$ for 1980, and so on.

As previously mentioned, the computations in table 14.2 are the same as in table 14.1. Notice, however, that because of the difference in coding the origins are not at the same part of the year. When there is an odd number of years, the origin is at the middle of the year (July 1), whereas, when there is an even number of years, the origin is at the end of the year (midnight December 31 or January 1).

The solutions shown in the preceding tables use the least-squares method based on the following set of normal equations:

$$\Sigma Y = na + b\Sigma X, \tag{14.5}$$

$$\Sigma XY = a\Sigma X + b\Sigma X^2, \tag{14.6}$$

where
$Y =$ the values of the dependent variable,
$n =$ the number of values of the independent (or dependent) variable,
$X =$ the values of the independent variable.

The values of a and b can always be found by solving simultane-ously the normal equations, but, if you code the independent variable so that $\Sigma x = 0$ as was done in our tables, the difficulty is greatly reduced. In this case the equations are reduced:

$$a = \frac{\Sigma Y}{n},$$ (14.7)

$$b = \frac{\Sigma xY}{\Sigma x^2}.$$ (14.8)

We have used x rather than X in the equations so that you will not forget that they are to be used only when the data are properly coded. Using the coded value (x), the straight-line trend equa-tion is

$$Y_c = a + bx,$$ (14.9)

where
Y_c = the trend value,
x = the coded value of X,
a = the intercept of the trend line (the value of Y_c when $x = 0$, the origin,
b = the slope of the trend line (the change in Y_c when x is changed one time period).

Finally look at the last calculation in tables 14.1 and 14.2. These last entries are both forecasts for the year 1982. Notice how the forecast is done. The appropriate value of x is determined for the year of the forecast, and the derived trend equation is used to make the forecast. Forecasting is one of the primary uses of trend analysis. The results, however, have to be used carefully, since we are basing the forecast for the future on past history. If the future holds numerous uncertainties due to changing economic conditions, product development, or technology changes, or other causes of variation, the forecast should be used with cau-tion and/or altered to reflect the changing situation.

Determining the Coefficients for a Logarithmic Straight-Line Trend Equation

In nearly the same way as fitting a straight line, you can trans-form the data and fit a straight line to the transformed data. For a discussion of transformation, read the appendix to chapter 12. Remember the logarithmic straight line is an exponential, or power, curve. The general equation is as follows:

$$Y_c = ab^X.$$ (14.10)

If the dependent variable data are transformed into logarithms, you can see the similarity to the arithmetic straight line:

$$\log Y_c = \log a + X \log b \qquad (14.11)$$

where

$\log Y_c$ = the logarithm of the trend value for any given value of the independent variable,

$\log a$ = the logarithm of the intercept of the trend line, $\log Y_c$ (the value of $\log Y_c$ when $X = 0$),

$\log b$ = the logarithm of the slope of the trend line (the change in $\log Y_c$ when X is changed one unit; the antilog of $\log b$ is the average annual rate of growth when $X = 1$ year),

X = any given value of the independent variable.

The equations used to solve for a and b using the method of least squares are

$$\Sigma(\log Y) = n(\log a) + \log b(\Sigma X), \qquad (14.12)$$

$$\Sigma(X \log Y) = \log a(\Sigma X) + \log b(\Sigma X^2). \qquad (14.13)$$

If we assume that the independent variable is coded so that $\Sigma x = 0$, these equations are reduced to the following:

$$\log a = \frac{\Sigma(\log Y)}{n}, \qquad (14.14)$$

$$\log b = \frac{\Sigma(x \log Y)}{\Sigma x^2}. \qquad (14.15)$$

An example of the type of data that fit the exponential curve is given in table 14.3. Notice that the change in the dependent variable is not constant from year to year, but that the increases get

Table 14.3
Data for the exponential trend equation

Year X	Sales Y
1973	50
1974	70
1975	90
1976	115
1977	150
1978	200
1979	265
1980	350
1981	450

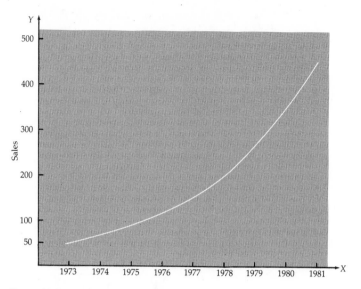

Figure 14.8   Arithmetic graph of the data for the exponential trend equation

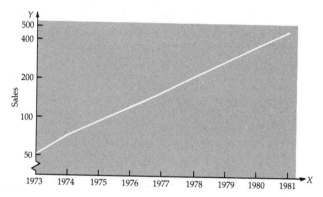

Figure 14.9   Semilogarithmic graph of the data for the exponential trend equation

larger as time goes on. The data presented in table 14.3 are plotted in figures 14.8 and 14.9.

In figure 14.8 the data are plotted on arithmetic paper so that you can see the continual increasing rate of Y. In figure 14.9 the data are plotted on semilogarithmic paper. Notice now that the graph approximates a straight line. Hence the nonlinear data (figure 14.8) are transformed to linear data (figure 14.9) and are easier to work with than the nonlinear data.

Now that you have seen graphs of the data, look at table 14.4 to see how the logarithmic trend equation is developed.

Notice at the end of table 14.4 sales for 1981 were calculated to

Table 14.4
Fitting a logarithmic straight line, least-squares method (exponential curve)

Year X	x	x^2	Sales Y	log Y	x log Y
1973	−4	16	50	1.69897	−6.79588
1974	−3	9	70	1.84510	−5.53530
1975	−2	4	90	1.95424	−3.90848
1976	−1	1	115	2.06070	−2.06070
1977	0	0	150	2.17609	0.0
1978	1	1	200	2.30103	2.30103
1979	2	4	265	2.42325	4.84650
1980	3	9	350	2.54407	7.63221
1981	4	16	450	2.65321	10.61284
	0	60		19.65666	7.09222

$$\log a = \frac{\Sigma \log Y}{n} = \frac{19.65666}{9} = 2.18407$$

$$\log b = \frac{\Sigma (x \log Y)}{\Sigma x^2} = \frac{7.09222}{60} = 0.11820$$

$$\log Y_c = \log a + x \log b$$
$$\log Y_c = 2.18407 + 0.11820x$$
$$x = 1 \text{ year}; Y = \text{annual sales}$$
$$\text{Origin} = \text{July 1, 1977}$$
$$\log Y_{c\ 1981} = 2.18407 + 0.11820(4)$$
$$\log Y_{c\ 1981} = 2.18407 + 0.47280 = 2.65687$$
$$Y_{c\ 1981} = 453.81$$

be $453.81 as compared to the actual value of $450.00. The difference represents an error of only 0.85 percent. Now let us use the logarithmic trend equation to forecast sales for 1983:

$$\log Y_{c\ 1983} = 2.18407 + 0.11820(6)$$
$$= 2.18407 + 0.70920 = 2.89327$$
$$Y_{c\ 1983} = \$782.15.$$

Determining the Coefficients for a Second-Degree Trend Equation

Sometimes the trend in time-series data appears to change direction. That is, it may have a positive slope over part of its length and a negative slope over the other part. When this happens, the trend might be appropriately expressed by a parabola, or second-degree curve. This can be done by the least-squares method but the procedure is slightly more complicated than the other two you have learned. A second-degree curve is described by the equation:

$$Y_c = a + bX + cX^2. \qquad (14.16)$$

Since you have three unknowns (a, b, and c), you need three equations to solve for them. Using the least-squares method of fitting the line, these three equations are

$$\Sigma Y = na + b\Sigma X + c\Sigma X^2, \tag{14.17}$$

$$\Sigma XY = a\Sigma X + b\Sigma X^2 + c\Sigma X^3, \tag{14.18}$$

$$\Sigma X^2 Y = a\Sigma X^2 + b\Sigma X^3 + c\Sigma X^4. \tag{14.19}$$

Fortunately, coding the independent variable so that $\Sigma x = 0$ reduces the complexity of solving for a, b, and c. With this form of coding the formulas are as follows:

$$\Sigma Y = na + c\Sigma x^2, \tag{14.20}$$

$$\Sigma xY = b\Sigma x^2, \tag{14.21}$$

$$\Sigma x^2 Y = a\Sigma x^2 + c\Sigma x^4. \tag{14.22}$$

(If $\Sigma x = 0$, Σx^3 is also equal to zero.) Now the value of b can be solved for directly, and a and c values can be found by using the two remaining equations:

$$b = \frac{\Sigma xY}{\Sigma x^2}. \tag{14.23}$$

On the other hand, you may prefer to use the following approach for determining the values of a and c. Either way, you will get the same results:

$$c = \frac{n\Sigma x^2 Y - \Sigma Y\Sigma x^2}{n\Sigma x^4 - (\Sigma x^2)^2}, \tag{14.24}$$

$$a = \frac{\Sigma Y - c\Sigma x^2}{n}. \tag{14.25}$$

Using the coded values of X, (x), the second-degree trend equation is

$$Y_c = a + bx + cx^2. \tag{14.26}$$

Now, look at table 14.5 for an example of fitting a second-degree curve.

Adjusting Data for Trend

The presence of trend in business and economic data may impair the data for evaluation purposes. For example, if there is a long-term (secular) growth trend of a 5 percent increase per year in a company's sales, extracting the trend from the data may facilitate the evaluation of the short-term performance of sales. In addition measures of other types of economic fluctuation, seasonal, cyclical, and random, must be free of "trend bias." Therefore either the trend must be removed from the data before these measures

Table 14.5
Fitting a second-degree curve, least-squares method

Year X	x	x^2	x^4	Sales Y	xY	x^2Y
1977	−2	4	16	$100	−200	400
1978	−1	1	1	80	− 80	80
1979	0	0	0	105	0	0
1980	+1	1	1	120	+120	120
1981	+2	4	16	145	+290	580
	0	10	34	$550	+130	1180

$$Y_c = a + bx + cx^2$$

$$b = \frac{\Sigma xY}{\Sigma x^2} = \frac{130}{10} = 13.0$$

$$c = \frac{n\Sigma x^2Y - \Sigma Y\Sigma x^2}{n\Sigma x^4 - (\Sigma x^2)^2} = \frac{5(1180) - (550)10}{5(34) - (10)^2} = \frac{400}{70} = 5.71$$

$$a = \frac{\Sigma Y - c\Sigma x^2}{n} = \frac{550 - (5.71)10}{5} = 98.58$$

$$Y_c = 98.58 + 13.0x + 5.71x^2$$

$x = 1$ year, $Y =$ annual sales

Origin = July 1, 1979

$$Y_{c\ 1981} = 98.58 + 13(+2) + 5.71(4)$$
$$= 98.58 + 26 + 22.84 = 147.43$$

are determined or the method of measurement must compensate for the presence of trend in the data.

Assuming the product model, trend is removed by dividing the actual data, Y, by the trend estimates, Y_c. The logic for this method is as follows:

IF $\quad Y = T \times S \times C \times I$

AND $\quad T \cong Y_c$,

THEN $\quad Y = Y_c \times S \times C \times I$

AND $\quad \dfrac{Y}{Y_c} = S \times C \times I.$

When trend is removed, what remains of the data is the influence of seasonal, cyclical, and irregular variations or cyclical and irregular if the data are annual data. Extracting trend in this manner leaves the residual $S \times C \times I$ in the form of a ratio rather than units (dollars, pounds, and so on). Study table 14.6 to see an example of how trend is removed in this manner.

Appendix

Trend equations can be manipulated to make them more useful. Two particular adjustments that we can do are (1) change the units and (2) shift the origin.

Table 14.6
Removing trend, product model

Year X	Sales Y	Y_c	Y/Y_c
1975	$ 80	$ 80.55	0.993
1976	95	91.80	1.035
1977	100	103.05	0.970
1978	110	114.30	0.962
1979	130	125.55	1.035
1980	140	136.80	1.023
1981	145	148.05	0.979

Changing the Units

Assume that the trend equation for the number of customers stopping at the Roadside Inn is

$Y_c = 1,800 + 60x,$
$x = 1$ year,
$Y =$ hundreds of customers,
Origin $=$ July 1, 1975.

Both a and b represent annual data; so, *to convert them to monthly values, we divide our trend equation by 12 as shown:*

$$Y_c = \frac{1,800}{12} + \frac{60}{12}x$$

$Y_c = 150 + 5x.$

This conversion has not, however, affected the x variable, which is still in years. Thus this trend equation reflects the monthly change for an annual period. That is, it gives the change for a given month in one year to the same month in the next year such as from July 1977 to July 1978. Consequently one more step is necessary to get our original trend equation into the form that reflects month-to-month changes. Now we must also *divide x by 12* as follows:

$$Y_c = \frac{1,800}{12} + \left(\frac{60}{12}\right)\left(\frac{x}{12}\right)$$

$Y_c = 150 + 0.417x,$
$x = 1$ month,
$Y =$ hundreds of customers,
Origin $=$ July 1, 1975.

In general, the conversion of an annual trend equation to a monthly trend equation is as shown as on page 411.

$$Y_c = a + bx,$$
$$x = 1 \text{ year,}$$
$$Y = \text{annual totals;}$$

$$Y_c = \frac{a}{12} + \left(\frac{b}{12}\right)\left(\frac{x}{12}\right),$$

$$x = 1 \text{ month,}$$
$$Y = \text{monthly totals.}$$

Conversion of an annual trend equation to a quarterly one follows the same procedure with a, b, and x divided by 4.

Shifting the Origin

Another adjustment may be necessary before two trend equations can be compared. Specifically, the origins of the equations must be equal. Thus, if for any reason the origins of two equations are not the same point in time, then you may wish to change one, or both, of the equations to reflect a different origin. This is done in the following manner. When the trend is an arithmetic straight line,

$$Y_c = a + b(x + (x_n)),$$

where x_n = the new origin.

Using the equation from table 14.1, the origin can be changed to 1980 by substituting in the formula:

$$Y_c = 115.71 + 12.14x \quad (\text{origin} = 1978)$$
$$= 115.71 + 12.14(x + 2) = 115.71 + 24.28 + 12.14x$$
$$Y_c = 139.99 + 12.14x \quad (\text{origin} = 1980).$$

If trend is expressed by a second-degree line, the procedure for changing the origin is much the same:

$$Y_c = a + bx + cx^2 \quad (\text{original origin})$$
$$Y_c = a + b(x + x_n) + c(x + x_n)^2 \quad (\text{new origin}).$$

Perhaps a more useful way to express the procedure is in its expanded form:

$$Y_c = a + b(x + x_n) + c[x^2 + 2x(x_n) + x_n^2].$$

The origin of the equation in table 14.5 is changed to 1977 in the following example:

$$Y_c = 98.58 + 13x + 5.71x^2 \quad (\text{origin} = 1979)$$
$$= 98.58 + 13(x - 2) + 5.71(x - 2)^2$$
$$= 98.58 + 13(x - 2) + 5.71(x^2 - 4x + 4)$$
$$= 98.58 + 13x - 26 + 5.71x^2 - 22.84x + 22.84$$
$$Y_c = 95.42 - 9.84x + 5.71x^2 \quad (\text{origin} = 1977).$$

Summary of Equations

14.1 Product time series model:

$$Y = T \times S \times C \times I$$

14.2 Additive time series model:

$$Y = T + S + C + I$$

14.3 Product times series model—without seasonal variation:

$$Y = T \times C \times I$$

14.4 Additive time series model—without seasonal variation:

$$Y = T + C + I$$

14.5 Normal equations for least-squares method—straight-line trend equation:

$$\Sigma Y = na + b\Sigma X$$

14.6 Normal equations for least-squares method—straight-line trend equation:

$$\Sigma XY = a\Sigma X + b\Sigma X^2$$

14.7 Intercept of trend line—coded independent variable:

$$a = \frac{\Sigma Y}{n}$$

14.8 Slope of the trend line—coded independent variable:

$$b = \frac{\Sigma xY}{\Sigma x^2}$$

14.9 Straight-line trend equation:

$$Y_c = a + bx$$

14.10 Exponential or power curve:

$$Y_c = ab^X$$

14.11   Logarithmic straight line:

$$\log Y_c = \log a + X \log b$$

14.12   Normal equation for least-squares method—logarithmic straight-line trend equation:

$$\Sigma(\log Y) = n(\log a) + \log b(\Sigma X)$$

14.13   Normal equation for least-squares method—logarithmic straight-line trend equation

$$\Sigma(X \log Y) = \log a(\Sigma X) + \log b(\Sigma X^2)$$

14.14   Result of simultaneous solution of preceding normal equations—coded independent variable:

$$\log a = \frac{\Sigma(\log Y)}{n}$$

14.15   Result of simultaneous solution of preceding normal equations—coded independent variable:

$$\log b = \frac{\Sigma(x \log Y)}{\Sigma x^2}$$

14.16   Second-degree trend equation:

$$Y_c = a + bX + cX^2$$

14.17   Normal equation for least-squares method—second-degree trend equation:

$$\Sigma Y = na + b\Sigma X + c\Sigma X^2$$

14.18   Normal equation for least-squares method—second-degree trend equation:

$$\Sigma XY = a\Sigma X + b\Sigma X^2 + c\Sigma X^3$$

14.19   Normal equation for least-squares method—second-degree trend equation:

$$\Sigma X^2 Y = a\Sigma X^2 + b\Sigma X^3 + c\Sigma X^4$$

14.20 Reduced normal equation—second-degree trend equation:

$$\Sigma Y = na + c\Sigma x^2$$

14.21 Reduced normal equation—second-degree trend equation:

$$\Sigma xY = b\Sigma x^2$$

14.22 Reduced normal equation—second-degree trend equation:

$$\Sigma x^2 Y = a\Sigma x^2 + c\Sigma x^4$$

14.23 Result of simultaneous solution of preceding reduced normal equations—second-degree trend equation:

$$b = \frac{\Sigma xY}{\Sigma x^2}$$

14.24 Result of simultaneous solution of preceding reduced normal equations—second-degree trend equation:

$$c = \frac{n\Sigma x^2 Y - \Sigma Y \Sigma x^2}{n\Sigma x^4 - (\Sigma x^2)^2}$$

14.25 Result of simultaneous solution of preceding reduced normal equations—second-degree trend equation:

$$a = \frac{\Sigma Y - c\Sigma x^2}{n}$$

14.26 Second-degree trend equation—coded independent variable:

$$Y_c = a + bx + cx^2$$

Review Questions

1. Explain the use of each equation given at the end of the chapter.

2. Define
(a) time series
(b) time series analysis

3. Name the four components of time series models.

4. Name the two types of time series models.

5. How do the representations of components of the two time series models differ?

Problems

1. The data are popultion data for the United States:
(*a*) Plot the data on graph paper.
(*b*) Fit a straight-line trend equation to the data using the least-squares method.
(*c*) Estimate the 1981 population.

Year	Total population (millions)
1970	204.9
1971	207.1
1972	208.8
1973	210.4
1974	211.9
1975	213.6
1976	215.2
1977	216.9
1978	218.5
1979	220.6

Source: *Statistical Abstract of the United States*, 1980, p. 6.

2. Assume the following data represent the corn production for a particular farm.

Year	Bushels
1976	200
1977	205
1978	215
1979	210
1980	230
1981	240
1982	245

(*a*) Plot the data on graph paper.
(*b*) Fit a straight-line trend equation to the data using the least-squares method.
(*c*) Estimate the 1983 production.

3. The following data are recent income security payments by the federal government:

Fiscal year	Income security payments (billions)
1973	73.0
1974	84.4
1975	108.6
1976	127.4
1977	137.9
1978	146.2
1979	160.2
1980	193.1
1981	225.1

Source: *The Budget of the United States Government, Fiscal Year 1983*. U.S. Government Printing Office, p. 9–54.

(*a*) Fit a linear trend equation to the data using the least-squares method.

(*b*) Estimate the 1982 expenditure.

(*c*) Fit a linear trend equation to the data using the freehand method.

4. The sales data for two organizations are given below.

Year	I Sales ($1,000)	II Sales ($1,000)
1976	356	420
1977	373	430
1978	388	438
1979	398	447
1980	417	453
1981	430	462
1982	445	470

(*a*) Use the method of least squares, and fit a linear trend equation to each set of data.

(*b*) Assuming that current trends continue, determine the year that I sales will first exceed II sales.

5. The following data are the debt of the federal government:

Fiscal year	Federal debt (billions)
1971	$409.5
1972	437.3
1973	468.4
1974	486.2
1975	544.1
1976	631.9
1977	709.1

Fiscal year	Federal debt (billions)
1978	$ 780.4
1979	833.8
1980	914.3
1981	1,003.9

Source: *The Budget of the United States Government, Fiscal Year 1983*. U.S. Government Printing Office, p. 9–60.

(a) Fit a linear trend equation to the data using the least-squares method.
(b) Estimate the 1982 debt, and comment on the estimate.

6. Consider the following data to be sales of New Store, Inc.:

Year	Sales (thousands)
1976	$10
1977	12
1978	18
1979	22
1980	30
1981	40
1982	52

(a) Plot the data on graph paper.
(b) Fit a linear trend equation to the data using the least-squares method.
(c) Fit a logarithmic trend equation to the data using the least-squares method.
(d) Which of the two curves appears to have the better fit?
(e) Estimate the sales for 1983 using the equation developed in part (c).

7. The following data represent a situation where the potential market is growing linearly but the sales are increasing exponentially:

Year	Potential market ($1,000)	Company sales ($1,000)
1976	88	33
1977	94	38
1978	101	45
1979	106	53
1980	111	63
1981	117	74
1982	125	87

(a) Determine the linear trend equation for the potential market using the least-squares method.
(b) Determine the exponential trend equation for the sales data using the least-squares method.
(c) Project the market and sales for 1983.

8. Our Favorite Publisher has published the following number of new books during the period 1974 to 1982:

Year	New books
1974	8
1975	10
1976	13
1977	17
1978	18
1979	21
1980	22
1981	26
1982	28

(a) Fit a linear regression equation to the data.
(b) Estimate the number of new books in 1983.

9. Use the following trend equation to answer the questions below:

$$Y_c = 284 + 14.4x,$$
origin $=$ July 1, 1974,
$x = 1$ year,
$Y =$ annual sales in thousands of dollars.

(a) If 1982 was the last year of the data, how many years of data were used to determine the trend equation assuming the origin has not been moved?
(b) Project sales for 1985.
(c) What is the annual dollar increase in sales?

10. Use the following trend equation to answer the questions below:

$$Y_c = 320 + 36x,$$
origin $=$ January 1, 1975,
$x = 1$ year,
$Y =$ annual sales in thousands of dollars.

(a) If 1982 was the last year of data, how many years of data were used to determine the trend equation if the origin has not been changed?
(b) Project sales for 1983.
(c) What is the annual dollar increase in sales?

11. Use the following trend equation to answer the questions below:

$$Y_c = 50 + 12x,$$
origin = December 31, 1977,
$$x = 1 \text{ year},$$
$$Y = \text{annual sales in thousands of dollars}.$$

(a) If 1982 was the last year of data, how many years of data were used to determine the trend equation if the origin has not been changed?
(b) What is the annual dollar increase in sales?
(c) Project sales for 1983.

12. Remove the trend from the data of problem 1 assuming the multiplicative model.

13. Fit a linear equation to the following data using the method of least squares:

Energy expenditures Fiscal year	Billions of dollars
1975	2.2
1976	3.1
1977	4.2
1978	5.9
1979	6.9
1980	6.3
1981	10.3

Source: *The Budget of the United States Government, Fiscal Year 1983*, U.S. Government Printing Office, p. 9–51.

14. Assume the following data represent employment in a large plant. Fit a second-degree least-squares equation to these data.

Year	Employment
1976	3,000
1977	3,500
1978	3,700
1979	4,000
1980	3,800
1981	3,600
1982	3,300

15. Fit a second-degree equation to the following data using the method of least squares:

Farm income stabilization payments Fiscal year	Billions of dollars
1972	4.6
1973	4.1
1974	1.5
1975	0.8
1976	1.6
1977	4.5
1978	6.9

Source: *The Budget of the United States Government, Fiscal Year 1983.* U.S. Government Printing Office, p. 9–51.

16. Fit a logarithmic trend equation, using the method of least squares, to the following data letting $x = 1$ year:

Sales of product Year	Billions of dollars
1975	4.6
1976	7.2
1977	8.9
1978	9.9
1979	11.2
1980	13.4
1981	14.1
1982	17.2

17. Use the hypothetical data to answer the questions below.

Gasoline sales (millions of gallons) Year	Quarter				Total
	I	II	III	IV	
1978	4.3	4.9	5.3	5.0	19.5
1979	4.4	4.9	5.4	5.1	19.8
1980	4.5	5.3	5.5	5.3	20.6
1981	4.6	5.3	5.6	5.4	20.9
1982	4.6	5.3	5.8	5.7	21.4

(a) Determine the straight-line trend equation for the yearly totals using the method of least squares. (Keep this work for chapter 15.)
(b) Convert the equation developed in part (a) to a quarterly equation with Y units expressed as quarterly sales.
(c) Shift the origin of the quarterly equation to the middle of the third quarter of 1980.

18. Estimate 1983 production using the following equation:

$$\log Y_c = 6.43921 + 0.23510x,$$
origin = July 1, 1977,
$\quad x = 1$ year,
$\quad Y =$ annual production in pounds.

19. Use the equation

$$Y_c = 100 + 10x,$$
origin = December 31, 1979,
$\quad x = 1$ year,
$\quad Y =$ annual sales in millions,

(a) to estimate sales for 1983.
(b) and shift the origin to December 31, 1981.

20. Using the trend equation in problem 9 of this chapter,
(a) shift the origin to July 1, 1978.
(b) convert the trend equation to monthly sales with $x = 1$ month, and set the origin at July 15, 1980.
(c) estimate sales for September 1983.

21. Use the equation

$$Y_c = 120 + 3.6x,$$
origin = December 31, 1978,
$\quad x = 1$ year,
$\quad Y =$ annual production in millions of pounds,

(a) and shift the origin to July 1, 1982.
(b) and convert the given trend equation to monthly data with $x = 1$ month.

Case Study

You will find twelve years of total personal income data presented in the case study at the end of chapter 12. Determine the linear trend for total personal income using first the original data and then logarithmic transformed data. Next, utilizing the additive model, determine $C + I$ using each trend equation, and present your results in graph form. Referring to the graph, does transformation of the data cause your measure of cyclical plus irregular variations $(C + I)$ to change? What part may inflation have attributed to your differing results?

Using each trend equation, project total personal income for 1982, 1983, 1984, and 1985, and construct a graph showing both the trends from 1970 to 1985. What is happening to the relative position of the two trend lines in the graph? Why?

Finally, in what year would the logarithmic trend equation project total personal income to be $4,000 billion? Do you believe that is reasonable? Why or why not?

15 Time Series: Seasonal and Cyclical Variation

Frequently the dependent variable in a time series is influenced by *cyclical variation* in addition to *trend*. Additionally, when the time units of the data are represented in periods less than a year (weekly, monthly, quarterly, and so on), the time-series data may also be affected by *seasonal variation*. The subject of this chapter is the measurement of *seasonal* and *cyclical* variation and the application of these measurements to data analysis.

The Nature and Importance of Seasonal Analysis

Seasonal variation may be attributable to such factors as holidays (for instance, Christmas), weather patterns, or established customs (such as graduations in May and football in the fall). Figure 15.1 contains a real-world example of seasonality. As you can see from the graph, the natural increase in population peaks near the

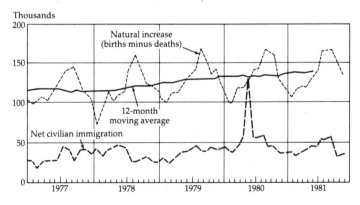

Figure 15.1 An example of seasonal variation: components of population change, by month, January 1, 1977, to December 1, 1981
Source: U.S. Department of Commerce, Bureau of the Census, *Population Estimates and Projections, Current Population Reports* (February 1982) p–25.

middle of each year and troughs near the beginning of each year.

As you may recall from the models discussed in chapter 14, seasonal variation may be expressed in actual units or as an index. *An index* is automatically used for the multiplicative model $(Y = T \times C \times S \times I)$, and it is ordinarily calculated even when the additive model $(Y = T + C + S + I)$ is used because it is with an index that seasonality is commonly measured. Also since customs and weather conditions are usually habitual, a seasonal index is a measure of the variation in time-series data that is attributable to annually recurring factors.

In order to measure seasonality over a year, there must be a series of seasonal index numbers. Monthly data require twelve indexes to express the seasonality—one for each month of the year. Similarly, quarterly data require a series of four indexes.

Like all indexes, a seasonal index must have a base. Generally speaking, in the time series the base used is either the *average year* or an *average period* (less than a year). When the base is an average period, the period may be a month, a quarter, or any other time segment of less than a year in length. The base (average year or average period) used in computing an index will cause minor differences in the methods of computation and, more important, differences in the interpretation and application of the index. The example in table 15.1 illustrates the use of alternative bases to measure seasonality.

When the base is an average period (quarters in table 15.1), the index expresses seasonal variation for each period in the year as a percent of the average period. (Notice, however, that the percent is *not stated*.) Referring to table 15.1, the index of 80 for quarter I means that sales in that quarter are typically 80 percent of average quarterly sales for a year. Since the base is assumed to be equal to 100, the difference between the index for any quarter and 100 (index minus 100) is the amount of variation attributable

Table 15.1
Seasonal index, alternative bases

| Quarter | Index | |
	Base-average year	Base-average quarter
I	20	80
II	30	120
III	35	140
IV	15	60
	100	400

to seasonality. Applying this to our example, sales in quarter II are typically 20 percent (120 − 100) above the average quarterly sales for a year.

If the average year is used as a base for the seasonal index, then the index for quarter I (shown in table 15.1) is 20. Stated in this manner, the index number means that typically 20 percent of a given year's sales occur in quarter I.

The choice of which base to use in measuring seasonality depends on the circumstances in which the seasonal index is being used. You may decide to use a particular base because it facilitates the interpretation, or communication, of the measures of seasonality. For example, if you are using monthly sales data in a report, the use of an average month base for a seasonal index of sales would retain the consistency of the data in your report. If you used an average year as the base, and your sales figures are in months, the reader might have a little trouble making the transition from months to years, and back to months. However, it is relatively simple to convert from one base to the other. Index numbers with the average quarter as a base are *four* times the size of quarterly index numbers computed using an average year as the base. Monthly indexes with the average month as the base are *twelve* times the size of monthly indexes with the average year as the base. Using this relationship, it should be relatively easy for you to determine how to convert a monthly seasonal index, a quarterly seasonal index, or one for other time periods from one base to another.

While we are considering alternative bases for a seasonal index, it may be worthwhile to take notice of a small technicality. The *sum* of the index numbers for time periods in a year will vary as the selection of a base varies. If an average year is used as the base, then the series of index numbers should have a sum of 100; if the average quarter is used, the sum should be 400; and when using the monthly average, the sum is 1,200. The logic of this is evident if you consider that the base is an arithmetic mean (average) equal to 100. All this may seem obvious, and perhaps trivial, to you, but it provides a ready method for checking the arithmetic accuracy of your calculations when computing a series of seasonal index numbers.

Adjusting for Seasonality

Given that you have a seasonal index, what can you do with it? In analyzing business and economic data, short-term fluctuations (seasonal variations) may camouflage, at least temporarily, more basic changes in the data. *Deseasonalizing* the actual data removes the distortion caused by seasonal variation. In addition, if a forecast of annual data (for instance, sales or output) is available, the seasonal index offers a capability to estimate how the series will behave during the year.

Example 15.1

▷ Assume that a toy shop has a November seasonal index of 150 and a December index equal to 200. If, in 1982, sales in November were $15,000 and in December they were $18,000, one might assume that the toy business was better in the month of December than it was in November. That is possible, for sales in December were actually 20 percent above November's sales. However, according to the index numbers, sales in December are normally 33.3 percent greater than sales in November! Perhaps business was not so good in December after all. ◁

Removing Seasonality

"Deseasonalizing data" is just another way of saying "removing the seasonal influence from the data." The procedure for deseasonalizing requires that the actual data be *divided* by the seasonal index, with the result multiplied by 100:

$$\frac{\text{actual data}}{\text{seasonal index}} \times 100 = \text{deseasonalized data.}$$

The origin of this procedure is evident from the product model. If

$$Y = T \times S \times C \times I,$$

then

$$\frac{Y}{S} = T \times C \times I = \text{deseasonalized data.}$$

Table 15.2 provides an example of applying a seasonal index to remove the seasonal influence from data. Theoretically, the de-

Table 15.2
Deseasonalizing data

Quarter	Actual sales	Seasonal index	Deseasonalized sales
I	$400	90	$444
II	540	110	491
III	630	120	525
IV	430	80	538

Sample calculation:

$$\frac{\$400 \text{ (actual sales)}}{90 \text{ (index)}} \times 100 = \$444 \text{ (deseasonalized sales)}$$

seasonalized data reflect only the influence of trend, cyclical, and random influences. Of course this is true only to the extent that the seasonal index is an accurate measure of seasonal influence.

It is difficult to decide, from observing actual sales, whether sales are increasing or decreasing because of seasonal variation that occurs within the year. However, deseasonalized sales data, as shown in the last column of table 15.2, allow the analyst to observe the trend of sales after the effect of seasonal variation has been removed. Actually, it should be remembered that the last column (sometimes called seasonally adjusted sales) of table 15.2 contains the TCI components of time-series data. However, the direction (increasing or decreasing) of the deseasonalized data is a reasonably good indicator of the trend.

Adding Seasonality

Frequently, you may have data (either actual or a forecast) available only in an annual form. For purposes of anticipating short-term (seasonal) fluctuations, you can apply a seasonal index to the data. If the index has as its base the average year, this requires only that you multiply the annual data by the seasonal index:

$$\text{annual data} \times \frac{\text{seasonal index}}{100} = \text{seasonalized data.}$$

If the base is an average time period, an intermediate step is necessary. In this case

$$\frac{\text{annual data}}{\begin{array}{c}\text{number of base time}\\\text{periods in a year}\end{array}} \times \frac{\text{seasonal index}}{100} = \text{seasonalized data.}$$

Both of these procedures are illustrated in table 15.3.

The preceding discussion is intended to cover the utilization of seasonal indexes in a very general sense. Perhaps you can already visualize how these techniques may prove useful as you approach problems in the functional areas of business administration. Knowledge of seasonal variations can assist the businessman in planning his personnel needs from month-to-month, in determining his inventory requirements and in anticipating changes in his cash flow. You will undoubtedly find that seasonal analysis is one of the most frequently used of all the statistical tools.

Determining the Existence of Seasonality

When we say that "data may contain, or are influenced by, seasonal variation" we imply that not all such dependent variables are subject to seasonal influences. This implication is correct.

Table 15.3
Seasonalizing data

Average year = base

Quarter	Index	Annual sales	Seasonalized data
I	20	$20,000	$ 4,000
II	30	20,000	6,000
III	35	20,000	7,000
IV	15	20,000	3,000
			$20,000

$$\$20{,}000 \text{ (annual sales)} \times \frac{20 \text{ (index)}}{100} = \$4{,}000 \text{ (seasonalized data)}$$

Average quarter = base

Quarter	Index	Annual sales ÷ 4	Seasonalized data
I	80	$5,000	$ 4,000
II	120	5,000	6,000
III	140	5,000	7,000
IV	60	5,000	3,000
			$20,000

$$\frac{\$20{,}000 \text{ (annual sales)}}{4 \text{ (number of periods)}} \times \frac{80 \text{ (index)}}{100} = \$4{,}000 \text{ (seasonalized data)}$$

Furthermore on occasion you may encounter a series of data where seasonality exists but its effect is so heterogeneous, or dispersed, that it is impossible, or at best misleading, to measure seasonality. An example of this is the monthly cash income of farmers. Variations in weather can cause "harvest time" to fluctuate between months. A farmer's cash income this year may be greater in August than in September. Next year, September's income may be the larger. Another possibility is that the seasonal pattern is changing through time. That is, there is a trend in the measures of seasonal influences.

How do you determine if seasonality exists in a set of data? The best way for you to find the answer to this question is to construct a set of scatter diagrams using the ratio of the actual value to the average for the year. A set is needed since you must prepare a separate one for each seasonal time period. That is, if you are working with monthly data, scatter diagrams are prepared for January, February, and so forth, for each of the twelve months. Figure 15.2 provides some examples of how these scatter diagrams might appear.

Graph (a) in the figure represents a situation where the values

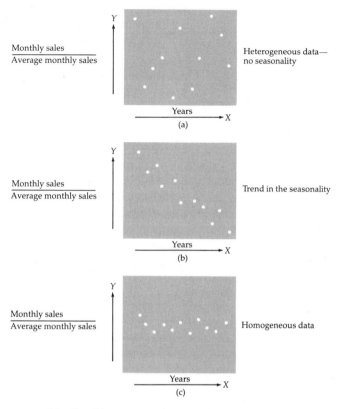

Figure 15.2   Possible variations for a particular month

for a hypothetical month are said to be heterogeneous, or dissimilar. Put another way, there appears to be little *central tendency* in the data. The measure of dispersion would be large regardless of the average you selected to repesent the data. Should you encounter data where this pattern exists for one or more months, it would be misleading to seek to compute, and use, a seasonal index.

In some cases the problem may be alleviated by lengthening the index time period, perhaps from monthly to quarterly. Failing to observe the requirement that there actually be some noticeable central tendency to the cluster of values for every time period will not prohibit you from continuing on to calculate a series of index numbers. However, since there would be no "typical" or "normal" seasonal influence, the resulting index numbers could hardly be construed as measuring the typical seasonal pattern.

Graph (b) represents another situation in which caution must be exercised when a seasonal index is computed. When you first examine the graph you may conclude that it is very much like graph (a). This is not the case, however. There is a distinct central tendency but it is *moving*. When a scatter of this nature occurs, there is said to be a *trend in the seasonality of the data*. Computation of an index requires that a trend equation be determined for each month or quarter. Thus the seasonal index for a time period would change each year. While this condition may exist quite often in business and economic data, in practice is it seldom recognized since (1) it generally requires a large number of years of data to detect the existence of trend and (2) the amount of computation required to determine the trend is so extensive as to be generally prohibitive in any except computerized procedures.

Graph (c) in figure 15.2 depicts the distribution of data most frequently assumed to exist when a seasonal index is computed. There is a definite central tendency to the data that can be adequately represented by an arithmetic mean. If there exists a few extreme deviations from the cluster at the midpoints, the influence of these values can be omitted by utilizing a modified arithmetic mean—assuming enough values are available to provide a reasonable estimate of the central tendency.

The methods presented on the following pages will carry an implicit assumption that there is a central tendency to the data and that this central tendency is reasonably fixed. In other words, the data are assumed to take on the approximate appearance of graph (c) in figure 15.2.

Measuring Seasonality

There are several variations to the general procedure of measuring seasonality. We have chosen to present only two methods that are fairly typical of the techniques that are generally utilized.

In measuring seasonality, however, a principle distinction in the methodology used depends upon the manner in which the influence of trend, cyclical, and random variations are removed from the data. For instance, the following circumstances would have a bearing on the methodology:

1. Ideally, the seasonal index (measure of seasonal variation) must be devoid of any influence from other types of fluctuations.

2. Trend may, or may not, have been measured prior to the computation of the seasonal index.

3. Cyclical variation and irregular variation are generally measured chronologically after seasonal analysis has taken place.

Moving Average Method

Look at tables 15.4 and 15.5. They illustrate the moving average method of computing a seasonal index. Using this technique, the index is determined from raw data. That is, the data in column 2 of table 15.4 represent actual sales data just as they appear in the accounting reports of the firm.

Procedurally, there are seven basic steps in the calculation of the seasonal index by this method. The first five steps are shown in table 15.4.

Table 15.4
Moving average method, quarterly data

(1) Quarter	(2) Sales	(3) 4-Quarter moving total	(4) Centered 8-quarter moving total	(5) Centered moving average (4)/8	(6) Specific seasonal [(2)/(5)][100]
1973 I	8				
II	12				
		42			
III	12		86	10.750	111.63
		44			
IV	10		91	11.375	87.91
		47			
1974 I	10		96	12.000	83.33
		49			
II	15		98	12.250	122.45
		49			
III	14		97	12.125	115.46
		48			
IV	10		95	11.875	84.21
		47			
1975 I	9		94	11.750	76.60
		47			
II	14		96	12.000	116.67
		49			
III	14		99	12.375	113.13
		50			
IV	12		102	12.750	94.12
		52			
1976 I	10		107	13.375	74.77
		55			
II	16		111	13.875	115.32
		56			
III	17		114	14.250	119.30
		58			
IV	13		118	14.750	88.14
		60			
1977 I	12		121	15.125	79.34
		61			
II	18		121	15.125	119.01
		60			
III	18		119	14.875	121.01
		59			
IV	12		118	14.750	81.36
		59			
1978 I	11		119	14.875	73.95
		60			
II	18		121	15.125	119.01
		61			
III	19		123	15.375	123.58
		62			
IV	13		125	15.625	83.20
		63			
1979 I	12		128	16.000	75.00
		65			
II	19		132	16.500	115.15
		67			
III	21		136	17.000	123.53
		69			
IV	15		143	17.875	83.92
		74			
1980 I	14		153	19.125	73.20
		79			
II	24		161	20.125	119.25
		82			
III	26		165	20.625	126.06
		83			
IV	18		167	20.875	86.23
		84			
1981 I	15		168	21.000	71.43
		84			
II	25		167	20.875	119.76
		83			
III	26				
IV	17				

Table 15.5
Calculating seasonal indexes

Year	I	II	III	IV	Total
1973			~~111.63~~	87.91	
1974	~~83.33~~	~~122.45~~	115.46	84.21	
1975	76.60	116.67	113.13	~~94.12~~	
1976	74.77	115.32	119.30	88.14	
1977	79.34	119.01	121.01	~~81.36~~	
1978	73.95	119.01	123.58	83.20	
1979	75.00	~~115.15~~	123.53	83.92	
1980	73.20	119.25	~~126.06~~	86.23	
1981	~~71.43~~	119.76			
Modified total	452.86	709.02	716.01	513.61	
Average specific seasonal (unadjusted)	75.48	118.17	119.34	85.60	398.59
Seasonal index (adjusted)	75.75	118.59	119.76	85.90	400.00

1. List the data in chronological order. This is shown in column 2 of table 15.4.

2. Take a 4-quarter moving total of the data. (If you are working with monthly data, it is a 12 month moving total.) The 4-quarter moving totals are taken for the entire time span of the data by dropping the sales figure for the earliest quarter and adding the next quarterly sales figure on the list. Put another way, when you drop quarter I for one year from the total you add quarter I of the next year to it. By looking at column 3 in table 15.4, this step in the procedure should become evident. The purpose of this step is to average out the seasonal and irregular components of the multiplicative model.

3. An 8-quarter (or 24-month) moving total is the next step. Actually, this is a moving total of the 4-quarter moving totals. This step is necessary because the 4-quarter totals are not centered on time periods—the middle of a 4-quarter period occurs between two quarters. A total of two successive 4-quarter totals is therefore represented as centered on a quarter.

4. Column 5 shows the fourth step. It is a centered quarterly moving average and is computed by dividing the 8-quarter moving totals by 8. When using monthly data, the monthly moving average is found by dividing the 24-month moving totals by 24. (Steps 2, 3, and 4 isolate the trend and cyclical components of the multiplicative model.)

5. The specific seasonals are shown in column 6. These are found by dividing the actual quarterly (monthly) data (column 2) by the

quarterly (monthly) moving averages (column 5) and multiplying by 100. (Dividing TCSI by TC leaves the seasonal and random components of the multiplicative model.)

The specific seasonal indexes shown in column 6 of table 15.4 identify whether sales in a given quarter were above or below the quarterly average for the year. For example, the third quarter sales of 1980 were 26.06 percent (126.06 − 100) above the quarterly average for the year. These specific seasonals, however, contain both seasonal and irregular variations. Table 15.5 presents the final two steps necessary to eliminate the irregular variation and determine a seasonal index for each quarter.

6. This step involves the computation of an average specific seasonal for each time period. In other words, when working with quarterly data, it is necessary to compute an average specific seasonal for quarter I, quarter II, and so on. In table 15.5 the "modified mean" approach is used, whereby an average is obtained from the modified total (the sum, after elimination of the high and low values, of the specific seasonals). The average specific seasonals are considered to be unadjusted indexes since they do not necessarily sum to 400 (1,200 for monthly indexes). Step 7 explains the adjustment process.

7. The final step involves computing the *adjusted seasonal index*. If you are working with quarterly data, the seasonal indexes should sum to 400. The difference between the unadjusted and adjusted indexes involves *forcing* the series to add to 400, or 1,200 in the case of monthly data. Thus simply sum the unadjusted seasonal indexes, and make the sum the denominator of a fraction whose numerator is 400 (1,200 for monthly indexes). Next multiply the unadjusted index for each quarter by the fraction. In table 15.5 the multiplier is $400 \div 398.59 = 1.00354$. The result is the *seasonal index* (adjusted).

The procedure described here yields an index with the average time period as a base. You might ask how the procedure removes the influence of trend, cyclical, and irregular variations from the data. This is assumed to be accomplished by using the 8-quarter moving average to determine the specific seasonals and averaging the specific seasonals. While this may not guarantee that the influence of the other types of fluctuations will be completely eliminated, it will almost certainly reduce their influence, and it is a widely accepted technique for computing seasonal indexes. Now look at tables 15.6 and 15.7 to see how monthly seasonal indexes are computed using the moving average method.

Table 15.6
Moving average method, monthly data

Year	(1) Month	(2) Sales	(3) 12-Month moving total	(4) Centered 24-month moving total	(5) = (4) ÷ 24 Centered 24-month moving average	(6) = [(2) ÷ (5)][100] Specific seasonal
1975	Jan.	30				
	Feb.	26				
	March	29				
	April	36				
	May	40				
	June	40	451			
	July	41	456	907	37.79	108.49
	Aug.	44	460	916	38.17	115.27
	Sept.	42	468	928	38.67	108.61
	Oct.	45	472	940	39.17	114.88
	Nov.	40	476	948	39.50	101.27
	Dec.	38	480	956	39.83	95.41
1976	Jan.	35	484	964	40.17	87.13
	Feb.	30	485	969	40.38	74.29
	March	37	488	973	40.54	91.27
	April	40	489	977	40.71	98.26
	May	44	490	979	40.79	107.87
	June	44	492	982	40.92	107.53
	July	45	491	983	40.96	109.86
	Aug.	45	494	985	41.04	109.65
	Sept.	45	493	987	41.12	109.44
	Oct.	46	493	986	41.08	111.98
	Nov.	41	493	986	41.08	99.81
	Dec.	40	491	984	41.00	97.56
1977	Jan.	34	488	979	40.79	83.35
	Feb.	33	490	978	40.75	80.98
	March	36	491	981	40.88	88.06
	April	40	493	984	41.00	97.56
	May	44	496	989	41.21	106.77
	June	42	498	994	41.42	101.40
	July	42	502	1,000	41.67	100.79
	Aug.	47	504	1,006	41.92	112.12
	Sept.	46	508	1,012	42.17	109.08
	Oct.	48	512	1,020	42.50	112.94
	Nov.	44	515	1,027	42.79	102.83
	Dec.	42	518	1,033	43.04	97.58
1978	Jan.	38	524	1,042	43.42	87.52
	Feb.	35	526	1,050	43.75	80.00
	March	40	528	1,054	43.92	91.07
	April	44	529	1,057	44.04	99.91
	May	47	530	1,059	44.12	106.53
	June	45	528	1,058	44.08	102.09

Table 15.6 (continued)

Year	(1) Month	(2) Sales	(3) 12-Month moving total	(4) Centered 24-month moving total	(5) = (4) ÷ 24 Centered 24-month moving average	(6) = [(2) ÷ (5)][100] Specific seasonal
	July	48	526	1,054	43.92	109.29
	Aug.	49	525	1,051	43.79	111.90
	Sept.	48	525	1,050	43.75	109.71
	Oct.	49	524	1,049	43.71	112.10
	Nov.	45	524	1,048	43.67	103.05
	Dec.	40	525	1,049	43.71	91.51
1979	Jan.	36	528	1,053	43.88	82.04
	Feb.	34	530	1,058	44.08	77.13
	March	40	533	1,063	44.29	90.31
	April	43	536	1,069	44.54	96.54
	May	47	538	1,074	44.75	105.03
	June	46	543	1,081	45.04	102.13
	July	51	547	1,090	45.42	112.29
	Aug.	51	548	1,095	45.62	111.79
	Sept.	51	553	1,101	45.88	111.16
	Oct.	52	560	1,113	46.38	112.12
	Nov.	47	567	1,127	46.96	100.09
	Dec.	45	575	1,142	47.58	94.58
1980	Jan.	40	579	1,154	48.08	83.19
	Feb.	35	584	1,163	48.46	72.22
	March	45	587	1,171	48.79	92.23
	April	50	591	1,178	49.08	101.87
	May	54	594	1,185	49.38	109.36
	June	54	595	1,189	49.54	109.00
	July	55	600	1,195	49.79	110.46
	Aug.	56	605	1,205	50.21	111.53
	Sept.	54	607	1,212	50.50	106.93
	Oct.	56	609	1,216	50.67	110.52
	Nov.	50	615	1,224	51.00	98.04
	Dec.	46	619	1,234	51.42	89.46
1981	Jan.	45	624	1,243	51.79	86.89
	Feb.	40	628	1,252	52.17	76.67
	March	47	631	1,259	52.46	89.59
	April	52	633	1,264	52.67	98.73
	May	60	635	1,268	52.83	113.57
	June	58	638	1,273	53.04	109.35
	July	60				
	Aug.	60				
	Sept.	57				
	Oct.	58				
	Nov.	52				
	Dec.	49				

Table 15.7
Calculating seasonal indexes

Year	January	February	March	April	May	June	July	August	September	October	November	December	Total
1975							108.49	~~115.27~~	108.61	~~114.88~~	101.27	95.41	
1976	87.13	74.29	91.27	98.26	107.87	107.53	109.86	~~109.65~~	109.44	111.98	99.81	97.56	
1977	83.35	~~80.98~~	~~88.06~~	97.56	106.77	~~101.40~~	~~100.79~~	112.12	109.08	112.94	102.83	~~97.58~~	
1978	~~87.52~~	80.00	91.07	99.91	106.53	102.09	109.29	111.90	109.71	112.10	~~103.05~~	91.51	
1979	~~82.04~~	77.13	90.31	~~96.54~~	~~105.03~~	102.13	~~112.29~~	111.79	~~111.16~~	112.12	100.09	94.58	
1980	83.19	~~72.22~~	~~92.23~~	~~101.87~~	109.36	109.00	110.46	111.53	~~106.93~~	~~110.52~~	~~98.04~~	~~89.46~~	
1981	86.89	76.67	89.59	98.73	~~113.57~~	~~109.35~~							
Modified total	340.56	308.09	362.24	394.46	430.53	420.75	438.10	447.34	436.84	449.14	404.00	379.06	
Average specific seasonal (unadjusted)	85.14	77.02	90.56	98.62	107.63	105.19	109.53	111.84	109.21	112.29	101.00	94.77	1202.80
Seasonal index (adjusted)	84.94	76.84	90.35	98.39	107.38	104.95	109.28	111.58	108.96	112.03	100.76	94.55	1200.01

Multiplier for adjustment $= \dfrac{1200}{1{,}202.80} = 0.9976721$

Detrended Method

It is possible to isolate the influence of seasonal variation other than by the moving average method. In chapter 14 you learned how to estimate trend. If a trend estimate is available, the trend can be removed from the data by using the method described near the end of chapter 14. One technique for doing this is illustrated in table 15.8.

Table 15.8
Detrending the data

Year	Quarter	Y Sales	$Y_c = 17.86 + 0.419x$ Trend	Y/Y_c
1977	I	12	13.670	0.878
	II	18	14.089	1.278
	III	18	14.508	1.241
	IV	12	14.927	0.804
1978	I	11	15.346	0.717
	II	18	15.765	1.142
	III	19	16.184	1.174
	IV	13	16.603	0.783
1979	I	12	17.022	0.705
	II	19	17.441	1.089
	III	21	17.860	1.176
	IV	15	18.279	0.821
1980	I	14	18.698	0.749
	II	24	19.117	1.255
	III	26	19.536	1.331
	IV	18	19.955	0.902
1981	I	15	20.374	0.736
	II	25	20.793	1.202
	III	26	21.212	1.226
	IV	17	21.631	0.786

Table 15.9
Calculating the seasonal indexes

| | Quarter | | | | |
Year	I	II	III	IV	Total
1977	~~0.878~~	~~1.278~~	1.241	0.804	
1978	0.717	1.142	~~1.174~~	~~0.783~~	
1979	~~0.705~~	~~1.089~~	1.176	0.821	
1980	0.749	1.255	~~1.331~~	~~0.902~~	
1981	0.736	1.202	1.226	0.786	
Modified total	2.202	3.599	3.643	2.411	
Unadjusted index	73.40	119.97	121.43	80.37	395.17
Adjusted index	74.30	121.44	122.91	81.35	400.00

$$\text{Multiplier for adjustment} = \frac{400}{395.17} = 1.0122226$$

After the data have been detrended as shown in table 15.8, then a modified mean *of the detrended values* for each time period can be determined in order to average out the random component of the multiplicative model. Each mean is adjusted to force the sum of the indexes to 400. This is the final step and yields the seasonal index for each period (table 15.9).

Detrending the data using the trend estimate offers a less time-consuming approach to time-series analysis than the moving average method. Furthermore the moving average method causes a loss of the first one-half year and the last one-half year of data from the index calculations. (This can be significant if you have data for only a few years.) On the other hand, the detrending method does not remove the cyclical component so effectively as the moving average method. Thus, unless you are sure that the cyclical effect is slight, the moving average method is probably the better choice.

Cyclical Variation—Characteristics and Effects

Until now we have examined the trend (T) and seasonal (S) components of a time series. There are two remaining components to consider, *cyclical variation* (C) and *irregular variation* (I). Our main concern in this section, however, is to determine the cyclical variation (C). Irregular variation simply may be regarded as a nuisance. It is the result of some *random event,* so we will attempt to eliminate it from time-series data by some averaging process. However, to see an illustration of a real irregular variation, look at the sharp peak in net civilian immigration in 1980 as shown in figure 15.1. This peak, or irregular, variation was caused by the large number of Cuban immigrants at that time.

Characteristics and Effects

Cyclical variation in the dependent variable of a time series is associated with changes in general business conditions of an economy—*the general business cycle*. For this reason cyclical variation usually shows graphically a wavelike movement in the data with alternating increases and declines.[1] Figure 15.3 is a graph of hypothetical data in which the effects of cyclical variation are exaggerated and obvious. In this illustration cyclical peaks occur in January 1963, July 1969, and January 1978. Cyclical troughs show in 1964 and 1973.

Over the past 200 years the economy of the United States has been characterized by changes in business activity resulting in cyclical peaks of prosperity and troughs of depression. Thus cyclical variation is a recurring condition. Its recurrence, however, is not with any sort of consistent pattern. Therefore cyclical movements may vary rather significantly in their amplitude (severity) and their magnitude (duration).

Not all time-series data are affected by cyclical variation, and some variables are more sensitive to cyclical influence than others. Figure 15.4 shows the index for industrial production in the United States for nondurable manufactures and durable manufactures from 1920 to 1980. If you examine the figure closely, there are several things you should notice regarding this time-series data:

1. The data have been "seasonally adjusted." That is, the influencing effects of seasonal variation (*S*) have been removed from the data.

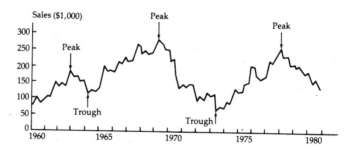

Figure 15.3   Hypothetical quarterly sales

---

1. For those familiar with trigonometry, it is a sine curve.

BASIC STATISTICS

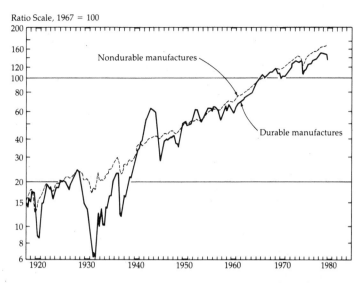

Ratio Scale, 1967 = 100

Figure 15.4   Index of industrial production in the United States by quarters, 1920–1980 (seasonally adjusted)

2. Durable manufactures is more sensitive to cyclical variation than nondurable manufactures. The cyclical troughs associated with known depressions and recessions are easily seen in the data for durable manufactures.

3. The severity of the cyclical movement was more pronounced prior to World War II. The amplitudes and magnitudes of the troughs are greater in the 1920s and 1930s.

4. Cyclical variation continues to recur. There were cyclical declines and upswings in the 1960s and 1970s. Notice the troughs in 1970 and 1976.

5. The upward trend in both series appears to be linear.

Computing the Cyclical Index

There are various methods for computing an index that measures the cyclical effect on a set of time-series data without considering the reason for this effect. In this text we will offer one simple method for computing the index when data are in quarterly units.

Determining the Cyclical Index for Quarterly Data

It is a fairly simple matter to determine a cyclical index series for quarterly data, particularly if you have already computed a seasonal index. Returning to the multiplicative time-series model, if the effects of trend and seasonal variation are removed from the data, the remaining variation must be attributable to cyclical and irregular variations ($C \times I$):

Table 15.10
Computing a cyclical index series

Year quarter		Sales (Y)	Trend (T)	Seasonal index (S)	Statistical normal $T \times S/100$	$[Y/(T \times S/100)]100 = (C \times I)$ (percent)	Cyclical index
1973	I	8	9.6306	75.75	7.2952	109.66	
	II	12	9.9517	118.59	11.8018	101.68	102.96
	III	12	10.2728	119.76	12.3028	97.54	103.04
	IV	10	10.5940	85.90	9.1002	109.89	109.46
1974	I	10	10.9151	75.75	8.2682	120.95	114.48
	II	15	11.2362	118.59	13.3250	112.59	111.56
	III	14	11.5573	119.76	13.8410	101.15	103.92
	IV	10	11.8784	85.90	10.2035	98.01	98.85
1975	I	9	12.1995	75.75	9.2411	97.39	96.56
	II	14	12.5206	118.59	14.8482	94.29	94.24
	III	14	12.8417	119.76	15.3792	91.03	97.15
	IV	12	13.1628	85.90	11.3069	106.13	98.35
1976	I	10	13.4839	75.75	10.2141	97.90	100.59
	II	16	13.8050	118.59	16.3714	97.73	98.71
	III	17	14.1261	119.76	16.9174	100.49	100.99
	IV	13	14.4472	85.90	12.4102	104.75	104.17
1977	I	12	14.7683	75.75	11.1870	107.27	104.20
	II	18	15.0894	118.50	17.8946	100.59	101.80
	III	18	15.4106	119.76	18.4557	97.53	95.64
	IV	12	15.7317	85.90	13.5135	88.80	92.26
1978	I	11	16.0528	75.75	12.1600	90.46	90.65
	II	18	16.3739	118.59	19.4178	92.70	92.73
	III	19	16.6950	119.76	19.9939	95.03	92.22
	IV	13	17.0161	85.90	14.6168	88.94	91.78
1979	I	12	17.3372	75.75	13.1329	91.37	90.35
	II	19	17.6583	118.59	20.9410	90.73	93.21
	III	21	17.9794	119.76	21.5321	97.53	94.56
	IV	15	18.3005	85.90	15.7201	95.42	97.40
1980	I	14	18.6216	75.75	14.1059	99.25	100.50
	II	24	18.9427	118.59	22.4642	106.84	106.26
	III	26	19.2638	119.76	23.0704	112.70	108.84
	IV	18	19.5849	85.90	16.8235	106.99	106.39
1981	I	15	19.9060	75.75	15.0788	99.48	103.56
	II	25	20.2272	118.59	23.9874	104.22	103.12
	III	26	20.5483	119.76	24.6086	105.65	101.57
	IV	17	20.8694	85.90	17.9268	94.83	

$$\frac{Y}{T \times S} = C \times I.$$

Then, if we adjust for irregular variations in some manner, the result will be a series measuring cyclical variation:

$$\frac{C \times I}{I} = C.$$

To illustrate how we can accomplish both these steps, we will again use the quarterly sales data first shown in table 15.4. Remember, the seasonal indexes for these data are shown in table 15.5. If you examine the data in table 15.10, you will see that the initial step in the procedure is to calculate a trend equation for the sales data. The linear least-squares trend equations for these data are

$$Y_c = 9.3095 + 0.3211x,$$
$$x = 1 \text{ quarter,}$$
$$\text{origin} = \text{QIV, 1972.}$$

Next, the products of the trend (T) values and the seasonal index values (S) are determined for each quarter. (This product, $T \times S$, is sometimes called *statistical normal.*) The results of these steps are shown graphically in figure 15.5. The shaded area in the figure represents $C \times I$, or that amount of variation in sales that is unexplained by the measures of trend and seasonal variation.

The third step in the procedure requires that we divide sales (Y) by statistical normal ($T \times S$). The result is a *crude cyclical index* because it may reflect some irregular variations. By smoothing out the irregularities, we refine the index.

One way of smoothing the crude index is to express it as a moving average. In table 15.10 the smoothing is done with a

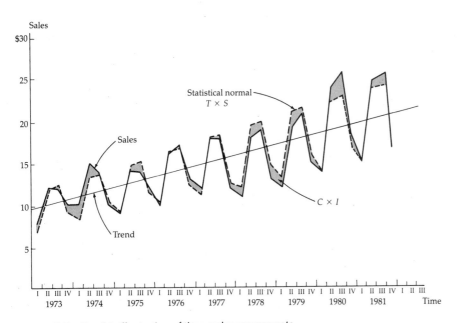

Figure 15.5 Graphic illustration of time-series components.
Source: Table 15.10

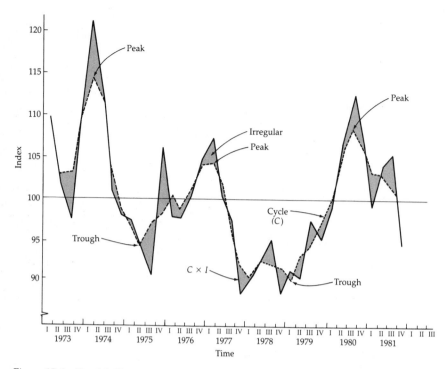

Figure 15.6   Graphic illustration of a smoothed cyclical index.
Source: Table 15.10

moving three-quarter average. It is through this smoothing process that we attempt to remove irregular variation, leaving only cyclical variation in our index number series. In figure 15.6, both the crude and smoothed series are shown so that you can see the results of smoothing. Notice in this figure that you identify the approximate amount of irregular variation at any point in time and also the peaks and troughs of the cycle.

Adjusting Data for Calendar Variation

There are many instances where calendar variations can influence the size of a variable. The monthly payroll of a company will vary if employees are paid weekly, simply because some months will contain more "paydays" than other months. Furthermore the number of "paydays" in a given month will vary from year to year. Monthly sales will vary for much the same reason—months contain varying numbers of "business days." Weekly sales will even vary for the same reasons. These calendar variations can cause significant distortions in the economic and business data you may encounter. Removing the influence of calendar variations is best illustrated by example 15.2.

Example 15.2

▷ A certain store is normally open 5 days a week, 52 weeks a year. However, the store takes eight holidays during the year. Thus it is open 252 days a year, for an average of approximately 4.81 days a week or 21 days a month. In February the store is open 20 days and has sales of $4,000. In the next month (March) it has sales of $5,000 and is open 24 days. The following adjustments should be made to the monthly sales before a comparison of sales in the two months is made:

February $4,000 \times \dfrac{21}{20} = \$4,200$ (adjusted sales),

March $5,000 \times \dfrac{21}{24} = \$4,375$ (adjusted sales). ◁

In many cases variations of this type have a negligible influence on data. However, you should assure yourself that this is the case before undertaking extensive analysis of time-series data. If the influence proves to be significant, the data should be adjusted before you attempt to measure seasonal, cyclical, or irregular variations.

1. Define seasonal index.

2. What is meant by deseasonalizing the data?

3. Why do we deseasonalize data?

4. How can seasonality be identified?

5. Discuss cyclical variation.

1. Calculate monthly seasonal indexes for the data that follows using a centered twelve-month moving average:

	Jan	Feb	Mar	Apr	May	June	July	Aug	Sep	Oct	Nov	Dec
1976	2	3	5	6	4	2	2	5	4	3	4	6
1977	3	4	6	7	5	3	3	7	5	4	5	7
1978	4	5	7	8	6	4	4	8	6	5	6	7
1979	5	6	8	9	7	5	5	9	7	6	7	9
1980	6	7	9	10	8	6	6	10	8	7	8	10
1981	7	8	10	11	9	7	7	11	9	8	9	11
1982	8	9	11	12	10	8	8	12	10	9	10	12

2. The following data represent the monthly electric bill (in dollars) for a residential customer. Calculate a monthly seasonal index for this customer using a centered twelve-month moving average.

	Jan	Feb	Mar	Apr	May	June	July	Aug	Sep	Oct	Nov	Dec
1976	128	129	125	81	79	85	90	150	145	153	80	69
1977	130	130	126	83	72	80	92	160	147	157	81	68
1978	134	134	131	89	78	84	94	157	153	160	85	71
1979	137	140	135	87	70	88	97	165	162	160	82	73
1980	139	141	140	85	70	82	98	171	165	165	80	79
1981	143	150	152	88	78	87	90	174	166	170	83	77
1982	156	158	155	86	79	82	95	178	172	175	80	74

3. Calculate monthly seasonal indexes using a centered twelve-month moving average.

Monthly sales
(in $1,000)

Year	Jan	Feb	Mar	Apr	May	June	July	Aug	Sep	Oct	Nov	Dec
1975	3	2	3	4	4	5	6	7	6	6	5	4
1976	4	3	5	7	8	8	9	10	10	8	8	6
1977	5	5	7	10	12	11	11	11	10	9	8	7
1978	6	5	6	6	7	8	9	10	10	10	9	9
1979	9	8	10	11	13	14	15	16	17	16	16	15
1980	13	13	14	15	16	17	17	17	17	16	14	12
1981	10	10	12	12	13	15	15	16	15	15	15	14
1982	14	13	15	15	14	15	16	18	19	19	18	18

4. Calculate monthly seasonal indexes using a centered twelve-month moving average.

Sales ($1,000)

Month	1975	1976	1977	1978	1979	1980	1981	1982
Jan	26.4	28.2	32.4	38.4	42.4	50.3	48.0	46.2
Feb	32.8	38.0	42.3	48.0	50.1	56.4	56.0	56.2
Mar	34.5	42.3	48.0	50.3	62.4	70.1	64.3	70.0
Apr	32.4	34.4	40.1	46.0	54.0	56.2	52.4	54.4
May	30.2	34.3	38.1	42.2	50.0	54.2	50.3	52.1
June	30.1	32.1	36.4	42.1	48.2	52.0	50.0	50.3
July	26.4	26.0	32.3	36.1	40.3	48.1	44.4	48.2
Aug	28.0	26.3	32.2	38.3	42.1	50.4	48.1	44.3
Sep	28.1	30.1	32.4	40.0	44.3	50.1	48.0	44.4
Oct	28.4	30.3	36.0	40.3	40.4	50.2	50.0	46.1
Nov	26.2	24.4	30.3	34.4	40.3	48.2	42.2	44.4
Dec	26.3	26.0	30.4	34.4	38.4	44.2	42.3	44.0

5. Calculate quarterly seasonal indexes using a centered four-quarter moving average.

Sales ($1,000)

Year	Quarter I	II	III	IV
1978	10	14	16	12
1979	12	21	26	17
1980	16	26	31	22
1981	22	31	36	27
1982	26	36	42	32

6. Calculate seasonal indexes for the data given below using a centered four-quarter moving average.

Gasoline sales (millions of gallons)

Year	Quarter I	II	III	IV	Total
1978	4.3	4.9	5.3	5.0	19.5
1979	4.4	4.9	5.4	5.1	19.8
1980	4.5	5.3	5.5	5.3	20.6
1981	4.6	5.3	5.6	5.4	20.9
1982	4.6	5.3	5.8	5.7	21.4

7. Calculate quarterly seasonal indexes using a centered four-quarter moving average.

Department store sales ($100,000)

	Quarter			
	I	II	III	IV
1978	1	2	3	4
1979	2	3	4	5
1980	3	4	5	6
1981	4	5	6	7
1982	5	6	7	8

8. Quarterly sales and seasonal indexes for Big Enterprise are given in the table that follows.

	I	II	III	IV
Quarterly sales	$110,560	$350,340	$322,220	$150,000
Seasonal index	49.7	115.4	125.9	109.0

Determine the seasonally adjusted sales.

9. Sales and seasonal indexes for the July Corporation are given below.

	Jan	Feb	Mar	Apr
Sales	$5,500	$5,200	$6,600	$8,500
Seasonal index	70	80	100	110

(a) What are the seasonally adjusted sales?
(b) The corporation forecaster believes that the average monthly sales will be $8,000. If the May seasonal index is 105, what will be forecast for May sales?

10. Sales of Bloudy Plastic, Ltd., were $84,000 for the first quarter of 1982.
(a) If the seasonal indexes by quarter are 75, 90, 125, 110, what are the deseasonalized sales for the first quarter?
(b) Assuming average quarterly sales of $100,000, estimate the sales for the third quarter.

11. A large state has determined the following seasonal indexes for sales tax collections:

	Quarter			
	I	II	III	IV
Seasonal index	70	120	80	130

If the annual sales tax collections are estimated to be $1.2 billion, how much should be collected in the second quarter of the year?

12. The bank has determined the following seasonal indexes for cash demands.

	June	July	Aug
Seasonal index	128	140	150

If the average monthly demand for cash is $4,000,000, how much cash should be available in July?

13. The trend equation for the sales of Lotta Manufacturing is $Y_c = 50 + 1.2x$ where Y is $1,000, $x = 1$ quarter, and the origin is the middle of the third quarter, 1978. If the seasonal indexes are 70, 80, 140, and 110 for each quarter, using the multiplicative model, estimate sales for each quarter of 1983, ignoring the cyclical and irregular fluctuations.

14. The following data represent the average of the specific seasonals for the months shown. Determine the adjusted seasonal index for each month.

Jan	89.50	May	99.85	Sep	114.75
Feb	72.46	June	105.36	Oct	106.84
Mar	81.28	July	125.43	Nov	93.96
Apr	95.63	Aug	115.60	Dec	88.88

15. Determine the seasonal indexes for the data in problem 6 using the detrending method. Use the trend equation developed in problem 17(c) of chapter 14.

16. The following data represent the average of the specific seasonals for the quarters shown. Determine the adjusted seasonal index for each quarter.

Quarter			
I	II	III	IV
114.86	94.63	98.48	92.80

17. Using the data and solution from problem 6 and the trend equation developed in problem 17(c) of chapter 14, determine the cyclical indexes.

18. Determine the cyclical indexes for the data given below.

	Quarter			
Year	I	II	III	IV
1975	12	16	16	14
1976	14	18	18	17
1977	19	23	23	21
1978	22	25	24	22
1979	18	22	22	20
1980	17	21	22	19
1981	18	22	22	20
1982	22	26	27	25

Case Study

Risk, Inc. is continuing to analyze the automotive dealer market. (Refer to the case studies at the end of chapters 12 and 13 for details concerning other analyses.) The management of the company is aware that the automobile manufacturing industry has encountered considerable economic difficulty in recent years, and they want you, the staff economist, to prepare a report that will provide some insight into what has been happening to automotive dealers. This report should include both tables and graphs, as well as a narrative describing your work and setting forth your conclusions. In addition, include a projection of monthly statistical normal (product model) through 1983. To help you get started, the data in the following table have been assembled from various issues of the *Survey of Current Business*:

Month	Monthly automotive dealer sales ($ millions)								
	1974	1975	1976	1977	1978	1979	1980	1981	1982
Jan.		6,732	8,299	9,612	9,976	12,805	13,366	13,351	12,118
Feb.		7,258	8,054	10,343	10,710	13,100	13,754	14,370	13,912
March		7,641	10,932	13,086	14,008	16,207	14,444	17,336	17,068
April		8,422	11,339	12,851	13,877	15,552	13,542	16,279	16,506
May	8,787	9,180	11,140	12,878	15,415	15,839	13,487	15,213	17,184
June	8,649	9,342	11,876	13,555	15,718	15,571	14,098	16,205	16,923
July	8,980	9,611	11,412	12,520	14,294	14,712	15,176	16,307	
Aug.	8,969	8,812	10,648	12,713	14,642	16,078	14,211	16,742	
Sept.	7,591	8,539	9,983	11,327	12,733	13,883	13,489	15,425	
Oct.	7,978	9,734	10,561	12,418	14,401	15,019	15,027	14,842	
Nov.	6,965	8,422	10,216	11,551	13,610	13,584	13,264	13,444	
Dec.	6,239	8,412	10,265	10,927	12,452	12,735	13,016	13,341	

Note: These data do not add to the annual totals given in the case study in chapter 12 as a result of revisions published at various times in the *Survey of Current Business*.

16

Index Numbers

"In the good old days things were a lot cheaper, and a dollar was worth something." Does that sound familiar? Probably so, for it is a common expression, particularly when the bills arrive each month. The phrase obviously implies that prices are higher today than in the "good old days." Is that really true, and, if it is, how can we measure the change?

In order to interpret a measurement meaningfully, we often use a *standard* or *yardstick*. You use a great many such standards—miles to measure distance, pounds to measure weight, quarts to measure volume, and so on. These standards remain fixed. That is, a mile is a mile yesterday, today, and tomorrow.

Change, however, is one of those things for which no stable standard of measurement exists. Thus, to handle this situation, the statistician uses *index numbers.* An index number is a ratio multiplied by 100 that can be used as a specialized form of a *yardstick.*

The purpose of this chapter is to show you how to construct and use index numbers. We will investigate simple and composite index numbers for both price and quantity changes. The primary effort, however, will concern price indexes.

Simple Price Index

A *simple index* is merely the ratio of each value in a series to one particular value in the series, which serves as the base or base period value. Table 16.1 shows hypothetical prices and quantities for three food items. A simple price index, *PI*, for each item can be constructed using equation (16.1):

$$(PI)_i = (p_i/p_b)(100), \qquad (16.1)$$

Table 16.1
Historical records

Item	Unit size	Price per unit			Quantity		
		1979	1980	1981	1979	1980	1981
Milk	Quart	$0.20	$0.25	$0.30	2,000	2,200	2,500
Eggs	Dozen	0.50	0.60	0.75	1,000	1,200	1,500
Bacon	Pound	2.00	2.10	2.30	1,500	1,750	2,000

Table 16.2
Construction of simple price index

	Index number		
Year	Milk (1979 = 100)	Eggs (1980 = 100)	Bacon (1981 = 100)
1979	(0.20/0.20)(100) = 100	(0.50/0.60)(100) = 83.3	(2.00/2.30)(100) = 87.0
1980	(0.25/0.20)(100) = 125	(0.60/0.60)(100) = 100	(2.10/2.30)(100) = 91.3
1981	(0.30/0.20)(100) = 150	(0.75/0.60)(100) = 125	(2.30/2.30)(100) = 100

where

$(PI)_i$ = index number for time period i,

p_i = price for time period i,

p_b = price for time period b (base period).

Table 16.2 shows the results of using equation (16.1) to construct a simple price index for each item in table 16.1. In table 16.2 a different base period year was used for each item. This is noted by the year that is set equal to 100. For milk, 1979 = 100 is the base period; for eggs, 1980 = 100 is the base period; and for bacon, 1981 = 100 is the base period.

Looking at the first column of table 16.2, we can see that the price of milk increased 25 percent in 1980 over the base period 1979, and the price of milk increased 50 percent in 1981 over the base period 1979. Notice that these increases can be read directly by subtracting 100 from the appropriate index. This method of determining the percentage increase is valid, however, only if the base period year is compared to some other year. That is, it would be incorrect to say that the percentage increase in the milk price from 1980 to 1981 is 25 (150 − 125). This is actually the percentage point increase. The percentage increase is determined as [(150 − 125)/125](100) = 20 percent.

All of the indexes in table 16.2 have as their base period a single year. At times, however, it is more convenient to use more than one year to form the base period. When this is the case, p_b in

Table 16.3
Simple price index for eggs

Year	Index numbers for eggs (1979–1980 = 100)
1979	$\dfrac{0.50}{\dfrac{(0.50 + 0.60)}{2}}(100) = 90.9$
1980	$\dfrac{0.60}{\dfrac{(0.50 + 0.60)}{2}}(100) = 109.1$
1981	$\dfrac{0.75}{\dfrac{(0.50 + 0.60)}{2}}(100) = 136.4$

equation (16.1) becomes the arithmetic mean price of the years chosen for the base period. Table 16.3 shows a price index for the egg prices with the base period as 1979–1980 = 100.

Notice in table 16.3 there is not an index number exactly equal to 100. This is so because the average price for the base period was $0.55, and no individual yearly price of eggs was exactly $0.55.

Composite Price Index

There are times when a simple price index is not sufficient. That is, suppose one really wants to know about the overall price change in the three items rather than the individual price changes. In this situation we must resort to the construction of a *composite index*. Four methods of constructing a composite price index will be presented.

Aggregate Method

An aggregate price index is constructed by summing the individual prices of a chosen year and dividing by the sum of prices in the base year as shown in equation (16.2). Table 16.4 shows the calculation of the aggregate price index using equation (16.2) for the three food items and setting 1979 = 100.

$$(PI)_i = \frac{\Sigma p_i}{\Sigma p_b}(100), \qquad (16.2)$$

where
$(PI)_i$ = index number for time period i,
p_i = prices in time period i,
p_b = prices in time period b (base period).

In the preceding method for calculating the index each price is

Table 16.4
Calculation of composite price index by the aggregate method

	Price per unit		
Year	Milk	Eggs	Bacon
1979	$0.20	$0.50	$2.00
1980	0.25	0.60	2.10
1981	0.30	0.75	2.30

$$(PI)_i = \frac{\Sigma p_i}{\Sigma p_b}(100)$$

$$PI_{1979} = \frac{0.20 + 0.50 + 2.00}{0.20 + 0.50 + 2.00}(100)$$

$$= \frac{2.70}{2.70}(100) = 100$$

$$PI_{1980} = \frac{0.25 + 0.60 + 2.10}{0.20 + 0.50 + 2.00}(100)$$

$$= \frac{2.95}{2.70}(100) = 109.3$$

$$PI_{1981} = \frac{0.30 + 0.75 + 2.30}{0.20 + 0.50 + 2.00}(100)$$

$$= \frac{3.35}{2.70}(100) = 124.1$$

treated without regard to units. Yet the unit of the product for which the price is given will make a difference in the numerical outcome of the calculated index. To overcome this deficiency, the weighted aggregate method is frequently used.

Weighted Aggregate Method

The *weighted aggregate method* utilizes some measure to *assign importance* to the individual prices. For our foods it would be appropriate to weight our items according to the quantities given in table 16.1.

When *quantities* are used as the weights, the quantities chosen for weights are *usually those in the base period or current period*. When base period weights are used, the index is referred to as a *Laspeyres index*, and, when current period weights are used, the index is referred to as a *Paasche index*.[1] In practice, the Laspeyres index is probably more prevalent because it requires only one

---

1. The indexes are named after the men who developed each concept.

period of quantity information. Equations (16.3) and (16.4) are used to calculate weighted aggregate price indexes:

$$(PI)_{Li} = \frac{\Sigma(p_iq_b)}{\Sigma(p_bq_b)}(100) \quad \text{(Laspeyres)}, \tag{16.3}$$

where

$(PI)_{Li}$ = price index for the time period i,
$\quad p_i$ = prices in time period i,
$\quad p_b$ = prices in time period b (base period),
$\quad q_b$ = quantities in time period b (base period);

$$(PI)_{Pi} = \frac{\Sigma(p_iq_i)}{\Sigma(p_bq_i)}(100) \quad \text{(Paasche)}, \tag{16.4}$$

where

$(PI)_{Pi}$ = price index for time period i,
$\quad p_i$ = prices in time period i,
$\quad p_b$ = prices in time period b (base period),
$\quad q_i$ = quantities in time period i.

Table 16.5 shows the calculation of the price index for food items using the weighted aggregate method and base period weights. That is, we have used equation (16.3) to construct table 16.5. Notice that only one year of quantity data is used. The index provides a measure of the price change focused on the base year quantities. Thus in 1980 it would take $1.09 to buy the same quantity that one dollar purchased in 1979.

Table 16.5
Calculation of composite price index by the weighted aggregate method, Laspeyres weighting (1979 = 100)

| Item | Price per unit | | | 1979 Base year |
	1979	1980	1981	quantity
Milk	$0.20	$0.25	$0.30	2,000
Eggs	0.50	0.60	0.75	1,000
Bacon	2.00	2.10	2.30	1,500

Price index $= \dfrac{\Sigma(p_iq_b)}{\Sigma(p_bq_b)}(100)$

$$(PI)_{1979} = \frac{(0.20)(2,000) + (0.50)(1,000) + (2.00)(1,500)}{(0.20)(2,000) + (0.50)(1,000) + (2.00)(1,500)}(100) = \frac{3,900}{3,900}(100) = 100$$

$$(PI)_{1980} = \frac{(0.25)(2,000) + (0.60)(1,000) + (2.10)(1,500)}{(0.20)(2,000) + (0.50)(1,000) + (2.00)(1,500)}(100) = \frac{4,250}{3,900}(100) = 109.0$$

$$(PI)_{1981} = \frac{(0.30)(2,000) + (0.75)(1,000) + (2.30)(1,500)}{(0.20)(2,000) + (0.50)(1,000) + (2.00)(1,500)}(100) = \frac{4,800}{3,900}(100) = 123.1$$

Table 16.6
Calculation of composite price index by the weighted aggregate method, Paasche weighting (1979 = 100)

Item	1979		1980		1981	
	Price p_b	Quantity q_b	Price p_1	Quantity q_1	Price p_2	Quantity q_2
Milk	$0.20	2,000	$0.25	2,200	$0.30	2,500
Eggs	0.50	1,000	0.60	1,200	0.75	1,500
Bacon	2.00	1,500	2.10	1,750	2.30	2,000

$$(PI)_i = \frac{\Sigma(p_i q_i)}{\Sigma(p_b q_i)}(100)$$

$$(PI)_{1979} = \frac{(0.20)(2,000) + (0.50)(1,000) + (2.00)(1,500)}{(0.20)(2,000) + (0.50)(1,000) + (2.00)(1,500)}(100) = \frac{3,900}{3,900}(100) = 100$$

$$(PI)_{1980} = \frac{(0.25)(2,200) + (0.60)(1,200) + (2.10)(1,750)}{(0.20)(2,200) + (0.50)(1,200) + (2.00)(1,750)}(100) = \frac{4,945}{4,540}(100) = 108.9$$

$$(PI)_{1981} = \frac{(0.30)(2,500) + (0.75)(1,500) + (2.30)(2,000)}{(0.20)(2,500) + (0.50)(1,500) + (2.00)(2,000)}(100) = \frac{6,475}{5,250}(100) = 123.3$$

Table 16.6 shows the calculation of the price index for the food items using the weighted aggregate method and current period weights. That is, we have used equation (16.4) to construct table 16.6. This index provides a measure of the price change focused on current quantities. That is, the quantities purchased in 1980 cost 8.9 percent more than if they had been purchased in 1979. Also it should be noted that even though the indexes in tables 16.5 and 16.6 are very nearly the same they do not have the same interpretation.

Average of Relatives Method

Another rather simple composite index that does not weight the individual prices is the *average of relatives method*. This method sums the simple price indexes and averages them by dividing by the number of items as shown in equation (16.5):

$$(PI)_i = \frac{\Sigma[(p_i/p_b)(100)]}{N}, \tag{16.5}$$

where
p_i = prices in period i,
p_b = prices in period b (base period),
N = number of items.

Table 16.7 shows how to calculate this composite index for the food items.

Table 16.7
Calculation of composite price index by the average of relatives method

	Simple price index			Average of relatives, $PI = \dfrac{\Sigma[(p_i/p_b)100]}{N}$
Year	Milk (1979 = 100)	Eggs (1979 = 100)	Bacon (1979 = 100)	
1979	(0.20/0.20)(100) = 100	(0.50/0.50)(100) = 100	(2.00/2.00)(100) = 100	300/3 = 100.0
1980	(0.25/0.20)(100) = 125	(0.60/0.50)(100) = 120	(2.10/2.00)(100) = 105	350/3 = 116.7
1981	(0.30/0.20)(100) = 150	(0.75/0.50)(100) = 150	(2.30/2.00)(100) = 115	415/3 = 138.3

The preceding method, however, has a weakness in that equal dollar changes in prices of different products may not have the same effect on the value of the index. To overcome this weakness, let us look at the weighted average of relatives method.

Weighted Average of Relatives Method

Equations (16.6) and (16.7) are used to calculate weighted average of relative price indexes according to the Laspeyres and Paasche methods, respectively. If you look closely, you will see that equation (16.6) is equivalent to equation (16.3), the weighted aggregate Laspeyres method, and equation (16.7) is equivalent to equation (16.4), the weighted aggregate Paasche method. Thus, when you apply either set of equations, you should get the same answers:

$$(PI)_{Li} = \frac{\Sigma[(p_i/p_b)(100)(p_b q_b)]}{\Sigma(p_b q_b)}, \tag{16.6}$$

$$(PI)_{Pi} = \frac{\Sigma[(p_i/p_b)(100)(p_b q_i)]}{\Sigma(p_b q_i)}. \tag{16.7}$$

The need for the weighted average of relatives method might then be questioned. It has an important place, however, because it can be applied to a set of index numbers already constructed. That is, in equations (16.6) and (16.7), the expression, $(p_i/p_b)(100)$, could be simple index numbers already calculated. Tables 16.8 and 16.9 show the index numbers resulting from using equations (16.6) and (16.7). As you can see, the results from the weighted average of relatives method are identical to the results from the weighted aggregate method for both the Laspeyres and Paasche weighting. And again, you should be reminded that, even though the Laspeyres and Paasche approaches may sometimes give very similar results, the interpretation of the two sets of index numbers is not the same.

Table 16.8
Calculation of composite price index by the weighted average of relatives method, Laspeyres weighting
(1979 = 100)

	Price per unit			1979 Base year quantity
Item	1979	1980	1981	
Milk	0.20	0.25	0.30	2,000
Eggs	0.50	0.60	0.75	1,000
Bacon	2.00	2.10	2.30	1,500

$$\text{Price index} = \frac{\Sigma[(p_i/p_b)(100)(p_bq_b)]}{\Sigma(p_bq_b)}$$

$$(PI)_{1979} = \frac{(0.20/0.20)(100)(0.20)(2,000) + (0.50/0.50)(100)(0.50)(1,000) + (2.00/2.00)(100)(2.00)(1,500)}{(0.20)(2,000) + (0.50)(1,000) + (2.00)(1,500)}$$

$$= \frac{40,000 + 50,000 + 300,000}{400 + 500 + 3,000} = \frac{390,000}{3,900} = 100$$

$$(PI)_{1980} = \frac{(0.25/0.20)(100)(0.20)(2,000) + (0.60/0.50)(100)(0.50)(1,000) + (2.10/2.00)(100)(2.00)(1,500)}{(0.20)(2,000) + (0.50)(1,000) + (2.00)(1,500)}$$

$$= \frac{50,000 + 60,000 + 315,000}{400 + 500 + 3,000} = \frac{425,000}{3,900} = 109.0$$

$$(PI)_{1981} = \frac{(0.30/0.20)(100)(0.20)(2,000) + (0.75/0.50)(100)(0.50)(1,000) + (2.30/2.00)(100)(2.00)(1,500)}{(0.20)(2,000) + (0.50)(1,000) + (2.00)(1,500)}$$

$$= \frac{60,000 + 75,000 + 345,000}{400 + 500 + 3,000} = \frac{480,000}{3,900} = 123.1$$

Specific Price Indexes

Two of the most commonly quoted price indexes are the Consumer Price Index and the Producer Price Index (formerly the Wholesale Price Index). Since these are indexes you have heard of, or probably will hear about, a brief description of each is presented below.

Consumer Price Index

Beginning in January 1978, the Bureau of Labor Statistics altered substantially the Consumer Price Index (CPI) that it publishes each month. As a matter of fact the bureau now publishes two CPI's. One is the traditional Consumer Price Index that is designed to measure the change in prices in a fixed market basket of goods and services purchased by urban wage earners and clerical workers, including their families. The second CPI published each month is for all urban consumers. This new index covers about 80 percent of the population, whereas the urban wage and

Table 16.9
Calculation of composite price index by weighted average of relatives method, Paasche weighting (1979 = 100)

Item	1979 Price p_b	1979 Quantity q_b	1980 Price p_1	1980 Quantity q_1	1981 Price p_2	1981 Quantity q_2
Milk	$0.20	2000	$0.25	2200	$0.30	2500
Eggs	0.50	1000	0.60	1200	0.75	1500
Bacon	2.00	1500	2.10	1750	2.30	2000

$$\text{Price index} = \frac{\Sigma[(p_i/p_b)(100)(p_b q_i)]}{\Sigma(p_b q_i)}$$

$$(PI)_{1979} = \frac{(0.20/0.20)(100)(0.20)(2{,}000) + (0.25/0.25)(100)(0.25)(1{,}000) + (2.00/2.00)(100)(2.00)(1{,}500)}{(0.20)(2{,}000) + (0.25)(1{,}000) + (2.00)(1{,}500)}$$

$$= \frac{40{,}000 + 50{,}000 + 300{,}000}{400 + 500 + 3{,}000} = \frac{390{,}000}{3{,}900} = 100$$

$$(PI)_{1980} = \frac{(0.25/0.20)(100)(0.20)(2{,}200) + (0.60/0.50)(100)(0.50)(1{,}200) + (2.10/2.00)(100)(2.00)(1{,}750)}{(0.20)(2{,}200) + (0.50)(1{,}200) + (2.00)(1{,}750)}$$

$$= \frac{55{,}000 + 72{,}000 + 367{,}500}{440 + 600 + 3{,}500} = \frac{494{,}500}{4{,}540} = 108.9$$

$$(PI)_{1981} = \frac{(0.30/0.20)(100)(0.20)(2{,}500) + (0.75/0.50)(100)(0.50)(1{,}500) + (2.30/2.00)(100)(2.00)(2{,}000)}{(0.20)(2{,}500) + (0.50)(1{,}500) + (2.00)(2{,}000)}$$

$$= \frac{75{,}000 + 112{,}500 + 460{,}000}{500 + 750 + 5{,}250} = \frac{647{,}500}{5{,}250} = 123.3$$

clerical worker CPI encompasses approximately 50 percent of the nation's population.

The CPI incorporates into its construction the price changes on 250 general items such as food, clothing, shelter, medical expenditures, transportation costs, recreation, and other goods and services purchased regularly by consumers included in the index. The index is an aggregative fixed-weights type constructed in essentially the same manner that we would construct a Laspeyres weighted average of relatives index. Currently the weights (the proportion of income devoted to each item) are based on a consumer expenditure survey that was conducted in 1972 and 1973.

The CPI has many uses. It is a measure of changes in prices; it can be used as a deflator for other economic series, and a third major use is to escalate income payments. However, it is not a cost-of-living index although it is frequently called that. There are several reasons why the CPI is not a true cost-of-living index. For example, the market basket is fixed so that is does not reflect changes in consumer spending patterns caused by changes in the relative prices of items. Moreover the CPI includes only those taxes directly associated with retail prices, such as sales taxes, and excludes income and social security taxes, among others.

One further point should be remembered when you use the CPI. Many prices vary on a seasonal basis. That is, the price of a hotel room in Miami increases each winter and then declines in the summer. Consequently the index is released in two forms—one in which the index has not been adjusted for seasonal price changes and the other where an adjustment has been made. While both reflect changes in the price level from month to month, you need to be sure you use the one that meets your needs.

Producer Price Index

The Producer Price Index (formerly known as the Wholesale Price Index) is also published monthly by the Bureau of Labor Statistics. The index is designed to measure changes in prices of commodities in primary markets in all stages of processing (essentially three main groups: crude materials, intermediate goods, and finished goods). The Producer Price Index is a weighted average of relatives index that includes about 2,800 commodities. Prices are obtained from representative producers and manufacturers and weighted by the values of shipments obtained from a 1972 study. Thus the Producer Price Index differs from the Consumer Price Index in that the Producer Price Index attempts to measure changes in prices for primary transactions that occur in other than retail markets whereas the Consumer Price Index attempts to measure changes in consumer (retail) prices.

Some Other Common Indexes

No doubt you have seen, or heard reference to, common stock price indexes. The Standard and Poor's composite index is computed daily using a group of 500 stocks, with a base period of 1940 to 1943 = 10. There are five New York Stock Exchange price indexes each with a base of December 31, 1965 = 50. These five indexes (composite, industrials, transportation, utilities, and finance) include the prices of all stocks listed on the New York Exchange and are developed from closing prices each day.

Other frequently used price indexes include those for prices received and paid by farmers. There are also price indexes, called *implicit price deflators,* for the gross national product, gross private domestic investment, and exports, among others.

All the price index series mentioned here are available in several different publications. However, one publication that includes all of them, as well as many other frequently used economic series, is *Economic Indicators.* This is a federal government publication that is released each month.

Working with Index Numbers

When working with index numbers that have already been constructed, it is possible that you may have to modify the index before you use it. Two useful modifications are shifting the base and splicing two index series.

Shifting the Base

You may encounter series of index numbers with *different base periods* or a series with a *rather old base year.* In either case, whether it is the need to put several series on the same base year or merely to update the base period, it is necessary to *shift the base.* Shifting the base is accomplished by taking the index number for the year desired as the base and dividing this index number into every other index number in the series. Table 16.10 shows a shift in the base of the hypothetical data for Product A. As you can see the base year was shifted from 1977 = 100 to 1979 = 100. This shift was accomplished by taking the 1979 index number of 108 and dividing it into the original price index producing the new price index with 1979 = 100.

Splicing Two Index Series

Sometimes, when using published index numbers, the period of interest will include the point in time where the base was changed. The result is that it becomes necessary to *splice* the two series, forming *one continuous series* of index numbers with a common base. Table 16.11 shows such a situation and the resulting spliced series.

Table 16.10
Calculations for shifting the base

Year	Product A price index 1977 = 100	Product A price index 1979 = 100
1977	100	(100/108)(100) = 92.6
1978	105	(105/108)(100) = 97.2
1979	108	(108/108)(100) = 100
1980	110	(110/108)(100) = 101.9

Table 16.11
Spliced series for hypothetical data

Year	Product A price index 1964 = 100	Product A price index 1972 = 100	Spliced series 1972 = 100
1976	130		(105)
1977	134		(108)
1978	139	112	112
1979		117	117
1980		122	122
1981		128	128

The circled index numbers in table 16.11 were obtained by using the following approach. If you look closely, you will see that new values for the spliced index are being constructed in such a manner as to maintain the same percentage change that existed in the 1964 = 100 index values:

$$I_{1977} = \frac{134}{139}(112) = 108,$$

$$I_{1976} = \frac{130}{134}(108) = 105.$$

Use of Price Index Numbers

In addition to showing the change in prices over time, price indexes are useful in the analysis of many sets of data expressed in money terms. Two common usages are (1) the determination of *real income* and (2) the *deflating of* sales data presented in the following two examples.

Example 16.1

▷ Suppose a particular wage earner has had salary figures of $11,000, $12,000, $12,500, $14,500, and $16,000 for the years 1975–1979. Let us also suppose that the Consumer Price Index (CPI) is representative of goods and services he must purchase each year. Now let us see what his purchasing power has actually been each year. Table 16.12 shows our wage earner's real wage in terms of 1967 dollars. As you can see, our wage earner's salary increased by $5,000 or 45.45 percent, but his buying power in terms of what could be purchased with 1967 dollars increased only $526 or 9.71 percent. Thus our wage earner is better off, but not by as much as you first might think. ◁

Example 16.2

▷ A manufacturer has a record of sales and of his product price index series for the years 1977 to 1981. In order to obtain a clearer

Table 16.12
Demonstration of determining real wage

Year	Money wage	CPI 1967 = 100[a]	1967 Dollars real wage
1975	$11,000	161.2	($11,000/161.2)(100) = $6,824
1976	12,000	170.5	($12,000/170.5)(100) = $7,038
1977	12,500	181.5	($12,500/181.5)(100) = $6,887
1978	14,500	195.3	($14,500/195.3)(100) = $7,424
1979	16,000	217.7	($16,000/217.7)(100) = $7,350

a. Source: U.S. Department of Labor, *Monthly Labor Review* (May 1980).

Table 16.13
Deflating sales

Year	Product sales	Product price index 1977 = 100	Deflated sales
1977	$14,000	100	$14,000
1978	15,000	103	($15,000/103)(100) = 14,563
1979	15,200	105	($15,200/105)(100) = 14,476
1980	17,000	108	($17,000/108)(100) = 15,740
1981	18,000	110	($18,000/110)(100) = 16,364

picture of the growth in sales, the manufacturer deflates the sales figures as shown in table 16.13. Now in terms of 1977 dollars the increase in sales was $2,364 rather than $4,000, or 16.9 percent rather than 28.6 percent.

The deflated sales figures reflect only the changes in sales that are the result of changes in the quantity of the product sold. Thus in terms of current dollars (column 2) the period 1978–1979 looks like a period where sales quantity increased, but the deflated sales column (constant dollars) shows the true picture. That is, the year 1979 actually had a small decline from the quantity sold in 1978. ◁

Quantity Index

Virtually all that has been said about price indexes in the preceding pages could be repeated with respect to quantity indexes. However, to expedite discussion, only the equations necessary for calculating quantity indexes are shown in table 16.14. The only difference in the equations involves a transposition of the p's and q's as can be seen in table 16.14. Each method of calculating an index has the same limitations or advantages whether applied to prices or quantities. And the quantity index is used to measure change in quantity over a period of time.

Table 16.14
Equations for price and quantity indexes

Method	Price index	Quantity index
Simple	$(p_i/p_b)100$	$(q_i/q_b)100$
Simple aggregate	$\dfrac{\Sigma p_i}{\Sigma p_b}(100)$	$\dfrac{\Sigma q_i}{\Sigma q_b}(100)^a$
Weighted aggregate		
Laspeyres	$\dfrac{\Sigma(p_i q_b)}{\Sigma(p_b q_b)}(100)$	$\dfrac{\Sigma(q_i p_b)}{\Sigma(q_b p_b)}(100)$
Paasche	$\dfrac{\Sigma(p_i q_i)}{\Sigma(p_b q_i)}(100)$	$\dfrac{\Sigma(q_i p_i)}{\Sigma(q_b p_i)}(100)$
Simple average of relatives	$\dfrac{\Sigma[(p_i/p_b)(100)]}{N}$	$\dfrac{\Sigma[(q_i/q_b)(100)]}{N}$
Weighted average of relatives		
Laspeyres	$\dfrac{\Sigma[(p_i/p_b)(100)(p_b q_b)]}{\Sigma(p_b q_b)}$	$\dfrac{\Sigma[(q_i/q_b)(100)(q_b p_b)]}{\Sigma(q_b p_b)}$
Paasche	$\dfrac{\Sigma[(p_i/p_b)(100)(p_b q_i)]}{\Sigma(p_b q_i)}$	$\dfrac{\Sigma[(q_i/q_b)(100)(q_b p_i)]}{\Sigma(q_b p_i)}$

a. Use of this equation requires homogeneous units.

16.1 Simple price index:

$$(PI)_i = (p_i/p_b)(100)$$

16.2 Composite aggregate price index:

$$(PI)_i = \frac{\Sigma p_i}{\Sigma p_b}(100)$$

16.3 Composite weighted aggregate price index—Laspeyres weighting:

$$(PI)_{L_i} = \frac{\Sigma(p_i q_b)}{\Sigma(p_b q_b)}(100)$$

16.4 Composite weighted aggregate price index—Paasche weighting:

$$(PI)_{P_i} = \frac{\Sigma(p_i q_i)}{\Sigma(p_b q_i)}(100)$$

16.5 Composite average of relatives price index:

$$(PI)_i = \frac{\Sigma[(p_i/p_b)(100)]}{N}$$

16.6 Composite weighted average of relatives price index—Laspeyres weighting:

$$(PI)_{L_i} = \frac{\Sigma[(p_i/p_b)(100)(p_b q_b)]}{\Sigma(p_b q_b)}$$

16.7 Composite weighted average of relatives price index—Paasche weighting:

$$(PI)_{P_i} = \frac{\Sigma[(p_i/p_b)(100)(p_b/q_i)]}{\Sigma(p_b q_i)}$$

Review Questions

1. Explain the use of each equation shown in table 16.14.

2. Define
(a) index number
(b) consumer price index (CPI)
(c) producer price index (PPI)

3. Explain the difference between Laspeyres and Paasche indexes.

4. What is meant by "shifting the base" of a series of index numbers?

5. What is meant by "splicing" two index series?

6. Distinguish between a simple index and a composite index.

Problems

1. Mr. Shopkeeper has kept a record on the costs of certain items purchased weekly.

Item	Unit size	Price per unit 1980	1981	1982	Quantity purchased 1980	1981	1982
Coffee	1 lb	$2.00	$2.10	$2.20	30	32	35
Cookies	1 lb	0.60	0.65	0.70	6	7	8
Fruit	1 lb	0.40	0.42	0.44	12	14	16
Sugar	5 lb	2.10	2.40	1.80	20	22	25

(a) Construct a simple price index for coffee with 1980 = 100.
(b) How many percentage points did the price of coffee increase from 1981 to 1982?
(c) What was the percentage increase in the price of coffee from 1981 to 1982?
(d) Construct a simple price index for sugar with 1982 = 100.
(e) Construct a simple quantity index for fruit with 1980 = 100.

2. Use the data in problem 1, and calculate a composite price index for each year using the simple aggregate method with 1980 = 100. Explain the confusion associated with using this index.

3. Use the data in problem 1, and calculate a composite quantity index for each year using the simple aggregate method with 1980 = 100.

4. Use the data in problem 1, and calculate a composite price index for each year using the average of relatives method with 1980 = 100.

5. Use the data in problem 1, and calculate a composite quantity index for each year using the average of relatives method with 1980 = 100.

6. Use the data in problem 1, and calculate a composite price index for each year using the weighted aggregate method with 1980 = 100.
(a) Use Laspeyres weights.
(b) Use Paasche weights.
(c) Interpret the results of parts (a) and (b).

7. Use the data in problem 1, and calculate a composite price index for each year using the weighted average of relatives method with 1980 = 100.
(a) Use Laspeyres weights.
(b) Use Paasche weights.
(c) Compare the answers in problems 6 and 7.

8. Use the data in problem 1, and calculate a composite quantity index for each year using the weighted aggregate method with 1980 = 100.
(a) Use Laspeyres weights.
(b) Use Paasche weights.

9. Use the data in problem 1, and calculate a composite quantity index for each year using the weighted average of relative method with 1980 = 100.
(a) Use Laspeyres weights.
(b) Use Paasche weights.
(c) Compare the answers in problems 8 and 9.

10.

Item	Unit size	Price per unit 1980	1981	1982	Quantity purchased 1980	1981	1982
Fish	1 lb	$1.25	$1.40	$1.50	200	250	300
Beef	1 lb	2.50	2.60	2.65	500	600	650
Chicken	1 lb	0.50	0.60	0.70	100	200	250

(a) Construct a composite price index for each year using the weighted aggregate approach and Laspeyres weights with 1980 = 100.
(b) What is the percentage increase in prices from 1980 to 1982 based on your index from part (a)?
(c) What is the percentage increase in prices from 1981 to 1982 based on your index from part (a)?

11. Repeat problem 10 for quantities rather than prices.

12. A small manufacturer has maintained records on certain basic expenses for two recent years.

Expense	Price/unit		Quantity used (units)	
	1981	1982	1981	1982
Electricity	$0.42	$0.63	1,000	1,200
Materials	4.00	4.50	250	280
Wages	2.00	2.25	3,000	4,000

(a) Determine the composite price index for 1982 using Laspeyres weights and 1981 = 100. Use the weighted average of relatives method.
(b) Interpret the answer in part (a).
(c) Determine the composite price index for 1982 using Paasche weights and 1981 = 100. Use the weighted average of relatives method.
(d) Interpret the answer in part (c).

13. An operator of a car rental agency has maintained the following records:

Item	Unit	Price per unit			Quantity used		
		1980	1981	1982	1980	1981	1982
Gasoline	Gallon	1.40	1.30	1.25	1,500	1,600	1,700
Tires	Radial	75.00	80.00	95.00	100	110	120
Oil	Quart	1.10	1.15	1.20	20	25	30

(a) Construct a simple price index for gasoline with 1980 = 100.
(b) Construct a composite price index for each year using the weighted aggregate method with 1980 = 100. Use Laspeyres weights.
(c) Construct a composite price index for each year using the weighted average of relatives method with 1980 = 100. Use Laspeyres weights.

14. Over several years, a manufacturer charged the following prices for a particular product.

Year	Price/unit
1977	$2.20
1978	2.06
1979	2.25
1980	2.30
1981	2.50
1982	2.69

(a) Construct a simple price index with 1977 = 100.
(b) Shift the base of the index constructed in part (a) to 1979.

15. Splice the following two series of price indexes together.

Year	1967 = 100	1972 = 100
1977	118	
1978	125	
1979	140	115
1980		129
1981		139
1982		146

(*a*) Maintain 1972 = 100.
(*b*) Maintain 1967 = 100.

16. From the data below, determine if the average real wage of the employees is keeping pace with increased living costs. If wages are falling behind price increases, what wage is necessary in 1980 to be equivalent to $175 in 1976?

Year	Average weekly wage	CPI[a] (1967 = 100)
1976	$175	170.5
1977	190	181.5
1978	200	195.3
1979	230	217.7
1980	270	247.0

[a] Source: *Monthly Labor Review,* U.S. Department of Labor, January, 1982.

17. Mr. Job Hopper is about to move again. He has been offered a new job with a salary of $23,500 per year in a city with a "cost of living" index of 162. If he presently earns $21,500 per year in a town with a "cost of living" index of 137, will he be better off financially in the new job?

18. New Products, Inc., has reported the following data for one product:

Year	Sales	Product price index 1978 = 100
1978	$27,200	100
1979	29,500	102
1980	29,900	106
1981	31,000	109
1982	32,500	112

Evaluate the growth recorded for this product.

19. If the consumer price index increased from 217.7 in 1979 to 247.0 in 1980, a clerical worker earning $2.90 per hour in 1979 must earn how much in 1980 in order to maintain a constant purchasing power?

20. Mr. Homeowner purchased his new home in 1975 at a cost of $51,000. It is now the year 1981, and Mr. Homeowner knows that residential construction costs have risen. In fact he has discovered that a residential construction cost index (1970 = 100), which was 132 in 1975, is now 215 . His home is currently insured for $65,000. Do you recommend he consider increasing his insurance? Why?

21. Assume that your real income, stated in 1967 dollars, increased from $8,000 in 1967 to $30,000 in 1980. If prices have increased 147 percent during the same period, what is your actual dollar income in 1980?

Case Study

A survey of voluntary nonprofit organizations in the State of Arkansas was conducted during the spring of 1981. The survey was directed, in cooperation with the Governor's Office of Voluntary Citizen Participation (Little Rock), by a Ph.D. candidate in the College of Business at the University of Arkansas (Fayetteville).

The responding organizations were classified into several cate-

Type of organization	Active members			Total funding		
	1978	1979	1980	1978	1979	1980
Social services	50	30	40	34,268	43,082	50,884
Sports/recreation	190	185	182	9,000	10,000	10,500
Environment	50	60	75	500	600	750
Religious	90	100	100	47,000	56,000	50,000
Arts & culture	25	25	25	3,780	3,900	4,350
Service	16	15	16	4,655	7,395	7,657
Civil	18	18	20	400	500	600
Education	5	8	7	185	300	200
Youth development	33	29	31	1,000	3,000	4,000
Social functions	16	18	20	240	250	300
Health services	25	45	75	13,000	50,000	32,000

CPI[a]
1967 = 100
1978 = 195.4
1979 = 217.4
1980 = 246.8

[a]Source: *U.S. Statistical Abstracts*, p. 467.

BASIC STATISTICS

gories by their primary organizational function. One organization has been randomly selected from each classification. The table shows the number of active members and total funding over a three-year period for each of the eleven selected organizations.

(*a*) Determine whether real dollars increased or decreased for each organization from 1978 to 1980.
(*b*) Determine the average funding per member (in real dollars) for each organization for 1978, 1979, 1980.

Contributed by William F. Crittenden, Florida State University and Vicky L. Crittenden, Florida State University/Tallahassee Community College

17

Expected Values and Decision Making

In this chapter we explain how to combine statistics with economic information in order to develop a framework for decision making. Situations where decisions are made may be divided into three categories: (1) *certainty*, (2) *risk*, and (3) *uncertainty*.

When there is perfect knowledge about which events will occur, we call this condition one of *certainty*. *Uncertainty* exists when there is no knowledge available about which of several events will occur. A third category, known as *risk*, is a condition between certainty and uncertainty. *Risk* is the condition where the occurrence of the possible events can be assigned probabilities or described with some probability distribution. That is, we know what events are possible, and the probability of each event can be specified. This chapter deals with decisions under the situation of risk.

Expected Values

As discussed at the end of chapter 7, one way to make decisions under risk conditions is to use *expected values*. The expected value of a random variable is an average value of the variable. The expected value of a discrete random variable can be calculated using equation 17.1:

$$E(X) = \Sigma[x \cdot P(X = x)], \tag{17.1}$$

where

(X) = expected value of the random variable, X,

x = the possible values (outcomes) of the random variable, X,

$P(X = x)$ = the probability that the random variable, X, is equal to x.

Expected value is sometimes called *mathematical expectation*. The concept is probably easiest understood through examples about games of chance.

Example 17.1

▷ Suppose you have been invited to toss a coin with the understanding that if a head occurs you win $6.00 but that if a tail occurs you lose $6.00. If the probability of a head is 1/2 and that of a tail is 1/2, what is the expected value of this game to you?

$$E(X) = \Sigma[x \cdot P(X = x)]$$

$$E(X) = \$6(1/2) + (-\$6)1/2$$
$$= \$3 - \$3$$
$$= 0.$$

Conclusion: If you play long enough, the expected value of the game to you is zero (0). A game or situation with an expected value of 0 is said to be a "fair game." ◁

Example 17.2

▷ Suppose we shift example 17.1 from the flip of a coin to the roll of a single fair die. Assume you win $6.00 if you roll a one or a two, but you lose $6.00 if you roll three, four, five, or six. The probability of each of the six possible outcomes is 1/6. Now, what is the expected value of this game to you?

$$E(X) = \Sigma[x \cdot P(X = x)]$$

$$= \$6 \cdot P(1 \text{ or } 2) + (-\$6)P(3 \text{ or } 4 \text{ or } 5 \text{ or } 6)$$
$$= \$6(1/6 + 1/6) + (-\$6)(1/6 + 1/6 + 1/6 + 1/6)$$
$$= \$6(2/6) + -\$6(4/6)$$
$$= \$2 - \$4$$
$$= -\$2.$$

Conclusion: This game is a long-run losing proposition to you, and your average loss will be $2 per play. ◁

Example 17.3

▷ A magazine distributor has been keeping records on the frequency of sales of his magazine as shown.

Number of magazines sold weekly	Frequency	Probability of sales
14	40	40/100 = 0.40
15	30	30/100 = 0.30
16	30	30/100 = 0.30
	100	

Based on this record of 100 sales, what have been the average weekly sales?

$$E(X) = \Sigma[x \cdot P(X = x)]$$
$$= 14(0.40) + 15(0.30) + 16(0.30)$$
$$= 5.60 + 4.50 + 4.80$$
$$E(X) = 14.90.$$

Conclusion: The average weekly sales have been 14.90 magazines. ◁

Payoff Tables

If we are trying to decide on which of several alternative choices to make, one way to analyze the result or outcome of the choice or decision is to construct a *payoff table*.

A payoff table shows the outcome for every action—event combination. *An action is a decision, an event is a state of nature, and an outcome is a result of the two,* usually a profit or loss. For instance, if you order two units of a product (the decision or choice), the payoff table will show how much profit (the outcome) you will make if you sell both, only one, or none of the units (the possible events). The table should also show your profit for other alternatives, such as ordering three units, one unit, or any other of several quantities.

Example 17.4

▷ Mr. Robert Mackey is a small farmer in southern California. Each planting season he must decide whether or not he will plant soybeans, sugar cane, or sorghum on a 10-acre tract reserved for this purpose. The amount of contribution to profit for each commodity planted depends on the season—dry, wet, or mixed. His alternative choices are which of three crops to plant. The events are the states of weather (dry, wet, or mixed), and the outcomes are the contribution to profit for each decision-event combination. Table 17.1 shows a payoff table for this example. From that table

Table 17.1
Payoff table of contribution to profit per acre of farm crop planted

Probability of state of nature	Events (i) States of nature	Actions (j) Decision alternative		
		Plant soybeans	Plant sugar cane	Plant sorghum
0.6	Dry	$600	$100	$400
0.1	Wet	200	800	100
0.3	Mixed	700	600	800

it can be seen that soybeans have the best payoff in dry weather, sugar cane is the best in wet weather, and sorghum has the best payoff in mixed weather. In the next section we will explain how to select a decision alternative using expected values. ◁

First, let us investigate in more detail the construction of a payoff table. Suppose an organization known as Tastree Partner has the sweets concession at the local Big Theater. A single item, known as Tastrees, is sold at the theater. Demand for the sweets has ranged from 30 to 50 dozen, so that the manager of Tastree Partner has to decide how many dozen Tastrees to make for a particular performance. She knows that a dozen Tastrees sell for $10.00 and cost $8.00. She also knows that any unsold Tastrees can be sold the next day (salvaged) for $4.00 a dozen. With this information she has decided to construct a payoff table to help with her decision.

Table 17.2 shows the payoff table the manager made. The entries in the table show the profits that *could* result from each decision she might make. If you examine the table closely, you will see that the first column represents the profit from preparing 30 dozen Tastrees. Since demand is always at least 30, there may be some unsatisfied demand (hungry customers), but there will never be any leftover Tastrees. The profit, $60.00, was found by using equation (17.2):

$$\text{payoff}_{ij} = (P - C) \cdot j, \qquad (17.2)$$

where

P = price per unit,
C = cost per unit,
j = number units prepared or produced,
i = number units demanded.

In the illustration the price is $10.00 per dozen, and the cost is

Table 17.2
A payoff ($) table for Tastree Partner

Events (i) Dozens of Tastrees demanded	Actions (j) Dozens of Tastrees prepared				
	30	35	40	45	50
30	60	40	20	0	−20
35	60	70	50	30	10
40	60	70	80	60	40
45	60	70	80	90	70
50	60	70	80	90	100

$8.00 per dozen. Thus, *when production is equal to or less than demand*, the potential profits are calculated as follows:

(for $i \geq 30$)
   payoff$_{i30}$ = ($10 − $8)30 = $60,
(for $i \geq 35$)
   payoff$_{i35}$ = ($10 − $8)35 = $70,
(for $i \geq 40$)
   payoff$_{i40}$ = ($10 − $8)40 = $80,
(for $i \geq 45$)
   payoff$_{i45}$ = ($10 − $8)45 = $90,
(for $i \geq 50$)
   payoff$_{i50}$ = ($10 − $8)50 = $100.

Table 17.2 shows these payoffs in the columns *below and on* the diagonal line. All the payoffs on and below this line are for situations where demand is equal to or greater than production. Thus there are never any Tastrees left when these combinations of actions and events occur.

Now consider those decisions that the manager can make that will cause some Tastrees to be left over. This will happen anytime *the number prepared is greater than the number demanded*. The payoffs for these action-event combinations can be found by using equation (17.3):

$$\text{payoff}_{ij} = (P − C)i + (S − C)(j − i), \tag{17.3}$$

where
P = price per unit,
C = cost per unit,
i = number of units demanded,
S = salvage value per unit,
j = number of units prepared.

In the payoff table all the conditions where demand is less than supply are above the diagonal line as positioned in table 17.2. The payoffs in the table for each possible action *when demand is 30* are calculated as follows:

(for $j = 35$)
   payoff$_{30,35}$ = ($10 − $8)30 + ($4 − $8)(35 − 30) = $40,
(for $j = 40$)
   payoff$_{30,40}$ = ($10 − $8)30 + ($4 − $8)(40 − 30) = $20,
(for $j = 45$)
   payoff$_{30,45}$ = ($10 − $8)30 + ($4 − $8)(45 − 30) = $0,
(for $j = 50$)
   payoff$_{30,35}$ = ($10 − $8)30 + ($4 − $8)(50 − 30) = −$20.

The remainder of the payoff table is completed by considering each level of demand and calculating the payoffs when supply is greater than demand using equation (17.3).

In looking at the payoff table in table 17.2, you can see that the most conservative decision would be to prepare (stock) 30 dozen Tastrees for every performance of the Big Theater and be assured of a $60.00 payoff (profit). On the other hand, the riskiest decision would be to prepare 50 dozen Tastrees and, depending on attendance, experience a profit as high as $100.00 or a loss of $20.00. The maximum gain or loss from each action-event combination is of interest to decision makers because of the marginal utility of money. Perhaps the easiest way to visualize the significance of this information is to assume that the business in our example has only $10.00 in assets. Thus a loss of more than $10.00 would bankrupt the firm. Under this condition the manager will probably not decide to produce 50 dozen Tastrees unless she is certain of selling 35 dozen or more. (See the last column of table 17.2.)

Decision Making Using Expected Money Values

Decision makers may find that information on the probability of the occurrence of each possible event would be helpful. That is, so far we have recognized only that different quantities may be demanded, but we have said nothing about the *likelihood* that each event (level of demand) will occur.

Table 17.3 shows the probability of each level of demand for Tastrees. Let's assume that the information in the table was derived from the sales records of the company. Thus 10 percent of the time 30 Tastrees are sold (demanded), 35 percent of the time 35 are demanded, and so on. With this information, we can now calculate the *expected money value* (EMV) of each production decision.

To see how the probabilities and payoffs are combined, look at table 17.4. This table illustrates the calculation of the expected

Table 17.3
Historical demand for Tastrees

Dozens of Tastrees	Probability of sales
30	0.10
35	0.35
40	0.30
45	0.15
50	0.10

Table 17.4
Determination of expected money value

Probability of demand	Profits associated with action: prepare 50 dozen	Expected value
0.10	$-20	$-2.00
0.35	10	3.50
0.30	40	12.00
0.15	70	10.50
0.10	100	10.00
		$34.00
Expected money value (prepare 50 dozen) = $34.00		

money value of producing 50 dozen Tastrees. As you can see, the payoff for each level of demand, given that 50 dozen are produced, is multiplied by the probability of that many Tastrees being demanded. The resulting expected values are then summed to give the expected money value of the action.

The preceding procedure is repeated for each of the actions (possible production levels). The expected money values are shown in the bottom row of table 17.5.

The manager now knows that the decision to produce 35 dozen Tastrees each night will yield the highest average profit. Thus, if the objective is to maximize profits, the company will always produce this quantity.

Expected Profit under Certainty

While producing the quantity that yields the highest expected money value will maximize average profits under conditions of risk; it may be possible to have a higher average profit. If decision makers had perfect information about how many units would be

Table 17.5
A payoff table including probabilities of demand and expected money values for Tastree partner

Probability of demand	Events (i) Dozens of Tastrees demanded	Actions (j) Dozens of Tastrees prepared				
		30	35	40	45	50
0.10	30	60	40	20	0	-20
0.35	35	60	70	50	30	10
0.30	40	60	70	80	60	40
0.15	45	60	70	80	90	70
0.10	50	60	70	80	90	100
Expected money value		$60	$67	$63.50	$51	$34

Table 17.6
A payoff table used to calculate the expected profit under certainty

Probability of demand	Events (i) Dozens of Tastrees demanded	Actions (j) Dozens of Tastrees prepared				
		30	35	40	45	50
0.10	30	60	40	20	0	−20
0.35	35	60	70	50	30	10
0.30	40	60	70	80	60	40
0.15	45	60	70	80	90	70
0.10	50	60	70	80	90	100

Expected profit under certainty
$$= (0.10)(60) + (0.35)(70) + (0.30)(80)$$
$$+(0.15)(90) + (0.10)100$$
$$= 6.00 + 24.50 + 24.00 + 13.50 + 10.00$$
$$= \$78.00$$

demanded, then they could always produce the exact quantity that was needed. This condition is a zero risk situation and, as you learned earlier, is called *certainty*.

The expected profit under certainty (EPUC) is the average payoff in a zero risk situation. Returning to our Tastree example, 10 percent of the time 30 dozen would be produced; 35 percent of the time the decision would be to produce 35 dozen, and so on. The circled numbers in table 17.6 are the profits that would be earned for each of the action-event combinations when exactly the quantity prepared would be demanded. To calculate the expected profit under certainty, it is necessary to multiply each payoff by the probability that the particular level of demand will occur. In our example, the long-run average profit for each night's operation, if the management has perfect information, is $78.00, as shown at the bottom of table 17.6.

Expected Value of Perfect Information

Now that you know how to determine expected money value under risk and the expected profit under certainty, you have the knowledge required to calculate the *expected value of perfect information* (EVPI). That is, the average amount of additional profit that could be earned by moving from a risk state to one of certainty. To calculate EVPI, you must subtract the maximum expected money value under risk from the expected profit under certainty. In our example, it is $78 − $67 = $11. This represents the maximum amount of money that should be spent to acquire additional information that would reduce risk. Thus, if the manager is going to purchase additional information, she should not spend more than the EVPI, which is $11.00.

Revised Probabilities

In the example in the preceding section, the probabilities were assumed to be historical in nature. That is, the probabilities reflected the likelihood of certain quantities being demanded and were determined from past sales records. These probabilities represent, however, a consolidation of sales for many performances. Thus for any one particular performance this historical probability distribution may be somewhat inadequate. It is the purpose of this section to demonstrate how sampling can be used to revise the historical probabilities so that more useful information will be available for making decisions.

Revising Probabilities Based on Sample Information

Continuing with our Tastree Partner manager, let us assume that the manager has obtained one additional piece of information about attendance at the Big Theater. Table 17.7 presents this information along with the historical probabilities that were given in table 17.3.

The additional information is contained in the first column of table 17.7 and is the probability of a season ticket holder attending a particular performance. These season ticket holder probabilities are associated with the Tastrees demanded in that the higher probabilities of a season ticket holder attending occur as demand becomes greater. Thus we are saying, when more season ticket holders attend a performance, overall attendance is likely to be higher, which means more Tastrees should be sold. Realizing this, the manager has decided to sample randomly the list of season ticket holders to find out what their attendance plans are.

To demonstrate, let us suppose that she takes a sample of 10 and learns that 4 are planning to attend the next performance. Under the assumption of sampling from a binomial distribution, let us see how the sample information can be used to revise the historical probabilities.

First, we can refer to appendix D to find the probabilities of a

Table 17.7
Big theater data

Proportion of season ticket holders attending	Dozens of Tastrees demanded	Probability of sales
0.05	30	0.10
0.15	35	0.35
0.30	40	0.30
0.60	45	0.15
0.90	50	0.10

sample of 10 showing that four will attend using the historical probabilities of season ticket-holder attendance. In other words, we are interested in the probability of finding 4 yes answers in a sample of 10 if, in fact, only 5 percent of the season ticket holders attend the performance, and so forth. As read from the appendix, the probabilities are presented as follows:

$P(X = 4)$ if $n = 10$ and $\pi = 0.05 = 0.001$,

$P(X = 4)$ if $n = 10$ and $\pi = 0.15 = 0.0401$,

$P(X = 4)$ if $n = 10$ and $\pi = 0.30 = 0.2001$,

$P(X = 4)$ if $n = 10$ and $\pi = 0.60 = P(n - X)$

$=$

$P(10 - 4)$ if $n = 10$ and $(1 - \pi) = 0.40 = 0.1115$,

$P(X = 4)$ if $n = 10$ and $\pi = 0.90 = P(n - X)$

$=$

$P(10 - 4)$ if $n = 10$ and $(1 - \pi) = 0.10 = 0.0001$.

You will also find these probabilities in column 2 of table 17.8.

The next step is to develop joint probabilities using the probabilities in column 2 and the historical probabilities for Tastree demand (column 3 of the table). For example, $(0.2001)(0.30) = 0.06003$ as shown in column 4.

The results of the final step in revising the probabilities based on the information from the sample are shown in column 5 of table 17.8. This step consists of dividing each of the joint probabilities by the sum of all the joint probabilities, or $(0.0001) \div (0.0909) = 0.0011$, for example.

In summary, now that we know that 4 out of a sample of 10 season ticket holders plan to attend the performance, the revised

Table 17.8
Revision of probabilities: sampling from a binomial distribution

(1)	(2)	(3)	(4) = (2)(3)	(5) = $\dfrac{(4)}{\sum(4)}$
Proportion (π) of season ticket holders attending	$P(X - 4)$ if $n = 10$ and π as given	(Prior) Historical probabilities	Joint probabilities	Revised probabilities (Posterior)
0.05	0.0010	0.10	0.000100	0.0011
0.15	0.0401	0.35	0.014035	0.1544
0.30	0.2001	0.30	0.060030	0.6604
0.60	0.1115	0.15	0.016725	0.1840
0.90	0.0001	0.10	0.000010	0.0001
			0.090900	1.0000

Table 17.9
Payoff table and calculations using revised probabilities

Probability of demand	Events (i) Dozens of Tastrees demanded	Actions (j) Dozens of Tastrees prepared				
		30	35	40	45	50
0.0011	30	60	40	20	0	−20
0.1544	35	60	70	50	30	10
0.6604	40	60	70	80	60	40
0.1840	45	60	70	80	90	70
0.0001	50	60	70	80	90	100
Expected money value		$60.00	$69.97	$75.30	$60.83	$40.83

Best action: prepare 40 dozen

Expected profit under certainty = (0.0011)($60) + (0.1544)($70) + (0.6604)($80)
$$+ (0.1840)($90) + (0.0001)($100)$$
$$= 0.066 + 10.808 + 52.832 + 16.56 + 0.01$$
$$= $80.28$$

Expected value of perfect information = $80.28 − $75.30
$$EVPI = $4.98$$

probability of selling 30 Tastrees is now 0.0011 instead of the original 0.10. This, and all other revisions, are shown in table 17.9. Using the improved information, we can now recompute the expected money values (EMV). In our example, the action that will yield the highest average profit is now the preparation of 40 dozen Tastrees.

Continuing with the revised probabilities, the expected profit under certainty for the next performance is $80.28. And by comparing the expected profit under certainty and the maximum expected money value, we find that the expected value of perfect information (EVPI) has been reduced to $4.98. In other words, the sample gave information that will result in a higher average profit, but we could still increase average profits. However, we should not spend more than $4.98 doing so.

It should be pointed out that the new information gained from the sample does not change the average expected profit under certainty ($78.00) shown in table 17.6. This amount ($78.00) is the maximum average profit possible if we never make an error. The data in table 17.9 are for *the next performance* and should be interpreted to mean that the average expected profit under certainty for those performances where 4 in a sample of 10 season ticket holders have said they will attend is $80.28.

At this point, you may wonder about the value of sample information as compared to its cost. In our example, it is obvious that the sample data increased the expected money value from $67.00

to $75.30, or by $8.30. If the sample costs less than $8.30, then deciding to "improve" the probabilities increased the firm's profits. But, is there a way we could have known this prior to taking the sample? To answer this question, you must know the *expected value of sample information* (EVSI).

Expected Value of Sample Information

The expected value of sample information is the difference between the expected return using sample information and the expected return under risk. The expected return under risk is the maximum profit (minimum cost) obtained from the payoff table analysis (see $67.00 in table 17.5) by using prior probabilities. The expected return using sample information is obtained from the revised probabilities. Then a comparison of the return (with and without sampling) is made to determine EVSI.

Earlier in this chapter you learned that the expected value of perfect information was the difference between expected profit under certainty and the maximum expected money value. In the Tastree Partner example, the expected value of perfect information is $11.00. Thus the decision maker would not pay more than $11.00 for information. If additional information, perhaps from a sample, is not available at a cost equal to or less than the expected value of perfect information, then the decision maker cannot improve the average profit of his firm.

Returning to our example, assume that it costs 50¢ per sample unit to sample the theater's season ticket holders. With an expected value of perfect information of $11.00, the manager could conceivably have a sample of 22 theater customers rather than 10. How, then, did she arrive at a sample size of 10 instead of 22, or 12, or perhaps only one? Once again, to answer this question you must know the expected value of sample information, but you also have to know the cost of each size sample.

Let's examine how to calculate EVSI. Actually, what is required is the construction of tables 17.8 and 17.9 for all possible responses for each of the possible sample sizes from 1 to 22. To serve as an example, with the sample of 10, we need to determine the total of the joint probabilities for every value of X (just like it was done for $X = 4$ in table 17.8) and the maximum expected money value. The next step requires that each joint probability be multiplied by the corresponding EMV as shown in table 17.10. These products are then summed, and the original EMV is subtracted from this sum to yield the expected value of sample information. Thus, in Table 17.10, $67.00 is subtracted from $72.80 to give EVSI = $5.80. Since the sample of 10 would cost $5.00 (50¢/unit), it is an economical decision to take the sample.

Table 17.10
Determination of the expected value of sample information ($n = 10$)

x	(1) $P(X = x)$ Probability	(2) Decision	(3) Maximum expected money value	(4) = (1)(3) $P(X = x) \cdot$ (max EMV)
0	0.1373	30	$60.00	$8.24
1	0.1897	35	65.02	12.33
2	0.1757	35	68.73	12.08
3	0.1329	35	69.76	9.27
4	0.0909[a]	40[b]	75.30[b]	6.85
5	0.0641	40	78.60	5.04
6	0.0502	45	82.88	4.16
7	0.0407	45	88.01	3.58
8	0.0379	45	89.67	3.40
9	0.0448	50	95.91	4.30
10	0.0358	50	99.24	3.55
	1.0000[c]			$72.80

EVSI = E (max EMV) − max (EMV under risk)
     = $P(X = x)$(max EMV) − max (EMV under risk)
EVSI = $72.80 − 67.00[d] = $5.80

[a] From table 17.8, the probability of obtaining 4 yes responses.
[b] From table 17.9.
[c] The probability of all possible outcomes from sample, $n = 10$.
[d] From table 17.5.

Actually, we still do not know if this is the "best" size sample to take. To answer this question, we must now repeat this process for all possible sample sizes. The optimum sample size is the one with the maximum difference between EVSI and the cost of sampling. We will not pursue this point any further, except to say that it could be done easily with the aid of a computer. However, before completing the discussion, let us look at one more example.

Example 17.5

▷ The executives of Supercard are considering accepting a proposal from a travel agency. The travel agency has proposed that Supercard offer all of its cardholders the opportunity to participate in a variety of vacation trips sponsored by the travel agency. The promotion and mailing expense incurred by Supercard would be $20,000. In return for this participation, Supercard would receive $5.00 per participant for the first 4,000 participants and $2.50 per participant for all participants above 4,000. At present, Supercard has 40,000 cardholders, and the executives have established the following anticipated participation rates and their probabilities.

Proportion of cardholders participating	Probability of participation
0.05	0.2
0.10	0.4
0.15	0.3
0.20	0.1

(a) Construct the appropriate payoff table.

First let us calculate the payoff for each participation rate:

(1) $(40,000)(0.05) = 2,000$

$(2,000)(\$5) - \$20,000 = -\$10,000,$

(2) $(40,000)(0.10) = 4,000$

$(4,000)(\$5) - \$20,000 = 0,$

(3) $(40,000)(0.15) = 6,000$

$(4,000)(\$5) + (\$2,000)(\$2.50) - \$20,000 =$
$\$20,000 + \$5,000 - \$20,000 = \$5,000,$

(4) $(40,000)(0.20) = 8,000$

$(4,000)(\$5) + (4,000)(\$2.50) - \$20,000 =$
$\$20,000 + \$10,000 - \$20,000 = \$10,000.$

(b) Determine the expected profit under certainty (EPUC):

EPUC $= (0.2)(0) + (0.4)(0) + (0.3)(5,000) + (0.10)(10,000)$

EPUC $= 0 + 0 + 1,500 + 1,000$

EPUC $= \$2,500.$

(c) Determine the expected value of perfect information (EVPI):

EVPI $=$ EPUC $-$ max (EMV)

$= \$2,500 - \$500$

EVPI $= \$2,000.$

(d) Determine the expected value of sample information (EVSI) using a sample size of three.

Table 17.11
Payoff table for marketing example

Probability of participation	Proportion of cardholders participating	Possible actions Accept proposal	Reject proposal
0.2	0.05	$-\$10,000$	0
0.4	0.10	0	0
0.3	0.15	5,000	0
0.1	0.20	10,000	0
Expected money value		$\$500$	0

Table 17.12
Probability revisions, $n = 3$, $X = 0$

(1)	(2)	(3)	(4) = (2)(3)	(5) = $\dfrac{(4)}{\Sigma(4)}$
Proportion of cardholders (π)	$P(X = 0)$ if $n = 3$ and π as given	Prior probabilities	Joint probabilities	Revised probabilities
0.05	0.8574	0.20	0.17480	0.24549
0.10	0.7290	0.40	0.29160	0.41746
0.15	0.6141	0.30	0.18423	0.26375
0.20	0.5120	0.10	0.05120	0.07330
			0.69851	

EMV (Accept proposal) = 0.24549(−$10,000) + 0.41746(0) + 0.26375($5,000)
$\qquad\qquad\qquad\quad$ + 0.07330($10,000)
$\qquad\qquad\qquad$ = −$2,454.90 + $1,318.75 + $733
$\qquad\qquad\qquad$ = −$403.15

EMV (Reject proposal) = 0.24549(0) + 0.41746(0) + 0.26375(0) + 0.07330(0)
$\qquad\qquad\qquad$ = 0

Table 17.13
Probability revisions, $n = 3$, $X = 1$

(1)	(2)	(3)	(4) = (2)(3)	(5) = $\dfrac{(4)}{\Sigma(4)}$
Proportion of cardholders (π)	$P(X = 1)$ if $n = 3$ and π as given	Prior probabilities	Joint probabilities	Revised probabilities
0.05	0.1354	0.20	0.02708	0.10407
0.10	0.2430	0.40	0.09720	0.37354
0.15	0.3251	0.30	0.09753	0.37481
0.20	0.3840	0.10	0.03840	0.14757
			0.26021	

EMV (Accept proposal) = 0.10407(−$10,000) + 0.37354(0) + 0.37481($5,000)
$\qquad\qquad\qquad\quad$ + 0.14757($10,000)
$\qquad\qquad\qquad$ = −$1,040.70 + 0 + $1,874.05 + $1,475.70
$\qquad\qquad\qquad$ = $2,309.05

EMV (Reject proposal) = 0.10407(0) + 0.37354(0) + 0.37481(0) + 0.14757(0)
$\qquad\qquad\qquad$ = 0

Obtaining the solution to EVSI necessitates the determination of the revised probabilities for each possible set of responses that might be obtained from the sample. That is, of the three people interviewed, 0, 1, 2, or 3 people might respond yes when asked if they would take one of the vacation trips. Tables 17.12 through 17.15 show the probability revisions and calculation of the expected money values.

Table 17.14
Probability revisions, $n = 3$, $X = 2$

(1)	(2)	(3)	(4) = (2)(3)	(5) = $\dfrac{(4)}{\Sigma(4)}$
Proportion of cardholders (π)	$P(X = 2)$ if $n = 3$ and π as given	Prior probabilities	Joint probabilities	Revised probabilities
0.05	0.0071	0.20	0.00142	0.03637
0.10	0.0270	0.40	0.01080	0.27664
0.15	0.0574	0.30	0.01722	0.44109
0.20	0.0960	0.10	0.00960	0.24590
			0.03904	

EMV (Accept proposal) = 0.03637(−$10,000) + 0.27664(0) + 0.44109($5,000)
$\qquad\qquad\qquad\qquad$ + 0.24590($10,000)
$\qquad\qquad\qquad\quad$ = −$363.70 + 0 + $2,205.45 + $2,459.00
$\qquad\qquad\qquad\quad$ = $4,300.75

EMV (Reject proposal) = 0.03637(0) + 0.27664(0) + 0.44109(0) + 0.24590(0)
$\qquad\qquad\qquad\quad$ = 0

Table 17.15
Probability revisions, $n = 3$, $X = 3$

(1)	(2)	(3)	(4) = (2)(3)	(5) = $\dfrac{(4)}{\Sigma(4)}$
Proportion of cardholders (π)	$P(X = 3)$ if $n = 3$ and π as given	Prior probabilities	Joint probabilities	Revised probabilities
0.05	0.0001	0.20	0.00002	0.00893
0.10	0.0010	0.40	0.00040	0.17857
0.15	0.0034	0.30	0.00102	0.45536
0.20	0.0080	0.10	0.00080	0.35714
			0.00224	

EMV (Accept proposal) = 0.00893(−$10,000) + 0.17857(0) + 0.45536($5,000)
$\qquad\qquad\qquad\qquad$ + 0.35714($10,000)
$\qquad\qquad\qquad\quad$ = −$89.30 + 0 + $2,276.80 + $3,571.40
$\qquad\qquad\qquad\quad$ = $5,758.90

EMV (Reject proposal) = 0.00893(0) + 0.17857(0) + 0.45536(0) + 0.35714(0)
$\qquad\qquad\qquad\quad$ = 0

Now, using the expected money values and the sum of the joint probabilities from the preceding tables and the original maximum expected money value from table 17.11, we can calculate the expected value of sample information (EVSI):

$$\text{EVSI} = P(X = x)(\max \text{EMV}) - \max(\text{EMV under risk})$$
$$= [0.69851(0) + 0.26021(\$2309.05) + 0.03904(\$4300.75)$$
$$+ 0.00224(\$5758.90)] - \$500$$
$$= [0 + \$600.84 + 167.90 \quad \$12.90] - \$500$$
$$= \$781.64 - \$500$$
$$\text{EVSI} = \$281.64.$$

Thus, if taking the sample of size three costs less than $281.64, then it is advantageous to take the sample. Some other sample size, however, may be a more optimum sample size. In order to decide on the optimum size, it would be necessary to find EVSI for all samples that would cost less than the EVPI ($2,000), and select the sample with the maximum difference between EVSI and cost of sampling. ◁

17.1   Expected value of a discrete random variable:

$$E(X) = \Sigma[x \cdot P(X = x)]$$

17.2   Profit when $i \geq j$:

$$\text{payoff}_{ij} = (P - C) \cdot j$$

17.3   Profit when $i < j$:

$$\text{payoff}_{ij} = (P - C)i + (S - C)(j - i)$$

Review Questions

1. Explain the meaning of each equation at the end of the chapter.

2. Define
(a) certainty
(b) risk
(c) uncertainty
(d) payoff table
(e) expected value of perfect information

3. Discuss the relationship between expected profit under certainty, expected money value, and expected value of perfect information.

4. What is the expected value of sample information and how is it used?

Problems

1. A study by the manager of a large store determined that the checkout times had the probability distribution shown below. Use the probability distribution to determine the average checkout time for a customer.

Checkout time (minutes)	Probability
2	0.30
4	0.45
6	0.15
8	0.10

2. A retailer has maintained the following records on weekly sales for Product M. Based on these records, estimate the average weekly sales for this product.

Weekly sales	Frequency
10	60
11	100
12	150
13	120
14	70

3. A Christmas tree lot owner anticipates the demand distribution for a particular size tree. If a tree costs $4.00 and is sold for $7.00:

Number of trees	Probability of sale
100	0.30
110	0.40
120	0.20
130	0.10

(a) Construct the appropriate payoff table.
(b) Calculate the expected money value for each action.
(c) Determine the best stocking action.
(d) Calculate the expected profit under certainty.
(e) Calculate EVPI.

4. The Film Shop has determined the following probability distribution for a particular film:

Demand	Probability
20	0.20
25	0.30
30	0.40
35	0.10

If each roll costs $3.00 and sells for $4.00:
(a) Construct the appropriate payoff table.
(b) Determine the best stocking action.
(c) Determine the expected profit under certainty.
(d) Calculate EVPI.

5. The Book Store expects the following demand distribution to exist for a certain book. If the book costs $10.00 per copy and sells for $12.00 per copy:

Number of books	Probability of sale
10	0.30
20	0.50
30	0.20

(a) Construct the appropriate payoff table.
(b) Calculate the expected money value for each action and determine the best stocking action.
(c) Calculate the expected profit under certainty.
(d) Calculate EVPI.

6. Repeat problem 5 if the books can be salvaged for $6.00 per copy.

7. Use the results of problem 5(*a*) to construct a loss table by subtracting every element in each row from the largest element in each row. (Hint: 0's should appear on the diagonal.)

8. Use the results of problem 7 to:
(*a*) Calculate the expected loss for each action.
(*b*) Determine the optimum stocking action by locating the minimum expected loss.
(*c*) Compare the minimum expected loss with the EVPI of problem 5(*d*).

9. A store owner has determined the following probability distribution for calendars. The calendars sell for $8.00 per dozen and cost $5.00 per dozen.

Calendars (dozens)	Probability
15	0.25
20	0.45
25	0.30

(*a*) Construct the appropriate payoff table.
(*b*) Determine the best stocking action.
(*c*) Determine the expected value of perfect information.

10. Repeat problem 9 if the calendars can be salvaged for $2.00 per dozen.

11. Use the results of problem 10 to establish a loss table by subtracting every element in each row from the largest element in each row. (Hint: The diagonal elements will be 0.)

12. Use the results of problem 11 to
(*a*) calculate the expected loss for each action.
(*b*) determine the optimum stocking by locating the minimum expected loss.
(*c*) compare your answer in problem 12(*b*) with your answer in problem 10(*c*).

13. A florist has recorded the following information concerning the demand for her famous Easter bouquets.

Bouquets	Probability
200	0.10
210	0.40
220	0.30
230	0.20

If the bouquets sell for $10.00 and cost $5.00 to make,
(a) construct the payoff table for the possible action-event combinations.
(b) determine the best stocking action.
(c) determine the expected profit under certainty.
(d) determine the EVPI.
(e) determine the expected demand.

14. Repeat problem 13 if the salvage value of each bouquet is $3.00.

15. Assume that the florist of problem 13 has acquired additional information as shown below:

Probability of repeat customers	Bouquets	Probability
0.05	200	0.10
0.25	210	0.40
0.35	220	0.30
0.50	230	0.20

(a) Based on a sample size of 2 previous customers who responded that they intended to purchase an Easter bouquet again this year, revise the prior probabilities.
(b) Determine the best stocking action based on the revised probabilities.

16. Calculate the expected value of sample information for the sample size of 2 in problem 15.

17. Manufacturing, Inc., receives a particular raw material in lot sizes of 1,000 units. The quality of previous shipments has had the following distribution.

Fraction defective	Probability
0.05	0.65
0.10	0.30
0.15	0.05

The company must decide between two procedures for accepting the materials. The first procedure involves 100 percent inspection at a fixed cost of $400.00. The second procedure is to place received lots directly into inventory and repair the defective parts as they are used at a cost of $5.00 per unit.
(a) Construct a table showing the cost of each action-event combination. (Hint: Do not use payoff table equations.)

(b) Determine which procedure is better by selecting the procedure with the lowest expected cost under risk.
(c) Determine the expected cost under certainty.
(d) Determine the expected value of perfect information.

18. Based on a sample of two items, both of which are perfect, revise the prior probability distribution of problem 17. Is it better to inspect 100 percent or use the material without inspection?

19. Determine the expected value of sample information for the situation in problems 17 and 18. If sampling costs are 20 cents per unit, is it profitable to sample?

20. Suppose EVSI = $75.00 and sampling costs are 30¢ per unit; what is the expected net gain from sampling (ENGS)? Assume sample size of 150.

21. A manufacturer of model airplanes is considering a new model. Market research has indicated the following possible annual demands. If the model cost $3.00 per unit and sells for $6.00 per unit:

Model airplanes (1,000's)	Probability
0	0.15
5	0.40
10	0.35
15	0.10

(a) Construct the appropriate payoff table.
(b) Determine the best production quantity.
(c) Determine the EVPI.
(d) Determine the expected demand for the new model.

22. An outboard motor distributor has established the following probability distribution of sales for the upcoming year:

Outboard motors	Probability
20	0.10
21	0.25
22	0.40
23	0.25

(a) What is the expected demand for outboard motors?
(b) If the profit per motor is $150.00, what is the expected profit from the sale of outboard motors?

23. Big Breakfast, Inc., is ready to introduce a new cereal, Toothbreakers. Market research has established the following demand information:

Demand (number of boxes)	Probability
75,000	0.19
100,000	0.28
125,000	0.25
150,000	0.17
175,000	0.11

If the cost of producing each box is 45¢, and each box sells for 75¢:
(a) Construct the appropriate payoff table.
(b) Determine the best quantity to produce.
(c) Calculate EVPI.

Case Study

There is a large producer of smoked chickens located in a small town in the Ozarks. This producer specializes in selling premium quality smoked chickens to be used as Christmas gifts for the customers of other business firms. The producer does, however, maintain a local outlet store where its surplus inventory of smoked chickens is sold at a discount. It is necessary to sell the product at a discount locally because the store is only open when the producer has a surplus and because of previous errors in estimating demand, the local consumer is somewhat tired of the product even though it is premium quality.

The producer must determine how many chickens will be smoked several months in advance of the Christmas season. For the forthcoming holiday period, the producer's accountants have estimated that the fixed cost of producing and marketing the product will be $50,000 and the variable cost per smoked chicken produced will be $2.00. The price of the chickens, when sold as Christmas gifts, is $3.50 each. If they are sold in the local outlet store, the price will be $2.50, but it is estimated that the maximum quantity that can be sold locally is 40,000 chickens. Surplus chickens in excess of 40,000 will be distributed free to local schools and charities, and the variable production costs will be written off against taxes. (The firm is in the fifty percent bracket.)

The technical problems associated with growing chickens and smoking them require that production be limited to increments of 20,000. The firm feels it must stay active in the market so it will produce at least 20,000 smoked chickens, but its capacity limit is 100,000. When the producer does not have enough inventory to meet demand, they spend an average of 50 cents per chicken demanded but not supplied to maintain customer goodwill.

Demand for product during twenty-year life of the producer

Quantity demanded	Years
100,000	3
80,000	4
60,000	7
40,000	5
20,000	1

Proportion of firms in Fortune 500 with stock prices forecast to be at record level high and quantity of chickens demanded

Quantity of chicken demand	Proportion
20,000	0.05
40,000	0.20
60,000	0.50
80,000	0.75
100,000	0.90

Just before making their production decision this year, the firm purchased forecasts of stock prices for five companies and found that three were forecast to have record level prices. With this information, what production level do you recommend? What is your expected profit? What is the maximum amount you would pay to acquire additional information at this point?

Appendixes

Appendixes

**Appendix A
Table of Common
Logarithms**

Common logs are exponents of the base 10. In other words, the common log of a number is that value such that when 10 is raised to that power, it is equal to the number. As a result the common log of 100 is equal to 2.00000, since $10^{2.00000}$ is equal to 100. Similarly,

log 1,000 = 3.00000, because $10^{3.00000}$ = 1,000,
 log 10 = 1.00000, because $10^{1.00000}$ = 10,
 log 0.1 = −1.00000, since 10^{-1} = 0.1 (this would
     normally be written as 9.00000 − 10.00000).

The part of the common log to the left of the decimal point, 2, 3, 1, and −1 in these examples, is called the characteristic of the log.

The part of the log to the right of the decimal point is the mantissa. It can be found from the tables in this appendix. For example, suppose we needed the common log of 4.325. Since there is one digit to the left of the decimal point, the characteristic of this number is equal to 0. Moving to the table we can find the mantissa of the log 4.325. We first move down the column headed by N until we find the number 432. Then the column is found having the value of 5 at its top, the 4th digit of the number. Reading from the column titled 5, we see the mantissa of the number is equal to 0.63599. Therefore

$$\log 4.325 = 0.63599.$$
Similarly, log 86.05 = 1.93475.

Due to the mathematical properties of logs, their use can be most helpful. Since $X^a \cdot X^b = X^{a+b}$, to find the product of two numbers, we merely need to add their common logs and find the number having the resulting log. This process is called finding the antilog, that number which has a log of a certain value.

For example, to find the product of 4.325 × 86.05, we need to add their common logs, 0.63599 + 1.93475 = 2.57074 and then find the antilog. Looking up 0.57074 in the body of the log table, we see the first four digits of the number would be 3722. Since a characteristic of 2 means there are 3 digits to the left of the decimal point, we see that the product of $4.325 \times 86.05 = 372.2$.

The mathematical property $X^a/X^b = X^{a-b}$ is also useful in working with logs. For example, suppose we needed to find the quotient of 86.05/4.325. Using logs, we would subtract the log of 4.325 from the log of 86.05 and find the antilog of the resulting value:

$$1.93475 - 0.63599 = 1.29876,$$
$$\text{antilog } 1.29876 = 19.90.$$

Thus the quotient of 86.05/4.325 is equal to 19.90.

Logs can also be used to raise numbers to powers (square, cube, and so on) and to find the roots of numbers:

$$\text{power} \equiv X^a = \text{antilog } (a \cdot \log X),$$
$$\text{root} \equiv X^{1/a} = \text{antilog } \frac{\log X}{a}.$$

As an illustration, suppose you want to raise 5 to the 6th power (5^6) and you also want to find the 3rd root of 9 or $9^{1/3}$:

$$5^6 = \text{antilog } (6)(0.69897)$$
$$= \text{antilog } 4.19382$$
$$5^6 = 15,625,$$
$$9^{1/3} = \text{antilog } \frac{0.95424}{3}$$
$$= \text{antilog } 0.31808$$
$$9^{1/3} = 2.0801.$$

Table of common logarithms

N	0	1	2	3	4	5	6	7	8	9
100	00000	00043	00087	00130	00173	00217	00260	00303	00346	00389
101	00432	00475	00518	00561	00604	00647	00689	00732	00775	00817
102	00860	00903	00945	00988	01030	01072	01115	01157	01199	01242
103	01284	01326	01368	01410	01452	01494	01536	01578	01620	01662
104	01703	01745	01787	01828	01870	01912	01953	01995	02036	02078
105	02119	02160	02202	02243	02284	02325	02366	02407	02449	02490
106	02531	02572	02612	02653	02694	02735	02776	02816	02857	02898
107	02938	02979	03019	03060	03100	03141	03181	03222	03262	03302
108	03342	03383	03423	03463	03503	03543	03583	03623	03663	03703
109	03743	03782	03822	03862	03902	03941	03981	04021	04060	04100
110	04139	04179	04218	04258	04297	04336	04376	04415	04454	04493
111	04532	04571	04610	04650	04689	04727	04766	04805	04844	04883
112	04922	04961	04999	05038	05077	05115	05154	05192	05231	05269
113	05308	05346	05385	05423	05461	05500	05538	05576	05614	05652
114	05690	05729	05767	05805	05843	05881	05918	05956	05994	06032
115	06070	06108	06145	06183	06221	06258	06296	06333	06371	06408
116	06446	06483	06521	06558	06595	06633	06670	06707	06744	06781
117	06819	06856	06893	06930	06967	07004	07041	07078	07115	07151
118	07188	07225	07262	07298	07335	07372	07408	07445	07482	07518
119	07555	07591	07628	07664	07700	07737	07773	07809	07846	07882
120	07918	07954	07990	08027	08063	08099	08135	08171	08207	08243
121	08279	08314	08350	08386	08422	08458	08493	08529	08565	08600
122	08636	08672	08707	08743	08778	08814	08849	08884	08920	08955
123	08991	09026	09061	09096	09132	09167	09202	09237	09272	09307
124	09342	09377	09412	09447	09482	09517	09552	09587	09621	09656
125	09691	09726	09760	09795	09830	09864	09899	09934	09968	10003
126	10037	10072	10106	10140	10175	10209	10243	10278	10312	10346
127	10380	10415	10449	10483	10517	10551	10585	10619	10653	10687
128	10721	10755	10789	10823	10857	10890	10924	10958	10992	11025
129	11059	11093	11126	11160	11193	11227	11261	11294	11327	11361
130	11394	11428	11461	11494	11528	11561	11594	11628	11661	11694
131	11727	11760	11793	11826	11860	11893	11926	11959	11992	12024
132	12057	12090	12123	12156	12189	12222	12254	12287	12320	12353
133	12385	12418	12450	12483	12516	12548	12581	12613	12646	12678
134	12710	12743	12775	12808	12840	12872	12905	12937	12969	13001
135	13033	13066	13098	13130	13162	13194	13226	13258	13290	13322
136	13354	13386	13418	13450	13481	13513	13545	13577	13609	13640
137	13672	13704	13735	13767	13799	13830	13862	13893	13925	13956
138	13988	14019	14051	14082	14114	14145	14176	14208	14239	14270
139	14301	14333	14364	14395	14426	14457	14489	14520	14551	14582
140	14613	14644	14675	14706	14737	14768	14799	14829	14860	14891
141	14922	14953	14983	15014	15045	15076	15106	15137	15168	15198
142	15229	15259	15290	15320	15351	15381	15412	15442	15473	15503
143	15534	15564	15594	15625	15655	15685	15715	15746	15776	15806
144	15836	15866	15897	15927	15957	15987	16017	16047	16077	16107
145	16137	16167	16197	16227	16256	16286	16316	16346	16376	16406
146	16435	16465	16495	16524	16554	16584	16613	16643	16673	16702
147	16732	16761	16791	16820	16850	16879	16909	16938	16967	16997
148	17026	17056	17085	17114	17143	17173	17202	17231	17260	17289
149	17319	17348	17377	17406	17435	17464	17493	17522	17551	17580
150	17609	17638	17667	17696	17725	17754	17783	17811	17840	17869
N	0	1	2	3	4	5	6	7	8	9

Table of common logarithms

N	0	1	2	3	4	5	6	7	8	9
150	17609	17638	17667	17696	17725	17754	17783	17811	17840	17869
151	17898	17926	17955	17984	18013	18041	18070	18099	18127	18156
152	18184	18213	18241	18270	18299	18327	18355	18384	18412	18441
153	18469	18498	18526	18554	18583	18611	18639	18667	18696	18724
154	18752	18780	18808	18837	18865	18893	18921	18949	18977	19005
155	19033	19061	19089	19117	19145	19173	19201	19229	19257	19285
156	19312	19340	19368	19396	19424	19451	19479	19507	19535	19562
157	19590	19618	19645	19673	19700	19728	19756	19783	19811	19838
158	19866	19893	19921	19948	19976	20003	20030	20058	20085	20112
159	20140	20167	20194	20222	20249	20276	20303	20330	20358	20385
160	20412	20439	20466	20493	20520	20548	20575	20602	20629	20656
161	20683	20710	20737	20763	20790	20817	20844	20871	20898	20925
162	20952	20978	21005	21032	21059	21085	21112	21139	21165	21192
163	21219	21245	21272	21299	21325	21352	21378	21405	21431	21458
164	21484	21511	21537	21564	21590	21617	21643	21669	21696	21722
165	21748	21775	21801	21827	21854	21880	21906	21932	21958	21985
166	22011	22037	22063	22089	22115	22141	22168	22194	22220	22246
167	22272	22298	22324	22350	22376	22401	22427	22453	22479	22505
168	22531	22557	22583	22608	22634	22660	22686	22712	22737	22763
169	22789	22814	22840	22866	22891	22917	22943	22968	22994	23019
170	23045	23070	23096	23121	23147	23172	23198	23223	23249	23274
171	23300	23325	23350	23376	23401	23426	23452	23477	23502	23528
172	23553	23578	23603	23629	23654	23679	23704	23729	23754	23780
173	23805	23830	23855	23880	23905	23930	23955	23980	24005	24030
174	24055	24080	24105	24130	24155	24180	24204	24229	24254	24279
175	24304	24329	24353	24378	24403	24428	24452	24477	24502	24527
176	24551	24576	24601	24625	24650	24674	24699	24724	24748	24773
177	24797	24822	24846	24871	24895	24920	24944	24969	24993	25018
178	25042	25066	25091	25115	25139	25164	25188	25212	25237	25261
179	25285	25310	25334	25358	25382	25406	25431	25455	25479	25503
180	25527	25551	25575	25600	25624	25648	25672	25696	25720	25744
181	25768	25792	25816	25840	25864	25888	25912	25935	25959	25983
182	26007	26031	26055	26079	26102	26126	26150	26174	26198	26221
183	26245	26269	26293	26316	26340	26364	26387	26411	26435	26458
184	26482	26505	26529	26553	26576	26600	26623	26647	26670	26694
185	26717	26741	26764	26788	26811	26834	26858	26881	26905	26928
186	26951	26975	26998	27021	27045	27068	27091	27114	27138	27161
187	27184	27207	27231	27254	27277	27300	27323	27346	27370	27393
188	27416	27439	27462	27485	27508	27531	27554	27577	27600	27623
189	27646	27669	27692	27715	27738	27761	27784	27807	27830	27853
190	27875	27898	27921	27944	27967	27990	28012	28035	28058	28081
191	28103	28126	28149	28172	28194	28217	28240	28262	28285	28308
192	28330	28353	28375	28398	28421	28443	28466	28488	28511	28533
193	28556	28578	28601	28623	28646	28668	28691	28713	28735	28758
194	28780	28803	28825	28847	28870	28892	28914	28937	28959	28981
195	29003	29026	29048	29070	29092	29115	29137	29159	29181	29203
196	29226	29248	29270	29292	29314	29336	29358	29380	29403	29425
197	29447	29469	29491	29513	29535	29557	29579	29601	29623	29645
198	29667	29688	29710	29732	29754	29776	29798	29820	29842	29863
199	29885	29907	29929	29951	29973	29994	30016	30038	30060	30081
200	30103	30125	30146	30168	30190	30211	30233	30255	30276	30298
N	0	1	2	3	4	5	6	7	8	9

Table of common logarithms

N	0	1	2	3	4	5	6	7	8	9
200	30103	30125	30146	30168	30190	30211	30233	30255	30276	30298
201	30320	30341	30363	30384	30406	30428	30449	30471	30492	30514
202	30535	30557	30578	30600	30621	30643	30664	30685	30707	30728
203	30750	30771	30792	30814	30835	30856	30878	30899	30920	30942
204	30963	30984	31006	31027	31048	31069	31091	31112	31133	31154
205	31175	31197	31218	31239	31260	31281	31302	31323	31345	31366
206	31387	31408	31429	31450	31471	31492	31513	31534	31555	31576
207	31597	31618	31639	31660	31681	31702	31723	31744	31765	31785
208	31806	31827	31848	31869	31890	31911	31931	31952	31973	31994
209	32015	32035	32056	32077	32098	32118	32139	32160	32181	32201
210	32222	32243	32263	32284	32305	32325	32346	32366	32387	32408
211	32428	32449	32469	32490	32511	32531	32552	32572	32593	32613
212	32634	32654	32675	32695	32715	32736	32756	32777	32797	32818
213	32838	32858	32879	32899	32919	32940	32960	32980	33001	33021
214	33041	33062	33082	33102	33122	33143	33163	33183	33203	33224
215	33244	33264	33284	33304	33325	33345	33365	33385	33405	33425
216	33445	33465	33486	33506	33526	33546	33566	33586	33606	33626
217	33646	33666	33686	33706	33726	33746	33766	33786	33806	33826
218	33846	33866	33885	33905	33925	33945	33965	33985	34005	34025
219	34044	34064	34084	34104	34124	34143	34163	34183	34203	34223
220	34242	34262	34282	34301	34321	34341	34361	34380	34400	34420
221	34439	34459	34479	34498	34518	34537	34557	34577	34596	34616
222	34635	34655	34674	34694	34713	34733	34753	34772	34792	34811
223	34830	34850	34869	34889	34908	34928	34947	34967	34986	35005
224	35025	35044	35064	35083	35102	35122	35141	35160	35180	35199
225	35218	35238	35257	35276	35295	35315	35334	35353	35372	35392
226	35411	35430	35449	35468	35488	35507	35526	35545	35564	35583
227	35603	35622	35641	35660	35679	35698	35717	35736	35755	35774
228	35793	35813	35832	35851	35870	35889	35908	35927	35946	35965
229	35984	36003	36021	36040	36059	36078	36097	36116	36135	36154
230	36173	36192	36211	36229	36248	36267	36286	36305	36324	36342
231	36361	36380	36399	36418	36436	36455	36474	36493	36511	36530
232	36549	36568	36586	36605	36624	36642	36661	36680	36698	36717
233	36736	36754	36773	36791	36810	36829	36847	36866	36884	36903
234	36922	36940	36959	36977	36996	37014	37033	37051	37070	37088
235	37107	37125	37144	37162	37181	37199	37218	37236	37254	37273
236	37291	37310	37328	37346	37365	37383	37401	37420	37438	37457
237	37475	37493	37511	37530	37548	37566	37585	37603	37621	37639
238	37658	37676	37694	37712	37731	37749	37767	37785	37803	37822
239	37840	37858	37876	37894	37912	37931	37949	37967	37985	38003
240	38021	38039	38057	38075	38093	38112	38130	38148	38166	38184
241	38202	38220	38238	38256	38274	38292	38310	38328	38346	38364
242	38382	38399	38417	38435	38453	38471	38489	38507	38525	38543
243	38561	38579	38596	38614	38632	38650	38668	38686	38703	38721
244	38739	38757	38775	38792	38810	38828	38846	38863	38881	38899
245	38917	38934	38952	38970	38987	39005	39023	39041	39058	39076
246	39094	39111	39129	39146	39164	39182	39199	39217	39235	39252
247	39270	39287	39305	39322	39340	39358	39375	39393	39410	39428
248	39445	39463	39480	39498	39515	39533	39550	39568	39585	39602
249	39620	39637	39655	39672	39690	39707	39724	39742	39759	39777
250	39794	39811	39829	39846	39863	39881	39898	39915	39933	39950
N	0	1	2	3	4	5	6	7	8	9

Table of common logarithms

N	0	1	2	3	4	5	6	7	8	9
250	39794	39811	39829	39846	39863	39881	39898	39915	39933	39950
251	39967	39985	40002	40019	40037	40054	40071	40088	40106	40123
252	40140	40157	40175	40192	40209	40226	40243	40261	40278	40295
253	40312	40329	40346	40364	40381	40398	40415	40432	40449	40466
254	40483	40500	40518	40535	40552	40569	40586	40603	40620	40637
255	40654	40671	40688	40705	40722	40739	40756	40773	40790	40807
256	40824	40841	40858	40875	40892	40909	40926	40943	40960	40976
257	40993	41010	41027	41044	41061	41078	41095	41111	41128	41145
258	41162	41179	41196	41212	41229	41246	41263	41280	41296	41313
259	41330	41347	41364	41380	41397	41414	41430	41447	41464	41481
260	41497	41514	41531	41547	41564	41581	41597	41614	41631	41647
261	41664	41681	41697	41714	41731	41747	41764	41780	41797	41814
262	41830	41847	41863	41880	41896	41913	41929	41946	41963	41979
263	41996	42012	42029	42045	42062	42078	42095	42111	42127	42144
264	42160	42177	42193	42210	42226	42243	42259	42275	42292	42308
265	42325	42341	42357	42374	42390	42406	42423	42439	42456	42472
266	42488	42504	42521	42537	42553	42570	42586	42602	42619	42635
267	42651	42667	42684	42700	42716	42732	42749	42765	42781	42797
268	42813	42830	42846	42862	42878	42894	42911	42927	42943	42959
269	42975	42991	43008	43024	43040	43056	43072	43088	43104	43120
270	43136	43152	43169	43185	43201	43217	43233	43249	43265	43281
271	43297	43313	43329	43345	43361	43377	43393	43409	43425	43441
272	43457	43473	43489	43505	43521	43537	43553	43569	43584	43600
273	43616	43632	43648	43664	43680	43696	43712	43727	43743	43759
274	43775	43791	43807	43823	43838	43854	43870	43886	43902	43917
275	43933	43949	43965	43981	43996	44012	44028	44044	44059	44075
276	44091	44107	44122	44138	44154	44170	44185	44201	44217	44232
277	44248	44264	44279	44295	44311	44326	44342	44358	44373	44389
278	44404	44420	44436	44451	44467	44483	44498	44514	44529	44545
279	44560	44576	44592	44607	44623	44638	44654	44669	44685	44700
280	44716	44731	44747	44762	44778	44793	44809	44824	44840	44855
281	44871	44886	44902	44917	44932	44948	44963	44979	44994	45010
282	45025	45040	45056	45071	45086	45102	45117	45133	45148	45163
283	45179	45194	45209	45225	45240	45255	45271	45286	45301	45317
284	45332	45347	45362	45378	45393	45408	45424	45439	45454	45469
285	45484	45500	45515	45530	45545	45561	45576	45591	45606	45621
286	45637	45652	45667	45682	45697	45712	45728	45743	45758	45773
287	45788	45803	45818	45834	45849	45864	45879	45894	45909	45924
288	45939	45954	45969	45984	46000	46015	46030	46045	46060	46075
289	46090	46105	46120	46135	46150	46165	46180	46195	46210	46225
290	46240	46255	46270	46285	46300	46315	46330	46345	46359	46374
291	46389	46404	46419	46434	46449	46464	46479	46494	46509	46523
292	46538	46553	46568	46583	46598	46613	46627	46642	46657	46672
293	46687	46702	46716	46731	46746	46761	46776	46790	46805	46820
294	46835	46850	46864	46879	46894	46909	46923	46938	46953	46967
295	46982	46997	47012	47026	47041	47056	47070	47085	47100	47115
296	47129	47144	47159	47173	47188	47202	47217	47232	47246	47261
297	47276	47290	47305	47319	47334	47349	47363	47378	47392	47407
298	47422	47436	47451	47465	47480	47494	47509	47524	47538	47553
299	47567	47582	47596	47611	47625	47640	47654	47669	47683	47698
300	47712	47727	47741	47756	47770	47784	47799	47813	47828	47842
N	0	1	2	3	4	5	6	7	8	9

Table of common logarithms

N	0	1	2	3	4	5	6	7	8	9
300	47712	47727	47741	47756	47770	47784	47799	47813	47828	47842
301	47857	47871	47886	47900	47914	47929	47943	47958	47972	47986
302	48001	48015	48029	48044	48058	48073	48087	48101	48116	48130
303	48144	48159	48173	48187	48202	48216	48230	48244	48259	48273
304	48287	48302	48316	48330	48344	48359	48373	48387	48402	48416
305	48430	48444	48458	48473	48487	48501	48515	48530	48544	48558
306	48572	48586	48601	48615	48629	48643	48657	48671	48686	48700
307	48714	48728	48742	48756	48770	48785	48799	48813	48827	48841
308	48855	48869	48883	48897	48911	48926	48940	48954	48968	48982
309	48996	49010	49024	49038	49052	49066	49080	49094	49108	49122
310	49136	49150	49164	49178	49192	49206	49220	49234	49248	49262
311	49276	49290	49304	49318	49332	49346	49360	49374	49388	49402
312	49415	49429	49443	49457	49471	49485	49499	49513	49527	49541
313	49554	49568	49582	49596	49610	49624	49638	49651	49665	49679
314	49693	49707	49721	49734	49748	49762	49776	49790	49803	49817
315	49831	49845	49859	49872	49886	49900	49914	49927	49941	49955
316	49969	49982	49996	50010	50024	50037	50051	50065	50079	50092
317	50106	50120	50133	50147	50161	50174	50188	50202	50215	50229
318	50243	50256	50270	50284	50297	50311	50325	50338	50352	50365
319	50379	50393	50406	50420	50433	50447	50461	50474	50488	50501
320	50515	50529	50542	50556	50569	50583	50596	50610	50623	50637
321	50651	50664	50678	50691	50705	50718	50732	50745	50759	50772
322	50786	50799	50813	50826	50840	50853	50866	50880	50893	50907
323	50920	50934	50947	50961	50974	50987	51001	51014	51028	51041
324	51055	51068	51081	51095	51108	51121	51135	51148	51162	51175
325	51188	51202	51215	51228	51242	51255	51268	51282	51295	51308
326	51322	51335	51348	51362	51375	51388	51402	51415	51428	51442
327	51455	51468	51481	51495	51508	51521	51534	51548	51561	51574
328	51587	51601	51614	51627	51640	51654	51667	51680	51693	51706
329	51720	51733	51746	51759	51772	51786	51799	51812	51825	51838
330	51851	51865	51878	51891	51904	51917	51930	51943	51957	51970
331	51983	51996	52009	52022	52035	52048	52061	52075	52088	52101
332	52114	52127	52140	52153	52166	52179	52192	52205	52218	52231
333	52244	52257	52271	52284	52297	52310	52323	52336	52349	52362
334	52375	52388	52401	52414	52427	52440	52453	52466	52479	52492
335	52504	52517	52530	52543	52556	52569	52582	52595	52608	52621
336	52634	52647	52660	52673	52686	52699	52711	52724	52737	52750
337	52763	52776	52789	52802	52815	52827	52840	52853	52866	52879
338	52892	52905	52917	52930	52943	52956	52969	52982	52994	53007
339	53020	53033	53046	53058	53071	53084	53097	53110	53122	53135
340	53148	53161	53173	53186	53199	53212	53224	53237	53250	53263
341	53275	53288	53301	53314	53326	53339	53352	53365	53377	53390
342	53403	53415	53428	53441	53453	53466	53479	53491	53504	53517
343	53529	53542	53555	53567	53580	53593	53605	53618	53631	53643
344	53656	53668	53681	53694	53706	53719	53732	53744	53757	53769
345	53782	53795	53807	53820	53832	53845	53857	53870	53883	53895
346	53908	53920	53933	53945	53958	53970	53983	53995	54008	54020
347	54033	54045	54058	54070	54083	54095	54108	54120	54133	54145
348	54158	54170	54183	54195	54208	54220	54233	54245	54258	54270
349	54283	54295	54307	54320	54332	54345	54357	54370	54382	54394
350	54407	54419	54432	54444	54456	54469	54481	54494	54506	54518
N	0	1	2	3	4	5	6	7	8	9

Table of common logarithms

N	0	1	2	3	4	5	6	7	8	9
350	54407	54419	54432	54444	54456	54469	54481	54494	54506	54518
351	54531	54543	54555	54568	54580	54593	54605	54617	54630	54642
352	54654	54667	54679	54691	54704	54716	54728	54741	54753	54765
353	54777	54790	54802	54814	54827	54839	54851	54864	54876	54888
354	54900	54913	54925	54937	54949	54962	54974	54986	54998	55011
355	55023	55035	55047	55060	55072	55084	55096	55108	55121	55133
356	55145	55157	55169	55182	55194	55206	55218	55230	55242	55255
357	55267	55279	55291	55303	55315	55328	55340	55352	55364	55376
358	55388	55400	55413	55425	55437	55449	55461	55473	55485	55497
359	55509	55522	55534	55546	55558	55570	55582	55594	55606	55618
360	55630	55642	55654	55666	55678	55691	55703	55715	55727	55739
361	55751	55763	55775	55787	55799	55811	55823	55835	55847	55859
362	55871	55883	55895	55907	55919	55931	55943	55955	55967	55979
363	55991	56003	56015	56027	56038	56050	56062	56074	56086	56098
364	56110	56122	56134	56146	56158	56170	56182	56194	56205	56217
365	56229	56241	56253	56265	56277	56289	56301	56313	56324	56336
366	56348	56360	56372	56384	56396	56407	56419	56431	56443	56455
367	56467	56478	56490	56502	56514	56526	56538	56549	56561	56573
368	56585	56597	56608	56620	56632	56644	56656	56667	56679	56691
369	56703	56714	56726	56738	56750	56761	56773	56785	56797	56808
370	56820	56832	56844	56855	56867	56879	56891	56902	56914	56926
371	56937	56949	56961	56973	56984	56996	57008	57019	57031	57043
372	57054	57066	57078	57089	57101	57113	57124	57136	57148	57159
373	57171	57183	57194	57206	57217	57229	57241	57252	57264	57276
374	57287	57299	57310	57322	57334	57345	57357	57368	57380	57392
375	57403	57415	57426	57438	57449	57461	57473	57484	57496	57507
376	57519	57530	57542	57553	57565	57577	57588	57600	57611	57623
377	57634	57646	57657	57669	57680	57692	57703	57715	57726	57738
378	57749	57761	57772	57784	57795	57807	57818	57830	57841	57852
379	57864	57875	57887	57898	57910	57921	57933	57944	57956	57967
380	57978	57990	58001	58013	58024	58035	58047	58058	58070	58081
381	58093	58104	58115	58127	58138	58149	58161	58172	58184	58195
382	58206	58218	58229	58240	58252	58263	58275	58286	58297	58309
383	58320	58331	58343	58354	58365	58377	58388	58399	58411	58422
384	58433	58444	58456	58467	58478	58490	58501	58512	58524	58535
385	58546	58557	58569	58580	58591	58602	58614	58625	58636	58647
386	58659	58670	58681	58692	58704	58715	58726	58737	58749	58760
387	58771	58782	58794	58805	58816	58827	58838	58850	58861	58872
388	58883	58894	58906	58917	58928	58939	58950	58961	58973	58984
389	58995	59006	59017	59028	59040	59051	59062	59073	59084	59095
390	59106	59118	59129	59140	59151	59162	59173	59184	59195	59207
391	59218	59229	59240	59251	59262	59273	59284	59295	59306	59318
392	59329	59340	59351	59362	59373	59384	59395	59406	59417	59428
393	59439	59450	59461	59472	59483	59494	59506	59517	59528	59539
394	59550	59561	59572	59583	59594	59605	59616	59627	59638	59649
395	59660	59671	59682	59693	59704	59715	59726	59737	59748	59759
396	59770	59780	59791	59802	59813	59824	59835	59846	59857	59868
397	59879	59890	59901	59912	59923	59934	59945	59956	59966	59977
398	59988	59999	60010	60021	60032	60043	60054	60065	60076	60086
399	60097	60108	60119	60130	60141	60152	60163	60173	60184	60195
400	60206	60217	60228	60239	60249	60260	60271	60282	60293	60304
N	0	1	2	3	4	5	6	7	8	9

Table of common logarithms

N	0	1	2	3	4	5	6	7	8	9
400	60206	60217	60228	60239	60249	60260	60271	60282	60293	60304
401	60314	60325	60336	60347	60358	60369	60379	60390	60401	60412
402	60423	60433	60444	60455	60466	60477	60487	60498	60509	60520
403	60531	60541	60552	60563	60574	60584	60595	60606	60617	60627
404	60638	60649	60660	60670	60681	60692	60703	60713	60724	60735
405	60746	60756	60767	60778	60788	60799	60810	60821	60831	60842
406	60853	60863	60874	60885	60895	60906	60917	60927	60938	60949
407	60959	60970	60981	60991	61002	61013	61023	61034	61045	61055
408	61066	61077	61087	61098	61109	61119	61130	61140	61151	61162
409	61172	61183	61194	61204	61215	61225	61236	61247	61257	61268
410	61278	61289	61300	61310	61321	61331	61342	61352	61363	61374
411	61384	61395	61405	61416	61426	61437	61448	61458	61469	61479
412	61490	61500	61511	61521	61532	61542	61553	61563	61574	61584
413	61595	61606	61616	61627	61637	61648	61658	61669	61679	61690
414	61700	61711	61721	61732	61742	61752	61763	61773	61784	61794
415	61805	61815	61826	61836	61847	61857	61868	61878	61888	61899
416	61909	61920	61930	61941	61951	61962	61972	61982	61993	62003
417	62014	62024	62034	62045	62055	62066	62076	62086	62097	62107
418	62118	62128	62138	62149	62159	62170	62180	62190	62201	62211
419	62221	62232	62242	62252	62263	62273	62284	62294	62304	62315
420	62325	62335	62346	62356	62366	62377	62387	62397	62408	62418
421	62428	62439	62449	62459	62469	62480	62490	62500	62511	62521
422	62531	62542	62552	62562	62572	62583	62593	62603	62614	62624
423	62634	62644	62655	62665	62675	62685	62696	62706	62716	62726
424	62737	62747	62757	62767	62778	62788	62798	62808	62818	62829
425	62839	62849	62859	62870	62880	62890	62900	62910	62921	62931
426	62941	62951	62961	62972	62982	62992	63002	63012	63022	63033
427	63043	63053	63063	63073	63083	63094	63104	63114	63124	63134
428	63144	63155	63165	63175	63185	63195	63205	63215	63225	63236
429	63246	63256	63266	63276	63286	63296	63306	63317	63327	63337
430	63347	63357	63367	63377	63387	63397	63407	63417	63428	63438
431	63448	63458	63468	63478	63488	63498	63508	63518	63528	63538
432	63548	63558	63568	63579	63589	63599	63609	63619	63629	63639
433	63649	63659	63669	63679	63689	63699	63709	63719	63729	63739
434	63749	63759	63769	63779	63789	63799	63809	63819	63829	63839
435	63849	63859	63869	63879	63889	63899	63909	63919	63929	63939
436	63949	63959	63969	63979	63988	63998	64008	64018	64028	64038
437	64048	64058	64068	64078	64088	64098	64108	64118	64128	64138
438	64147	64157	64167	64177	64187	64197	64207	64217	64227	64237
439	64246	64256	64266	64276	64286	64296	64306	64316	64326	64335
440	64345	64355	64365	64375	64385	64395	64404	64414	64424	64434
441	64444	64454	64464	64473	64483	64493	64503	64513	64523	64532
442	64542	64552	64562	64572	64582	64591	64601	64611	64621	64631
443	64640	64650	64660	64670	64680	64689	64699	64709	64719	64729
444	64738	64748	64758	64768	64777	64787	64797	64807	64816	64826
445	64836	64846	64856	64865	64875	64885	64895	64904	64914	64924
446	64933	64943	64953	64963	64972	64982	64992	65002	65011	65021
447	65031	65040	65050	65060	65070	65079	65089	65099	65108	65118
448	65128	65138	65147	65157	65167	65176	65186	65196	65205	65215
449	65225	65234	65244	65254	65263	65273	65283	65292	65302	65312
450	65321	65331	65341	65350	65360	65369	65379	65389	65398	65408
N	0	1	2	3	4	5	6	7	8	9

Table of common logarithms

N	0	1	2	3	4	5	6	7	8	9
450	65321	65331	65341	65350	65360	65369	65379	65389	65398	65408
451	65418	65427	65437	65447	65456	65466	65475	65485	65495	65504
452	65514	65523	65533	65543	65552	65562	65571	65581	65591	65600
453	65610	65619	65629	65639	65648	65658	65667	65677	65686	65696
454	65706	65715	65725	65734	65744	65753	65763	65773	65782	65792
455	65801	65811	65820	65830	65839	65849	65858	65868	65877	65887
456	65896	65906	65916	65925	65935	65944	65954	65963	65973	65982
457	65992	66001	66011	66020	66030	66039	66049	66058	66068	66077
458	66087	66096	66106	66115	66124	66134	66143	66153	66162	66172
459	66181	66191	66200	66210	66219	66229	66238	66247	66257	66266
460	66276	66285	66295	66304	66314	66323	66332	66342	66351	66361
461	66370	66380	66389	66398	66408	66417	66427	66436	66445	66455
462	66464	66474	66483	66492	66502	66511	66521	66530	66539	66549
463	66558	66567	66577	66586	66596	66605	66614	66624	66633	66642
464	66652	66661	66671	66680	66689	66699	66708	66717	66727	66736
465	66745	66755	66764	66773	66783	66792	66801	66811	66820	66829
466	66839	66848	66857	66867	66876	66885	66894	66904	66913	66922
467	66932	66941	66950	66960	66969	66978	66987	66997	67006	67015
468	67025	67034	67043	67052	67062	67071	67080	67090	67099	67108
469	67117	67127	67136	67145	67154	67164	67173	67182	67191	67201
470	67210	67219	67228	67238	67247	67256	67265	67274	67284	67293
471	67302	67311	67321	67330	67339	67348	67357	67367	67376	67385
472	67394	67403	67413	67422	67431	67440	67449	67459	67468	67477
473	67486	67495	67504	67514	67523	67532	67541	67550	67560	67569
474	67578	67587	67596	67605	67614	67624	67633	67642	67651	67660
475	67669	67679	67688	67697	67706	67715	67724	67733	67742	67752
476	67761	67770	67779	67788	67797	67806	67815	67825	67834	67843
477	67852	67861	67870	67879	67888	67897	67906	67916	67925	67934
478	67943	67952	67961	67970	67979	67988	67997	68006	68015	68024
479	68034	68043	68052	68061	68070	68079	68088	68097	68106	68115
480	68124	68133	68142	68151	68160	68169	68178	68187	68196	68205
481	68215	68224	68233	68242	68251	68260	68269	68278	68287	68296
482	68305	68314	68323	68332	68341	68350	68359	68368	68377	68386
483	68395	68404	68413	68422	68431	68440	68449	68458	68467	68476
484	68485	68494	68502	68511	68520	68529	68538	68547	68556	68565
485	68574	68583	68592	68601	68610	68619	68628	68637	68646	68655
486	68664	68673	68682	68690	68699	68708	68717	68726	68735	68744
487	68753	68762	68771	68780	68789	68797	68806	68815	68824	68833
488	68842	68851	68860	68869	68878	68886	68895	68904	68913	68922
489	68931	68940	68949	68958	68966	68975	68984	68993	69002	69011
490	69020	69028	69037	69046	69055	69064	69073	69082	69090	69099
491	69108	69117	69126	69135	69144	69152	69161	69170	69179	69188
492	69197	69205	69214	69223	69232	69241	69249	69258	69267	69276
493	69285	69294	69302	69311	69320	69329	69338	69346	69355	69364
494	69373	69381	69390	69399	69408	69417	69425	69434	69443	69452
495	69461	69469	69478	69487	69496	69504	69513	69522	69531	69539
496	69548	69557	69566	69574	69583	69592	69601	69609	69618	69627
497	69636	69644	69653	69662	69671	69679	69688	69697	69705	69714
498	69723	69732	69740	69749	69758	69767	69775	69784	69793	69801
499	69810	69819	69827	69836	69845	69854	69862	69871	69880	69888
500	69897	69906	69914	69923	69932	69940	69949	69958	69966	69975
N	0	1	2	3	4	5	6	7	8	9

Table of common logarithms

N	0	1	2	3	4	5	6	7	8	9
500	69897	69906	69914	69923	69932	69940	69949	69958	69966	69975
501	69984	69992	70001	70010	70018	70027	70036	70044	70053	70062
502	70070	70079	70088	70096	70105	70114	70122	70131	70140	70148
503	70157	70165	70174	70183	70191	70200	70209	70217	70226	70234
504	70243	70252	70260	70269	70278	70286	70295	70303	70312	70321
505	70329	70338	70346	70355	70364	70372	70381	70389	70398	70406
506	70415	70424	70432	70441	70449	70458	70467	70475	70484	70492
507	70501	70509	70518	70526	70535	70544	70552	70561	70569	70578
508	70586	70595	70603	70612	70621	70629	70638	70646	70655	70663
509	70672	70680	70689	70697	70706	70714	70723	70731	70740	70749
510	70757	70766	70774	70783	70791	70800	70808	70817	70825	70834
511	70842	70851	70859	70868	70876	70885	70893	70902	70910	70919
512	70927	70935	70944	70952	70961	70969	70978	70986	70995	71003
513	71012	71020	71029	71037	71046	71054	71063	71071	71079	71088
514	71096	71105	71113	71122	71130	71139	71147	71155	71164	71172
515	71181	71189	71198	71206	71214	71223	71231	71240	71248	71257
516	71265	71273	71282	71290	71299	71307	71315	71324	71332	71341
517	71349	71357	71366	71374	71383	71391	71399	71408	71416	71425
518	71433	71441	71450	71458	71467	71475	71483	71492	71500	71508
519	71517	71525	71533	71542	71550	71559	71567	71575	71584	71592
520	71600	71609	71617	71625	71634	71642	71650	71659	71667	71675
521	71684	71692	71700	71709	71717	71725	71734	71742	71750	71759
522	71767	71775	71784	71792	71800	71809	71817	71825	71834	71842
523	71850	71858	71867	71875	71883	71892	71900	71908	71917	71925
524	71933	71941	71950	71958	71966	71975	71983	71991	71999	72008
525	72016	72024	72032	72041	72049	72057	72066	72074	72082	72090
526	72099	72107	72115	72123	72132	72140	72148	72156	72165	72173
527	72181	72189	72198	72206	72214	72222	72230	72239	72247	72255
528	72263	72272	72280	72288	72296	72305	72313	72321	72329	72337
529	72346	72354	72362	72370	72378	72387	72395	72403	72411	72419
530	72428	72436	72444	72452	72460	72469	72477	72485	72493	72501
531	72509	72518	72526	72534	72542	72550	72559	72567	72575	72583
532	72591	72599	72607	72616	72624	72632	72640	72648	72656	72665
533	72673	72681	72689	72697	72705	72713	72722	72730	72738	72746
534	72754	72762	72770	72779	72787	72795	72803	72811	72819	72827
535	72835	72844	72852	72860	72868	72876	72884	72892	72900	72908
536	72916	72925	72933	72941	72949	72957	72965	72973	72981	72989
537	72997	73006	73014	73022	73030	73038	73046	73054	73062	73070
538	73078	73086	73094	73102	73111	73119	73127	73135	73143	73151
539	73159	73167	73175	73183	73191	73199	73207	73215	73223	73231
540	73239	73247	73255	73264	73272	73280	73288	73296	73304	73312
541	73320	73328	73336	73344	73352	73360	73368	73376	73384	73392
542	73400	73408	73416	73424	73432	73440	73448	73456	73464	73472
543	73480	73488	73496	73504	73512	73520	73528	73536	73544	73552
544	73560	73568	73576	73584	73592	73600	73608	73616	73624	73632
545	73640	73648	73656	73664	73672	73679	73687	73695	73703	73711
546	73719	73727	73735	73743	73751	73759	73767	73775	73783	73791
547	73799	73807	73815	73823	73830	73838	73846	73854	73862	73870
548	73878	73886	73894	73902	73910	73918	73926	73934	73941	73949
549	73957	73965	73973	73981	73989	73997	74005	74013	74020	74028
550	74036	74044	74052	74060	74068	74076	74084	74092	74099	74107
N	0	1	2	3	4	5	6	7	8	9

Table of common logarithms

N	0	1	2	3	4	5	6	7	8	9
550	74036	74044	74052	74060	74068	74076	74084	74092	74099	74107
551	74115	74123	74131	74139	74147	74155	74162	74170	74178	74186
552	74194	74202	74210	74218	74225	74233	74241	74249	74257	74265
553	74273	74280	74288	74296	74304	74312	74320	74327	74335	74343
554	74351	74359	74367	74374	74382	74390	74398	74406	74414	74421
555	74429	74437	74445	74453	74461	74468	74476	74484	74492	74500
556	74507	74515	74523	74531	74539	74547	74554	74562	74570	74578
557	74586	74593	74601	74609	74617	74624	74632	74640	74648	74656
558	74663	74671	74679	74687	74695	74702	74710	74718	74726	74733
559	74741	74749	74757	74764	74772	74780	74788	74796	74803	74811
560	74819	74827	74834	74842	74850	74858	74865	74873	74881	74889
561	74896	74904	74912	74920	74927	74935	74943	74950	74958	74966
562	74974	74981	74989	74997	75005	75012	75020	75028	75035	75043
563	75051	75059	75066	75074	75082	75089	75097	75105	75113	75120
564	75128	75136	75143	75151	75159	75166	75174	75182	75189	75197
565	75205	75213	75220	75228	75236	75243	75251	75259	75266	75274
566	75282	75289	75297	75305	75312	75320	75328	75335	75343	75351
567	75358	75366	75374	75381	75389	75397	75404	75412	75420	75427
568	75435	75442	75450	75458	75465	75473	75481	75488	75496	75504
569	75511	75519	75526	75534	75542	75549	75557	75565	75572	75580
570	75587	75595	75603	75610	75618	75626	75633	75641	75648	75656
571	75664	75671	75679	75686	75694	75702	75709	75717	75724	75732
572	75740	75747	75755	75762	75770	75778	75785	75793	75800	75808
573	75815	75823	75831	75838	75846	75853	75861	75868	75876	75884
574	75891	75899	75906	75914	75921	75929	75937	75944	75952	75959
575	75967	75974	75982	75989	75997	76005	76012	76020	76027	76035
576	76042	76050	76057	76065	76072	76080	76087	76095	76103	76110
577	76118	76125	76133	76140	76148	76155	76163	76170	76178	76185
578	76193	76200	76208	76215	76223	76230	76238	76245	76253	76260
579	76268	76275	76283	76290	76298	76305	76313	76320	76328	76335
580	76343	76350	76358	76365	76373	76380	76388	76395	76403	76410
581	76418	76425	76433	76440	76448	76455	76462	76470	76477	76485
582	76492	76500	76507	76515	76522	76530	76537	76545	76552	76559
583	76567	76574	76582	76589	76597	76604	76612	76619	76626	76634
584	76641	76649	76656	76664	76671	76678	76686	76693	76701	76708
585	76716	76723	76730	76738	76745	76753	76760	76768	76775	76782
586	76790	76797	76805	76812	76819	76827	76834	76842	76849	76856
587	76864	76871	76879	76886	76893	76901	76908	76916	76923	76930
588	76938	76945	76953	76960	76967	76975	76982	76989	76997	77004
589	77012	77019	77026	77034	77041	77048	77056	77063	77070	77078
590	77085	77093	77100	77107	77115	77122	77129	77137	77144	77151
591	77159	77166	77173	77181	77188	77195	77203	77210	77218	77225
592	77232	77240	77247	77254	77262	77269	77276	77284	77291	77298
593	77305	77313	77320	77327	77335	77342	77349	77357	77364	77371
594	77379	77386	77393	77401	77408	77415	77423	77430	77437	77444
595	77452	77459	77466	77474	77481	77488	77495	77503	77510	77517
596	77525	77532	77539	77546	77554	77561	77568	77576	77583	77590
597	77597	77605	77612	77619	77627	77634	77641	77648	77656	77663
598	77670	77677	77685	77692	77699	77706	77714	77721	77728	77735
599	77743	77750	77757	77764	77772	77779	77786	77793	77801	77808
600	77815	77822	77830	77837	77844	77851	77859	77866	77873	77880
N	0	1	2	3	4	5	6	7	8	9

APPENDIX A

Table of common logarithms

N	0	1	2	3	4	5	6	7	8	9
600	77815	77822	77830	77837	77844	77851	77859	77866	77873	77880
601	77887	77895	77902	77909	77916	77924	77931	77938	77945	77952
602	77960	77967	77974	77981	77989	77996	78003	78010	78017	78025
603	78032	78039	78046	78053	78061	78068	78075	78082	78089	78097
604	78104	78111	78118	78125	78132	78140	78147	78154	78161	78168
605	78176	78183	78190	78197	78204	78211	78219	78226	78233	78240
606	78247	78254	78262	78269	78276	78283	78290	78297	78305	78312
607	78319	78326	78333	78340	78347	78355	78362	78369	78376	78383
608	78390	78398	78405	78412	78419	78426	78433	78440	78447	78455
609	78462	78469	78476	78483	78490	78497	78505	78512	78519	78526
610	78533	78540	78547	78554	78561	78569	78576	78583	78590	78597
611	78604	78611	78618	78625	78633	78640	78647	78654	78661	78668
612	78675	78682	78689	78696	78704	78711	78718	78725	78732	78739
613	78746	78753	78760	78767	78774	78781	78789	78796	78803	78810
614	78817	78824	78831	78838	78845	78852	78859	78866	78873	78880
615	78888	78895	78902	78909	78916	78923	78930	78937	78944	78951
616	78958	78965	78972	78979	78986	78993	79000	79007	79014	79021
617	79029	79036	79043	79050	79057	79064	79071	79078	79085	79092
618	79099	79106	79113	79120	79127	79134	79141	79148	79155	79162
619	79169	79176	79183	79190	79197	79204	79211	79218	79225	79232
620	79239	79246	79253	79260	79267	79274	79281	79288	79295	79302
621	79309	79316	79323	79330	79337	79344	79351	79358	79365	79372
622	79379	79386	79393	79400	79407	79414	79421	79428	79435	79442
623	79449	79456	79463	79470	79477	79484	79491	79498	79505	79512
624	79518	79525	79532	79539	79546	79553	79560	79567	79574	79581
625	79588	79595	79602	79609	79616	79623	79630	79637	79644	79651
626	79657	79664	79671	79678	79685	79692	79699	79706	79713	79720
627	79727	79734	79741	79748	79754	79761	79768	79775	79782	79789
628	79796	79803	79810	79817	79824	79831	79837	79844	79851	79858
629	79865	79872	79879	79886	79893	79900	79906	79913	79920	79927
630	79934	79941	79948	79955	79962	79969	79975	79982	79989	79996
631	80003	80010	80017	80024	80030	80037	80044	80051	80058	80065
632	80072	80079	80085	80092	80099	80106	80113	80120	80127	80134
633	80140	80147	80154	80161	80168	80175	80182	80188	80195	80202
634	80209	80216	80223	80229	80236	80243	80250	80257	80264	80271
635	80277	80284	80291	80298	80305	80312	80318	80325	80332	80339
636	80346	80353	80359	80366	80373	80380	80387	80393	80400	80407
637	80414	80421	80428	80434	80441	80448	80455	80462	80468	80475
638	80482	80489	80496	80502	80509	80516	80523	80530	80536	80543
639	80550	80557	80564	80570	80577	80584	80591	80598	80604	80611
640	80618	80625	80632	80638	80645	80652	80659	80665	80672	80679
641	80686	80693	80699	80706	80713	80720	80726	80733	80740	80747
642	80754	80760	80767	80774	80781	80787	80794	80801	80808	80814
643	80821	80828	80835	80841	80848	80855	80862	80868	80875	80882
644	80889	80895	80902	80909	80916	80922	80929	80936	80943	80949
645	80956	80963	80969	80976	80983	80990	80996	81003	81010	81017
646	81023	81030	81037	81043	81050	81057	81064	81070	81077	81084
647	81090	81097	81104	81111	81117	81124	81131	81137	81144	81151
648	81158	81164	81171	81178	81184	81191	81198	81204	81211	81218
649	81224	81231	81238	81245	81251	81258	81265	81271	81278	81285
650	81291	81298	81305	81311	81318	81325	81331	81338	81345	81351
N	0	1	2	3	4	5	6	7	8	9

Table of common logarithms

N	0	1	2	3	4	5	6	7	8	9
650	81291	81298	81305	81311	81318	81325	81331	81338	81345	81351
651	81358	81365	81371	81378	81385	81391	81398	81405	81411	81418
652	81425	81431	81438	81445	81451	81458	81465	81471	81478	81485
653	81491	81498	81505	81511	81518	81525	81531	81538	81545	81551
654	81558	81564	81571	81578	81584	81591	81598	81604	81611	81618
655	81624	81631	81637	81644	81651	81657	81664	81671	81677	81684
656	81690	81697	81704	81710	81717	81723	81730	81737	81743	81750
657	81757	81763	81770	81776	81783	81790	81796	81803	81809	81816
658	81823	81829	81836	81842	81849	81856	81862	81869	81875	81882
659	81889	81895	81902	81908	81915	81921	81928	81935	81941	81948
660	81954	81961	81968	81974	81981	81987	81994	82000	82007	82014
661	82020	82027	82033	82040	82046	82053	82060	82066	82073	82079
662	82086	82092	82099	82105	82112	82119	82125	82132	82138	82145
663	82151	82158	82164	82171	82178	82184	82191	82197	82204	82210
664	82217	82223	82230	82236	82243	82250	82256	82263	82269	82276
665	82282	82289	82295	82302	82308	82315	82321	82328	82334	82341
666	82347	82354	82360	82367	82374	82380	82387	82393	82400	82406
667	82413	82419	82426	82432	82439	82445	82452	82458	82465	82471
668	82478	82484	82491	82497	82504	82510	82517	82523	82530	82536
669	82543	82549	82556	82562	82569	82575	82582	82588	82595	82601
670	82607	82614	82620	82627	82633	82640	82646	82653	82659	82666
671	82672	82679	82685	82692	82698	82705	82711	82718	82724	82730
672	82737	82743	82750	82756	82763	82769	82776	82782	82789	82795
673	82802	82808	82814	82821	82827	82834	82840	82847	82853	82860
674	82866	82872	82879	82885	82892	82898	82905	82911	82918	82924
675	82930	82937	82943	82950	82956	82963	82969	82975	82982	82988
676	82995	83001	83008	83014	83020	83027	83033	83040	83046	83052
677	83059	83065	83072	83078	83085	83091	83097	83104	83110	83117
678	83123	83129	83136	83142	83149	83155	83161	83168	83174	83181
679	83187	83193	83200	83206	83213	83219	83225	83232	83238	83245
680	83251	83257	83264	83270	83276	83283	83289	83296	83302	83308
681	83315	83321	83327	83334	83340	83347	83353	83359	83366	83372
682	83378	83385	83391	83398	83404	83410	83417	83423	83429	83436
683	83442	83448	83455	83461	83468	83474	83480	83487	83493	83499
684	83506	83512	83518	83525	83531	83537	83544	83550	83556	83563
685	83569	83575	83582	83588	83594	83601	83607	83613	83620	83626
686	83632	83639	83645	83651	83658	83664	83670	83677	83683	83689
687	83696	83702	83708	83715	83721	83727	83734	83740	83746	83753
688	83759	83765	83771	83778	83784	83790	83797	83803	83809	83816
689	83822	83828	83835	83841	83847	83853	83860	83866	83872	83879
690	83885	83891	83898	83904	83910	83916	83923	83929	83935	83942
691	83948	83954	83960	83967	83973	83979	83986	83992	83998	84004
692	84011	84017	84023	84029	84036	84042	84048	84055	84061	84067
693	84073	84080	84086	84092	84098	84105	84111	84117	84123	84130
694	84136	84142	84148	84155	84161	84167	84173	84180	84186	84192
695	84198	84205	84211	84217	84223	84230	84236	84242	84248	84255
696	84261	84267	84273	84280	84286	84292	84298	84305	84311	84317
697	84323	84330	84336	84342	84348	84354	84361	84367	84373	84379
698	84386	84392	84398	84404	84410	84417	84423	84429	84435	84442
699	84448	84454	84460	84466	84473	84479	84485	84491	84497	84504
700	84510	84516	84522	84528	84535	84541	84547	84553	84559	84566
N	0	1	2	3	4	5	6	7	8	9

Table of common logarithms

N	0	1	2	3	4	5	6	7	8	9
700	84510	84516	84522	84528	84535	84541	84547	84553	84559	84566
701	84572	84578	84584	84590	84597	84603	84609	84615	84621	84628
702	84634	84640	84646	84652	84658	84665	84671	84677	84683	84689
703	84696	84702	84708	84714	84720	84726	84733	84739	84745	84751
704	84757	84763	84770	84776	84782	84788	84794	84800	84807	84813
705	84819	84825	84831	84837	84844	84850	84856	84862	84868	84874
706	84880	84887	84893	84899	84905	84911	84917	84924	84930	84936
707	84942	84948	84954	84960	84967	84973	84979	84985	84991	84997
708	85003	85009	85016	85022	85028	85034	85040	85046	85052	85059
709	85065	85071	85077	85083	85089	85095	85101	85107	85114	85120
710	85126	85132	85138	85144	85150	85156	85163	85169	85175	85181
711	85187	85193	85199	85205	85211	85218	85224	85230	85236	85242
712	85248	85254	85260	85266	85272	85278	85285	85291	85297	85303
713	85309	85315	85321	85327	85333	85339	85345	85352	85358	85364
714	85370	85376	85382	85388	85394	85400	85406	85412	85418	85425
715	85431	85437	85443	85449	85455	85461	85467	85473	85479	85485
716	85491	85497	85503	85510	85516	85522	85528	85534	85540	85546
717	85552	85558	85564	85570	85576	85582	85588	85594	85600	85606
718	85612	85619	85625	85631	85637	85643	85649	85655	85661	85667
719	85673	85679	85685	85691	85697	85703	85709	85715	85721	85727
720	85733	85739	85745	85751	85757	85763	85769	85775	85781	85788
721	85794	85800	85806	85812	85818	85824	85830	85836	85842	85848
722	85854	85860	85866	85872	85878	85884	85890	85896	85902	85908
723	85914	85920	85926	85932	85938	85944	85950	85956	85962	85968
724	85974	85980	85986	85992	85998	86004	86010	86016	86022	86028
725	86034	86040	86046	86052	86058	86064	86070	86076	86082	86088
726	86094	86100	86106	86112	86118	86124	86130	86136	86142	86147
727	86153	86159	86165	86171	86177	86183	86189	86195	86201	86207
728	86213	86219	86225	86231	86237	86243	86249	86255	86261	86267
729	86273	86279	86285	86291	86297	86303	86308	86314	86320	86326
730	86332	86338	86344	86350	86356	86362	86368	86374	86380	86386
731	86392	86398	86404	86410	86416	86421	86427	86433	86439	86445
732	86451	86457	86463	86469	86475	86481	86487	86493	86499	86504
733	86510	86516	86522	86528	86534	86540	86546	86552	86558	86564
734	86570	86576	86581	86587	86593	86599	86605	86611	86617	86623
735	86629	86635	86641	86646	86652	86658	86664	86670	86676	86682
736	86688	86694	86700	86705	86711	86717	86723	86729	86735	86741
737	86747	86753	86759	86764	86770	86776	86782	86788	86794	86800
738	86806	86812	86817	86823	86829	86835	86841	86847	86853	86859
739	86864	86870	86876	86882	86888	86894	86900	86906	86911	86917
740	86923	86929	86935	86941	86947	86953	86958	86964	86970	86976
741	86982	86988	86994	86999	87005	87011	87017	87023	87029	87035
742	87040	87046	87052	87058	87064	87070	87076	87081	87087	87093
743	87099	87105	87111	87116	87122	87128	87134	87140	87146	87151
744	87157	87163	87169	87175	87181	87186	87192	87198	87204	87210
745	87216	87221	87227	87233	87239	87245	87251	87256	87262	87268
746	87274	87280	87286	87291	87297	87303	87309	87315	87320	87326
747	87332	87338	87344	87350	87355	87361	87367	87373	87379	87384
748	87390	87396	87402	87408	87413	87419	87425	87431	87437	87442
749	87448	87454	87460	87466	87471	87477	87483	87489	87495	87500
750	87506	87512	87518	87524	87529	87535	87541	87547	87552	87558
N	0	1	2	3	4	5	6	7	8	9

Table of common logarithms

N	0	1	2	3	4	5	6	7	8	9
750	87506	87512	87518	87524	87529	87535	87541	87547	87552	87558
751	87564	87570	87576	87581	87587	87593	87599	87604	87610	87616
752	87622	87628	87633	87639	87645	87651	87656	87662	87668	87674
753	87680	87685	87691	87697	87703	87708	87714	87720	87726	87731
754	87737	87743	87749	87754	87760	87766	87772	87777	87783	87789
755	87795	87800	87806	87812	87818	87823	87829	87835	87841	87846
756	87852	87858	87864	87869	87875	87881	87887	87892	87898	87904
757	87910	87915	87921	87927	87933	87938	87944	87950	87955	87961
758	87967	87973	87978	87984	87990	87996	88001	88007	88013	88018
759	88024	88030	88036	88041	88047	88053	88059	88064	88070	88076
760	88081	88087	88093	88099	88104	88110	88116	88121	88127	88133
761	88138	88144	88150	88156	88161	88167	88173	88178	88184	88190
762	88196	88201	88207	88213	88218	88224	88230	88235	88241	88247
763	88252	88258	88264	88270	88275	88281	88287	88292	88298	88304
764	88309	88315	88321	88326	88332	88338	88343	88349	88355	88360
765	88366	88372	88378	88383	88389	88395	88400	88406	88412	88417
766	88423	88429	88434	88440	88446	88451	88457	88463	88468	88474
767	88480	88485	88491	88497	88502	88508	88514	88519	88525	88530
768	88536	88542	88547	88553	88559	88564	88570	88576	88581	88587
769	88593	88598	88604	88610	88615	88621	88627	88632	88638	88643
770	88649	88655	88660	88666	88672	88677	88683	88689	88694	88700
771	88705	88711	88717	88722	88728	88734	88739	88745	88750	88756
772	88762	88767	88773	88779	88784	88790	88795	88801	88807	88812
773	88818	88824	88829	88835	88840	88846	88852	88857	88863	88868
774	88874	88880	88885	88891	88897	88902	88908	88913	88919	88925
775	88930	88936	88941	88947	88953	88958	88964	88969	88975	88981
776	88986	88992	88997	89003	89009	89014	89020	89025	89031	89037
777	89042	89048	89053	89059	89064	89070	89076	89081	89087	89092
778	89098	89104	89109	89115	89120	89126	89131	89137	89143	89148
779	89154	89159	89165	89170	89176	89182	89187	89193	89198	89204
780	89209	89215	89221	89226	89232	89237	89243	89248	89254	89260
781	89265	89271	89276	89282	89287	89293	89298	89304	89310	89315
782	89321	89326	89332	89337	89343	89348	89354	89360	89365	89371
783	89376	89382	89387	89393	89398	89404	89409	89415	89421	89426
784	89432	89437	89443	89448	89454	89459	89465	89470	89476	89481
785	89487	89493	89498	89504	89509	89515	89520	89526	89531	89537
786	89542	89548	89553	89559	89564	89570	89575	89581	89586	89592
787	89597	89603	89609	89614	89620	89625	89631	89636	89642	89647
788	89653	89658	89664	89669	89675	89680	89686	89691	89697	89702
789	89708	89713	89719	89724	89730	89735	89741	89746	89752	89757
790	89763	89768	89774	89779	89785	89790	89796	89801	89807	89812
791	89818	89823	89829	89834	89840	89845	89851	89856	89862	89867
792	89873	89878	89883	89889	89894	89900	89905	89911	89916	89922
793	89927	89933	89938	89944	89949	89955	89960	89966	89971	89977
794	89982	89988	89993	89998	90004	90009	90015	90020	90026	90031
795	90037	90042	90048	90053	90059	90064	90069	90075	90080	90086
796	90091	90097	90102	90108	90113	90119	90124	90129	90135	90140
797	90146	90151	90157	90162	90168	90173	90179	90184	90189	90195
798	90200	90206	90211	90217	90222	90228	90233	90238	90244	90249
799	90255	90260	90266	90271	90276	90282	90287	90293	90298	90304
800	90309	90314	90320	90325	90331	90336	90342	90347	90352	90358
N	0	1	2	3	4	5	6	7	8	9

Table of common logarithms

N	0	1	2	3	4	5	6	7	8	9
800	90309	90314	90320	90325	90331	90336	90342	90347	90352	90358
801	90363	90369	90374	90380	90385	90390	90396	90401	90407	90412
802	90417	90423	90428	90434	90439	90445	90450	90455	90461	90466
803	90472	90477	90482	90488	90493	90499	90504	90509	90515	90520
804	90526	90531	90536	90542	90547	90553	90558	90563	90569	90574
805	90580	90585	90590	90596	90601	90607	90612	90617	90623	90628
806	90634	90639	90644	90650	90655	90660	90666	90671	90677	90682
807	90687	90693	90698	90704	90709	90714	90720	90725	90730	90736
808	90741	90747	90752	90757	90763	90768	90773	90779	90784	90789
809	90795	90800	90806	90811	90816	90822	90827	90832	90838	90843
810	90849	90854	90859	90865	90870	90875	90881	90886	90891	90897
811	90902	90907	90913	90918	90924	90929	90934	90940	90945	90950
812	90956	90961	90966	90972	90977	90982	90988	90993	90998	91004
813	91009	91014	91020	91025	91030	91036	91041	91046	91052	91057
814	91062	91068	91073	91078	91084	91089	91094	91100	91105	91110
815	91116	91121	91126	91132	91137	91142	91148	91153	91158	91164
816	91169	91174	91180	91185	91190	91196	91201	91206	91212	91217
817	91222	91228	91233	91238	91243	91249	91254	91259	91265	91270
818	91275	91281	91286	91291	91297	91302	91307	91312	91318	91323
819	91328	91334	91339	91344	91350	91355	91360	91366	91371	91376
820	91381	91387	91392	91397	91403	91408	91413	91418	91424	91429
821	91434	91440	91445	91450	91455	91461	91466	91471	91477	91482
822	91487	91492	91498	91503	91508	91514	91519	91524	91529	91535
823	91540	91545	91551	91556	91561	91566	91572	91577	91582	91587
824	91593	91598	91603	91609	91614	91619	91624	91630	91635	91640
825	91645	91651	91656	91661	91666	91672	91677	91682	91687	91693
826	91698	91703	91709	91714	91719	91724	91730	91735	91740	91745
827	91751	91756	91761	91766	91772	91777	91782	91787	91793	91798
828	91803	91808	91814	91819	91824	91829	91835	91840	91845	91850
829	91855	91861	91866	91871	91876	91882	91887	91892	91897	91903
830	91908	91913	91918	91924	91929	91934	91939	91944	91950	91955
831	91960	91965	91971	91976	91981	91986	91991	91997	92002	92007
832	92012	92018	92023	92028	92033	92038	92044	92049	92054	92059
833	92065	92070	92075	92080	92085	92091	92096	92101	92106	92111
834	92117	92122	92127	92132	92137	92143	92148	92153	92158	92163
835	92169	92174	92179	92184	92189	92195	92200	92205	92210	92215
836	92221	92226	92231	92236	92241	92247	92252	92257	92262	92267
837	92273	92278	92283	92288	92293	92298	92304	92309	92314	92319
838	92324	92330	92335	92340	92345	92350	92355	92361	92366	92371
839	92376	92381	92387	92392	92397	92402	92407	92412	92418	92423
840	92428	92433	92438	92443	92449	92454	92459	92464	92469	92474
841	92480	92485	92490	92495	92500	92505	92511	92516	92521	92526
842	92531	92536	92542	92547	92552	92557	92562	92567	92572	92578
843	92583	92588	92593	92598	92603	92609	92614	92619	92624	92629
844	92634	92639	92645	92650	92655	92660	92665	92670	92675	92681
845	92686	92691	92696	92701	92706	92711	92717	92722	92727	92732
846	92737	92742	92747	92752	92758	92763	92768	92773	92778	92783
847	92788	92793	92799	92804	92809	92814	92819	92824	92829	92834
848	92840	92845	92850	92855	92860	92865	92870	92875	92881	92886
849	92891	92896	92901	92906	92911	92916	92921	92927	92932	92937
850	92942	92947	92952	92957	92962	92967	92973	92978	92983	92988
N	0	1	2	3	4	5	6	7	8	9

Table of common logarithms

N	0	1	2	3	4	5	6	7	8	9
850	92942	92947	92952	92957	92962	92967	92973	92978	92983	92988
851	92993	92998	93003	93008	93013	93018	93024	93029	93034	93039
852	93044	93049	93054	93059	93064	93069	93075	93080	93085	93090
853	93095	93100	93105	93110	93115	93120	93125	93131	93136	93141
854	93146	93151	93156	93161	93166	93171	93176	93181	93186	93192
855	93197	93202	93207	93212	93217	93222	93227	93232	93237	93242
856	93247	93252	93258	93263	93268	93273	93278	93283	93288	93293
857	93298	93303	93308	93313	93318	93323	93328	93334	93339	93344
858	93349	93354	93359	93364	93369	93374	93379	93384	93389	93394
859	93399	93404	93409	93414	93420	93425	93430	93435	93440	93445
860	93450	93455	93460	93465	93470	93475	93480	93485	93490	93495
861	93500	93505	93510	93515	93520	93526	93531	93536	93541	93546
862	93551	93556	93561	93566	93571	93576	93581	93586	93591	93596
863	93601	93606	93611	93616	93621	93626	93631	93636	93641	93646
864	93651	93656	93661	93666	93671	93677	93682	93687	93692	93697
865	93702	93707	93712	93717	93722	93727	93732	93737	93742	93747
866	93752	93757	93762	93767	93772	93777	93782	93787	93792	93797
867	93802	93807	93812	93817	93822	93827	93832	93837	93842	93847
868	93852	93857	93862	93867	93872	93877	93882	93887	93892	93897
869	93902	93907	93912	93917	93922	93927	93932	93937	93942	93947
870	93952	93957	93962	93967	93972	93977	93982	93987	93992	93997
871	94002	94007	94012	94017	94022	94027	94032	94037	94042	94047
872	94052	94057	94062	94067	94072	94077	94082	94087	94091	94096
873	94101	94106	94111	94116	94121	94126	94131	94136	94141	94146
874	94151	94156	94161	94166	94171	94176	94181	94186	94191	94196
875	94201	94206	94211	94216	94221	94226	94231	94236	94241	94245
876	94250	94255	94260	94265	94270	94275	94280	94285	94290	94295
877	94300	94305	94310	94315	94320	94325	94330	94335	94340	94345
878	94349	94354	94359	94364	94369	94374	94379	94384	94389	94394
879	94399	94404	94409	94414	94419	94424	94429	94433	94438	94443
880	94448	94453	94458	94463	94468	94473	94478	94483	94488	94493
881	94498	94503	94507	94512	94517	94522	94527	94532	94537	94542
882	94547	94552	94557	94562	94567	94571	94576	94581	94586	94591
883	94596	94601	94606	94611	94616	94621	94626	94630	94635	94640
884	94645	94650	94655	94660	94665	94670	94675	94680	94685	94689
885	94694	94699	94704	94709	94714	94719	94724	94729	94734	94738
886	94743	94748	94753	94758	94763	94768	94773	94778	94783	94787
887	94792	94797	94802	94807	94812	94817	94822	94827	94832	94836
888	94841	94846	94851	94856	94861	94866	94871	94876	94880	94885
889	94890	94895	94900	94905	94910	94915	94919	94924	94929	94934
890	94939	94944	94949	94954	94959	94963	94968	94973	94978	94983
891	94988	94993	94998	95002	95007	95012	95017	95022	95027	95032
892	95036	95041	95046	95051	95056	95061	95066	95071	95075	95080
893	95085	95090	95095	95100	95105	95109	95114	95119	95124	95129
894	95134	95139	95143	95148	95153	95158	95163	95168	95173	95177
895	95182	95187	95192	95197	95202	95207	95211	95216	95221	95226
896	95231	95236	95241	95245	95250	95255	95260	95265	95270	95274
897	95279	95284	95289	95294	95299	95303	95308	95313	95318	95323
898	95328	95332	95337	95342	95347	95352	95357	95361	95366	95371
899	95376	95381	95386	95390	95395	95400	95405	95410	95415	95419
900	95424	95429	95434	95439	95444	95448	95453	95458	95463	95468
N	0	1	2	3	4	5	6	7	8	9

Table of common logarithms

N	0	1	2	3	4	5	6	7	8	9
900	95424	95429	95434	95439	95444	95448	95453	95458	95463	95468
901	95472	95477	95482	95487	95492	95497	95501	95506	95511	95516
902	95521	95525	95530	95535	95540	95545	95550	95554	95559	95564
903	95569	95574	95578	95583	95588	95593	95598	95602	95607	95612
904	95617	95622	95626	95631	95636	95641	95646	95650	95655	95660
905	95665	95670	95674	95679	95684	95689	95694	95698	95703	95708
906	95713	95718	95722	95727	95732	95737	95742	95746	95751	95756
907	95761	95766	95770	95775	95780	95785	95789	95794	95799	95804
908	95809	95813	95818	95823	95828	95833	95837	95842	95847	95852
909	95856	95861	95866	95871	95876	95880	95885	95890	95895	95899
910	95904	95909	95914	95918	95923	95928	95933	95938	95942	95947
911	95952	95957	95961	95966	95971	95976	95980	95985	95990	95995
912	95999	96004	96009	96014	96019	96023	96028	96033	96038	96042
913	96047	96052	96057	96061	96066	96071	96076	96080	96085	96090
914	96095	96099	96104	96109	96114	96118	96123	96128	96133	96137
915	96142	96147	96152	96156	96161	96166	96171	96175	96180	96185
916	96190	96194	96199	96204	96209	96213	96218	96223	96227	96232
917	96237	96242	96246	96251	96256	96261	96265	96270	96275	96280
918	96284	96289	96294	96298	96303	96308	96313	96317	96322	96327
919	96332	96336	96341	96346	96350	96355	96360	96365	96369	96374
920	96379	96384	96388	96393	96398	96402	96407	96412	96417	96421
921	96426	96431	96435	96440	96445	96450	96454	96459	96464	96468
922	96473	96478	96483	96487	96492	96497	96501	96506	96511	96515
923	96520	96525	96530	96534	96539	96544	96548	96553	96558	96563
924	96567	96572	96577	96581	96586	96591	96595	96600	96605	96609
925	96614	96619	96624	96628	96633	96638	96642	96647	96652	96656
926	96661	96666	96670	96675	96680	96685	96689	96694	96699	96703
927	96708	96713	96717	96722	96727	96731	96736	96741	96745	96750
928	96755	96759	96764	96769	96774	96778	96783	96788	96792	96797
929	96802	96806	96811	96816	96820	96825	96830	96834	96839	96844
930	96848	96853	96858	96862	96867	96872	96876	96881	96886	96890
931	96895	96900	96904	96909	96914	96918	96923	96928	96932	96937
932	96942	96946	96951	96956	96960	96965	96970	96974	96979	96984
933	96988	96993	96997	97002	97007	97011	97016	97021	97025	97030
934	97035	97039	97044	97049	97053	97058	97063	97067	97072	97077
935	97081	97086	97090	97095	97100	97104	97109	97114	97118	97123
936	97128	97132	97137	97142	97146	97151	97155	97160	97165	97169
937	97174	97179	97183	97188	97193	97197	97202	97206	97211	97216
938	97220	97225	97230	97234	97239	97243	97248	97253	97257	97262
939	97267	97271	97276	97280	97285	97290	97294	97299	97304	97308
940	97313	97317	97322	97327	97331	97336	97341	97345	97350	97354
941	97359	97364	97368	97373	97377	97382	97387	97391	97396	97400
942	97405	97410	97414	97419	97424	97428	97433	97437	97442	97447
943	97451	97456	97460	97465	97470	97474	97479	97483	97488	97493
944	97497	97502	97506	97511	97516	97520	97525	97529	97534	97539
945	97543	97548	97552	97557	97562	97566	97571	97575	97580	97585
946	97589	97594	97598	97603	97607	97612	97617	97621	97626	97630
947	97635	97640	97644	97649	97653	97658	97663	97667	97672	97676
948	97681	97685	97690	97695	97699	97704	97708	97713	97717	97722
949	97727	97731	97736	97740	97745	97750	97754	97759	97763	97768
950	97772	97777	97782	97786	97791	97795	97800	97804	97809	97813
N	0	1	2	3	4	5	6	7	8	9

Table of common logarithms

N	0	1	2	3	4	5	6	7	8	9
950	97772	97777	97782	97786	97791	97795	97800	97804	97809	97813
951	97818	97823	97827	97832	97836	97841	97845	97850	97855	97859
952	97864	97868	97873	97877	97882	97887	97891	97896	97900	97905
953	97909	97914	97918	97923	97928	97932	97937	97941	97946	97950
954	97955	97959	97964	97969	97973	97978	97982	97987	97991	97996
955	98000	98005	98009	98014	98019	98023	98028	98032	98037	98041
956	98046	98050	98055	98059	98064	98069	98073	98078	98082	98087
957	98091	98096	98100	98105	98109	98114	98118	98123	98127	98132
958	98137	98141	98146	98150	98155	98159	98164	98168	98173	98177
959	98182	98186	98191	98195	98200	98205	98209	98214	98218	98223
960	98227	98232	98236	98241	98245	98250	98254	98259	98263	98268
961	98272	98277	98281	98286	98290	98295	98299	98304	98308	98313
962	98318	98322	98327	98331	98336	98340	98345	98349	98354	98358
963	98363	98367	98372	98376	98381	98385	98390	98394	98399	98403
964	98408	98412	98417	98421	98426	98430	98435	98439	98444	98448
965	98453	98457	98462	98466	98471	98475	98480	98484	98489	98493
966	98498	98502	98507	98511	98516	98520	98525	98529	98534	98538
967	98543	98547	98552	98556	98561	98565	98570	98574	98579	98583
968	98588	98592	98597	98601	98605	98610	98614	98619	98623	98628
969	98632	98637	98641	98646	98650	98655	98659	98664	98668	98673
970	98677	98682	98686	98691	98695	98700	98704	98709	98713	98717
971	98722	98726	98731	98735	98740	98744	98749	98753	98758	98762
972	98767	98771	98776	98780	98785	98789	98793	98798	98802	98807
973	98811	98816	98820	98825	98829	98834	98838	98843	98847	98851
974	98856	98860	98865	98869	98874	98878	98883	98887	98892	98896
975	98900	98905	98909	98914	98918	98923	98927	98932	98936	98941
976	98945	98949	98954	98958	98963	98967	98972	98976	98981	98985
977	98989	98994	98998	99003	99007	99012	99016	99021	99025	99029
978	99034	99038	99043	99047	99052	99056	99061	99065	99069	99074
979	99078	99083	99087	99092	99096	99100	99105	99109	99114	99118
980	99123	99127	99131	99136	99140	99145	99149	99154	99158	99162
981	99167	99171	99176	99180	99185	99189	99193	99198	99202	99207
982	99211	99216	99220	99224	99229	99233	99238	99242	99247	99251
983	99255	99260	99264	99269	99273	99277	99282	99286	99291	99295
984	99300	99304	99308	99313	99317	99322	99326	99330	99335	99339
985	99344	99348	99352	99357	99361	99366	99370	99374	99379	99383
986	99388	99392	99397	99401	99405	99410	99414	99419	99423	99427
987	99432	99436	99441	99445	99449	99454	99458	99463	99467	99471
988	99476	99480	99484	99489	99493	99498	99502	99506	99511	99515
989	99520	99524	99528	99533	99537	99542	99546	99550	99555	99559
990	99564	99568	99572	99577	99581	99585	99590	99594	99599	99603
991	99607	99612	99616	99621	99625	99629	99634	99638	99642	99647
992	99651	99656	99660	99664	99669	99673	99677	99682	99686	99691
993	99695	99699	99704	99708	99712	99717	99721	99726	99730	99734
994	99739	99743	99747	99752	99756	99760	99765	99769	99774	99778
995	99782	99787	99791	99795	99800	99804	99808	99813	99817	99822
996	99826	99830	99835	99839	99843	99848	99852	99856	99861	99865
997	99870	99874	99878	99883	99887	99891	99896	99900	99904	99909
998	99913	99917	99922	99926	99930	99935	99939	99944	99948	99952
999	99957	99961	99965	99970	99974	99978	99983	99987	99991	99996
1000	100000	100004	100009	100013	100017	100022	100026	100030	100035	100039
N	0	1	2	3	4	5	6	7	8	9

**Appendix B
Table of Squares and
Square Roots**

This appendix can be most useful whenever an electronic calculator is not available. It gives N^2, $\sqrt{N}$, and $\sqrt{10N}$ for various values of N. If your value of N is an integer between 1 and 1,000, merely find that number under the N column. Then by reading across the row, you obtain the values of N^2, $\sqrt{N}$, and $\sqrt{10N}$. For example, 50^2 is equal to 2,500, $\sqrt{50}$ is equal to 7.071, and $\sqrt{10(50)}$ or $\sqrt{500}$ is equal to 22.36.

By moving the decimal point, similar information can be obtained for many other numbers. For example, 0.50^2 is equal to 0.2500, $\sqrt{0.50}$ is equal to 0.7071, and $\sqrt{10(0.5)}$ or $\sqrt{5}$ is equal to 2.236. Likewise, $5{,}000^2$ is equal to 25,000,000, $\sqrt{5{,}000}$ is equal to 70.71, and $\sqrt{10(5{,}000)}$ or $\sqrt{50{,}000}$ is equal to 223.607.

Table of squares and square roots

N	N^2	$\sqrt{N}$	$\sqrt{10N}$	N	N^2	$\sqrt{N}$	$\sqrt{10N}$
1	1	1.0000	3.1623	51	2 601	7.1414	22.5832
2	4	1.4142	4.4721	52	2 704	7.2111	22.8035
3	9	1.7321	5.4772	53	2 809	7.2801	23.0217
4	16	2.0000	6.3246	54	2 916	7.3485	23.2379
5	25	2.2361	7.0711	55	3 025	7.4162	23.4521
6	36	2.4495	7.7460	56	3 136	7.4833	23.6643
7	49	2.6458	8.3666	57	3 249	7.5498	23.8747
8	64	2.8284	8.9443	58	3 364	7.6158	24.0832
9	81	3.0000	9.4868	59	3 481	7.6811	24.2899
10	100	3.1623	10.0000	60	3 600	7.7460	24.4949
11	121	3.3166	10.4881	61	3 721	7.8102	24.6982
12	144	3.4641	10.9545	62	3 844	7.8740	24.8998
13	169	3.6056	11.4018	63	3 969	7.9373	25.0998
14	196	3.7417	11.8322	64	4 096	8.0000	25.2982
15	225	3.8730	12.2474	65	4 225	8.0623	25.4951
16	256	4.0000	12.6491	66	4 356	8.1240	25.6905
17	289	4.1231	13.0384	67	4 489	8.1854	25.8844
18	324	4.2426	13.4164	68	4 624	8.2462	26.0768
19	361	4.3589	13.7840	69	4 761	8.3066	26.2679
20	400	4.4721	14.1421	70	4 900	8.3666	26.4575
21	441	4.5826	14.4914	71	5 041	8.4261	26.6458
22	484	4.6904	14.8324	72	5 184	8.4853	26.8328
23	529	4.7958	15.1658	73	5 329	8.5440	27.0185
24	576	4.8990	15.4919	74	5 476	8.6023	27.2029
25	625	5.0000	15.8114	75	5 625	8.6603	27.3861
26	676	5.0990	16.1245	76	5 776	8.7178	27.5681
27	729	5.1962	16.4317	77	5 929	8.7750	27.7489
28	784	5.2915	16.7332	78	6 084	8.8318	27.9285
29	841	5.3852	17.0294	79	6 241	8.8882	28.1069
30	900	5.4772	17.3205	80	6 400	8.9443	28.2843
31	961	5.5678	17.6068	81	6 561	9.0000	28.4605
32	1 024	5.6569	17.8885	82	6 724	9.0554	28.6356
33	1 089	5.7446	18.1659	83	6 889	9.1104	28.8097
34	1 156	5.8310	18.4391	84	7 056	9.1652	28.9828
35	1 225	5.9161	18.7083	85	7 225	9.2195	29.1548
36	1 296	6.0000	18.9737	86	7 396	9.2736	29.3258
37	1 369	6.0828	19.2354	87	7 569	9.3274	29.4958
38	1 444	6.1644	19.4936	88	7 744	9.3808	29.6648
39	1 521	6.2450	19.7484	89	7 921	9.4340	29.8329
40	1 600	6.3246	20.0000	90	8 100	9.4868	30.0000
41	1 681	6.4031	20.2485	91	8 281	9.5394	30.1662
42	1 764	6.4807	20.4939	92	8 464	9.5917	30.3315
43	1 849	6.5574	20.7364	93	8 649	9.6437	30.4959
44	1 936	6.6332	20.9762	94	8 836	9.6954	30.6594
45	2 025	6.7082	21.2132	95	9 025	9.7468	30.8221
46	2 116	6.7823	21.4476	96	9 216	9.7980	30.9839
47	2 209	6.8557	21.6795	97	9 409	9.8489	31.1448
48	2 304	6.9282	21.9089	98	9 604	9.8995	31.3049
49	2 401	7.0000	22.1359	99	9 801	9.9499	31.4643
50	2 500	7.0711	22.3607	100	10 000	10.0000	31.6228
N	N^2	$\sqrt{N}$	$\sqrt{10N}$	N	N^2	$\sqrt{N}$	$\sqrt{10N}$

Table of squares and square roots

N	N^2	$\sqrt{N}$	$\sqrt{10N}$	N	N^2	$\sqrt{N}$	$\sqrt{10N}$
101	10 201	10.0499	31.7805	151	22 801	12.2882	38.8587
102	10 404	10.0995	31.9374	152	23 104	12.3288	38.9872
103	10 609	10.1489	32.0936	153	23 409	12.3693	39.1152
104	10 816	10.1980	32.2490	154	23 716	12.4097	39.2428
105	11 025	10.2470	32.4037	155	24 025	12.4499	39.3700
106	11 236	10.2956	32.5576	156	24 336	12.4900	39.4968
107	11 449	10.3441	32.7109	157	24 649	12.5300	39.6232
108	11 664	10.3923	32.8634	158	24 964	12.5698	39.7492
109	11 881	10.4403	33.0152	159	25 281	12.6095	39.8748
110	12 100	10.4881	33.1662	160	25 600	12.6491	40.0000
111	12 321	10.5357	33.3167	161	25 921	12.6886	40.1248
112	12 544	10.5830	33.4664	162	26 244	12.7279	40.2492
113	12 769	10.6301	33.6155	163	26 569	12.7671	40.3733
114	12 996	10.6771	33.7639	164	26 896	12.8062	40.4969
115	13 225	10.7238	33.9117	165	27 225	12.8452	40.6202
116	13 456	10.7703	34.0588	166	27 556	12.8841	40.7431
117	13 689	10.8167	34.2053	167	27 889	12.9228	40.8656
118	13 924	10.8628	34.3511	168	28 224	12.9615	40.9878
119	14 161	10.9087	34.4964	169	28 561	13.0000	41.1096
120	14 400	10.9545	34.6410	170	28 900	13.0384	41.2311
121	14 641	11.0000	34.7850	171	29 241	13.0767	41.3521
122	14 884	11.0454	34.9285	172	29 584	13.1149	41.4729
123	15 129	11.0905	35.0714	173	29 929	13.1529	41.5933
124	15 376	11.1355	35.2136	174	30 276	13.1909	41.7133
125	15 625	11.1803	35.3553	175	30 625	13.2288	41.8330
126	15 876	11.2250	35.4965	176	30 976	13.2665	41.9523
127	16 129	11.2694	35.6371	177	31 329	13.3041	42.0714
128	16 384	11.3137	35.7771	178	31 684	13.3417	42.1900
129	16 641	11.3578	35.9166	179	32 041	13.3791	42.3084
130	16 900	11.4018	36.0555	180	32 400	13.4164	42.4264
131	17 161	11.4455	36.1939	181	32 761	13.4536	42.5441
132	17 424	11.4891	36.3318	182	33 124	13.4907	42.6615
133	17 689	11.5326	36.4692	183	33 489	13.5277	42.7785
134	17 956	11.5758	36.6060	184	33 856	13.5647	42.8952
135	18 225	11.6189	36.7423	185	34 225	13.6015	43.0116
136	18 496	11.6619	36.8782	186	34 596	13.6382	43.1277
137	18 769	11.7047	37.0135	187	34 969	13.6748	43.2435
138	19 044	11.7473	37.1483	188	35 344	13.7113	43.3590
139	19 321	11.7898	37.2827	189	35 721	13.7477	43.4741
140	19 600	11.8322	37.4166	190	36 100	13.7840	43.5890
141	19 881	11.8743	37.5500	191	36 481	13.8203	43.7036
142	20 164	11.9164	37.6829	192	36 864	13.8564	43.8178
143	20 449	11.9583	37.8153	193	37 249	13.8924	43.9318
144	20 736	12.0000	37.9473	194	37 636	13.9284	44.0454
145	21 025	12.0416	38.0789	195	38 025	13.9642	44.1588
146	21 316	12.0830	38.2099	196	38 416	14.0000	44.2719
147	21 609	12.1244	38.3406	197	38 809	14.0357	44.3847
148	21 904	12.1655	38.4708	198	39 204	14.0712	44.4972
149	22 201	12.2066	38.6005	199	39 601	14.1067	44.6094
150	22 500	12.2474	38.7298	200	40 000	14.1421	44.7214
N	N^2	$\sqrt{N}$	$\sqrt{10N}$	N	N^2	$\sqrt{N}$	$\sqrt{10N}$

Table of squares and square roots

N	N²	√N	√10N	N	N²	√N	√10N
201	40 401	14.1774	44.8330	251	63 001	15.8430	50.0999
202	40 804	14.2127	44.9444	252	63 504	15.8745	50.1996
203	41 209	14.2478	45.0555	253	64 009	15.9060	50.2991
204	41 616	14.2829	45.1664	254	64 516	15.9374	50.3984
205	42 025	14.3178	45.2769	255	65 025	15.9687	50.4975
206	42 436	14.3527	45.3872	256	65 536	16.0000	50.5964
207	42 849	14.3875	45.4973	257	66 049	16.0312	50.6952
208	43 264	14.4222	45.6070	258	66 564	16.0624	50.7937
209	43 681	14.4568	45.7165	259	67 081	16.0935	50.8920
210	44 100	14.4914	45.8258	260	67 600	16.1245	50.9902
211	44 521	14.5258	45.9347	261	68 121	16.1555	51.0882
212	44 944	14.5602	46.0435	262	68 644	16.1864	51.1859
213	45 369	14.5945	46.1519	263	69 169	16.2173	51.2835
214	45 796	14.6287	46.2601	264	69 696	16.2481	51.3809
215	46 225	14.6629	46.3681	265	70 225	16.2788	51.4781
216	46 656	14.6969	46.4758	266	70 756	16.3095	51.5752
217	47 089	14.7309	46.5833	267	71 289	16.3401	51.6720
218	47 524	14.7648	46.6905	268	71 824	16.3707	51.7687
219	47 961	14.7986	46.7974	269	72 361	16.4012	51.8652
220	48 400	14.8324	46.9042	270	72 900	16.4317	51.9615
221	48 841	14.8661	47.0106	271	73 441	16.4621	52.0577
222	49 284	14.8997	47.1169	272	73 984	16.4924	52.1536
223	49 729	14.9332	47.2229	273	74 529	16.5227	52.2494
224	50 176	14.9666	47.3286	274	75 076	16.5529	52.3450
225	50 625	15.0000	47.4342	275	75 625	16.5831	52.4404
226	51 076	15.0333	47.5395	276	76 176	16.6133	52.5357
227	51 529	15.0665	47.6445	277	76 729	16.6433	52.6308
228	51 984	15.0997	47.7493	278	77 284	16.6733	52.7257
229	52 441	15.1327	47.8539	279	77 841	16.7033	52.8204
230	52 900	15.1658	47.9583	280	78 400	16.7332	52.9150
231	53 361	15.1987	48.0625	281	78 961	16.7631	53.0094
232	53 824	15.2315	48.1664	282	79 524	16.7929	53.1037
233	54 289	15.2643	48.2701	283	80 089	16.8226	53.1977
234	54 756	15.2971	48.3736	284	80 656	16.8523	53.2917
235	55 225	15.3297	48.4768	285	81 225	16.8819	53.3854
236	55 696	15.3623	48.5798	286	81 796	16.9115	53.4790
237	56 169	15.3948	48.6826	287	82 369	16.9411	53.5724
238	56 644	15.4272	48.7852	288	82 944	16.9706	53.6656
239	57 121	15.4596	48.8876	289	83 521	17.0000	53.7587
240	57 600	15.4919	48.9898	290	84 100	17.0294	53.8517
241	58 081	15.5242	49.0918	291	84 681	17.0587	53.9444
242	58 564	15.5563	49.1935	292	85 264	17.0880	54.0370
243	59 049	15.5885	49.2950	293	85 849	17.1172	54.1295
244	59 536	15.6205	49.3964	294	86 436	17.1464	54.2218
245	60 025	15.6525	49.4975	295	87 025	17.1756	54.3139
246	60 516	15.6844	49.5984	296	87 616	17.2047	54.4059
247	61 009	15.7162	49.6991	297	88 209	17.2337	54.4977
248	61 504	15.7480	49.7996	298	88 804	17.2627	54.5894
249	62 001	15.7797	49.8999	299	89 401	17.2916	54.6809
250	62 500	15.8114	50.0000	300	90 000	17.3205	54.7723
N	N²	√N	√10N	N	N²	√N	√10N

Table of squares and square roots

N	N²	√N	√10N	N	N²	√N	√10N
301	90 601	17.3493	54.8635	351	123 201	18.7350	59.2453
302	91 204	17.3781	54.9545	352	123 904	18.7617	59.3296
303	91 809	17.4069	55.0454	353	124 609	18.7883	59.4138
304	92 416	17.4356	55.1362	354	125 316	18.8149	59.4979
305	93 025	17.4642	55.2268	355	126 025	18.8414	59.5819
306	93 636	17.4929	55.3173	356	126 736	18.8680	59.6657
307	94 249	17.5214	55.4076	357	127 449	18.8944	59.7495
308	94 864	17.5499	55.4977	358	128 164	18.9209	59.8331
309	95 481	17.5784	55.5878	359	128 881	18.9473	59.9166
310	96 100	17.6068	55.6776	360	129 600	18.9737	60.0000
311	96 721	17.6352	55.7674	361	130 321	19.0000	60.0833
312	97 344	17.6635	55.8570	362	131 044	19.0263	60.1664
313	97 969	17.6918	55.9464	363	131 769	19.0526	60.2495
314	98 596	17.7200	56.0357	364	132 496	19.0788	60.3324
315	99 225	17.7482	56.1249	365	133 225	19.1050	60.4152
316	99 856	17.7764	56.2139	366	133 956	19.1311	60.4979
317	100 489	17.8045	56.3027	367	134 689	19.1572	60.5805
318	101 124	17.8326	56.3915	368	135 424	19.1833	60.6630
319	101 761	17.8606	56.4801	369	136 161	19.2094	60.7454
320	102 400	17.8885	56.5685	370	136 900	19.2354	60.8276
321	103 041	17.9165	56.6569	371	137 641	19.2614	60.9098
322	103 684	17.9444	56.7450	372	138 384	19.2873	60.9918
323	104 329	17.9722	56.8331	373	139 129	19.3132	61.0737
324	104 976	18.0000	56.9210	374	139 876	19.3391	61.1555
325	105 625	18.0278	57.0088	375	140 625	19.3649	61.2372
326	106 276	18.0555	57.0964	376	141 376	19.3907	61.3188
327	106 929	18.0831	57.1839	377	142 129	19.4165	61.4003
328	107 584	18.1108	57.2713	378	142 884	19.4422	61.4817
329	108 241	18.1384	57.3585	379	143 641	19.4679	61.5630
330	108 900	18.1659	57.4456	380	144 400	19.4936	61.6441
331	109 561	18.1934	57.5326	381	145 161	19.5192	61.7252
332	110 224	18.2209	57.6194	382	145 924	19.5448	61.8062
333	110 889	18.2483	57.7061	383	146 689	19.5704	61.8870
334	111 556	18.2757	57.7927	384	147 456	19.5959	61.9677
335	112 225	18.3030	57.8792	385	148 225	19.6214	62.0484
336	112 896	18.3303	57.9655	386	148 996	19.6469	62.1289
337	113 569	18.3576	58.0517	387	149 769	19.6723	62.2093
338	114 244	18.3848	58.1378	388	150 544	19.6977	62.2896
339	114 921	18.4120	58.2237	389	151 321	19.7231	62.3699
340	115 600	18.4391	58.3095	390	152 100	19.7484	62.4500
341	116 281	18.4662	58.3952	391	152 881	19.7737	62.5300
342	116 964	18.4932	58.4808	392	153 664	19.7990	62.6099
343	117 649	18.5203	58.5662	393	154 449	19.8242	62.6897
344	118 336	18.5472	58.6515	394	155 236	19.8494	62.7694
345	119 025	18.5742	58.7367	395	156 025	19.8746	62.8490
346	119 716	18.6011	58.8218	396	156 816	19.8997	62.9285
347	120 409	18.6279	58.9067	397	157 609	19.9249	63.0079
348	121 104	18.6548	58.9915	398	158 404	19.9499	63.0872
349	121 801	18.6815	59.0762	399	159 201	19.9750	63.1664
350	122 500	18.7083	59.1608	400	160 000	20.0000	63.2456
N	N²	√N	√10N	N	N²	√N	√10N

Table of squares and square roots

N	N^2	$\sqrt{N}$	$\sqrt{10N}$	N	N^2	$\sqrt{N}$	$\sqrt{10N}$
401	160 801	20.0250	63.3246	451	203 401	21.2368	67.1565
402	161 604	20.0499	63.4035	452	204 304	21.2603	67.2309
403	162 409	20.0749	63.4823	453	205 209	21.2838	67.3053
404	163 216	20.0997	63.5610	454	206 116	21.3073	67.3795
405	164 025	20.1246	63.6396	455	207 025	21.3307	67.4537
406	164 836	20.1494	63.7181	456	207 936	21.3542	67.5278
407	165 649	20.1742	63.7966	457	208 849	21.3776	67.6018
408	166 464	20.1990	63.8749	458	209 764	21.4009	67.6757
409	167 281	20.2238	63.9531	459	210 681	21.4243	67.7495
410	168 100	20.2485	64.0313	460	211 600	21.4476	67.8233
411	168 921	20.2731	64.1093	461	212 521	21.4709	67.8970
412	169 744	20.2978	64.1872	462	213 444	21.4942	67.9706
413	170 569	20.3224	64.2651	463	214 369	21.5174	68.0441
414	171 396	20.3470	64.3428	464	215 296	21.5407	68.1175
415	172 225	20.3716	64.4205	465	216 225	21.5639	68.1909
416	173 056	20.3961	64.4981	466	217 156	21.5870	68.2642
417	173 889	20.4206	64.5755	467	218 089	21.6102	68.3374
418	174 724	20.4451	64.6529	468	219 024	21.6333	68.4105
419	175 561	20.4695	64.7302	469	219 961	21.6564	68.4836
420	176 400	20.4939	64.8074	470	220 900	21.6795	68.5565
421	177 241	20.5183	64.8845	471	221 841	21.7025	68.6294
422	178 084	20.5426	64.9615	472	222 784	21.7256	68.7023
423	178 929	20.5670	65.0385	473	223 729	21.7486	68.7750
424	179 776	20.5913	65.1153	474	224 676	21.7715	68.8477
425	180 625	20.6155	65.1920	475	225 625	21.7945	68.9202
426	181 476	20.6398	65.2687	476	226 576	21.8174	68.9928
427	182 329	20.6640	65.3452	477	227 529	21.8403	69.0652
428	183 184	20.6882	65.4217	478	228 484	21.8632	69.1375
429	184 041	20.7123	65.4981	479	229 441	21.8861	69.2098
430	184 900	20.7364	65.5744	480	230 400	21.9089	69.2820
431	185 761	20.7605	65.6506	481	231 361	21.9317	69.3542
432	186 624	20.7846	65.7267	482	232 324	21.9545	69.4262
433	187 489	20.8087	65.8027	483	233 289	21.9773	69.4982
434	188 356	20.8327	65.8787	484	234 256	22.0000	69.5701
435	189 225	20.8567	65.9545	485	235 225	22.0227	69.6419
436	190 096	20.8806	66.0303	486	236 196	22.0454	69.7137
437	190 969	20.9045	66.1060	487	237 169	22.0681	69.7854
438	191 844	20.9285	66.1816	488	238 144	22.0907	69.8570
439	192 721	20.9523	66.2571	489	239 121	22.1133	69.9285
440	193 600	20.9762	66.3325	490	240 100	22.1359	70.0000
441	194 481	21.0000	66.4078	491	241 081	22.1585	70.0714
442	195 364	21.0238	66.4831	492	242 064	22.1811	70.1427
443	196 249	21.0476	66.5582	493	243 049	22.2036	70.2140
444	197 136	21.0713	66.6333	494	244 036	22.2261	70.2851
445	198 025	21.0950	66.7083	495	245 025	22.2486	70.3562
446	198 916	21.1187	66.7832	496	246 016	22.2711	70.4273
447	199 809	21.1424	66.8581	497	247 009	22.2935	70.4982
448	200 704	21.1660	66.9328	498	248 004	22.3159	70.5691
449	201 601	21.1896	67.0075	499	249 001	22.3383	70.6399
450	202 500	21.2132	67.0820	500	250 000	22.3607	70.7107
N	N^2	$\sqrt{N}$	$\sqrt{10N}$	N	N^2	$\sqrt{N}$	$\sqrt{10N}$

APPENDIX B

Table of squares and square roots

N	N²	√N	√10N	N	N²	√N	√10N
501	251 001	22.3830	70.7814	551	303 601	23.4734	74.2294
502	252 004	22.4053	70.8520	552	304 704	23.4947	74.2967
503	253 009	22.4277	70.9225	553	305 809	23.5159	74.3640
504	254 016	22.4500	70.9930	554	306 916	23.5372	74.4312
505	255 025	22.4722	71.0634	555	308 025	23.5584	74.4983
506	256 036	22.4944	71.1337	556	309 136	23.5797	74.5654
507	257 049	22.5167	71.2039	557	310 249	23.6008	74.6324
508	258 064	22.5388	71.2741	558	311 364	23.6220	74.6994
509	259 081	22.5610	71.3442	559	312 481	23.6432	74.7663
510	260 100	22.5832	71.4143	560	313 600	23.6643	74.8331
511	261 121	22.6053	71.4843	561	314 721	23.6854	74.8999
512	262 144	22.6274	71.5542	562	315 844	23.7065	74.9667
513	263 169	22.6495	71.6240	563	316 969	23.7276	75.0333
514	264 196	22.6716	71.6938	564	318 096	23.7487	75.0999
515	265 225	22.6936	71.7635	565	319 225	23.7697	75.1665
516	266 256	22.7156	71.8331	566	320 356	23.7908	75.2330
517	267 289	22.7376	71.9027	567	321 489	23.8118	75.2994
518	268 324	22.7596	71.9722	568	322 624	23.8327	75.3658
519	269 361	22.7816	72.0417	569	323 761	23.8537	75.4321
520	270 400	22.8035	72.1110	570	324 900	23.8747	75.4983
521	271 441	22.8254	72.1803	571	326 041	23.8956	75.5645
522	272 484	22.8473	72.2496	572	327 184	23.9165	75.6307
523	273 529	22.8692	72.3187	573	328 329	23.9374	75.6968
524	274 576	22.8911	72.3878	574	329 476	23.9583	75.7628
525	275 625	22.9129	72.4569	575	330 625	23.9792	75.8288
526	276 676	22.9347	72.5259	576	331 776	24.0000	75.8947
527	277 729	22.9565	72.5948	577	332 929	24.0208	75.9605
528	278 784	22.9783	72.6636	578	334 084	24.0416	76.0263
529	279 841	23.0000	72.7324	579	335 241	24.0624	76.0921
530	280 900	23.0217	72.8011	580	336 400	24.0832	76.1577
531	281 961	23.0434	72.8698	581	337 561	24.1039	76.2234
532	283 024	23.0651	72.9383	582	338 724	24.1247	76.2889
533	284 089	23.0868	73.0069	583	339 889	24.1454	76.3544
534	285 156	23.1084	73.0753	584	341 056	24.1661	76.4199
535	286 225	23.1301	73.1437	585	342 225	24.1868	76.4853
536	287 296	23.1517	73.2120	586	343 396	24.2074	76.5506
537	288 369	23.1733	73.2803	587	344 569	24.2281	76.6159
538	289 444	23.1948	73.3485	588	345 744	24.2487	76.6812
539	290 521	23.2164	73.4166	589	346 921	24.2693	76.7463
540	291 600	23.2379	73.4847	590	348 100	24.2899	76.8115
541	292 681	23.2594	73.5527	591	349 281	24.3105	76.8765
542	293 764	23.2809	73.6207	592	350 464	24.3311	76.9415
543	294 849	23.3024	73.6885	593	351 649	24.3516	77.0065
544	295 936	23.3238	73.7564	594	352 836	24.3721	77.0714
545	297 025	23.3452	73.8241	595	354 025	24.3926	77.1362
546	298 116	23.3666	73.8918	596	355 216	24.4131	77.2010
547	299 209	23.3880	73.9594	597	356 409	24.4336	77.2658
548	300 304	23.4094	74.0270	598	357 604	24.4540	77.3305
549	301 401	23.4308	74.0945	599	358 801	24.4745	77.3951
550	302 500	23.4521	74.1620	600	360 000	24.4949	77.4597
N	N²	√N	√10N	N	N²	√N	√10N

Table of squares and square roots

N	N^2	$\sqrt{N}$	$\sqrt{10N}$	N	N^2	$\sqrt{N}$	$\sqrt{10N}$
601	361 201	24.5153	77.5242	651	423 801	25.5147	80.6846
602	362 404	24.5357	77.5887	652	425 104	25.5343	80.7465
603	363 609	24.5561	77.6531	653	426 409	25.5539	80.8084
604	364 816	24.5764	77.7174	654	427 716	25.5734	80.8703
605	366 025	24.5967	77.7818	655	429 025	25.5930	80.9321
606	367 236	24.6171	77.8460	656	430 336	25.6125	80.9938
607	368 449	24.6374	77.9102	657	431 649	25.6320	81.0555
608	369 664	24.6577	77.9743	658	432 964	25.6515	81.1172
609	370 881	24.6779	78.0385	659	434 281	25.6710	81.1788
610	372 100	24.6982	78.1025	660	435 600	25.6905	81.2404
611	373 321	24.7184	78.1665	661	436 921	25.7099	81.3019
612	374 544	24.7386	78.2304	662	438 244	25.7294	81.3634
613	375 769	24.7588	78.2943	663	439 569	25.7488	81.4248
614	376 996	24.7790	78.3582	664	440 896	25.7682	81.4862
615	378 225	24.7992	78.4219	665	442 225	25.7876	81.5475
616	379 456	24.8194	78.4857	666	443 556	25.8070	81.6088
617	380 689	24.8395	78.5493	667	444 889	25.8263	81.6701
618	381 924	24.8596	78.6130	668	446 224	25.8457	81.7313
619	383 161	24.8797	78.6766	669	447 561	25.8650	81.7924
620	384 400	24.8998	78.7401	670	448 900	25.8844	81.8535
621	385 641	24.9199	78.8036	671	450 241	25.9037	81.9146
622	386 884	24.9399	78.8670	672	451 584	25.9230	81.9756
623	388 129	24.9600	78.9303	673	452 929	25.9422	82.0366
624	389 376	24.9800	78.9937	674	454 276	25.9615	82.0975
625	390 625	25.0000	79.0569	675	455 625	25.9808	82.1584
626	391 876	25.0200	79.1202	676	456 976	26.0000	82.2192
627	393 129	25.0400	79.1833	677	458 329	26.0192	82.2800
628	394 384	25.0599	79.2464	678	459 684	26.0384	82.3408
629	395 641	25.0799	79.3095	679	461 041	26.0576	82.4015
630	396 900	25.0998	79.3725	680	462 400	26.0768	82.4621
631	398 161	25.1197	79.4355	681	463 761	26.0960	82.5227
632	399 424	25.1396	79.4984	682	465 124	26.1151	82.5833
633	400 689	25.1595	79.5613	683	466 489	26.1343	82.6438
634	401 956	25.1794	79.6241	684	467 856	26.1534	82.7043
635	403 225	25.1992	79.6869	685	469 225	26.1725	82.7647
636	404 496	25.2190	79.7496	686	470 596	26.1916	82.8251
637	405 769	25.2389	79.8123	687	471 969	26.2107	82.8855
638	407 044	25.2587	79.8749	688	473 344	26.2298	82.9458
639	408 321	25.2784	79.9375	689	474 721	26.2488	83.0060
640	409 600	25.2982	80.0000	690	476 100	26.2679	83.0662
641	410 881	25.3180	80.0625	691	477 481	26.2869	83.1264
642	412 164	25.3377	80.1249	692	478 864	26.3059	83.1865
643	413 449	25.3575	80.1873	693	480 249	26.3249	83.2466
644	414 736	25.3772	80.2496	694	481 636	26.3439	83.3067
645	416 025	25.3969	80.3119	695	483 025	26.3629	83.3667
646	417 316	25.4165	80.3741	696	484 416	26.3818	83.4266
647	418 609	25.4362	80.4363	697	485 809	26.4008	83.4865
648	419 904	25.4558	80.4984	698	487 204	26.4197	83.5464
649	421 201	25.4755	80.5605	699	488 601	26.4386	83.6062
650	422 500	25.4951	80.6226	700	490 000	26.4575	83.6660
N	N^2	$\sqrt{N}$	$\sqrt{10N}$	N	N^2	$\sqrt{N}$	$\sqrt{10N}$

Table of squares and square roots

N	N²	√N	√10N	N	N²	√N	√10N
701	491 401	26.4764	83.7257	751	564 001	27.4044	86.6603
702	492 804	26.4953	83.7854	752	565 504	27.4226	86.7179
703	494 209	26.5141	83.8451	753	567 009	27.4408	86.7756
704	495 616	26.5330	83.9047	754	568 516	27.4591	86.8332
705	497 025	26.5518	83.9643	755	570 025	27.4773	86.8907
706	498 436	26.5707	84.0238	756	571 536	27.4955	86.9483
707	499 849	26.5895	84.0833	757	573 049	27.5136	87.0058
708	501 264	26.6083	84.1427	758	574 564	27.5318	87.0632
709	502 681	26.6271	84.2021	759	576 081	27.5500	87.1206
710	504 100	26.6458	84.2615	760	577 600	27.5681	87.1780
711	505 521	26.6646	84.3208	761	579 121	27.5862	87.2353
712	506 944	26.6833	84.3801	762	580 644	27.6044	87.2926
713	508 369	26.7021	84.4393	763	582 169	27.6225	87.3499
714	509 796	26.7208	84.4985	764	583 696	27.6405	87.4071
715	511 225	26.7395	84.5577	765	585 225	27.6586	87.4643
716	512 656	26.7582	84.6168	766	586 756	27.6767	87.5214
717	514 089	26.7769	84.6759	767	588 289	27.6948	87.5785
718	515 524	26.7955	84.7349	768	589 824	27.7128	87.6356
719	516 961	26.8142	84.7939	769	591 361	27.7309	87.6926
720	518 400	26.8328	84.8528	770	592 900	27.7489	87.7496
721	519 841	26.8514	84.9117	771	594 441	27.7669	87.8066
722	521 284	26.8701	84.9706	772	595 984	27.7849	87.8635
723	522 729	26.8887	85.0294	773	597 529	27.8029	87.9204
724	524 176	26.9072	85.0882	774	599 076	27.8209	87.9773
725	525 625	26.9258	85.1469	775	600 625	27.8388	88.0341
726	527 076	26.9444	85.2056	776	602 176	27.8568	88.0909
727	528 529	26.9629	85.2643	777	603 729	27.8747	88.1476
728	529 984	26.9815	85.3229	778	605 284	27.8927	88.2043
729	531 441	27.0000	85.3815	779	606 841	27.9106	88.2610
730	532 900	27.0185	85.4400	780	608 400	27.9285	88.3176
731	534 361	27.0370	85.4985	781	609 961	27.9464	88.3742
732	535 824	27.0555	85.5570	782	611 524	27.9643	88.4308
733	537 289	27.0740	85.6154	783	613 089	27.9821	88.4873
734	538 756	27.0924	85.6738	784	614 656	28.0000	88.5438
735	540 225	27.1109	85.7321	785	616 225	28.0179	88.6002
736	541 696	27.1293	85.7905	786	617 796	28.0357	88.6566
737	543 169	27.1478	85.8487	787	619 369	28.0535	88.7130
738	544 644	27.1662	85.9069	788	620 944	28.0713	88.7694
739	546 121	27.1846	85.9651	789	622 521	28.0891	88.8257
740	547 600	27.2029	86.0233	790	624 100	28.1069	88.8819
741	549 081	27.2213	86.0814	791	625 681	28.1247	88.9382
742	550 564	27.2397	86.1394	792	627 264	28.1425	88.9944
743	552 049	27.2580	86.1974	793	628 849	28.1603	89.0506
744	553 536	27.2764	86.2554	794	630 436	28.1780	89.1067
745	555 025	27.2947	86.3134	795	632 025	28.1957	89.1628
746	556 516	27.3130	86.3713	796	633 616	28.2135	89.2188
747	558 009	27.3313	86.4292	797	635 209	28.2312	89.2749
748	559 504	27.3496	86.4870	798	636 804	28.2489	89.3308
749	561 001	27.3679	86.5448	799	638 401	28.2666	89.3868
750	562 500	27.3861	86.6025	800	640 000	28.2843	89.4427
N	N²	√N	√10N	N	N²	√N	√10N

Table of squares and square roots

N	N²	√N	√10N	N	N²	√N	√10N
801	641 601	28.3019	89.4986	851	724 201	29.1719	92.2497
802	643 204	28.3196	89.5545	852	725 904	29.1890	92.3038
803	644 809	28.3372	89.6103	853	727 609	29.2062	92.3580
804	646 416	28.3549	89.6660	854	729 316	29.2233	92.4121
805	648 025	28.3725	89.7218	855	731 025	29.2404	92.4662
806	649 636	28.3901	89.7775	856	732 736	29.2575	92.5203
807	651 249	28.4077	89.8332	857	734 449	29.2746	92.5743
808	652 864	28.4253	89.8888	858	736 164	29.2916	92.6283
809	654 481	28.4429	89.9444	859	737 881	29.3087	92.6823
810	656 100	28.4605	90.0000	860	739 600	29.3258	92.7362
811	657 721	28.4781	90.0555	861	741 321	29.3428	92.7901
812	659 344	28.4956	90.1110	862	743 044	29.3598	92.8440
813	660 969	28.5132	90.1665	863	744 769	29.3769	92.8978
814	662 596	28.5307	90.2220	864	746 496	29.3939	92.9516
815	664 225	28.5482	90.2774	865	748 225	29.4109	93.0054
816	665 856	28.5657	90.3327	866	749 956	29.4279	93.0591
817	667 489	28.5832	90.3880	867	751 689	29.4449	93.1128
818	669 124	28.6007	90.4434	868	753 424	29.4618	93.1665
819	670 761	28.6182	90.4986	869	755 161	29.4788	93.2202
820	672 400	28.6356	90.5538	870	756 900	29.4958	93.2738
821	674 041	28.6531	90.6091	871	758 641	29.5127	93.3274
822	675 684	28.6705	90.6642	872	760 384	29.5296	93.3809
823	677 329	28.6880	90.7193	873	762 129	29.5466	93.4345
824	678 976	28.7054	90.7744	874	763 876	29.5635	93.4880
825	680 625	28.7228	90.8295	875	765 625	29.5804	93.5414
826	682 276	28.7402	90.8845	876	767 376	29.5973	93.5949
827	683 929	28.7576	90.9395	877	769 129	29.6142	93.6483
828	685 584	28.7750	90.9945	878	770 884	29.6311	93.7017
829	687 241	28.7924	91.0494	879	772 641	29.6479	93.7550
830	688 900	28.8097	91.1043	880	774 400	29.6648	93.8083
831	690 561	28.8271	91.1592	881	776 161	29.6816	93.8616
832	692 224	28.8444	91.2140	882	777 924	29.6985	93.9149
833	693 889	28.8617	91.2688	883	779 689	29.7153	93.9681
834	695 556	28.8791	91.3236	884	781 456	29.7321	94.0213
835	697 225	28.8964	91.3783	885	783 225	29.7489	94.0744
836	698 896	28.9137	91.4330	886	784 996	29.7657	94.1276
837	700 569	28.9310	91.4877	887	786 769	29.7825	94.1807
838	702 244	28.9482	91.5423	888	788 544	29.7993	94.2337
839	703 921	28.9655	91.5969	889	790 321	29.8161	94.2868
840	705 600	28.9828	91.6515	890	792 100	29.8329	94.3398
841	707 281	29.0000	91.7061	891	793 881	29.8496	94.3928
842	708 964	29.0172	91.7606	892	795 664	29.8664	94.4458
843	710 649	29.0345	91.8150	893	797 449	29.8831	94.4987
844	712 336	29.0517	91.8695	894	799 236	29.8998	94.5516
845	714 025	29.0689	91.9239	895	801 025	29.9165	94.6044
846	715 716	29.0861	91.9783	896	802 816	29.9333	94.6573
847	717 409	29.1033	92.0326	897	804 609	29.9500	94.7101
848	719 104	29.1204	92.0869	898	806 404	29.9666	94.7629
849	720 801	29.1376	92.1412	899	808 201	29.9833	94.8156
850	722 500	29.1548	92.1954	900	810 000	30.0000	94.8683
N	N²	√N	√10N	N	N²	√N	√10N

Table of squares and square roots

N	N²	√N	√10N	N	N²	√N	√10N
901	811 801	30.0167	94.9210	951	904 401	30.8383	97.5192
902	813 604	30.0333	94.9737	952	906 304	30.8545	97.5705
903	815 409	30.0500	95.0263	953	908 209	30.8707	97.6217
904	817 216	30.0666	95.0789	954	910 116	30.8869	97.6729
905	819 025	30.0832	95.1315	955	912 025	30.9031	97.7241
906	820 836	30.0998	95.1840	956	913 936	30.9193	97.7753
907	822 649	30.1164	95.2365	957	915 849	30.9354	97.8264
908	824 464	30.1330	95.2890	958	917 764	30.9516	97.8775
909	826 281	30.1496	95.3415	959	919 681	30.9677	97.9285
910	828 100	30.1662	95.3939	960	921 600	30.9839	97.9796
911	829 921	30.1828	95.4463	961	923 521	31.0000	98.0306
912	831 744	30.1993	95.4987	962	925 444	31.0161	98.0816
913	833 569	30.2159	95.5510	963	927 369	31.0322	98.1326
914	835 396	30.2324	95.6033	964	929 296	31.0484	98.1835
915	837 225	30.2490	95.6556	965	931 225	31.0645	98.2344
916	839 056	30.2655	95.7079	966	933 156	31.0805	98.2853
917	840 889	30.2820	95.7601	967	935 089	31.0966	98.3362
918	842 724	30.2985	95.8123	968	937 024	31.1127	98.3870
919	844 561	30.3150	95.8645	969	938 961	31.1288	98.4378
920	846 400	30.3315	95.9166	970	940 900	31.1448	98.4886
921	848 241	30.3480	95.9688	971	942 841	31.1609	98.5393
922	850 084	30.3645	96.0208	972	944 784	31.1769	98.5901
923	851 929	30.3809	96.0729	973	946 729	31.1929	98.6408
924	853 776	30.3974	96.1249	974	948 676	31.2090	98.6914
925	855 625	30.4138	96.1769	975	950 625	31.2250	98.7421
926	857 476	30.4303	96.2289	976	952 576	31.2410	98.7927
927	859 329	30.4467	96.2808	977	954 529	31.2570	98.8433
928	861 184	30.4631	96.3328	978	956 484	31.2730	98.8939
929	863 041	30.4795	96.3846	979	958 441	31.2890	98.9444
930	864 900	30.4959	96.4365	980	960 400	31.3049	98.9949
931	866 761	30.5123	96.4883	981	962 361	31.3209	99.0454
932	868 624	30.5287	96.5401	982	964 324	31.3369	99.0959
933	870 489	30.5450	96.5919	983	966 289	31.3528	99.1464
934	872 356	30.5614	96.6437	984	968 256	31.3688	99.1968
935	874 225	30.5778	96.6954	985	970 225	31.3847	99.2472
936	876 096	30.5941	96.7471	986	972 196	31.4006	99.2975
937	877 969	30.6105	96.7988	987	974 169	31.4165	99.3479
938	879 844	30.6268	96.8504	988	976 144	31.4325	99.3982
939	881 721	30.6431	96.9020	989	978 121	31.4484	99.4485
940	883 600	30.6594	96.9536	990	980 100	31.4643	99.4987
941	885 481	30.6757	97.0052	991	982 081	31.4801	99.5490
942	887 364	30.6920	97.0567	992	984 064	31.4960	99.5992
943	889 249	30.7083	97.1082	993	986 049	31.5119	99.6494
944	891 136	30.7246	97.1597	994	988 036	31.5278	99.6996
945	893 025	30.7408	97.2111	995	990 025	31.5436	99.7497
946	894 916	30.7571	97.2625	996	992 016	31.5595	99.7998
947	896 809	30.7734	97.3139	997	994 009	31.5753	99.8499
948	898 704	30.7896	97.3653	998	996 004	31.5911	99.8999
949	900 601	30.8058	97.4166	999	998 001	31.6070	99.9500
950	902 500	30.8221	97.4679	1000	1000 000	31.6228	100.0000
N	N²	√N	√10N	N	N²	√N	√10N

Appendix C
Table of Combinatorials

The table in this appendix gives the values for the combination of n items taken x at a time up to n and x equal to 20. It should be remembered that, in using combinations, the order in which the items are chosen is not of importance and no item can be chosen more than once.

To illustrate the use of this table, suppose we had the following combination problem:

$$C_4^{15},$$

where 15 is n (the total number of items) and 4 is x (the number of items drawn). To find the total number of possible combinations, first find the correct column, in this case the column headed by $\binom{n}{4}$. Next, this column is moved down until the row is found that represents the value of n, in this case 15. As seen from the table, the C_4^{15} is equal to 1,365.

For values of x greater than 10, the identity

$$C_x^n = C_{n-x}^n$$

can be used to determine the total number of combinations. For example, the determination of the total number of possible combinations for n equal to 15 and x equal to 11 is given as follows:

$$C_{11}^{15} = C_4^{15}$$

which as seen earlier is equal to 1,365.

Table of combinatorials

n	C_0^n	C_1^n	C_2^n	C_3^n	C_4^n	C_5^n	C_6^n	C_7^n	C_8^n	C_9^n	C_{10}^n
0	1	0	0	0	0	0	0	0	0	0	0
1	1	1	0	0	0	0	0	0	0	0	0
2	1	2	1	0	0	0	0	0	0	0	0
3	1	3	3	1	0	0	0	0	0	0	0
4	1	4	6	4	1	0	0	0	0	0	0
5	1	5	10	10	5	1	0	0	0	0	0
6	1	6	15	20	15	6	1	0	0	0	0
7	1	7	21	35	35	21	7	1	0	0	0
8	1	8	28	56	70	56	28	8	1	0	0
9	1	9	36	84	126	126	84	36	9	1	0
10	1	10	45	120	210	252	210	120	45	10	1
11	1	11	55	165	330	462	462	330	165	55	11
12	1	12	66	220	495	792	924	792	495	220	66
13	1	13	78	286	715	1287	1716	1716	1287	715	286
14	1	14	91	364	1001	2002	3003	3432	3003	2002	1001
15	1	15	105	455	1365	3003	5005	6435	6435	5005	3003
16	1	16	120	560	1820	4368	8008	11440	12870	11440	8008
17	1	17	136	680	2380	6188	12376	19448	24310	24310	19448
18	1	18	153	816	3060	8568	18564	31824	43758	48620	43758
19	1	19	171	969	3876	11628	27132	50388	75582	92378	92378
20	1	20	190	1140	4845	15504	38760	77520	125970	167960	184756

Appendix D
The Binomial Probability Distribution

This appendix is actually a collection of the values of several binomial probability distributions. There is a distribution included for each value of n from 1 to 20.

As an illustration of how the tables are used, consider the following problem:

$$C_2^3(0.50)^2(0.50)^1.$$

Remember that the general form of this equation is

$$C_x^n \pi^x(1 - \pi)^{n-x}.$$

Since $n = 3$, refer to the section of the table headed by $n = 3$. Across the top of the table are a series of column headings for specific values of π. Find the one labeled 0.50. Now, referring to the column headed by x, find the row designating $x = 2$. Reading in that row of the 0.50 column, you will find the value 0.3750. This is the solution to the problem.

Notice that the values of π terminate at 0.5 in the table. To determine probabilities associated with values of π greater than 0.5, simply subtract π from one $(1 - \pi)$, and let $x = (n - x)$, then proceed as before. For example,

$$C_4^5(0.6)^4(0.4)^1.$$

Using the table for $n = 5$, look under the column headed 0.4 $(1 - 0.6)$ to the row corresponding with $x = 1$ $(5 - 4)$. The answer to the problem is 0.2592.

The binomial probability distribution

n	x	.05	.10	.15	.20	.25	π .30	.35	.40	.45	.50
1	0	.9500	.9000	.8500	.8000	.7500	.7000	.6500	.6000	.5500	.5000
	1	.0500	.1000	.1500	.2000	.2500	.3000	.3500	.4000	.4500	.5000
2	0	.9025	.8100	.7225	.6400	.5625	.4900	.4225	.3600	.3025	.2500
	1	.0950	.1800	.2550	.3200	.3750	.4200	.4550	.4800	.4950	.5000
	2	.0025	.0100	.0225	.0400	.0625	.0900	.1225	.1600	.2025	.2500
3	0	.8574	.7290	.6141	.5120	.4219	.3430	.2746	.2160	.1664	.1250
	1	.1354	.2430	.3251	.3840	.4219	.4410	.4436	.4320	.4084	.3750
	2	.0071	.0270	.0574	.0960	.1406	.1890	.2389	.2880	.3341	.3750
	3	.0001	.0010	.0034	.0080	.0156	.0270	.0429	.0640	.0911	.1250
4	0	.8145	.6561	.5220	.4096	.3164	.2401	.1785	.1296	.0915	.0625
	1	.1715	.2916	.3685	.4096	.4219	.4116	.3845	.3456	.2995	.2500
	2	.0135	.0486	.0975	.1536	.2109	.2646	.3105	.3456	.3675	.3750
	3	.0005	.0036	.0115	.0256	.0469	.0756	.1115	.1536	.2005	.2500
	4	.0000	.0001	.0005	.0016	.0039	.0081	.0150	.0256	.0410	.0625
5	0	.7738	.5905	.4437	.3277	.2373	.1681	.1160	.0778	.0503	.0313
	1	.2036	.3280	.3915	.4096	.3955	.3601	.3124	.2592	.2059	.1563
	2	.0214	.0729	.1382	.2048	.2637	.3087	.3364	.3456	.3369	.3125
	3	.0011	.0081	.0244	.0512	.0879	.1323	.1811	.2304	.2757	.3125
	4	.0000	.0004	.0022	.0064	.0146	.0283	.0488	.0768	.1128	.1562
	5	.0000	.0000	.0001	.0003	.0010	.0024	.0053	.0102	.0185	.0312
6	0	.7351	.5314	.3771	.2621	.1780	.1176	.0754	.0467	.0277	.0156
	1	.2321	.3543	.3993	.3932	.3560	.3025	.2437	.1866	.1359	.0938
	2	.0305	.0984	.1762	.2458	.2966	.3241	.3280	.3110	.2780	.2344
	3	.0021	.0146	.0415	.0819	.1318	.1852	.2355	.2765	.3032	.3125
	4	.0001	.0012	.0055	.0154	.0330	.0595	.0951	.1382	.1861	.2344
	5	.0000	.0001	.0004	.0015	.0044	.0102	.0205	.0369	.0609	.0937
	6	.0000	.0000	.0000	.0001	.0002	.0007	.0018	.0041	.0083	.0156
7	0	.6983	.4783	.3206	.2097	.1335	.0824	.0490	.0280	.0152	.0078
	1	.2573	.3720	.3960	.3670	.3115	.2471	.1848	.1306	.0872	.0547
	2	.0406	.1240	.2097	.2753	.3115	.3177	.2985	.2613	.2140	.1641
	3	.0036	.0230	.0617	.1147	.1730	.2269	.2679	.2903	.2918	.2734
	4	.0002	.0026	.0109	.0287	.0577	.0972	.1442	.1935	.2388	.2734
	5	.0000	.0002	.0012	.0043	.0115	.0250	.0466	.0774	.1172	.1641
	6	.0000	.0000	.0001	.0004	.0013	.0036	.0084	.0172	.0320	.0547
	7	.0000	.0000	.0000	.0000	.0001	.0002	.0006	.0016	.0037	.0078
8	0	.6634	.4305	.2725	.1678	.1001	.0576	.0319	.0168	.0084	.0039
	1	.2793	.3826	.3847	.3355	.2670	.1977	.1373	.0896	.0548	.0313
	2	.0515	.1488	.2376	.2936	.3115	.2965	.2587	.2090	.1569	.1094
	3	.0054	.0331	.0839	.1468	.2076	.2541	.2786	.2787	.2568	.2188
	4	.0004	.0046	.0185	.0459	.0865	.1361	.1875	.2322	.2627	.2734
	5	.0000	.0004	.0026	.0092	.0231	.0467	.0808	.1239	.1719	.2188
	6	.0000	.0000	.0002	.0011	.0038	.0100	.0217	.0413	.0703	.1094
	7	.0000	.0000	.0000	.0001	.0004	.0012	.0033	.0079	.0164	.0312
	8	.0000	.0000	.0000	.0000	.0000	.0001	.0002	.0007	.0017	.0039
9	0	.6302	.3874	.2316	.1342	.0751	.0404	.0207	.0101	.0046	.0020
	1	.2985	.3874	.3679	.3020	.2253	.1556	.1004	.0605	.0339	.0176
	2	.0629	.1722	.2597	.3020	.3003	.2668	.2162	.1612	.1110	.0703
	3	.0077	.0446	.1069	.1762	.2336	.2668	.2716	.2508	.2119	.1641
	4	.0006	.0074	.0283	.0661	.1168	.1715	.2194	.2508	.2600	.2461
	5	.0000	.0008	.0050	.0165	.0389	.0735	.1181	.1672	.2128	.2461
	6	.0000	.0001	.0006	.0028	.0087	.0210	.0424	.0743	.1160	.1641
	7	.0000	.0000	.0000	.0003	.0012	.0039	.0098	.0212	.0407	.0703
	8	.0000	.0000	.0000	.0000	.0001	.0004	.0013	.0035	.0083	.0176
	9	.0000	.0000	.0000	.0000	.0000	.0000	.0001	.0003	.0008	.0020

The binomial probability distribution

n	x	.05	.10	.15	.20	.25	π .30	.35	.40	.45	.50
10	0	.5987	.3487	.1969	.1074	.0563	.0282	.0135	.0060	.0025	.0010
	1	.3151	.3874	.3474	.2684	.1877	.1211	.0725	.0403	.0207	.0098
	2	.0746	.1937	.2759	.3020	.2816	.2335	.1757	.1209	.0763	.0439
	3	.0105	.0574	.1298	.2013	.2503	.2668	.2522	.2150	.1665	.1172
	4	.0010	.0112	.0401	.0881	.1460	.2001	.2377	.2508	.2384	.2051
	5	.0001	.0015	.0085	.0264	.0584	.1029	.1536	.2007	.2340	.2461
	6	.0000	.0001	.0012	.0055	.0162	.0368	.0689	.1115	.1596	.2051
	7	.0000	.0000	.0001	.0008	.0031	.0090	.0212	.0425	.0746	.1172
	8	.0000	.0000	.0000	.0001	.0004	.0014	.0043	.0106	.0229	.0439
	9	.0000	.0000	.0000	.0000	.0000	.0001	.0005	.0016	.0042	.0098
	10	.0000	.0000	.0000	.0000	.0000	.0000	.0000	.0001	.0003	.0010
11	0	.5688	.3138	.1673	.0859	.0422	.0198	.0088	.0036	.0014	.0005
	1	.3293	.3835	.3248	.2362	.1549	.0932	.0518	.0266	.0125	.0054
	2	.0867	.2131	.2866	.2953	.2581	.1998	.1395	.0887	.0513	.0269
	3	.0137	.0710	.1517	.2215	.2581	.2568	.2254	.1774	.1259	.0806
	4	.0014	.0158	.0536	.1107	.1721	.2201	.2428	.2365	.2060	.1611
	5	.0001	.0025	.0132	.0388	.0803	.1321	.1830	.2207	.2360	.2256
	6	.0000	.0003	.0023	.0097	.0268	.0566	.0985	.1471	.1931	.2256
	7	.0000	.0000	.0003	.0017	.0064	.0173	.0379	.0701	.1128	.1611
	8	.0000	.0000	.0000	.0002	.0011	.0037	.0102	.0234	.0462	.0806
	9	.0000	.0000	.0000	.0000	.0001	.0005	.0018	.0052	.0126	.0269
	10	.0000	.0000	.0000	.0000	.0000	.0000	.0002	.0007	.0021	.0054
	11	.0000	.0000	.0000	.0000	.0000	.0000	.0000	.0000	.0002	.0005
12	0	.5404	.2824	.1422	.0687	.0317	.0138	.0057	.0022	.0008	.0002
	1	.3413	.3766	.3012	.2062	.1267	.0712	.0368	.0174	.0075	.0029
	2	.0988	.2301	.2924	.2835	.2323	.1678	.1088	.0639	.0339	.0161
	3	.0173	.0852	.1720	.2362	.2581	.2397	.1954	.1419	.0923	.0537
	4	.0021	.0213	.0683	.1329	.1936	.2311	.2367	.2128	.1700	.1208
	5	.0002	.0038	.0193	.0532	.1032	.1585	.2039	.2270	.2225	.1934
	6	.0000	.0005	.0040	.0155	.0401	.0792	.1281	.1766	.2124	.2256
	7	.0000	.0000	.0006	.0033	.0115	.0291	.0591	.1009	.1489	.1934
	8	.0000	.0000	.0001	.0005	.0024	.0078	.0199	.0420	.0762	.1208
	9	.0000	.0000	.0000	.0001	.0004	.0015	.0048	.0125	.0277	.0537
	10	.0000	.0000	.0000	.0000	.0000	.0002	.0008	.0025	.0068	.0161
	11	.0000	.0000	.0000	.0000	.0000	.0000	.0001	.0003	.0010	.0029
	12	.0000	.0000	.0000	.0000	.0000	.0000	.0000	.0000	.0001	.0002
13	0	.5133	.2542	.1209	.0550	.0238	.0097	.0037	.0013	.0004	.0001
	1	.3512	.3672	.2774	.1787	.1029	.0540	.0259	.0113	.0045	.0016
	2	.1109	.2448	.2937	.2680	.2059	.1388	.0836	.0453	.0220	.0095
	3	.0214	.0997	.1900	.2457	.2516	.2181	.1651	.1107	.0660	.0349
	4	.0028	.0277	.0838	.1535	.2097	.2337	.2222	.1845	.1350	.0873
	5	.0003	.0055	.0266	.0691	.1258	.1803	.2154	.2214	.1989	.1571
	6	.0000	.0008	.0063	.0230	.0559	.1030	.1546	.1968	.2169	.2095
	7	.0000	.0001	.0011	.0058	.0186	.0442	.0833	.1312	.1775	.2095
	8	.0000	.0000	.0001	.0011	.0047	.0142	.0336	.0656	.1089	.1571
	9	.0000	.0000	.0000	.0001	.0009	.0034	.0101	.0243	.0495	.0873
	10	.0000	.0000	.0000	.0000	.0001	.0006	.0022	.0065	.0162	.0349
	11	.0000	.0000	.0000	.0000	.0000	.0001	.0003	.0012	.0036	.0095
	12	.0000	.0000	.0000	.0000	.0000	.0000	.0000	.0001	.0005	.0016
	13	.0000	.0000	.0000	.0000	.0000	.0000	.0000	.0000	.0000	.0001
14	0	.4877	.2288	.1028	.0440	.0178	.0068	.0024	.0008	.0002	.0001
	1	.3593	.3559	.2539	.1539	.0832	.0407	.0181	.0073	.0027	.0009
	2	.1229	.2570	.2912	.2501	.1802	.1134	.0634	.0317	.0141	.0056
	3	.0259	.1142	.2056	.2501	.2402	.1943	.1366	.0845	.0462	.0222
	4	.0037	.0349	.0998	.1720	.2202	.2290	.2022	.1549	.1040	.0611

The binomial probability distribution

n	x	.05	.10	.15	.20	.25	π .30	.35	.40	.45	.50
14	5	.0004	.0078	.0352	.0860	.1468	.1963	.2178	.2066	.1701	.1222
	6	.0000	.0013	.0093	.0322	.0734	.1262	.1759	.2066	.2088	.1833
	7	.0000	.0002	.0019	.0092	.0280	.0618	.1082	.1574	.1952	.2095
	8	.0000	.0000	.0003	.0020	.0082	.0232	.0510	.0918	.1398	.1833
	9	.0000	.0000	.0000	.0003	.0018	.0066	.0183	.0408	.0762	.1222
	10	.0000	.0000	.0000	.0000	.0003	.0014	.0049	.0136	.0312	.0611
	11	.0000	.0000	.0000	.0000	.0000	.0002	.0010	.0033	.0093	.0222
	12	.0000	.0000	.0000	.0000	.0000	.0000	.0001	.0005	.0019	.0056
	13	.0000	.0000	.0000	.0000	.0000	.0000	.0000	.0001	.0002	.0009
	14	.0000	.0000	.0000	.0000	.0000	.0000	.0000	.0000	.0000	.0001
15	0	.4633	.2059	.0874	.0352	.0134	.0047	.0016	.0005	.0001	.0000
	1	.3658	.3432	.2312	.1319	.0668	.0305	.0126	.0047	.0016	.0005
	2	.1348	.2669	.2856	.2309	.1559	.0916	.0476	.0219	.0090	.0032
	3	.0307	.1285	.2184	.2501	.2252	.1700	.1110	.0634	.0318	.0139
	4	.0049	.0428	.1156	.1876	.2252	.2186	.1792	.1268	.0780	.0417
	5	.0006	.0105	.0449	.1032	.1651	.2061	.2123	.1859	.1404	.0916
	6	.0000	.0019	.0132	.0430	.0917	.1472	.1906	.2066	.1914	.1527
	7	.0000	.0003	.0030	.0138	.0393	.0811	.1319	.1771	.2013	.1964
	8	.0000	.0000	.0005	.0035	.0131	.0348	.0710	.1181	.1647	.1964
	9	.0000	.0000	.0001	.0007	.0034	.0116	.0298	.0612	.1048	.1527
	10	.0000	.0000	.0000	.0001	.0007	.0030	.0096	.0245	.0515	.0916
	11	.0000	.0000	.0000	.0000	.0001	.0006	.0024	.0074	.0191	.0417
	12	.0000	.0000	.0000	.0000	.0000	.0001	.0004	.0016	.0052	.0139
	13	.0000	.0000	.0000	.0000	.0000	.0000	.0001	.0003	.0010	.0032
	14	.0000	.0000	.0000	.0000	.0000	.0000	.0000	.0000	.0001	.0005
	15	.0000	.0000	.0000	.0000	.0000	.0000	.0000	.0000	.0000	.0000
16	0	.4401	.1853	.0743	.0281	.0100	.0033	.0010	.0003	.0001	.0000
	1	.3706	.3294	.2096	.1126	.0535	.0228	.0087	.0030	.0009	.0002
	2	.1463	.2745	.2775	.2111	.1336	.0732	.0353	.0150	.0056	.0018
	3	.0359	.1423	.2285	.2463	.2079	.1465	.0888	.0468	.0215	.0085
	4	.0061	.0514	.1311	.2001	.2252	.2040	.1553	.1014	.0572	.0278
	5	.0008	.0137	.0555	.1201	.1802	.2099	.2008	.1623	.1123	.0666
	6	.0001	.0028	.0180	.0550	.1101	.1649	.1982	.1983	.1684	.1222
	7	.0000	.0004	.0045	.0197	.0524	.1010	.1524	.1889	.1969	.1746
	8	.0000	.0001	.0009	.0055	.0197	.0487	.0923	.1417	.1812	.1964
	9	.0000	.0000	.0001	.0012	.0058	.0185	.0442	.0839	.1318	.1746
	10	.0000	.0000	.0000	.0002	.0014	.0056	.0167	.0392	.0755	.1222
	11	.0000	.0000	.0000	.0000	.0002	.0013	.0049	.0142	.0337	.0666
	12	.0000	.0000	.0000	.0000	.0000	.0002	.0011	.0040	.0115	.0278
	13	.0000	.0000	.0000	.0000	.0000	.0000	.0002	.0008	.0029	.0085
	14	.0000	.0000	.0000	.0000	.0000	.0000	.0000	.0001	.0005	.0018
	15	.0000	.0000	.0000	.0000	.0000	.0000	.0000	.0000	.0001	.0002
	16	.0000	.0000	.0000	.0000	.0000	.0000	.0000	.0000	.0000	.0000
17	0	.4181	.1668	.0631	.0225	.0075	.0023	.0007	.0002	.0000	.0000
	1	.3741	.3150	.1893	.0957	.0426	.0169	.0060	.0019	.0005	.0001
	2	.1575	.2800	.2673	.1914	.1136	.0581	.0260	.0102	.0035	.0010
	3	.0415	.1556	.2359	.2393	.1893	.1245	.0701	.0341	.0144	.0052
	4	.0076	.0605	.1457	.2093	.2209	.1868	.1320	.0796	.0411	.0182
	5	.0010	.0175	.0668	.1361	.1914	.2081	.1849	.1379	.0875	.0472
	6	.0001	.0039	.0236	.0680	.1276	.1784	.1991	.1839	.1432	.0944
	7	.0000	.0007	.0065	.0267	.0668	.1201	.1685	.1927	.1841	.1484
	8	.0000	.0001	.0014	.0084	.0279	.0644	.1134	.1606	.1883	.1855
	9	.0000	.0000	.0003	.0021	.0093	.0276	.0611	.1070	.1540	.1855
	10	.0000	.0000	.0000	.0004	.0025	.0095	.0263	.0571	.1008	.1484
	11	.0000	.0000	.0000	.0001	.0005	.0026	.0090	.0242	.0525	.0944
	12	.0000	.0000	.0000	.0000	.0001	.0006	.0024	.0081	.0215	.0472

The binomial probability distribution

n	x	.05	.10	.15	.20	.25	π .30	.35	.40	.45	.50
17	13	.0000	.0000	.0000	.0000	.0000	.0001	.0005	.0021	.0068	.0182
	14	.0000	.0000	.0000	.0000	.0000	.0000	.0001	.0004	.0016	.0052
	15	.0000	.0000	.0000	.0000	.0000	.0000	.0000	.0001	.0003	.0010
	16	.0000	.0000	.0000	.0000	.0000	.0000	.0000	.0000	.0000	.0001
	17	.0000	.0000	.0000	.0000	.0000	.0000	.0000	.0000	.0000	.0000
18	0	.3972	.1501	.0536	.0180	.0056	.0016	.0004	.0001	.0000	.0000
	1	.3763	.3002	.1704	.0811	.0338	.0126	.0042	.0012	.0003	.0001
	2	.1684	.2835	.2556	.1723	.0958	.0458	.0190	.0069	.0022	.0006
	3	.0473	.1680	.2406	.2297	.1704	.1046	.0547	.0246	.0095	.0031
	4	.0093	.0700	.1592	.2153	.2130	.1681	.1104	.0614	.0291	.0117
	5	.0014	.0218	.0787	.1507	.1988	.2017	.1664	.1146	.0666	.0327
	6	.0002	.0052	.0301	.0816	.1436	.1873	.1941	.1655	.1181	.0708
	7	.0000	.0010	.0091	.0350	.0820	.1376	.1792	.1892	.1657	.1214
	8	.0000	.0002	.0022	.0120	.0376	.0811	.1327	.1734	.1864	.1669
	9	.0000	.0000	.0004	.0033	.0139	.0386	.0794	.1284	.1694	.1855
	10	.0000	.0000	.0001	.0008	.0042	.0149	.0385	.0771	.1248	.1669
	11	.0000	.0000	.0000	.0001	.0010	.0046	.0151	.0374	.0742	.1214
	12	.0000	.0000	.0000	.0000	.0002	.0012	.0047	.0145	.0354	.0708
	13	.0000	.0000	.0000	.0000	.0000	.0002	.0012	.0045	.0134	.0327
	14	.0000	.0000	.0000	.0000	.0000	.0000	.0002	.0011	.0039	.0117
	15	.0000	.0000	.0000	.0000	.0000	.0000	.0000	.0002	.0009	.0031
	16	.0000	.0000	.0000	.0000	.0000	.0000	.0000	.0000	.0001	.0006
	17	.0000	.0000	.0000	.0000	.0000	.0000	.0000	.0000	.0000	.0001
	18	.0000	.0000	.0000	.0000	.0000	.0000	.0000	.0000	.0000	.0000
19	0	.3774	.1351	.0456	.0144	.0042	.0011	.0003	.0001	.0000	.0000
	1	.3773	.2852	.1529	.0685	.0268	.0093	.0029	.0008	.0002	.0000
	2	.1787	.2852	.2428	.1540	.0803	.0358	.0138	.0046	.0013	.0003
	3	.0533	.1796	.2428	.2182	.1517	.0869	.0422	.0175	.0062	.0018
	4	.0112	.0798	.1714	.2182	.2023	.1491	.0909	.0467	.0203	.0074
	5	.0018	.0266	.0907	.1636	.2023	.1916	.1468	.0933	.0497	.0222
	6	.0002	.0069	.0374	.0955	.1574	.1916	.1844	.1451	.0949	.0518
	7	.0000	.0014	.0122	.0443	.0974	.1525	.1844	.1797	.1443	.0961
	8	.0000	.0002	.0032	.0166	.0487	.0981	.1489	.1797	.1771	.1442
	9	.0000	.0000	.0007	.0051	.0198	.0514	.0980	.1464	.1771	.1762
	10	.0000	.0000	.0001	.0013	.0066	.0220	.0528	.0976	.1449	.1762
	11	.0000	.0000	.0000	.0003	.0018	.0077	.0233	.0532	.0970	.1442
	12	.0000	.0000	.0000	.0000	.0004	.0022	.0083	.0237	.0529	.0961
	13	.0000	.0000	.0000	.0000	.0001	.0005	.0024	.0085	.0233	.0517
	14	.0000	.0000	.0000	.0000	.0000	.0001	.0006	.0024	.0082	.0222
	15	.0000	.0000	.0000	.0000	.0000	.0000	.0001	.0005	.0022	.0074
	16	.0000	.0000	.0000	.0000	.0000	.0000	.0000	.0001	.0005	.0018
	17	.0000	.0000	.0000	.0000	.0000	.0000	.0000	.0000	.0001	.0003
	18	.0000	.0000	.0000	.0000	.0000	.0000	.0000	.0000	.0000	.0000
	19	.0000	.0000	.0000	.0000	.0000	.0000	.0000	.0000	.0000	.0000
20	0	.3585	.1216	.0388	.0115	.0032	.0008	.0002	.0000	.0000	.0000
	1	.3774	.2702	.1368	.0576	.0211	.0068	.0020	.0005	.0001	.0000
	2	.1887	.2852	.2293	.1369	.0669	.0278	.0100	.0031	.0008	.0002
	3	.0596	.1901	.2428	.2054	.1339	.0716	.0323	.0123	.0040	.0011
	4	.0133	.0898	.1821	.2182	.1897	.1304	.0738	.0350	.0139	.0046
	5	.0022	.0319	.1028	.1746	.2023	.1789	.1272	.0746	.0365	.0148
	6	.0003	.0089	.0454	.1091	.1686	.1916	.1712	.1244	.0746	.0370
	7	.0000	.0020	.0160	.0545	.1124	.1643	.1844	.1659	.1221	.0739
	8	.0000	.0004	.0046	.0222	.0609	.1144	.1613	.1797	.1623	.1201
	9	.0000	.0001	.0011	.0074	.0271	.0654	.1158	.1597	.1771	.1602
	10	.0000	.0000	.0002	.0020	.0099	.0308	.0686	.1171	.1593	.1762

The binomial probability distribution

n	x						π				
		.05	.10	.15	.20	.25	.30	.35	.40	.45	.50
20	11	.0000	.0000	.0000	.0005	.0030	.0120	.0336	.0710	.1185	.1602
	12	.0000	.0000	.0000	.0001	.0008	.0039	.0136	.0355	.0727	.1201
	13	.0000	.0000	.0000	.0000	.0002	.0010	.0045	.0146	.0366	.0739
	14	.0000	.0000	.0000	.0000	.0000	.0002	.0012	.0049	.0150	.0370
	15	.0000	.0000	.0000	.0000	.0000	.0000	.0003	.0013	.0049	.0148
	16	.0000	.0000	.0000	.0000	.0000	.0000	.0000	.0003	.0013	.0046
	17	.0000	.0000	.0000	.0000	.0000	.0000	.0000	.0000	.0002	.0011
	18	.0000	.0000	.0000	.0000	.0000	.0000	.0000	.0000	.0000	.0002
	19	.0000	.0000	.0000	.0000	.0000	.0000	.0000	.0000	.0000	.0000
	20	.0000	.0000	.0000	.0000	.0000	.0000	.0000	.0000	.0000	.0000

This appendix is actually a series of tables that gives the cumulative values of the binomial probability distribution. It should be remembered that it is appropriate to use the binomial distribution whenever the simple events are independent, the outcome of a simple event can be classified into a dichotomy, and the number of successful simple events is a discrete value.

To illustrate the use of these tables, suppose we have a problem in which the probability of success on any simple event π is equal to 0.4, the number of simple events observed n is equal to 5, and the number of successful simple events X is equal to either 0, 1, 2, or 3. In other words, we want to find

$$\sum_{x=0}^{3} C_x^5 (0.4)^x (0.6)^{5-x}.$$

First move to the table headed by $n = 5$. Upon finding this table, locate the column with a π of 0.4. Moving down this column, find the row representing 3 successful simple events. Reading from this intersection, we see that the cumulative probability of obtaining a 0, 1, 2, or 3 in this problem is equal to 0.9130. The probability of obtaining 4 or 5 successful simple events in this problem is equal to $1.0000 - 0.9130$ or 0.0870, since a probability distribution sums to one.

If a problem is encountered where $\pi > 0.50$, this table can also be used in its solution. For example, suppose we have a problem in which $n = 5$, $\pi = 0.6$, and $X \le 2$. For our illustration, the probability of 0, 1 or 2 successes is the same as the probability of 3, 4, or 5 failures. In the table we can find the probability of $X \le 2$ failures and subtract from 1.0 to determine the probability of $X > 2$. From the table headed $n = 5$ under the 0.40 column,

$P(X \le 2) = 0.6826;$

then

$P(X > 2) = 1 - 0.6826 = 0.3174.$

Thus the probability of $X \le 2$ when $\pi = 0.60$ is 0.3174, which is the same as the probability of $X > 2$ when $(1 - \pi) = 0.40$.

Table of cumulative binomial

n = 1

π	.01	.05	.10	.20	.30	.40	.50
x							
0	0.9900	0.9500	0.9000	0.8000	0.7000	0.6000	0.5000
1	1.0000	1.0000	1.0000	1.0000	1.0000	1.0000	1.0000

n = 2

π	.01	.05	.10	.20	.30	.40	.50
x							
0	0.9801	0.9025	0.8100	0.6400	0.4900	0.3600	0.2500
1	0.9999	0.9975	0.9900	0.9600	0.9100	0.8400	0.7500
2	1.0000	1.0000	1.0000	1.0000	1.0000	1.0000	1.0000

n = 3

π	.01	.05	.10	.20	.30	.40	.50
x							
0	0.9703	0.8574	0.7290	0.5120	0.3430	0.2160	0.1250
1	0.9997	0.9927	0.9720	0.8960	0.7840	0.6480	0.5000
2	1.0000	0.9999	0.9990	0.9920	0.9730	0.9360	0.8750
3		1.0000	1.0000	1.0000	1.0000	1.0000	1.0000

n = 4

π	.01	.05	.10	.20	.30	.40	.50
x							
0	0.9606	0.8145	0.6561	0.4096	0.2401	0.1296	0.0625
1	0.9994	0.9860	0.9477	0.8192	0.6517	0.4752	0.3125
2	1.0000	0.9995	0.9963	0.9728	0.9163	0.8208	0.6875
3		1.0000	0.9999	0.9984	0.9919	0.9744	0.9375
4			1.0000	1.0000	1.0000	1.0000	1.0000

n = 5

π	.01	.05	.10	.20	.30	.40	.50
x							
0	0.9510	0.7738	0.5905	0.3277	0.1681	0.0778	0.0313
1	0.9990	0.9774	0.9185	0.7373	0.5282	0.3370	0.1875
2	1.0000	0.9988	0.9914	0.9421	0.8369	0.6826	0.5000
3		1.0000	0.9995	0.9933	0.9692	0.9130	0.8125
4			1.0000	0.9997	0.9976	0.9898	0.9688
5				1.0000	1.0000	1.0000	1.0000

n = 6

π	.01	.05	.10	.20	.30	.40	.50
x							
0	0.9415	0.7351	0.5314	0.2621	0.1176	0.0467	0.0156
1	0.9985	0.9672	0.8857	0.6554	0.4202	0.2333	0.1094
2	1.0000	0.9978	0.9841	0.9011	0.7443	0.5443	0.3438
3		0.9999	0.9987	0.9830	0.9295	0.8208	0.6563
4		1.0000	0.9999	0.9984	0.9891	0.9590	0.8906
5			1.0000	0.9999	0.9993	0.9959	0.9844
6			1.0000	1.0000	1.0000	1.0000	

Table of cumulative binomial

n = 7

π	.01	.05	.10	.20	.30	.40	.50
x							
0	0.9321	0.6983	0.4783	0.2097	0.0824	0.0280	0.0078
1	0.9980	0.9556	0.8503	0.5767	0.3294	0.1586	0.0625
2	1.0000	0.9962	0.9743	0.8520	0.6471	0.4199	0.2266
3		0.9998	0.997.	0.9667	0.8740	0.7102	0.5000
4		1.0000	0.9998	0.9953	0.9712	0.9037	0.7734
5			1.0000	0.9996	0.9962	0.9812	0.9375
6				1.0000	0.9998	0.9984	0.9922
7					1.0000	1.0000	1.0000

n = 8

π	.01	.05	.10	.20	.30	.40	.50
x							
0	0.9227	0.6634	0.4305	0.1678	0.0576	0.0168	0.0039
1	0.9973	0.9428	0.8131	0.5033	0.2553	0.1064	0.0352
2	0.9999	0.9942	0.9619	0.7969	0.5518	0.3154	0.1445
3	1.0000	0.9996	0.9950	0.9437	0.8059	0.5941	0.3633
4		1.0000	0.9996	0.9896	0.9420	0.8263	0.6367
5			1.0000	0.9988	0.9887	0.9502	0.8555
6				0.9999	0.9987	0.9915	0.9648
7				1.0000	0.9999	0.9993	0.9961
8					1.0000	1.0000	1.0000

n = 9

π	.01	.05	.10	.20	.30	.40	.50
x							
0	0.9135	0.6302	0.3874	0.1342	0.0404	0.0101	0.0020
1	0.9966	0.9288	0.7748	0.4362	0.1960	0.0705	0.0195
2	0.9999	0.9916	0.9470	0.7382	0.4628	0.2318	0.0898
3	1.0000	0.9994	0.9917	0.9144	0.7297	0.4826	0.2539
4		1.0000	0.9991	0.9804	0.9012	0.7334	0.5000
5			0.9999	0.9969	0.9747	0.9006	0.7461
6			1.0000	0.9997	0.9957	0.9750	0.9102
7				1.0000	0.9996	0.9962	0.9805
8					1.0000	0.9997	0.9980
9						1.0000	1.0000

n = 10

π	.01	.05	.10	.20	.30	.40	.50
x							
0	0.9044	0.5987	0.3487	0.1074	0.0282	0.0060	0.0010
1	0.9957	009139	0.7361	0.3758	0.1493	0.0464	0.0107
2	0.9999	0.9885	0.9298	0.6778	0.3828	0.1973	0.0547
3	1.0000	0.9990	0.9872	0.8791	0.6496	0.3823	0.1719
4		0.9999	0.9984	0.9670	0.8497	0.6331	0.3770
5		1.0000	0.9999	0.9936	0.9526	0.8338	0.6230
6			1.0000	0.9991	0.9894	0.9120	0.8281
7				0.9999	0.9999	0.9877	0.9453
8				1.0000	1.0000	0.9983	0.9893
9						0.9999	0.9990
10						1.0000	1.0000

Table of cumulative binomial

n = 11

π	.01	.05	.10	.20	.30	.40	.50
x							
0	0.8953	0.5688	0.3138	0.0859	0.0198	0.0036	0.0005
1	0.9948	0.8981	0.6974	0.3221	0.1130	0.0302	0.0059
2	0.9998	0.9848	0.9104	0.6174	0.3127	0.1189	0.0327
3	1.0000	0.9984	0.9815	0.8369	0.5696	0.2963	0.1133
4		0.9999	0.9972	0.9496	0.7897	0.5328	0.2744
5		1.0000	0.9997	0.9883	0.9218	0.7535	0.5000
6			1.0000	0.9980	0.9784	0.9006	0.7256
7				0.9998	0.9957	0.9707	0.8867
8				1.0000	0.9994	0.9941	0.9673
9					1.0000	0.9993	0.9941
10						1.0000	0.9995
11							1.0000

n = 12

π	.01	.05	.10	.20	.30	.40	.50
x							
0	0.8864	0.5404	0.2824	0.0687	0.0138	0.0022	0.0002
1	0.9938	0.8816	0.6590	0.2749	0.0850	0.0196	0.0032
2	0.9998	0.9804	0.8891	0.5583	0.2528	0.0837	0.0193
3	1.0000	0.9978	0.9744	0.7946	0.4925	0.2253	0.0730
4		0.9998	0.9957	0.9274	0.7237	0.4382	0.1938
5		1.0000	0.9995	0.9806	0.8821	0.6652	0.3872
6			0.9999	0.9961	0.9614	0.8418	0.6128
7			1.0000	0.9994	0.9905	0.9427	0.8062
8				0.9999	0.9983	0.9847	0.9270
9				1.0000	0.9998	0.9970	0.9807
10					1.0000	0.9997	0.9968
11						1.0000	0.9998
12							1.0000

n = 13

π	.01	.05	.10	.20	.30	.40	.50
x							
0	0.8775	0.5133	0.2542	0.0550	0.0097	0.0013	0.0001
1	0.9928	0.8646	0.6213	0.2336	0.0637	0.0126	0.0017
2	0.9997	0.9755	0.8661	0.5017	0.2025	0.0579	0.0112
3	1.0000	0.9969	0.9658	0.7473	0.4206	0.1686	0.0461
4		0.9997	0.9935	0.9009	0.6543	0.3530	0.1334
5		1.0000	0.9991	0.9700	0.8346	0.5744	0.2905
6			0.9999	0.9930	0.9376	0.7712	0.5000
7			1.0000	0.9988	0.9818	0.9023	0.7095
8				0.9998	0.9960	0.9679	0.8666
9				1.0000	0.9993	0.9922	0.9539
10					0.9999	0.9987	0.9888
11					1.0000	0.9999	0.9983
12						1.0000	0.9999
13							1.0000

Table of cumulative binomial

π	.01	.05	n = 14 .10	.20	.30	.40	.50
x							
0	0.8687	0.4877	0.2288	0.0440	0.0068	0.0008	0.0001
1	0.9916	0.8470	0.5846	0.1979	0.0475	0.0081	0.0009
2	0.9997	0.9699	0.8416	0.4481	0.1608	0.0398	0.0065
3	1.0000	0.9958	0.9559	0.6982	0.3552	0.1243	0.0287
4		0.9996	0.9908	0.8702	0.5842	0.2793	0.0898
5		1.0000	0.9985	0.9561	0.7805	0.4859	0.2120
6			0.9998	0.9884	0.9067	0.6925	0.3953
7			1.0000	0.9976	0.9685	0.8499	0.6047
8				0.9996	0.9917	0.9417	0.7880
9				1.0000	0.9983	0.9825	0.9102
10					0.9998	0.9961	0.9713
11					1.0000	0.9994	0.9935
12						0.9999	0.9991
13						1.0000	0.9999
14							1.0000

π	.01	.05	n = 15 .10	.20	.30	.40	.50
x							
0	0.8601	0.4633	0.2059	0.0352	0.0047	0.0005	0.0000
1	0.9904	0.8290	0.5490	0.1671	0.0353	0.0052	0.0005
2	0.9996	0.9638	0.8159	0.3980	0.1268	0.0271	0.0037
3	1.0000	0.9945	0.9444	0.6482	0.2969	0.0905	0.0176
4		0.9994	0.9873	0.8358	0.5155	0.2173	0.0592
5		0.9999	0.9978	0.9389	0.7216	0.4032	0.1509
6		1.0000	0.9997	0.9819	0.9689	0.6098	0.3036
7			1.0000	0.9958	0.9500	0.7869	0.5000
8				0.9992	0.9848	0.9050	0.6964
9				0.9999	0.9963	0.9662	0.8491
10				1.0000	0.9993	0.9907	0.9408
11					0.9999	0.9981	0.9824
12					1.0000	0.9997	0.9963
13						1.0000	0.9995
14							1.0000

π	.01	.05	n = 16 .10	.20	.30	.40	.50
x							
0	0.8515	0.4401	0.1853	0.0281	0.003	.00003	0.0003
1	0.9891	0.8108	0.5147	0.1407	0.0261	0.0000	0.0003
2	0.9995	0.9571	0.7892	0.3518	0.0994	0.0183	0.0021
3	1.0000	0.9930	0.9316	0.5981	0.2459	0.0651	0.0106
4		0.9991	0.9830	0.7982	0.4499	0.1666	0.0384
5		0.9999	0.9967	0.9183	0.6598	0.3288	0.1051
6		1.0000	0.9995	0.9733	0.8247	0.5272	0.2272
7			0.9999	0.9930	0.9256	0.7161	0.4018
8			1.0000	0.9985	0.9743	0.8577	0.5982
9				0.9998	0.9929	0.9417	0.7728
10				1.0000	0.9984	0.9809	0.8949
11					0.9997	0.9951	0.9616
12					1.0000	0.9991	0.9894
13						0.9999	0.9979
14						1.0000	0.9997
15							1.0000

Table of cumulative binomial

π	.01	.05	n = 17 .10	.20	.30	.40	.50
x							
0	0.8429	0.4181	0.1668	0.0225	0.0023	0.0002	0.0000
1	0.9877	0.7922	0.4818	0.1182	0.0193	0.0021	0.0001
2	0.9994	0.9497	0.7618	0.3096	0.0774	0.0123	0.0012
3	1.0000	0.9912	0.9174	0.5489	0.2019	0.0464	0.0064
4		0.9988	0.9779	0.7582	0.3887	0.1260	0.0245
5		0.9999	0.9953	0.8943	0.5968	0.2639	0.0717
6		1.0000	0.9992	0.9623	0.7752	0.4478	0.1662
7			0.9999	0.9891	0.8954	0.6405	0.3145
8			1.0000	0.9974	0.9597	0.8011	0.5000
9				0.9995	0.9873	0.9081	0.6855
10				0.9999	0.9968	0.9652	0.8338
11				1.0000	0.9993	0.9894	0.9283
12					0.9999	0.9975	0.9755
13					1.0000	0.9995	0.9936
14						0.9999	0.9988
15						1.0000	0.9999
16							1.0000

π	.01	.05	n = 18 .10	.20	.30	.40	.50
x							
0	0.8345	0.3972	0.1501	0.0180	0.0016	0.0001	0.0000
1	0.9862	0.7735	0.4503	0.0991	0.0142	0.0013	0.0001
2	0.9993	0.9419	0.7338	0.2713	0.0600	0.0082	0.0007
3	1.0000	0.9891	0.9018	0.5010	0.1646	0.0328	0.0038
4		0.9985	0.9718	0.7064	0.3327	0.0942	0.0154
5		0.9998	0.9936	0.8671	0.5344	0.2088	0.0481
6		1.0000	0.9988				
7			0.9988	0.9487	0.7217	0.3743	0.1189
8			0.9998	0.9837	0.8593	0.5634	0.2403
9			1.0000	0.9957	0.9404	0.7368	0.4073
10				0.9991	0.9790	0.8653	0.5927
11				0.9998	0.9939	0.9424	0.7597
12				1.0000	0.9986	0.9797	0.8811
13					0.9997	0.9942	0.9846
14					1.0000	0.9998	0.9962
15						1.0000	0.9993
16							0.9999
17							1.0000

Table of cumulative binomial

π	.01	.05	.10	.20	.30	.40	.50
x							
0	0.8262	0.3774	0.1351	0.0144	0.0011	0.0001	0.0000
1	0.9847	0.7547	0.4203	0.0829	0.0104	0.0008	0.0000
2	0.9991	0.9335	0.7054	0.2369	0.0462	0.0055	0.0004
3	1.0000	0.9868	0.8850	0.4551	0.1332	0.0230	0.0022
4		0.9980	0.9648	0.6733	0.2822	0.0696	0.0096
5		0.9998	0.9914	0.8369	0.4739	0.1629	0.0318
6		1.0000	0.9983	.09324	0.6655	0.3081	0.0835
7			0.9997	0.9767	0.8180	0.4878	0.1796
8			1.0000	0.9933	0.9161	0.6675	0.3238
9				0.9984	0.9674	0.8139	0.5000
10				0.9997	0.9895	0.9115	0.6762
11				0.9999	0.9972	0.9648	0.8204
12				1.0000	0.9994	0.9884	0.9165
13					0.9999	0.9969	0.9682
14					1.0000	0.9994	0.9904
15						0.9999	0.9978
16						1.0000	0.9996
17							1.0000

π	.01	.05	.10	.20	.30	.40	.50
x							
0	0.8179	0.3585	0.1216	0.0115	0.0008	0.0000	0.0000
1	0.9831	0.7358	0.3917	0.0692	0.0076	0.0005	0.0000
2	0.9999	0.9245	0.6769	0.2061	0.0355	0.0036	0.0002
3	1.0000	0.9841	0.8670	0.4114	0.1071	0.0160	0.0013
4		0.9974	0.9568	0.6296	0.2375	0.0510	0.0059
5		0.9997	0.9887	0.8042	0.4164	0.1256	0.0207
6		1.0000	0.9976	0.9133	0.6080	0.2500	0.0577
7			0.9996	0.9679	0.7703	0.4159	0.1316
8			0.9999	0.9900	0.8867	0.5956	0.2517
9			1.0000	0.9974	0.9520	0.7553	0.4119
10				0.9994	0.9829	0.8725	0.5881
11				0.9999	0.9949	0.9435	0.7483
12				1.0000	0.9987	0.9790	0.8684
13					0.9997	0.9935	0.9423
14					1.0000	0.9984	0.9793
15						0.9997	0.9941
16						1.0000	0.9987
17							0.9998
18							1.0000

APPENDIX E

Appendix F
Poisson Distribution

As pointed out in the text, the Poisson distribution can be most useful in determining certain probabilities whenever the probability, π, for a simple event is relatively small. The table in this appendix gives the probability of various values of x and μ, with the Poisson formula being

$$P(X = x) = \frac{\mu^x e^{-\mu}}{x!}$$

Suppose we have a problem where the probability of the simple event occurring is 0.05, and we are to observe 10 such simple events. In addition, suppose we are interested in finding the probability of 4 of these simple events being successful. Since the mean μ of this problem is equal to $(0.05)(10) = 0.5$, we first locate the column headed by 0.5 in the table. Moving down the column, we find the row representing the number of successful simple events in which we are interested, in this case 4. Reading from the intersection of the 0.5 column and 4 row, we see that the answer to this problem is 0.0016.

Poisson distribution

x	0.1	0.2	0.3	0.4	μ 0.5	0.6	0.7	0.8	0.9	1.0
0	.9048	.8187	.7408	.6703	.6065	.5488	.4966	.4493	.4066	.3679
1	.0905	.1637	.2222	.2681	.3033	.3293	.3476	.3595	.3659	.3679
2	.0045	.0164	.0333	.0536	.0758	.0988	.1217	.1438	.1647	.1839
3	.0002	.0011	.0033	.0072	.0126	.0198	.0284	.0383	.0494	.0613
4	.0000	.0001	.0003	.0007	.0016	.0030	.0050	.0077	.0111	.0153
5	.0000	.0000	.0000	.0001	.0002	.0004	.0007	.0012	.0020	.0031
6	.0000	.0000	.0000	.0000	.0000	.0000	.0001	.0002	.0003	.0005
7	.0000	.0000	.0000	.0000	.0000	.0000	.0000	.0000	.0000	.0001

x	1.1	1.2	1.3	1.4	μ 1.5	1.6	1.7	1.8	1.9	2.0
0	.3329	.3012	.2725	.2466	.2231	.2019	.1827	.1653	.1496	.1353
1	.3662	.3614	.3543	.3452	.3347	.3230	.3106	.2975	.2842	.2707
2	.2014	.2169	.2303	.2417	.2510	.2584	.2640	.2678	.2700	.2707
3	.0738	.0867	.0998	.1128	.1255	.1378	.1496	.1607	.1710	.1804
4	.0203	.0260	.0324	.0395	.0471	.0551	.0636	.0723	.0812	.0902
5	.0045	.0062	.0084	.0111	.0141	.0176	.0216	.0260	.0309	.0361
6	.0008	.0012	.0018	.0026	.0035	.0047	.0061	.0078	.0098	.0120
7	.0001	.0002	.0003	.0005	.0008	.0011	.0015	.0020	.0027	.0034
8	.0000	.0000	.0001	.0001	.0001	.0002	.0003	.0005	.0006	.0009
9	.0000	.0000	.0000	.0000	.0000	.0000	.0001	.0001	.0001	.0002

x	2.1	2.2	2.3	2.4	μ 2.5	2.6	2.7	2.8	2.9	3.0
0	.1225	.1108	.1003	.0907	.0821	.0743	.0672	.0608	.0550	.0498
1	.2572	.2438	.2306	.2177	.2052	.1931	.1815	.1703	.1596	.1494
2	.2700	.2681	.2652	.2613	.2565	.2510	.2450	.2384	.2314	.2240
3	.1890	.1966	.2033	.2090	.2138	.2176	.2205	.2225	.2237	.2240
4	.0992	.1082	.1169	.1254	.1336	.1414	.1488	.1557	.1622	.1680
5	.0417	.0476	.0538	.0602	.0668	.0735	.0804	.0872	.0940	.1008
6	.0146	.0174	.0206	.0241	.0278	.0319	.0362	.0407	.0455	.0504
7	.0044	.0055	.0068	.0083	.0099	.0118	.0139	.0163	.0188	.0216
8	.0011	.0015	.0019	.0025	.0031	.0038	.0047	.0057	.0068	.0081
9	.0003	.0004	.0005	.0007	.0009	.0011	.0014	.0018	.0022	.0027
10	.0001	.0001	.0001	.0002	.0002	.0003	.0004	.0005	.0006	.0008
11	.0000	.0000	.0000	.0000	.0000	.0001	.0001	.0001	.0002	.0002
12	.0000	.0000	.0000	.0000	.0000	.0000	.0000	.0000	.0000	.0001

x	3.1	3.2	3.3	3.4	μ 3.5	3.6	3.7	3.8	3.9	4.0
0	.0450	.0408	.0369	.0334	.0302	.0273	.0247	.0224	.0202	.0183
1	.1397	.1304	.1217	.1135	.1057	.0984	.0915	.0850	.0789	.0733
2	.2165	.2087	.2008	.1929	.1850	.1771	.1692	.1615	.1539	.1465
3	.2237	.2226	.2209	.2186	.2158	.2125	.2087	.2046	.2001	.1954
4	.1733	.1781	.1823	.1858	.1888	.1912	.1931	.1944	.1951	.1954
5	.1075	.1140	.1203	.1264	.1322	.1377	.1429	.1477	.1522	.1563
6	.0555	.0608	.0662	.0716	.0771	.0826	.0881	.0936	.0989	.1042
7	.0246	.0278	.0312	.0348	.0385	.0425	.0466	.0508	.0551	.0595
8	.0095	.0111	.0129	.0148	.0169	.0191	.0215	.0241	.0269	.0298
9	.0033	.0040	.0047	.0056	.0066	.0076	.0089	.0102	.0116	.0132

Poisson distribution

x	μ 3.1	3.2	3.3	3.4	3.5	3.6	3.7	3.8	3.9	4.0
10	.0010	.0013	.0016	.0019	.0023	.0028	.0033	.0039	.0045	.0053
11	.0003	.0004	.0005	.0006	.0007	.0009	.0011	.0013	.0016	.0019
12	.0001	.0001	.0001	.0002	.0002	.0003	.0003	.0004	.0005	.0006
13	.0000	.0000	.0000	.0000	.0001	.0001	.0001	.0001	.0002	.0002
14	.0000	.0000	.0000	.0000	.0000	.0000	.0000	.0000	.0000	.0001

x	μ 4.1	4.2	4.3	4.4	4.5	4.6	4.7	4.8	4.9	5.0
0	.0166	.0150	.0136	.0123	.0111	.0101	.0091	.0082	.0074	.0067
1	.0679	.0630	.0583	.0540	.0500	.0462	.0427	.0395	.0365	.0337
2	.1393	.1323	.1254	.1188	.1125	.1063	.1005	.0948	.0894	.0842
3	.1904	.1852	.1798	.1743	.1687	.1631	.1574	.1517	.1460	.1404
4	.1951	.1944	.1933	.1917	.1898	.1875	.1849	.1820	.1789	.1755
5	.1600	.1633	.1662	.1687	.1708	.1725	.1738	.1747	.1753	.1755
6	.1093	.1143	.1191	.1237	.1281	.1323	.1362	.1398	.1432	.1462
7	.0640	.0686	.0732	.0778	.0824	.0869	.0914	.0959	.1002	.1044
8	.0328	.0360	.0393	.0428	.0463	.0500	.0537	.0575	.0614	.0653
9	.0150	.0168	.0188	.0209	.0232	.0255	.0280	.0307	.0334	.0363
10	.0061	.0071	.0081	.0092	.0104	.0118	.0132	.0147	.0164	.0181
11	.0023	.0027	.0032	.0037	.0043	.0049	.0056	.0064	.0073	.0082
12	.0008	.0009	.0011	.0013	.0016	.0019	.0022	.0026	.0030	.0034
13	.0002	.0003	.0004	.0005	.0006	.0007	.0008	.0009	.0011	.0013
14	.0001	.0001	.0001	.0001	.0002	.0002	.0003	.0003	.0004	.0005
15	.0000	.0000	.0000	.0000	.0001	.0001	.0001	.0001	.0001	.0002

x	μ 5.1	5.2	5.3	5.4	5.5	5.6	5.7	5.8	5.9	6.0
0	.0061	.0055	.0050	.0045	.0041	.0037	.0033	.0030	.0027	.0025
1	.0311	.0287	.0265	.0244	.0225	.0207	.0191	.0176	.0162	.0149
2	.0793	.0746	.0701	.0659	.0618	.0580	.0544	.0509	.0477	.0446
3	.1348	.1293	.1239	.1185	.1133	.1082	.1033	.0985	.0938	.0892
4	.1719	.1681	.1641	.1600	.1558	.1515	.1472	.1428	.1383	.1339
5	.1753	.1748	.1740	.1728	.1714	.1697	.1678	.1656	.1632	.1606
6	.1490	.1515	.1537	.1555	.1571	.1584	.1594	.1601	.1605	.1606
7	.1086	.1125	.1163	.1200	.1234	.1267	.1298	.1326	.1353	.1377
8	.0692	.0731	.0771	.0810	.0849	.0887	.0925	.0962	.0998	.1033
9	.0392	.0423	.0454	.0486	.0519	.0552	.0586	.0620	.0654	.0688
10	.0200	.0220	.0241	.0262	.0285	.0309	.0334	.0359	.0386	.0413
11	.0093	.0104	.0116	.0129	.0143	.0157	.0173	.0190	.0207	.0225
12	.0039	.0045	.0051	.0058	.0065	.0073	.0082	.0092	.0102	.0113
13	.0015	.0018	.0021	.0024	.0028	.0032	.0036	.0041	.0046	.0052
14	.0006	.0007	.0008	.0009	.0011	.0013	.0015	.0017	.0019	.0022
15	.0002	.0002	.0003	.0003	.0004	.0005	.0006	.0007	.0008	.0009
16	.0001	.0001	.0001	.0001	.0001	.0002	.0002	.0002	.0003	.0003
17	.0000	.0000	.0000	.0000	.0000	.0001	.0001	.0001	.0001	.0001

x	μ 6.1	6.2	6.3	6.4	6.5	6.6	6.7	6.8	6.9	7.0
0	.0022	.0020	.0018	.0017	.0015	.0014	.0012	.0011	.0010	.0009
1	.0137	.0126	.0116	.0106	.0098	.0090	.0082	.0076	.0070	.0064
2	.0417	.0390	.0364	.0340	.0318	.0296	.0276	.0258	.0240	.0223
3	.0848	.0806	.0765	.0726	.0688	.0652	.0617	.0584	.0552	.0521
4	.1294	.1249	.1205	.1162	.1118	.1076	.1034	.0992	.0952	.0912

Poisson distribution

x	6.1	6.2	6.3	6.4	μ 6.5	6.6	6.7	6.8	6.9	7.0
5	.1579	.1549	.1519	.1487	.1454	.1420	.1385	.1349	.1314	.1277
6	.1605	.1601	.1595	.1586	.1575	.1562	.1546	.1529	.1511	.1490
7	.1399	.1418	.1435	.1450	.1462	.1472	.1480	.1486	.1489	.1490
8	.1066	.1099	.1130	.1160	.1188	.1215	.1240	.1263	.1284	.1304
9	.0723	.0757	.0791	.0825	.0858	.0891	.0923	.0954	.0985	.1014
10	.0441	.0469	.0498	.0528	.0558	.0588	.0618	.0649	.0679	.0710
11	.0244	.0265	.0285	.0307	.0330	.0353	.0377	.0401	.0426	.0452
12	.0124	.0137	.0150	.0164	.0179	.0194	.0210	.0227	.0245	.0263
13	.0058	.0065	.0073	.0081	.0089	.0099	.0108	.0119	.0130	.0142
14	.0025	.0029	.0033	.0037	.0041	.0046	.0052	.0058	.0064	.0071
15	.0010	.0012	.0014	.0016	.0018	.0020	.0023	.0026	.0029	.0033
16	.0004	.0005	.0005	.0006	.0007	.0008	.0010	.0011	.0013	.0014
17	.0001	.0002	.0002	.0002	.0003	.0003	.0004	.0004	.0005	.0006
18	.0000	.0001	.0001	.0001	.0001	.0001	.0001	.0002	.0002	.0002
19	.0000	.0000	.0000	.0000	.0000	.0000	.0001	.0001	.0001	.0001

x	7.1	7.2	7.3	7.4	μ 7.5	7.6	7.7	7.8	7.9	8.0
0	.0008	.0007	.0007	.0006	.0006	.0005	.0005	.0004	.0004	.0003
1	.0059	.0054	.0049	.0045	.0041	.0038	.0035	.0032	.0029	.0027
2	.0208	.0194	.0180	.0167	.0156	.0145	.0134	.0125	.0116	.0107
3	.0492	.0464	.0438	.0413	.0389	.0366	.0345	.0324	.0305	.0286
4	.0874	.0836	.0799	.0764	.0729	.0696	.0663	.0632	.0602	.0573
5	.1241	.1204	.1167	.1130	.1094	.1057	.1021	.0986	.0951	.0916
6	.1468	.1445	.1420	.1394	.1367	.1339	.1311	.1281	.1252	.1221
7	.1489	.1486	.1481	.1474	.1465	.1454	.1442	.1428	.1413	.1396
8	.1321	.1337	.1351	.1363	.1373	.1381	.1388	.1392	.1395	.1396
9	.1042	.1070	.1096	.1121	.1144	.1167	.1187	.1207	.1224	.1241
10	.0740	.0770	.0800	.0829	.0858	.0887	.0914	.0941	.0967	.0993
11	.0478	.0504	.0531	.0558	.0585	.0613	.0640	.0667	.0695	.0722
12	.0283	.0302	.0323	.0344	.0366	.0388	.0411	.0434	.0457	.0481
13	.0154	.0168	.0181	.0196	.0211	.0227	.0243	.0260	.0278	.0296
14	.0078	.0086	.0095	.0104	.0113	.0123	.0134	.0145	.0157	.0169
15	.0037	.0041	.0046	.0051	.0057	.0062	.0069	.0075	.0083	.0090
16	.0016	.0019	.0021	.0024	.0026	.0030	.0033	.0037	.0041	.0045
17	.0007	.0008	.0009	.0010	.0012	.0013	.0015	.0017	.0019	.0021
18	.0003	.0003	.0004	.0004	.0005	.0006	.0006	.0007	.0008	.0009
19	.0001	.0001	.0001	.0002	.0002	.0002	.0003	.0003	.0003	.0004
20	.0000	.0000	.0001	.0001	.0001	.0001	.0001	.0001	.0001	.0002
21	.0000	.0000	.0000	.0000	.0000	.0000	.0000	.0000	.0001	.0001

x	8.1	8.2	8.3	8.4	μ 8.5	8.6	8.7	8.8	8.9	9.0
0	.0003	.0003	.0002	.0002	.0002	.0002	.0002	.0002	.0001	.0001
1	.0025	.0023	.0021	.0019	.0017	.0016	.0014	.0013	.0012	.0011
2	.0100	.0092	.0086	.0079	.0074	.0068	.0063	.0058	.0054	.0050
3	.0269	.0252	.0237	.0222	.0208	.0195	.0183	.0171	.0160	.0150
4	.0544	.0517	.0491	.0466	.0443	.0420	.0398	.0377	.0357	.0337
5	.0882	.0848	.0816	.0784	.0752	.0722	.0692	.0663	.0635	.0607
6	.1191	.1160	.1128	.1097	.1066	.1034	.1003	.0972	.0941	.0911
7	.1378	.1358	.1338	.1317	.1294	.1271	.1247	.1222	.1197	.1171
8	.1395	.1392	.1388	.1382	.1375	.1366	.1356	.1344	.1332	.1318
9	.1255	.1269	.1280	.1290	.1299	.1305	.1311	.1315	.1317	.1318

APPENDIX F

Poisson distribution

x	8.1	8.2	8.3	8.4	μ 8.5	8.6	8.7	8.8	8.9	9.0
10	.1017	.1040	.1063	.1084	.1104	.1123	.1140	.1157	.1172	.1186
11	.0749	.0775	.0802	.0828	.0853	.0878	.0902	.0925	.0948	.0970
12	.0505	.0530	.0555	.0579	.0604	.0629	.0654	.0679	.0703	.0728
13	.0315	.0334	.0354	.0374	.0395	.0416	.0438	.0459	.0481	.0504
14	.0182	.0196	.0210	.0225	.0240	.0256	.0272	.0289	.0306	.0324
15	.0098	.0107	.0116	.0126	.0136	.0147	.0158	.0169	.0182	.0194
16	.0050	.0055	.0060	.0066	.0072	.0079	.0086	.0093	.0101	.0109
17	.0024	.0026	.0029	.0033	.0036	.0040	.0044	.0048	.0053	.0058
18	.0011	.0012	.0014	.0015	.0017	.0019	.0021	.0024	.0026	.0029
19	.0005	.0005	.0006	.0007	.0008	.0009	.0010	.0011	.0012	.0014
20	.0002	.0002	.0002	.0003	.0003	.0004	.0004	.0005	.0005	.0006
21	.0001	.0001	.0001	.0001	.0001	.0002	.0002	.0002	.0002	.0003
22	.0000	.0000	.0000	.0000	.0001	.0001	.0001	.0001	.0001	.0001

x	9.1	9.2	9.3	9.4	μ 9.5	9.6	9.7	9.8	9.9	10.0
0	.0001	.0001	.0001	.0001	.0001	.0001	.0001	.0001	.0001	.0000
1	.0010	.0009	.0009	.0008	.0007	.0007	.0006	.0005	.0005	.0005
2	.0046	.0043	.0040	.0037	.0034	.0031	.0029	.0027	.0025	.0023
3	.0140	.0131	.0123	.0115	.0107	.0100	.0093	.0087	.0081	.0076
4	.0319	.0302	.0285	.0269	.0254	.0240	.0226	.0213	.0201	.0189
5	.0581	.0555	.0530	.0506	.0483	.0460	.0439	.0418	.0398	.0378
6	.0881	.0851	.0822	.0793	.0764	.0736	.0709	.0682	.0656	.0631
7	.1145	.1118	.1091	.1064	.1037	.1010	.0982	.0955	.0928	.0901
8	.1302	.1286	.1269	.1251	.1232	.1212	.1191	.1170	.1148	.1126
9	.1317	.1315	.1311	.1306	.1300	.1293	.1284	.1274	.1263	.1251
10	.1198	.1209	.1219	.1228	.1235	.1241	.1245	.1248	.1250	.1251
11	.0991	.1012	.1031	.1049	.1067	.1083	.1098	.1112	.1125	.1137
12	.0752	.0776	.0799	.0822	.0844	.0866	.0888	.0908	.0928	.0948
13	.0526	.0549	.0572	.0594	.0617	.0640	.0662	.0685	.0707	.0729
14	.0342	.0361	.0380	.0399	.0419	.0439	.0459	.0479	.0500	.0521
15	.0208	.0221	.0235	.0250	.0265	.0281	.0297	.0313	.0330	.0347
16	.0118	.0127	.0137	.0147	.0157	.0168	.0180	.0192	.0204	.0217
17	.0063	.0069	.0075	.0081	.0088	.0095	.0103	.0111	.0119	.0128
18	.0032	.0035	.0039	.0042	.0046	.0051	.0055	.0060	.0065	.0071
19	.0015	.0017	.0019	.0021	.0023	.0026	.0028	.0031	.0034	.0037
20	.0007	.0008	.0009	.0010	.0011	.0012	.0014	.0015	.0017	.0019
21	.0003	.0003	.0004	.0004	.0005	.0006	.0006	.0007	.0008	.0009
22	.0001	.0001	.0002	.0002	.0002	.0002	.0003	.0003	.0004	.0004
23	.0000	.0001	.0001	.0001	.0001	.0001	.0001	.0001	.0002	.0002
24	.0000	.0000	.0000	.0000	.0000	.0000	.0000	.0001	.0001	.0001

x	11.0	12.0	13.0	14.0	μ 15.0	16.0	17.0	18.0	19.0	20.0
0	.0000	.0000	.0000	.0000	.0000	.0000	.0000	.0000	.0000	.0000
1	.0002	.0001	.0000	.0000	.0000	.0000	.0000	.0000	.0000	.0000
2	.0010	.0004	.0002	.0001	.0000	.0000	.0000	.0000	.0000	.0000
3	.0037	.0018	.0008	.0004	.0002	.0001	.0000	.0000	.0000	.0000
4	.0102	.0053	.0027	.0013	.0006	.0003	.0001	.0001	.0000	.0000
5	.0224	.0127	.0070	.0037	.0019	.0010	.0005	.0002	.0001	.0001
6	.0411	.0255	.0152	.0087	.0048	.0026	.0014	.0007	.0004	.0002
7	.0646	.0437	.0281	.0174	.0104	.0060	.0034	.0018	.0010	.0005
8	.0888	.0655	.0457	.0304	.0194	.0120	.0072	.0042	.0024	.0013
9	.1085	.0874	.0661	.0473	.0324	.0213	.0135	.0083	.0050	.0029

Poisson distribution

x	11.0	12.0	13.0	14.0	μ 15.0	16.0	17.0	18.0	19.0	20.0
10	.1194	.1048	.0859	.0663	.0486	.0341	.0230	.0150	.0095	.0058
11	.1194	.1144	.1015	.0844	.0663	.0496	.0355	.0245	.0164	.0106
12	.1094	.1144	.1099	.0984	.0829	.0661	.0504	.0368	.0259	.0176
13	.0926	.1056	.1099	.1060	.0956	.0814	.0658	.0509	.0378	.0271
14	.0728	.0905	.1021	.1060	.1024	.0930	.0800	.0655	.0513	.0387
15	.0534	.0724	.0885	.0989	.1024	.0992	.0906	.0786	.0650	.0516
16	.0367	.0543	.0719	.0866	.0960	.0992	.0963	.0884	.0772	.0646
17	.0237	.0383	.0550	.0713	.0847	.0934	.0963	.0936	.0863	.0760
18	.0145	.0255	.0397	.0554	.0706	.0830	.0909	.0936	.0911	.0844
19	.0084	.0161	.0272	.0409	.0557	.0699	.0814	.0887	.0911	.0888
20	.0046	.0097	.0177	.0286	.0418	.0559	.0692	.0798	.0866	.0888
21	.0024	.0055	.0109	.0191	.0299	.0426	.0560	.0684	.0783	.0846
22	.0012	.0030	.0065	.0121	.0204	.0310	.0433	.0560	.0676	.0769
23	.0006	.0016	.0037	.0074	.0133	.0216	.0320	.0438	.0559	.0669
24	.0003	.0008	.0020	.0043	.0083	.0144	.0226	.0328	.0442	.0557
25	.0001	.0004	.0010	.0024	.0050	.0092	.0154	.0237	.0336	.0446
26	.0000	.0002	.0005	.0013	.0029	.0057	.0101	.0164	.0246	.0343
27	.0000	.0001	.0002	.0007	.0016	.0034	.0063	.0109	.0173	.0254
28	.0000	.0000	.0001	.0003	.0009	.0019	.0038	.0070	.0117	.0181
29	.0000	.0000	.0001	.0002	.0004	.0011	.0023	.0044	.0077	.0125
30	.0000	.0000	.0000	.0001	.0002	.0006	.0013	.0026	.0049	.0083
31	.0000	.0000	.0000	.0000	.0001	.0003	.0007	.0015	.0030	.0054
32	.0000	.0000	.0000	.0000	.0001	.0001	.0004	.0009	.0018	.0034
33	.0000	.0000	.0000	.0000	.0000	.0001	.0002	.0005	.0010	.0020
34	.0000	.0000	.0000	.0000	.0000	.0000	.0001	.0002	.0006	.0012
35	.0000	.0000	.0000	.0000	.0000	.0000	.0000	.0001	.0003	.0007
36	.0000	.0000	.0000	.0000	.0000	.0000	.0000	.0001	.0002	.0004
37	.0000	.0000	.0000	.0000	.0000	.0000	.0000	.0000	.0001	.0002
38	.0000	.0000	.0000	.0000	.0000	.0000	.0000	.0000	.0000	.0001
39	.0000	.0000	.0000	.0000	.0000	.0000	.0000	.0000	.0000	.0001

Appendix G
Values of e^{-x}

The table in this appendix provides values for the expression e^{-x}. To use the table, locate x on the left side of the column where opposite the value of x is the value of e^{-x}. For instance, if x equals 0.40, then e^{-x} equals 0.67032.

Values of x between 1 and 10 not given in the table can be obtained in the following manner:

$$e^{-1.3} = (e^{-1})(e^{-0.3}) = (0.36788)(0.74082)$$
$$e^{-1.3} = 0.27253.$$

Values of e^{-x}

x	e^{-x}	x	e^{-x}	x	e^{-x}	x	e^{-x}
.00	1.00000	.30	.74082	.60	.54881	.90	.40657
.01	.99005	.31	.73345	.61	.54335	.91	.40252
.02	.98020	.32	.72615	.62	.53794	.92	.39852
.03	.97045	.33	.71892	.63	.53259	.93	.39455
.04	.96079	.34	.71177	.64	.52729	.94	.39063
.05	.95123	.35	.70469	.65	.52205	.95	.38674
.06	.94176	.36	.69768	.66	.51865	.96	.38289
.07	.93239	.37	.69073	.67	.51171	.97	.37908
.08	.92312	.38	.68386	.68	.50662	.98	.37531
.09	.91393	.39	.67706	.69	.50158	.99	.37158
.10	.90484	.40	.67032	.70	.49659	1.00	.36788
.11	.89583	.41	.66365	.71	.49164		
.12	.88692	.42	.65705	.72	.48675	2.00	.13534
.13	.87810	.43	.65051	.73	.48191		
.14	.86936	.44	.64404	.74	.47711	3.00	.04979
.15	.86071	.45	.63763	.75	.47237		
.16	.85214	.46	.63128	.76	.46767	4.00	.01832
.17	.84366	.47	.62500	.77	.46301		
.18	.83527	.48	.61878	.78	.45841	5.00	.00674
.19	.82696	.49	.61263	.79	.45384		
.20	.81873	.50	.60653	.80	.44933	6.00	.00248
.21	.81058	.51	.60050	.81	.44486		
.22	.80252	.52	.59452	.82	.44043	7.00	.00091
.23	.79453	.53	.58860	.83	.43605		
.24	.78663	.54	.58275	.84	.43171	8.00	.00034
.25	.77880	.55	.57695	.85	.42741		
.26	.77105‾	.56	.57121	.86	.42316	9.00	.00012
.27	.76338	.57	.56553	.87	.41895		
.28	.75578	.58	.55990	.88	.41478	10.00	.00005
.29	.74826	.59	.55433	.89	.41066		

**Appendix H
Areas under the Normal
Curve from 0 to Z**

The table in this appendix gives areas under the unit normal curve, this curve having a mean equal to 0 and a standard deviation equal to 1. It is constructed so that it will give the area between the mean of the normal curve and a point represented by Z, the number of standard deviations the point is away from the mean. For example, to find the area under the unit normal curve between the mean and a point 1.96σ to the right of the mean, the column on the left side of the table is moved down to the row having a Z of 1.9. Then this row is moved across to the 0.06 column, thus representing the point 1.96σ from the mean. As read from the intersection of the 1.9 row and the 0.06 column, the area between the mean of the unit normal curve and the point 1.96σ to the right of the mean is 0.4750.

Due to the fact that the area under the normal curve is equal to 1, and the normal curve is symmetrical, this table can be used to solve other types of problems than that outlined here. For example, due to the symmetry of the distribution, the area between the mean and $-Z\sigma$ is equal to the area between the mean and $+Z\sigma$. To illustrate, the area between the mean and $+1.96\sigma$ is equal to the area between the mean and -1.96σ, or 0.4750.

A second type of problem that can be solved is the determination of the area in the tail of the distribution. Since the area on either side of the mean is equal to 0.5000, the area to the right of 1.96σ is equal to $0.5000 - 0.4750$, or 0.0250.

Other problems involving the normal curve can be solved by similar methods.

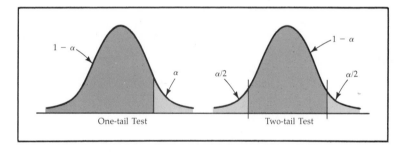

Areas under the normal curve from 0 to Z

Z	0.0	0.01	0.02	0.03	0.04	0.05	0.06	0.07	0.08	0.09
0.0	.0000	.0040	.0080	.0120	.0160	.0199	.0239	.0279	.0319	.0359
0.1	.0398	.0438	.0478	.0517	.0557	.0596	.0636	.0675	.0714	.0753
0.2	.0793	.0832	.0871	.0910	.0948	.0987	.1026	.1064	.1103	.1141
0.3	.1179	.1217	.1255	.1293	.1331	.1368	.1406	.1443	.1480	.1517
0.4	.1554	.1591	.1628	.1664	.1700	.1736	.1772	.1808	.1844	.1879
0.5	.1915	.1950	.1985	.2019	.2054	.2088	.2123	.2157	.2190	.2224
0.6	.2257	.2291	.2324	.2357	.2389	.2422	.2454	.2486	.2517	.2549
0.7	.2580	.2611	.2642	.2673	.2704	.2734	.2764	.2794	.2823	.2852
0.8	.2881	.2910	.2939	.2967	.2995	.3023	.3051	.3078	.3106	.3133
0.9	.3159	.3186	.3212	.3238	.3264	.3289	.3315	.3340	.3365	.3389
1.0	.3413	.3438	.3461	.3485	.3508	.3531	.3554	.3577	.3599	.3621
1.1	.3643	.4665	.4686	.3708	.3729	.3749	.3770	.3790	.3810	.3830
1.2	.3849	.3869	.3888	.3907	.3925	.3944	.3962	.3980	.3997	.4015
1.3	.4032	.4049	.4066	.4082	.4099	.4115	.4131	.4147	.4162	.4177
1.4	.4192	.4207	.4222	.4236	.4251	.4265	.4279	.4292	.4306	.4319
1.5	.4332	.4345	.4357	.4370	.4382	.4394	.4406	.4418	.4429	.4441
1.6	.4452	.4463	.4474	.4484	.4495	.4505	.4515	.4525	.4535	.4545
1.7	.4554	.4564	.4573	.4582	.4591	.4599	.4608	.4616	.4625	.4633
1.8	.4641	.4649	.4656	.4664	.4671	.4678	.4686	.4693	.4699	.4706
1.9	.4713	.4719	.4726	.4732	.4738	.4744	.4750	.4756	.4761	.4767
2.0	.4772	.4778	.4783	.4788	.4793	.4798	.4803	.4808	.4812	.4817
2.1	.4821	.4826	.4830	.4834	.4838	.4842	.4846	.4850	.4854	.4857
2.2	.4861	.4864	.4868	.4871	.4875	.4878	.4881	.4884	.4887	.4890
2.3	.4893	.4896	.4898	.4901	.4904	.4906	.4909	.4911	.4913	.4916
2.4	.2918	.4920	.4922	.4925	.4927	.4929	.4931	.4932	.4934	.4936
2.5	.4938	.4940	.4941	.4943	.4945	.4946	.4948	.4949	.4951	.4952
2.6	.4953	.4955	.4956	.4957	.4959	.4960	.4961	.4962	.4963	.4964
2.7	.4965	.4966	.4967	.4968	.4969	.4970	.4971	.4972	.4973	.4974
2.8	.4974	.4975	.4976	.4977	.4977	.4978	.4979	.4979	.4980	.4981
2.9	.4981	.4982	.4982	.4983	.4984	.4984	.4985	.4985	.4986	.4986
3.0	.4987	.4987	.4987	.4988	.4988	.4989	.4989	.4989	.4990	.4990
3.1	.4990	.4991	.4991	.4991	.4992	.4992	.4992	.4992	.4993	.4993
3.2	.4993	.4993	.4994	.4994	.4994	.4994	.4994	.4995	.4995	.4995
3.3	.4995	.4995	.4995	.4996	.4996	.4996	.4996	.4996	.4996	.4997
3.4	.4997	.4997	.4997	.4997	.4997	.4997	.4997	.4997	.4997	.4998
3.6	.4998	.4998	.4999	.4999	.4999	.4999	.4999	.4999	.4999	.4999
3.9	.5000	.5000	.5000	.5000	.5000	.5000	.5000	.5000	.5000	.5000

Source: Reprinted by permission from *Statistical Methods* by George W. Snedecor and William G. Cochran, sixth edition © 1967 by Iowa State University Press, Ames, Iowa.

APPENDIX H

Appendix I
Student _t_ Distribution

This appendix gives values for t of the Student's distribution, with the value being determined by the number of degrees of freedom and the level of confidence being used. The Student's distribution is used whenever the standard deviation of a sample is used as an estimator for the standard deviation of the population and the sample size is less than 31. It should be remembered that, in order to use this distribution, the population from which the sample is taken must be assumed to be normally distributed.

To illustrate the use of the Student's distribution, suppose there was a sample of size 25 whose standard deviation was to be used as an estimator of the population's standard deviation. In addition, suppose that a confidence interval was being constructed at a 95 percent level of confidence. To determine the appropriate t value, the column for a two-tail 95 percent level of confidence situation is located, representing an α of 0.05. This column is then moved down until the row representing 24 degrees of freedom is found, since the number of degrees of freedom is equal to the sample size minus one. As read from the table, the appropriate t value for this situation is 2.064.

If this same sample was to be used in a one-tail hypothesis testing situation, the 95 percent one-tail column would instead be used. The resulting t value would be 1.711.

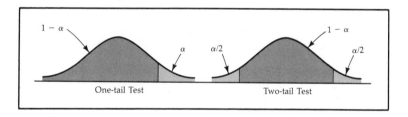

Student t-distribution

	Level of confidence (1-α)								
	0.20	0.50	0.80	0.90	0.95	0.98	0.99	0.999	Two-tail test
	Level of significance (α)								
n-1 Degrees of freedom	0.4	0.25	0.1	0.05	0.025	0.01	0.005	0.0005	One-tail test
	0.8	0.5	0.2	0.10	0.05	0.02	0.01	0.001	Two-tail test
1	0.325	1.000	3.078	6.314	12.706	31.821	63.657	636.619	
2	0.289	0.816	1.886	2.920	4.303	6.965	9.925	31.598	
3	0.277	0.765	1.638	2.353	3.182	4.541	5.841	12.924	
4	0.271	0.741	1.533	2.132	2.776	3.747	4.604	8.610	
5	0.267	0.727	1.476	2.015	2.571	3.365	4.032	6.869	
6	0.265	0.718	1.440	1.943	2.447	3.143	3.707	5.959	
7	0.263	0.711	1.415	1.895	2.365	2.998	3.499	5.408	
8	0.262	0.706	1.397	1.860	2.306	2.896	3.355	5.041	
9	0.261	0.703	1.383	1.833	2.262	2.821	3.250	4.781	
10	0.260	0.700	1.372	1.812	2.228	2.764	3.169	4.587	
11	0.260	0.697	1.363	1.796	2.201	2.718	3.106	4.437	
12	0.259	0.695	1.356	1.782	2.179	2.681	3.055	4.318	
13	0.259	0.694	1.350	1.771	2.160	2.650	3.012	4.221	
14	0.258	0.692	1.345	1.761	2.145	2.624	2.977	4.140	
15	0.258	0.691	1.341	1.753	2.131	2.602	2.947	4.073	
16	0.258	0.690	1.337	1.746	2.120	2.583	2.921	4.015	
17	0.257	0.689	1.333	1.740	2.110	2.567	2.898	3.965	
18	0.257	0.688	1.330	1.734	2.101	2.552	2.878	3.922	
19	0.257	0.688	1.328	1.729	2.093	2.539	2.861	3.883	
20	0.257	0.687	1.325	1.725	2.086	2.528	2.845	3.850	
21	0.257	0.686	1.323	1.721	2.080	2.518	2.831	3.819	
22	0.256	0.686	1.321	1.717	2.074	2.508	2.819	3.792	
23	0.256	0.685	1.319	1.714	2.069	2.500	2.807	3.767	
24	0.256	0.685	1.318	1.711	2.064	2.492	2.797	3.745	
25	0.256	0.684	1.316	1.708	2.060	2.485	2.787	3.725	
26	0.256	0.684	1.315	1.706	2.056	2.479	2.779	3.707	
27	0.256	0.684	1.314	1.703	2.052	2.473	2.771	3.690	
28	0.256	0.683	1.313	1.701	2.048	2.467	2.763	3.674	
29	0.256	0.683	1.311	1.699	2.045	2.462	2.756	3.659	
30	0.256	0.683	1.310	1.697	2.042	2.457	2.750	3.646	
40	0.255	0.681	1.303	1.684	2.021	2.423	2.704	3.551	
60	0.254	0.679	1.296	1.671	2.000	2.390	2.660	3.460	
120	0.254	0.677	1.289	1.658	1.980	2.358	2.617	3.373	
∞	0.253	0.674	1.282	1.645	1.960	2.326	2.576	3.291	

Source: Adapted from Table 12, Percentage points of the t-distribution, in E. S. Pearson and H. O. Hartley, *Biometrika Tables for Statisticians*, Vol. I, third edition, 1970, p. 146. Used with permission of Biometrika Trustees

Appendix J
F Distribution

The table in this appendix gives the values of F for various degrees of freedom and α's of 0.01 and 0.05. The F values for an α of 0.05 are given in lightface type, while the F values for an α of 0.01 are given in the boldface type.

To illustrate the use of this table, suppose three samples of size 8 each were chosen and an α value 0.05 is to be used. The column corresponding to the number of degrees of freedom in the numerator is located. For the numerator the number of degrees of freedom is equal to the number of samples minus one. In this problem the number of degrees of freedom in the numerator is $3 - 1 = 2$. After locating the correct column, move down to the row that represents the number of degrees of freedom in the denominator. For the denominator the number of degrees of freedom is equal to the combined sample size minus the number of samples. In this problem the number of degrees of freedom in the denominator is $24 - 3 = 21$. Reading from the table, with an α of 0.05, the appropriate F value is 3.47. If an α of 0.01 were used, the appropriate F value would be 5.78.

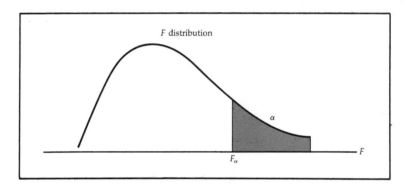

F Distribution.
$\alpha = 0.05$ (roman type) and $\alpha = 0.01$ (bold face type)

f_1 Degrees of freedom

f_2	1	2	3	4	5	6	7	8	9	10	11	12	14	16	20	24	30	40	50	∞
1	161. **4052.**	200. **4999.**	216. **5403.**	225. **5625.**	230. **5764.**	234. **5859.**	237. **5928.**	239. **5981.**	241. **6022.**	242. **6056.**	243. **6082.**	244. **6106.**	245. **6142.**	246. **6169.**	248. **6208.**	249. **6234.**	250. **6261.**	251. **6286.**	252. **6302.**	254. **6366.**
2	18.51 **98.49**	19.00 **99.00**	19.16 **99.17**	19.25 **99.25**	19.30 **99.30**	19.33 **99.33**	19.36 **99.36**	19.37 **99.37**	19.38 **99.39**	19.39 **99.40**	19.40 **99.41**	19.41 **99.42**	19.42 **99.43**	19.43 **99.44**	19.44 **99.45**	19.45 **99.46**	19.46 **99.47**	19.47 **99.48**	19.47 **99.48**	19.50 **99.50**
3	10.13 **34.12**	9.55 **30.82**	9.28 **29.46**	9.12 **28.71**	9.01 **28.24**	8.94 **27.91**	8.88 **27.67**	8.84 **27.49**	8.81 **27.34**	8.78 **27.23**	8.76 **27.13**	8.74 **27.05**	8.71 **26.92**	8.69 **26.83**	8.66 **26.69**	8.64 **26.60**	8.62 **26.50**	8.60 **26.41**	8.58 **26.35**	8.53 **26.12**
4	7.71 **21.20**	6.94 **18.00**	6.59 **16.69**	6.39 **15.98**	6.26 **15.52**	6.16 **15.21**	6.09 **14.98**	6.04 **14.80**	6.00 **14.66**	5.96 **14.54**	5.93 **14.45**	5.91 **14.37**	5.87 **14.24**	5.84 **14.15**	5.80 **14.02**	5.77 **13.93**	5.74 **13.83**	5.71 **13.74**	5.70 **13.69**	5.63 **13.46**
5	6.61 **16.26**	5.79 **13.27**	5.41 **12.06**	5.19 **11.39**	5.05 **10.97**	4.95 **10.67**	4.88 **10.45**	4.82 **10.29**	4.78 **10.15**	4.74 **10.05**	4.70 **9.96**	4.68 **9.89**	4.64 **9.77**	4.60 **9.68**	4.56 **9.55**	4.53 **9.47**	4.50 **9.38**	4.46 **9.29**	4.44 **9.24**	4.36 **9.02**
6	5.99 **13.74**	5.14 **10.92**	4.76 **9.78**	4.53 **9.15**	4.39 **8.75**	4.28 **8.47**	4.21 **8.26**	4.15 **8.10**	4.10 **7.98**	4.06 **7.87**	4.03 **7.79**	4.00 **7.72**	3.96 **7.60**	3.92 **7.52**	3.87 **7.39**	3.84 **7.31**	3.81 **7.23**	3.77 **7.14**	3.75 **7.09**	3.67 **6.88**
7	5.59 **12.25**	4.74 **9.55**	4.35 **8.45**	4.12 **7.85**	3.97 **7.46**	3.87 **7.19**	3.79 **7.00**	3.73 **6.84**	3.68 **6.71**	3.63 **6.62**	3.60 **6.54**	3.57 **6.47**	3.52 **6.35**	3.49 **6.27**	3.44 **6.15**	3.41 **6.07**	3.38 **5.98**	3.34 **5.90**	3.32 **5.85**	3.23 **5.65**
8	5.32 **11.26**	4.46 **8.65**	4.07 **7.59**	3.84 **7.01**	3.69 **6.63**	3.58 **6.37**	3.50 **6.19**	3.44 **6.03**	3.39 **5.91**	3.34 **5.82**	3.31 **5.74**	3.28 **5.67**	3.23 **5.56**	3.20 **5.48**	3.15 **5.36**	3.12 **5.28**	3.08 **5.20**	3.05 **5.11**	3.03 **5.06**	2.93 **4.86**
9	5.12 **10.56**	4.26 **8.02**	3.86 **6.99**	3.63 **6.42**	3.48 **6.06**	3.37 **5.80**	3.29 **5.62**	3.23 **5.47**	3.18 **5.35**	3.13 **5.26**	3.10 **5.18**	3.07 **5.11**	3.02 **5.00**	2.98 **4.92**	2.93 **4.80**	2.90 **4.73**	2.86 **4.64**	2.82 **4.56**	2.80 **4.51**	2.71 **4.31**
10	4.96 **10.04**	4.10 **7.56**	3.71 **6.55**	3.48 **5.99**	3.33 **5.64**	3.22 **5.39**	3.14 **5.21**	3.07 **5.06**	3.02 **4.95**	2.97 **4.85**	2.94 **4.78**	2.91 **4.71**	2.86 **4.60**	2.82 **4.52**	2.77 **4.41**	2.74 **4.33**	2.70 **4.25**	2.67 **4.17**	2.64 **4.12**	2.54 **3.91**
11	4.84 **9.65**	3.98 **7.20**	3.59 **6.22**	3.36 **5.67**	3.20 **5.32**	3.09 **5.07**	3.01 **4.88**	2.95 **4.74**	2.90 **4.63**	2.86 **4.54**	2.82 **4.46**	2.79 **4.40**	2.74 **4.29**	2.70 **4.21**	2.65 **4.10**	2.61 **4.02**	2.57 **3.94**	2.53 **3.86**	2.50 **3.80**	2.40 **3.60**
12	4.75 **9.33**	3.88 **6.93**	3.49 **5.95**	3.26 **5.41**	3.11 **5.06**	3.00 **4.82**	2.92 **4.65**	2.85 **4.50**	2.80 **4.39**	2.76 **4.30**	2.72 **4.22**	2.69 **4.16**	2.64 **4.05**	2.60 **3.98**	2.54 **3.86**	2.50 **3.78**	2.46 **3.70**	2.42 **3.61**	2.40 **3.56**	2.30 **3.36**
13	4.67 **9.07**	3.80 **6.70**	3.41 **5.74**	3.18 **5.20**	3.02 **4.86**	2.92 **4.62**	2.84 **4.44**	2.77 **4.30**	2.72 **4.19**	2.67 **4.10**	2.63 **4.02**	2.60 **3.96**	2.55 **3.85**	2.51 **3.78**	2.46 **3.67**	2.42 **3.59**	2.38 **3.51**	2.34 **3.42**	2.32 **3.37**	2.21 **3.16**

F Distribution.
$\alpha = 0.05$ (roman type) and $\alpha = 0.01$ (bold face type)

f_1 Degrees of freedom

f_2	1	2	3	4	5	6	7	8	9	10	11	12	14	16	20	24	30	40	50	∞
14	4.60 **8.86**	3.74 **6.51**	3.34 **5.56**	3.11 **5.03**	2.96 **4.69**	2.85 **4.46**	2.77 **4.28**	2.70 **4.14**	2.65 **4.03**	2.60 **3.94**	2.56 **3.86**	2.53 **3.80**	2.48 **3.70**	2.44 **3.62**	2.39 **3.51**	2.35 **3.43**	2.31 **3.34**	2.27 **3.26**	2.24 **3.21**	2.13 **3.00**
15	4.54 **8.68**	3.68 **6.36**	3.29 **5.42**	3.06 **4.89**	2.90 **4.56**	2.79 **4.32**	2.70 **4.14**	2.64 **4.00**	2.59 **3.89**	2.55 **3.80**	2.51 **3.73**	2.48 **3.67**	2.43 **3.56**	2.39 **3.48**	2.33 **3.36**	2.29 **3.29**	2.25 **3.20**	2.21 **3.12**	2.18 **3.07**	2.07 **2.87**
16	4.49 **8.53**	3.63 **6.23**	3.24 **5.29**	3.01 **4.77**	2.85 **4.44**	2.74 **4.20**	2.66 **4.03**	2.59 **3.89**	2.54 **3.78**	2.49 **3.69**	2.45 **3.61**	2.42 **3.55**	2.37 **3.45**	2.33 **3.37**	2.28 **3.25**	2.24 **3.18**	2.20 **3.10**	2.16 **3.01**	2.13 **2.96**	2.01 **2.75**
17	4.45 **8.40**	3.59 **6.11**	3.20 **5.18**	2.96 **4.67**	2.81 **4.34**	2.70 **4.10**	2.62 **3.93**	2.55 **3.79**	2.50 **3.68**	2.45 **3.59**	2.41 **3.52**	2.38 **3.45**	2.33 **3.35**	2.29 **3.27**	2.23 **3.16**	2.19 **3.08**	2.15 **3.00**	2.11 **2.92**	2.08 **2.86**	1.96 **2.65**
18	4.41 **8.28**	3.55 **6.01**	3.16 **5.09**	2.93 **4.58**	2.77 **4.25**	2.66 **4.01**	2.58 **3.85**	2.51 **3.71**	2.46 **3.60**	2.41 **3.51**	2.37 **3.44**	2.34 **3.37**	2.29 **3.27**	2.25 **3.19**	2.19 **3.07**	2.15 **3.00**	2.11 **2.91**	2.07 **2.83**	2.04 **2.78**	1.92 **2.57**
19	4.38 **8.18**	3.52 **5.93**	3.13 **5.01**	2.90 **4.50**	2.74 **4.17**	2.63 **3.94**	2.55 **3.77**	2.48 **3.63**	2.43 **3.52**	2.38 **3.43**	2.34 **3.36**	2.31 **3.30**	2.26 **3.19**	2.21 **3.12**	2.15 **3.00**	2.11 **2.92**	2.07 **2.84**	2.02 **2.76**	2.00 **2.70**	1.88 **2.49**
20	4.35 **8.10**	3.49 **5.85**	3.10 **4.94**	2.87 **4.43**	2.71 **4.10**	2.60 **3.87**	2.52 **3.71**	2.45 **3.56**	2.40 **3.45**	2.35 **3.37**	2.31 **3.30**	2.28 **3.23**	2.23 **3.13**	2.18 **3.05**	2.12 **2.94**	2.08 **2.86**	2.04 **2.77**	1.99 **2.69**	1.96 **2.63**	1.84 **2.42**
21	4.32 **8.02**	3.47 **5.78**	3.07 **4.87**	2.84 **4.37**	2.68 **4.04**	2.57 **3.81**	2.49 **3.65**	2.42 **3.51**	2.37 **3.40**	2.32 **3.31**	2.28 **3.24**	2.25 **3.17**	2.20 **3.07**	2.15 **2.99**	2.09 **2.88**	2.05 **2.80**	2.00 **2.72**	1.96 **2.63**	1.93 **2.58**	1.81 **2.36**
22	4.30 **7.94**	3.44 **5.72**	3.05 **4.82**	2.82 **4.31**	2.66 **3.99**	2.55 **3.76**	2.47 **3.59**	2.40 **3.45**	2.35 **3.35**	2.30 **3.26**	2.26 **3.18**	2.23 **3.12**	2.18 **3.02**	2.13 **2.94**	2.07 **2.83**	2.03 **2.75**	1.98 **2.67**	1.93 **2.58**	1.91 **2.53**	1.78 **2.31**
23	4.28 **7.88**	3.42 **5.66**	3.03 **4.76**	2.80 **4.26**	2.64 **3.94**	2.53 **3.71**	2.45 **3.54**	2.38 **3.41**	2.32 **3.30**	2.28 **3.21**	2.24 **3.14**	2.20 **3.07**	2.14 **2.97**	2.10 **2.89**	2.04 **2.78**	2.00 **2.70**	1.96 **2.62**	1.91 **2.53**	1.88 **2.48**	1.76 **2.26**
24	4.26 **7.82**	3.40 **5.61**	3.01 **4.72**	2.78 **4.22**	2.62 **3.90**	2.51 **3.67**	2.43 **3.50**	2.36 **3.36**	2.30 **3.25**	2.26 **3.17**	2.22 **3.09**	2.18 **3.03**	2.13 **2.93**	2.09 **2.85**	2.02 **2.74**	1.98 **2.66**	1.94 **2.58**	1.89 **2.49**	1.86 **2.44**	1.73 **2.21**
25	4.24 **7.77**	3.38 **5.57**	2.99 **4.68**	2.76 **4.18**	2.60 **3.86**	2.49 **3.63**	2.41 **3.46**	2.34 **3.32**	2.28 **3.21**	2.24 **3.13**	2.20 **3.05**	2.16 **2.99**	2.11 **2.89**	2.06 **2.81**	2.00 **2.70**	1.96 **2.62**	1.92 **2.54**	1.87 **2.45**	1.84 **2.40**	1.71 **2.17**
26	4.22 **7.72**	3.37 **5.53**	2.98 **4.64**	2.74 **4.14**	2.59 **3.82**	2.47 **3.59**	2.39 **3.42**	2.32 **3.29**	2.27 **3.17**	2.22 **3.09**	2.18 **3.02**	2.15 **2.96**	2.10 **2.86**	2.05 **2.77**	1.99 **2.66**	1.95 **2.58**	1.90 **2.50**	1.85 **2.41**	1.82 **2.36**	1.69 **2.13**

F Distribution.
α = 0.05 (roman type) and α = 0.01 (bold face type)

f₁ Degrees of freedom

f_2	1	2	3	4	5	6	7	8	9	10	11	12	14	16	20	24	30	40	50	∞
27	4.21 **7.68**	3.35 **5.49**	2.96 **4.60**	2.73 **4.11**	2.57 **3.79**	2.46 **3.56**	2.37 **3.39**	2.30 **3.26**	2.25 **3.14**	2.20 **3.06**	2.16 **2.98**	2.13 **2.93**	2.08 **2.83**	2.03 **2.74**	1.97 **2.63**	1.93 **2.55**	1.88 **2.47**	1.84 **2.38**	1.80 **2.33**	1.67 **2.10**
28	4.20 **7.64**	3.34 **5.45**	2.95 **4.57**	2.71 **4.07**	2.56 **3.76**	2.44 **3.53**	2.36 **3.36**	2.29 **3.23**	2.24 **3.11**	2.19 **3.03**	2.15 **2.95**	2.12 **2.90**	2.06 **2.80**	2.02 **2.71**	1.96 **2.60**	1.91 **2.52**	1.87 **2.44**	1.81 **2.35**	1.78 **2.30**	1.65 **2.06**
29	4.18 **7.60**	3.33 **5.42**	2.93 **4.54**	2.70 **4.04**	2.54 **3.73**	2.43 **3.50**	2.35 **3.33**	2.28 **3.20**	2.22 **3.08**	2.18 **3.00**	2.14 **2.92**	2.10 **2.87**	2.05 **2.77**	2.00 **2.68**	1.94 **2.57**	1.90 **2.49**	1.85 **2.41**	1.80 **2.32**	1.77 **2.27**	1.64 **2.03**
30	4.17 **7.56**	3.32 **5.39**	2.92 **4.51**	2.69 **4.02**	2.53 **3.70**	2.42 **3.47**	2.34 **3.30**	2.27 **3.17**	2.21 **3.06**	2.16 **2.98**	2.12 **2.90**	2.09 **2.84**	2.04 **2.74**	1.99 **2.66**	1.93 **2.55**	1.89 **2.47**	1.84 **2.38**	1.79 **2.29**	1.76 **2.24**	1.62 **2.01**
32	4.15 **7.50**	3.30 **5.34**	2.90 **4.46**	2.67 **3.97**	2.51 **3.66**	2.40 **3.42**	2.32 **3.25**	2.25 **3.12**	2.19 **3.01**	2.14 **2.94**	2.10 **2.86**	2.07 **2.80**	2.02 **2.70**	1.97 **2.62**	1.91 **2.51**	1.86 **2.42**	1.82 **2.34**	1.76 **2.25**	1.74 **2.20**	1.59 **1.96**
34	4.13 **7.44**	3.28 **5.29**	2.88 **4.42**	2.65 **3.93**	2.49 **3.61**	2.38 **3.38**	2.30 **3.21**	2.23 **3.08**	2.17 **2.97**	2.12 **2.89**	2.08 **2.82**	2.05 **2.76**	2.00 **2.66**	1.95 **2.58**	1.89 **2.47**	1.84 **2.38**	1.80 **2.30**	1.74 **2.21**	1.71 **2.15**	1.57 **1.91**
36	4.11 **7.39**	3.26 **5.25**	2.86 **4.38**	2.63 **3.89**	2.48 **3.58**	2.36 **3.35**	2.28 **3.18**	2.21 **3.04**	2.15 **2.94**	2.10 **2.86**	2.06 **2.78**	2.03 **2.72**	1.98 **2.62**	1.93 **2.54**	1.87 **2.43**	1.82 **2.35**	1.78 **2.26**	1.72 **2.17**	1.69 **2.12**	1.55 **1.87**
38	4.10 **7.35**	3.25 **5.21**	2.85 **4.34**	2.62 **3.86**	2.46 **3.54**	2.35 **3.32**	2.26 **3.15**	2.19 **3.02**	2.14 **2.91**	2.09 **2.82**	2.05 **2.75**	2.02 **2.69**	1.96 **2.59**	1.92 **2.51**	1.85 **2.40**	1.80 **2.32**	1.76 **2.22**	1.71 **2.14**	1.67 **2.08**	1.53 **1.84**
40	4.08 **7.31**	3.23 **5.18**	2.84 **4.31**	2.61 **3.83**	2.45 **3.51**	2.34 **3.29**	2.25 **3.12**	2.18 **2.99**	2.12 **2.88**	2.07 **2.80**	2.04 **2.73**	2.00 **2.66**	1.95 **2.56**	1.90 **2.49**	1.84 **2.37**	1.79 **2.29**	1.74 **2.20**	1.69 **2.11**	1.66 **2.05**	1.51 **1.81**
42	4.07 **7.27**	3.22 **5.15**	2.83 **4.29**	2.59 **3.80**	2.44 **3.49**	2.32 **3.26**	2.24 **3.10**	2.17 **2.96**	2.11 **2.86**	2.06 **2.77**	2.02 **2.70**	1.99 **2.64**	1.94 **2.54**	1.89 **2.46**	1.82 **2.35**	1.78 **2.26**	1.73 **2.17**	1.68 **2.08**	1.64 **2.02**	1.49 **1.78**
44	4.06 **7.24**	3.21 **5.12**	2.82 **4.26**	2.58 **3.78**	2.43 **3.46**	2.31 **3.24**	2.23 **3.07**	2.16 **2.94**	2.10 **2.84**	2.05 **2.75**	2.01 **2.68**	1.98 **2.62**	1.92 **2.52**	1.88 **2.44**	1.81 **2.32**	1.76 **2.24**	1.72 **2.15**	1.66 **2.06**	1.63 **2.00**	1.48 **1.75**
46	4.05 **7.21**	3.20 **5.10**	2.81 **4.24**	2.57 **3.76**	2.42 **3.44**	2.30 **3.22**	2.22 **3.05**	2.14 **2.92**	2.09 **2.82**	2.04 **2.73**	2.00 **2.66**	1.97 **2.60**	1.91 **2.50**	1.87 **2.42**	1.80 **2.30**	1.75 **2.22**	1.71 **2.13**	1.65 **2.04**	1.62 **1.98**	1.46 **1.72**
48	4.04 **7.19**	3.19 **5.08**	2.80 **4.22**	2.56 **3.74**	2.41 **3.42**	2.30 **3.20**	2.21 **3.04**	2.14 **2.90**	2.08 **2.80**	2.03 **2.71**	1.99 **2.64**	1.96 **2.58**	1.90 **2.48**	1.86 **2.40**	1.79 **2.28**	1.74 **2.20**	1.70 **2.11**	1.64 **2.02**	1.61 **1.96**	1.45 **1.70**

F Distribution.

$\alpha = 0.05$ (roman type) and $\alpha = 0.01$ (bold face type)

									f_1 Degrees of freedom											
f_2	1	2	3	4	5	6	7	8	9	10	11	12	14	16	20	24	30	40	50	∞
50	4.03 **7.17**	3.18 **5.06**	2.79 **4.20**	2.56 **3.72**	2.40 **3.41**	2.29 **3.18**	2.20 **3.02**	2.13 **2.88**	2.07 **2.78**	2.02 **2.70**	1.98 **2.62**	1.95 **2.56**	1.90 **2.46**	1.85 **2.39**	1.78 **2.26**	1.74 **2.18**	1.69 **2.10**	1.63 **2.00**	1.60 **1.94**	1.44 **1.68**
55	4.02 **7.12**	3.17 **5.01**	2.78 **4.16**	2.54 **3.68**	2.38 **3.37**	2.27 **3.15**	2.18 **2.98**	2.11 **2.85**	2.05 **2.75**	2.00 **2.66**	1.97 **2.59**	1.93 **2.53**	1.88 **2.43**	1.83 **2.35**	1.76 **2.23**	1.72 **2.15**	1.67 **2.06**	1.61 **1.96**	1.58 **1.90**	1.41 **1.64**
60	4.00 **7.08**	3.15 **4.98**	2.76 **4.13**	2.52 **3.65**	2.37 **3.34**	2.25 **3.12**	2.17 **2.95**	2.10 **2.82**	2.04 **2.72**	1.99 **2.63**	1.95 **2.56**	1.92 **2.50**	1.86 **2.40**	1.81 **2.32**	1.75 **2.20**	1.70 **2.12**	1.65 **2.03**	1.59 **1.93**	1.56 **1.87**	1.39 **1.60**
65	3.99 **7.04**	3.14 **4.95**	2.75 **4.10**	2.51 **3.62**	2.36 **3.31**	2.24 **3.09**	2.15 **2.93**	2.08 **2.79**	2.02 **2.70**	1.98 **2.61**	1.94 **2.54**	1.90 **2.47**	1.85 **2.37**	1.80 **2.30**	1.73 **2.18**	1.68 **2.09**	1.63 **2.00**	1.57 **1.90**	1.54 **1.84**	1.37 **1.56**
70	3.98 **7.01**	3.13 **4.92**	2.74 **4.08**	2.50 **3.60**	2.35 **3.29**	2.23 **3.07**	2.14 **2.91**	2.07 **2.77**	2.01 **2.67**	1.97 **2.59**	1.93 **2.51**	1.89 **2.45**	1.84 **2.35**	1.79 **2.28**	1.72 **2.15**	1.67 **2.07**	1.62 **1.98**	1.56 **1.88**	1.53 **1.82**	1.35 **1.53**
80	3.96 **6.96**	3.11 **4.88**	2.72 **4.04**	2.48 **3.56**	2.33 **3.25**	2.21 **3.04**	2.12 **2.87**	2.05 **2.74**	1.99 **2.64**	1.95 **2.55**	1.91 **2.48**	1.88 **2.41**	1.82 **2.32**	1.77 **2.24**	1.70 **2.11**	1.65 **2.03**	1.60 **1.94**	1.54 **1.84**	1.51 **1.78**	1.32 **1.49**
100	3.94 **6.90**	3.09 **4.82**	2.70 **3.98**	2.46 **3.51**	2.30 **3.20**	2.19 **2.99**	2.10 **2.82**	2.03 **2.69**	1.97 **2.59**	1.92 **2.51**	1.88 **2.43**	1.85 **2.36**	1.79 **2.26**	1.75 **2.19**	1.68 **2.06**	1.63 **1.98**	1.57 **1.89**	1.51 **1.79**	1.48 **1.73**	1.28 **1.43**
125	3.92 **6.84**	3.07 **4.78**	2.68 **3.94**	2.44 **3.47**	2.29 **3.17**	2.17 **2.95**	2.08 **2.79**	2.01 **2.65**	1.95 **2.56**	1.90 **2.47**	1.86 **2.40**	1.83 **2.33**	1.77 **2.23**	1.72 **2.15**	1.65 **2.03**	1.60 **1.94**	1.55 **1.85**	1.49 **1.75**	1.45 **1.68**	1.25 **1.37**
150	3.91 **6.81**	3.06 **4.75**	2.67 **3.91**	2.43 **3.44**	2.27 **3.14**	2.16 **2.92**	2.07 **2.76**	2.00 **2.62**	1.94 **2.53**	1.89 **2.44**	1.85 **2.37**	1.82 **2.30**	1.76 **2.20**	1.71 **2.12**	1.64 **2.00**	1.59 **1.91**	1.54 **1.83**	1.47 **1.72**	1.44 **1.66**	1.22 **1.33**
200	3.89 **6.76**	3.04 **4.71**	2.65 **3.88**	2.41 **3.41**	2.26 **3.11**	2.14 **2.90**	2.05 **2.73**	1.98 **2.60**	1.92 **2.50**	1.87 **2.41**	1.83 **2.34**	1.80 **2.28**	1.74 **2.17**	1.69 **2.09**	1.62 **1.97**	1.57 **1.88**	1.52 **1.79**	1.45 **1.69**	1.42 **1.62**	1.19 **1.28**
400	3.86 **6.70**	3.02 **4.66**	2.62 **3.83**	2.39 **3.36**	2.23 **3.06**	2.12 **2.85**	2.03 **2.69**	1.96 **2.55**	1.90 **2.46**	1.85 **2.37**	1.81 **2.29**	1.78 **2.23**	1.72 **2.12**	1.67 **2.04**	1.60 **1.92**	1.54 **1.84**	1.49 **1.74**	1.42 **1.64**	1.38 **1.57**	1.13 **1.19**
1000	3.85 **6.66**	3.00 **4.62**	2.61 **3.80**	2.38 **3.34**	2.22 **3.04**	2.10 **2.82**	2.02 **2.66**	1.95 **2.53**	1.89 **2.43**	1.84 **2.34**	1.80 **2.26**	1.76 **2.20**	1.70 **2.09**	1.65 **2.01**	1.58 **1.89**	1.53 **1.81**	1.47 **1.71**	1.41 **1.61**	1.36 **1.54**	1.08 **1.11**
∞	3.84 **6.63**	2.99 **4.60**	2.60 **3.78**	2.37 **3.32**	2.21 **3.02**	2.09 **2.80**	2.01 **2.64**	1.94 **2.51**	1.88 **2.41**	1.83 **2.32**	1.79 **2.24**	1.75 **2.18**	1.69 **2.07**	1.64 **1.99**	1.57 **1.87**	1.52 **1.79**	1.46 **1.69**	1.40 **1.59**	1.35 **1.52**	1.00 **1.00**

Source: Reprinted by permission from *Statistical Methods* by George W. Snedecor and William G. Cochran, sixth edition © 1967 by Iowa State University Press, Ames, Iowa.

**Appendix K
Chi Square
Distribution**

The table in this appendix gives values for χ^2 for various degrees of freedom and levels of confidence. The values down the left side of the table represent the number of degrees of freedom in the problem. The values across the top of this table represent the area in the tail of the distribution.

To illustrate the use of this table, suppose we have an experiment in which the observations are classified into a table of 4 rows and 3 columns. In addition, a 95 percent level of confidence is to be used in the problem. The number of degrees of freedom in this type of problem is equal to $(r - 1)(c - 1)$, where r is the number of rows in the table of observations and c is the number of columns in the table of observations. To find the appropriate χ^2 value, the column headed by 0.05 is located since the level of confidence is 0.95. This column is then moved down until the row representing the correct number of degrees of freedom is located, $(4 - 1)(3 - 1) = 6$ in this case. Reading from the intersection of this column and row, the appropriate χ^2 value is seen to be 12.5916.

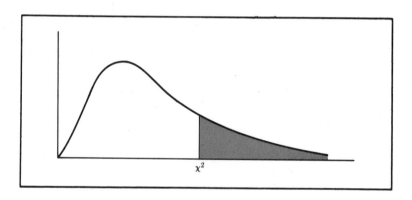

Chi square distribution

df \ α	0.995	0.975	0.500	0.100	0.050	0.010	0.001
1	0.00003927	0.00098201	0.454936	2.70554	3.84146	6.6349	10.828
2	0.0100251	0.0506356	1.38629	4.60517	5.99146	9.21034	13.816
3	0.0717208	0.215795	2.36597	6.25139	7.81473	11.3449	16.266
4	0.206989	0.484419	3.35669	7.77944	9.48773	13.2767	18.467
5	0.411742	0.831212	4.35146	9.23636	11.0705	15.0863	20.515
6	0.675727	1.23734	5.34812	10.6446	12.5916	16.8119	22.458
7	0.989266	1.68987	6.34581	12.0170	14.0671	18.4753	24.322
8	1.34441	2.17973	7.34412	13.3616	15.5073	20.0902	26.125
9	1.73493	2.70039	8.34283	14.6837	16.9190	21.6660	27.877
10	2.15586	3.24697	9.34182	15.9872	18.3070	23.2093	29.588
11	2.60322	3.81575	10.3410	17.2750	19.6751	24.7250	31.264
12	3.07382	4.40379	11.3403	18.5493	21.0261	26.2170	32.909
13	3.56503	5.00875	12.3398	19.8119	22.3620	27.6882	34.528
14	4.07467	5.62873	13.3393	21.0641	23.6848	29.1412	36.123
15	4.60092	6.26214	14.3389	22.3071	24.9958	30.5779	37.697
16	5.14221	6.90766	15.3385	23.5418	26.2962	31.9999	39.252
17	5.69722	7.56419	16.3382	24.7690	27.5871	33.4087	40.790
18	6.26480	8.23075	17.3379	25.9894	28.8693	34.8053	42.312
19	6.84397	8.90652	18.3377	27.2036	30.1435	36.1909	43.820
20	7.43384	9.59078	19.3374	28.4120	31.4104	37.5662	45.315
21	8.03365	10.28293	20.3372	29.6151	32.6706	38.9322	46.797
22	8.64272	10.9823	21.3370	30.8133	33.9244	40.2894	48.268
23	9.26043	11.6886	22.3369	32.0069	35.1725	41.6384	49.728
24	9.88623	12.4012	23.3367	33.1962	36.4150	42.9798	51.179
25	10.5197	13.1197	24.3366	34.3816	37.6525	44.3141	52.618
26	11.1602	13.8439	25.3365	35.5632	38.8851	45.6417	54.052
27	11.8076	14.5734	26.3363	36.7412	40.1133	46.9629	55.476
28	12.4613	15.3079	27.3362	37.9159	41.3371	48.2782	56.892
29	13.1211	16.0471	28.3361	39.0875	42.5570	49.5879	58.301
30	13.7867	16.7908	29.3360	40.2560	43.7730	50.8922	59.703
40	20.7065	24.4330	39.3353	51.8051	55.7585	63.6907	73.402
50	27.9907	32.3574	49.3349	63.1671	67.5048	76.1539	86.661
60	35.5345	40.4817	59.3347	74.3970	79.0819	88.3794	99.607
70	43.2752	48.7576	69.3345	85.5270	90.5312	100.425	112.317
80	51.1719	57.1532	79.3343	96.5782	101.879	112.329	124.839
90	59.1963	65.6466	89.3342	107.565	113.145	124.116	137.208
100	67.3276	74.2219	99.3341	118.498	124.342	135.807	149.449

Source: Adapted from Table 8, Percentage points of the χ^2-distribution, in E. S. Pearson and H. O. Hartley, *Biometrika Tables for Statisticians*, Vol. I, third edition, 1970, pp. 136–137. Used with permission of Biometrika Trustees.

Table of random digits

	00-04	05-09	10-14	15-19	20-24	25-29	30-34	35-39	40-44	45-49
0	70577	42866	24969	61210	76046	67699	42054	12696	93758	03283
1	08244	27647	33851	44705	94211	46716	11738	55784	95374	72655
2	98409	66162	95763	47420	20792	61527	20441	39435	11859	41567
3	44560	38750	83635	56540	64900	42912	13953	79149	18710	68618
4	54463	22662	65905	70639	79365	67382	29085	69831	47058	08186
5	96887	12479	80621	66223	86085	78285	02432	53342	42846	94771
6	40950	84820	29881	85966	62800	70326	84740	62660	77379	90279
7	07521	56898	12236	60277	39102	62315	12239	07105	11844	01117
8	15389	85205	18850	39226	42249	90669	96325	23248	60933	26927
9	85941	40756	82414	02015	13858	78030	16269	65978	01385	15345
10	61149	69440	11286	88218	58925	03638	52862	62733	33451	77455
11	05219	81619	10651	67079	92511	59888	84502	72095	83463	75577
12	41417	98326	87719	92294	46614	50948	64886	20002	97365	30976
13	28357	94070	20652	35774	16249	75019	21145	05217	47286	76305
14	17783	00015	10806	83091	91530	36466	39981	62481	49177	75779
15	82995	64157	66164	41180	10089	41757	78258	96488	88629	37231
16	96754	17676	55659	44105	47361	34833	86679	23930	53249	27083
17	34357	88040	53364	71726	45690	66334	60332	22554	90600	71113
18	06318	37403	49927	57715	50423	67372	63116	48888	21505	80182
19	62111	52820	07243	79931	89292	84767	85693	73947	22278	11551
20	47534	09243	67879	00544	23410	12740	02540	54440	32949	13491
21	98614	75993	84460	62846	58944	14922	48730	73443	48167	34770
22	24856	03648	44898	09351	98795	18644	39765	71058	90368	44104
23	90801	21472	42815	77408	37390	76766	52615	32141	30268	18106
24	55165	77312	83666	36028	28420	70219	81369	41943	47366	41067
25	75884	12952	84318	95108	72305	64620	91318	89872	45375	85436
26	16777	37116	58550	42958	21460	43910	01175	87894	81378	10620
27	46230	43877	80207	88877	89380	32992	91380	03164	98656	59337
28	42902	66892	46134	01432	94710	23474	20423	60137	60609	13119
29	81007	00333	39693	28039	10154	95425	39220	19774	31782	49037
30	68089	02211	51111	72373	06902	74373	96199	97017	41273	21546
31	20411	67081	89950	16944	93054	87687	96693	87236	77054	33848
32	58212	13160	06468	15718	82627	76999	05999	58680	96739	63700
33	94522	74358	71659	62038	79643	79169	44741	05437	39038	13163
34	42626	86819	85651	88678	17401	03252	99547	32404	17918	62880
35	16501	33763	57194	16752	54450	19031	58580	47629	54132	60631
36	59497	04392	09419	89964	51211	04894	72882	17805	21896	83864
37	97155	13428	40293	09985	58434	01412	69124	82171	59058	82859
38	45476	84882	65109	96597	25930	66790	65706	61203	53634	22557
39	89300	69700	50741	30329	11658	23166	05400	66669	48708	03887
40	50051	95137	91631	66315	91428	12275	24816	68091	71710	33258
41	31753	85178	31310	89642	98364	02306	24617	09609	83942	22716
42	79152	53829	77250	20190	56535	18760	69942	77448	33278	48805
43	68328	83378	63369	71381	39564	05615	42451	64559	97501	65747
44	46939	38689	58625	08342	30459	85863	20781	09284	26333	91777
45	91896	67126	04151	03795	59077	11848	12630	98375	52068	60142
46	55751	62515	21108	80830	02263	29303	37204	96926	30506	09808
47	85156	87689	95493	88842	00664	55017	55539	17771	69488	87530
48	83944	86141	15707	96256	23068	13782	08467	89469	93842	55349
49	91621	00881	04900	54224	46177	55309	17852	27491	89415	23466

APPENDIX L

	50-54	55-59	60-64	65-69	70-74	75-79	80-84	85-89	90-94	95-99
0	95068	88628	35911	14530	33020	80428	39936	31855	34334	64865
1	16874	62677	57412	13215	31389	62233	80827	73917	82802	84420
2	36013	97518	51400	25670	98342	61891	27101	37855	06235	33316
3	86859	19558	64432	16706	99612	59798	32803	67708	15297	28612
4	11258	24591	36863	55368	31721	94335	34936	02566	80972	08188
5	99567	76364	77204	04615	27062	96621	43918	01896	83991	51141
6	54463	47237	73800	91017	36239	71824	83671	39892	60518	37092
7	92494	63157	76593	91316	03505	72389	96363	52887	01087	66091
8	59391	58030	52098	82718	87024	82848	01490	96574	90464	29065
9	15669	56689	35682	40844	53256	81872	35213	09840	34471	74441
10	99116	75486	84989	23476	52967	67104	39495	39100	17217	74073
11	15696	10703	65178	90637	63110	17622	53988	71087	84148	11670
12	97720	15369	51269	69620	03388	13699	33423	67453	43269	56720
13	11666	13841	71681	98000	35979	39719	81899	07449	47985	46967
14	71628	73130	78783	75691	41632	09847	61547	18707	85489	69944
15	40501	51089	99943	91843	41995	88931	73631	69361	05375	15417
16	22518	55576	98215	82068	10798	86211	36584	67466	69373	40054
17	75112	30485	62173	02132	14878	92879	22281	16783	86352	00077
18	80327	02671	98191	84342	90813	49268	95441	15496	20168	09271
19	60251	45548	02146	05597	48228	81366	34598	72856	66762	17002
20	57430	82270	10421	00540	43648	75888	66049	21511	47676	33444
21	73528	39559	34434	88596	54086	71693	43132	14414	79949	85193
22	25991	65959	70769	64721	86413	33475	42740	06175	82758	66248
23	78388	16638	09134	59980	63806	48472	39318	35434	24057	74739
24	12477	09965	96657	57994	59439	76330	24596	77515	09577	91871
25	83266	32883	42451	15579	38155	29793	40914	65990	16255	17777
26	76970	80876	10237	39515	79152	74798	39357	09054	73579	92359
27	37074	65198	44785	68624	98336	84481	97610	78735	46703	98265
28	83712	06514	30101	78295	54656	85417	43189	60048	72781	72606
29	20287	56862	69727	94443	64936	08366	27227	05158	50326	59566
30	74261	32592	86538	27041	65172	85532	07571	80609	39285	65340
31	64081	49863	08478	96001	18888	14810	70545	89755	59064	07210
32	05617	75818	47750	67814	29575	10526	66192	44464	27058	40467
33	26793	74951	95466	74307	13330	42664	85515	20632	05497	33625
34	65988	72850	48737	54719	52056	01596	03845	35067	03134	70322
35	27366	42271	44300	73399	21105	03280	73457	43093	05192	48657
36	56760	10909	98147	34736	33863	95256	12731	66598	50771	83665
37	72880	43338	93643	58904	59543	23943	11231	83268	65938	81581
38	86878	38100	03062	58103	47961	83841	25878	23746	55903	44115
39	63525	94441	77033	12147	51054	49955	58312	76923	96071	05813
40	47606	93410	16359	89033	89696	47231	64498	31776	05383	39902
41	28440	07819	21580	51459	47971	29882	13990	29226	23608	15873
42	52669	45030	96279	14709	52372	87832	02735	50803	72744	88208
43	16738	60159	07425	62369	07515	82721	37875	71153	21315	00132
44	59348	11695	45751	15865	74739	05572	32688	20271	65128	14551
45	12900	71775	29845	60774	94924	21810	38636	33717	67598	82521
46	75086	23537	49939	33595	13484	97588	28617	17979	70749	34234
47	26075	31671	45386	36583	93459	48599	52022	41330	60651	91321
48	66361	93596	23377	51133	95126	61496	42474	45141	46660	83342
49	49599	51434	29181	09993	38190	42553	68922	52125	91077	79014

Answers to Odd-Numbered Problems

Chapter 1

5. (a) Gross national product
(Billions of dollars)

Sectors	1980	1981
Durable Goods	$ 458.6	$ 507.0
Nondurable Goods	671.9	764.2
Services	1229.6	1370.3
Structures	266.0	280.7
Total	$2626.1	$2922.2

Source: *Survey of Current Business,* January
1982, p. 9, Tables 1.3–1.4.

7. (c) The bar chart provides actual amounts whereas the pie chart provides only percent of total. However, the pie chart provides an easier-to-read presentation of the relative share of each expense item.

15. (c) Both graphs are acceptable. But, because of the wide range of the data, the semi-log graph depicts the information in a somewhat more balanced fashion.

17. (a) 18, 21, 24, 25, 28, 29, 30, 31, 32, 32, 33, 34, 35, 35, 36, 36, 36, 37, 38, 38, 39, 39, 40, 40, 41, 41, 42, 42, 42, 43, 44, 45, 45, 46, 47, 47, 49, 50, 50, 50, 51, 52, 52, 55, 55, 57, 58, 59, 62, 64

(a)&(d) Ages	f	Relative frequency
18 and under 24	2	.04
24 and under 30	4	.08
30 and under 36	8	.16
36 and under 42	12	.24
42 and under 48	10	.20
48 and under 54	7	.14
54 and under 60	5	.10
60 and under 66	2	.04
	50	1.00

19. (a)

Ages	f
18 or more	50
24 or more	48
30 or more	44
36 or more	36
42 or more	24
48 or more	14
54 or more	7
60 or more	2
66 or more	0

(c)

Ages	f
less than 18	0
less than 24	2
less than 30	6
less than 36	14
less than 42	26
less than 48	36
less than 54	43
less than 60	48
less than 66	50

21. (a) ($60000)(0.10) = $6000   (b) ($60000)(0.05) = $3000

(c) $70000 − 40000 = $30000

(d) $\dfrac{\$70000 - 40000}{40000}(100) = 75\%$

Chapter 2

1. (a) 87.8   (b) 97   (c) 97   (d) 27.81

3. (a) 83.8   (b) 90   (c) 92   (d) 12.3   (e) 35

5. (a) 6.7   (b) 4.5   (c) 8   (d) 24   (e) 6.94   (f) 3

(g) The arithmetic mean is overly influenced by the extreme value of 25.

7. $\overline{X} = 5.1$; Median = 5; Mode = 6; Half of the employees work 5 or more hours of overtime with the most frequent number of overtime hours equaling 6.

9. (a) $148.57   (b) $125   (c) There is no mode

11. (a) Median, because of the extreme value ($100000)

(b) $7800

13. You would opt for Investment A if you were looking for the investment with the lower risk (V = 6.1% for A versus 14.3% for B). But, you would opt for B if you were looking for the investment with the greater potential value ($22350 + 3(3200) = $31950 versus $18000 + 3(1100) = $21300).

15. (a) 12.27   (b) 13.83   (c) 98.19   (d) 71.66

(e) $Q_1 = 4.6875$; $Q_3 = 22$

17. (a) 13.50   (b) 13.00   (c) 37.58   (d) 6.13

19. (a) $53.40   (b) $52.14

21. 34.4¢

23. Assistant Professors have the greater relative dispersion at 33.3%.

25. (a) 31   (b) 11.45   (c) 131.10   (d) 24   (e) 30.83

(f) 39.17   (g) 36.94%   (h) +0.045

27. 5.83%

29. (a) 41   (b) 56.4   (c) Half of the employees are between the ages of 41 and 56.4.   (d) +0.1062

Chapter 3

1. 60   3. 96   5. 20   7. 15   9. (a) $^{2}\!/_{90}$ (b) $^{72}\!/_{90}$
11. (a) 150   (b) 315
13. (a) 0.9875   (b) 0.1625
15. (a) 0.125   (b) 0.083   (c) 0.375
17. (a) 0.00009   (b) 0.000135
19. 0.000000000002
21. (a) $^{49}\!/_{100}$ (b) $^{42}\!/_{100}$
23. 0.38
25. 0.0004
27. 0.923   (b) 0.22
29. (a) 0.688   (b) 0.563   (c) 0.438   (d) No

Chapter 4

1. (a) Yes   (b) No   (c) No   (d) Yes   (e) No
   (f) Yes   (g) Yes   (h) Yes

3.

Number of heads (X)	$P(X \leq x)$
0	0.125
1	0.500
2	0.875
3	1.000

5. 0.25
7. 0.2031
9. (a) 0.2508   (b) 0.6177   (c) 0.0123   (d) 0.0464
11. 0.0486
13. 0.0574
15. (a) 0.1353   (b) 0.2707   (c) 0.1606
17. (a) 0.0498   (b) 0.2240   (c) 0.0337
19. (a) 0.1465   (b) 0.1221   (c) 0.7852
21. 0.0467
23. 0.24303
25. 0.3106
27. (a) 0.2119   (b) 0.2743   (c) 0.1464   (d) 0.0228
    (e) 0.1151   (f) 0.3811   (g) 0.2620
29. (a) 0.7881   (b) 0.3674   (c) 28.35 inches
31. (a) 12.54 months   (b) 11.925 months
33. (a) 0.0116   (b) 0.0869
35. 0.2019
37. (a) 0.7358   (b) 0.6915

Chapter 5

1. (a) $4^{2} = 16$ (b) $P_{2}^{4} = 12$ (c) $C_{2}^{4} = 6$
3. Possible Samples (3,3), (3,5), (3,7), (3,9), (5,3), (5,5), (5,7),
   (5,9), (7,3), (7,5), (7,7), (7,9), (9,3), (9,5), (9,7), (9,9)
5. (a) 6   (b) 1.58   (c) 1.58

7.

$\overline{X}$	f
4	2
5	2
6	4
7	2
8	2

9. (a) $P_2^4 = 12$ (b) Possible Samples (5,8), (5,11), (5,14), (8,5), (8,11), (8,14), (11,5), (11,8), (11,14), (14,5), (14,8), (14,11)   (c) 9.5   (d) 1.936   (e) 1.936

11. 0.25

13. (a) 0.6826   (b) 0.2514   (c) 0.9544   (d) 0.0918

15. (a) 1.5   (b) 1.466   (c) b is a more conservative figure due to the use of the finite population correction factor

17. (a) 0.9216   (b) 0.2005   (c) 0.121

19. 0.0316

21. 0.0283

23. (a) 0.8   (b) Yes   (c) because of small samples, normal population, and unknown standard deviation

25. 0.0002

27. 0.0401

Chapter 6

1. $40.9024 \leq \mu \leq 43.0976$

3. $56.578 \leq \mu \leq 63.422$

5. $20.216 \leq \mu \leq 21.784$

7. $45.01 \leq \mu \leq 58.99$

9. $36.2848 \leq \mu \leq 43.7152$

11. $27.7852 \leq \mu \leq 32.2148$

13. $42.9468 \leq \mu \leq 47.0532$

15. $75.0464 \leq \mu \leq 84.9536$

17. $81.20 \leq \mu \leq 118.80$

19. $72.23 \leq \mu \leq 77.77$

21. $2.9314 \leq \mu \leq 3.0686$

23. $530.65 \leq \mu \leq 569.35$

25. $0.526 \leq \pi \leq 0.774$

27. $0.17 \leq \pi \leq 0.33$

29. $0.5845 \leq \pi \leq 0.6955$

31. $0.139 \leq \pi \leq 0.361$

33. $0.1323 \leq \pi \leq 0.4677$

Chapter 7

1. 4888, 3214, 4194, 0316, 1977

3. $\overline{X} = 10$

5. $n_1 = 91, n_2 = 136, n_3 = 159, n_4 = 114$

7. $n_1 = 159, n_2 = 79, n_3 = 40, n_4 = 22$

9. $n_1 = 20$, $n_2 = 30$, $n_3 = 40$, $n_4 = 10$

11. $n_1 = 133$ urban residents; $n_2 = 267$ rural residents

13. 384

15. 420

17. 403

19. 864

21. 1067

23. \$7/9

25. 32

Chapter 8

1. $Z_{test} = 2$, Reject H_0; Reject H_0 if $Z_{test} < -1.96$ or $Z_{test} > 1.96$

2. $Z_{test} = 5$, Reject H_0; conclude that the machine is not filling properly; Reject H_0 if $Z_{test} < -2.58$ or $Z_{test} > 2.58$

5. $Z_{test} = 1.79$, still accept (do not Reject) H_0; Reject H_0 if $Z_{test} > 2.33$

7. $t_{test} = -2$, Reject H_0; The manager's claim is not supported; Reject H_0 if t_{test} is < -1.711

9. $t_{test} = 2.5$, Accept (do not Reject) H_0; Reject H_0 if $t_{test} \le -1.711$

11. (a) If fills are normally distributed (b) $t_{test} = -0.57$, still accept H_0; Reject H_0 if $t_{test} < -2.131$ or $t_{test} > 2.131$

13. $Z_{test} = 6.67$, Reject H_0; conclude that the average balance is increasing; Reject H_0 if $Z_{test} > 1.64$

15. $Z_{test} = 3.2$, Reject H_0; conclude that time for service calls exceeds 25 minutes; Reject H_0 if $Z_{test} > 1.64$

17. $Z_{test} = 9.52$, Reject H_0; conclude that the average number of days has increased; Reject H_0 if $Z_{test} > 2.05$

19. $Z_{test} = 2.07$, Reject H_0; conclude that the company should market nationally; Reject H_0 if $Z_{test} > 1.64$

21. $Z_{test} = 1.54$, Accept (do not Reject) H_0; conclude that the sample does not indicate that the proportion of claims is increasing; Reject H_0 if $Z_{test} > 2.33$

23. $Z_{test} = -0.327$, still Reject H_0; conclude that the shipment should not be accepted; Reject H_0 if $Z_{test} < -2.33$

25. $Z_{test} = -2.5$, Reject H_0; conclude that product acceptance is not high enough for mass production; Reject H_0 if $Z_{test} \le -1.64$

27. $Z_{test} = 0.73$, Accept (do not Reject) H_0; conclude that more than 50% are in favor of the issue; Reject H_0 if $Z_{test} \le -1.64$

29. $\beta = 0.484$

31. $\beta = 0.7352$

33. $n = 8.45$ or 88

35. 190

Chapter 9

1. A partial listing of the 81 sample pairs is given below.

Samples from N_1	$\overline{X}_1$	Samples from N_2	$\overline{X}_2$	$\overline{X}_1 - \overline{X}_2$
4,4	4.0	1,1	1.0	3.0
4,4	4.0	1,2	1.5	2.5
4,4	4.0	1,3	2.0	2.0
4,4	4.0	2,1	1.5	2.5
4,4	4.0	2,2	2.0	2.0
4,4	4.0	2,3	2.5	1.5
4,4	4.0	3,1	2.0	2.0
4,4	4.0	3,2	2.5	1.5
4,4	4.0	3,3	3.0	1.0
4,5	4.5	1,1	1.0	3.5
4,5	4.5	1,2	1.5	3.0
4,5	4.5	1,3	2.0	2.5
.	.	.	.	.
.	.	.	.	.
.	.	.	.	.
6,6	6.0	1,1	1.0	5.0
6,6	6.0	1,2	1.5	4.5
6,6	6.0	1,3	2.0	4.0
6,6	6.0	2,1	1.5	4.5
6,6	6.0	2,2	2.0	4.0
6,6	6.0	2,3	2.5	3.5
6,6	6.0	3,1	2.0	4.0
6,6	6.0	3,2	2.5	3.5
6,6	6.0	3,3	3.0	3.0

3. (a) 3   (b) 0.816, yes

5. $Z_{\text{test}} = 0.775$, Accept (do not Reject) H_0; Reject H_0 if $Z_{\text{test}} < -1.96$ or $Z_{\text{test}} > 1.96$

7. $t_{\text{test}} = -3.21$, Reject H_0; conclude that the operations are different; Reject H_0 if $t_{\text{test}} < -2.069$ or $t_{\text{test}} > 2.069$

9. $t_{\text{test}} = -2.867$, Reject H_0; conclude that mornings and afternoons are different; Reject H_0 if $t_{\text{test}} < -2.064$ or $t_{\text{test}} > 2.064$

11. $t_{\text{test}} = -3.26$, Reject H_0; conclude that the tear strengths are different; Reject H_0 if $t_{\text{test}} < -2.306$ or $t_{\text{test}} > 2.306$

13. $Z_{\text{test}} = 3.44$, Accept (do not Reject) H_0; Reject H_0 if $Z_{\text{test}} < -1.28$

15. $Z_{\text{test}} = -9.78$, Reject H_0; conclude that difference in income is significant; Reject H_0 if $Z_{\text{test}} < -1.96$ or $Z_{\text{test}} > 1.96$

17. (a) $t_{\text{test}} = -0.92$, Accept (do not Reject) H_0; conclude that the average expenditures are not different; Reject H_0 if $t_{\text{test}} < -2.807$ or $t_{\text{test}} > 2.807$ (b) Assume normal population and $\sigma_1 = \sigma_2$

19. $Z_{\text{test}} = 6.07$, Reject H_0; conclude that new paint provides greater coverage; Reject H_0 if $Z_{\text{test}} > 2.33$

21. 90% C.I. for $\pi_1 - \pi_2 = 0.02 \pm 0.09348$

23. $Z_{\text{test}} = -0.873$, Accept (do not Reject) H_0; conclude that the samples came from the same population; Reject H_0 if Z_{test} is < -1.96 or $Z_{\text{test}} > 1.96$

25. $Z_{\text{test}} = 1.56$, Accept (do not Reject) H_0; conclude that there is no difference in proportion of calls

Chapter 10

1. $F_{\text{test}} = 1.96$, Reject H_0; conclude that the variances are not equal; Reject H_0 if $F_{\text{test}} > 1.39$

3. $F_{\text{test}} = 2.25$, Reject H_0; conclude that the variances are not equal; Reject H_0 if $F_{\text{test}} > 1.85$

5. $F_{\text{test}} = 6.75$, Reject H_0; conclude that at least two means are different; Reject H_0 if $F_{\text{test}} > 3.88$

7. $F_{\text{test}} = 1.114$, Accept (do not Reject) H_0; conclude that the means are equal; Reject H_0 if $F_{\text{test}} > 3.47$

9. $F_{\text{test}} = 28.57$, Reject H_0; conclude that the three categories have different payment times; Reject H_0 if $F_{\text{test}} > 6.93$

Chapter 11

1. $X^2_{\text{test}} = 39.9$, Reject H_0; conclude that the grades are not uniformly distributed; Reject H_0 if $X^2_{\text{test}} > 13.2767$

3. $X^2_{\text{test}} = 1.142$, Accept (do not Reject) H_0; conclude that the arrivals are uniformly distributed; Reject H_0 if $X^2_{\text{test}} > 15.0863$

5. $X^2_{\text{test}} = 9.619$, Accept (do not Reject) H_0; conclude that the number of people waiting in line is normally distributed; Reject H_0 if $X^2_{\text{test}} > 11.0705$

7. $X^2_{\text{test}} = 5.64$, Accept (do not Reject) H_0; conclude that envelopes make no difference; Reject H_0 if $X^2_{\text{test}} > 6.63490$

9. $X^2_{\text{test}} = 38.11$, Reject H_0; conclude that racket sales are not independent of store location; Reject H_0 if $X^2_{\text{test}} > 13.2767$

11. $X^2_{\text{test}} = 31.34$, Reject H_0; conclude that purchase and type of campaign are not independent; Reject H_0 if $X^2_{\text{test}} > 3.84146$

13. $Z_{\text{test}} = 1.55$, Accept (do not Reject) H_0; conclude that digits are random; Reject H_0 if $Z_{\text{test}} < -1.96$ or $Z_{\text{test}} > 1.96$

15. $Z_{\text{test}} = -0.157$, Accept (do not Reject) H_0; conclude that the departures are random; Reject H_0 if $Z_{\text{test}} < -2.58$ or $Z_{\text{test}} > 2.58$

17. $Z_{\text{test}} = 1.74$, Reject H_0; apparently the new plan does not increase the dollar amount of the commission of the majority of the employees; Reject H_0 if $Z_{\text{test}} > 1.64$

19. $Z_{\text{test}} = -0.45$, Accept (do not Reject) H_0; conclude that the chefs are equally popular; Reject H_0 if $Z_{\text{test}} < -1.96$ or $Z_{\text{test}} > 1.96$

21. $H = 6.9$, Reject H_0; conclude that there is a difference in academic performances; Reject H_0 if $H > 5.99147$

Chapter 12

1. (b) $Y_c = -0.38 + 0.133X$ (c) $t_{test} = 12.09$, Reject H_0; Reject H_0 if $t_{test} < -2.447$ or $t_{test} > 2.447$
   (d) $r = 0.98$ (e) 99% C.I. for $Y_c = 6.27 \pm 0.6721$
   (f) 99% C.I. for $Y = 9.196 \pm 2.16$

3. (a) $Y_c = 3.7501 + 3.0729X$ (c) $t_{test} = 3.998$, Reject H_0; Reject H_0 if $t_{test} < -2.776$ or $t_{test} > 2.776$
   (d) $Y_c = 40.6249$ (e) 95% C.I. for $Y_c = 40.6249 \pm 12.0703$ (f) 95% C.I. for $Y = 40.6249 \pm 24.139$ (g) $r = 0.8944$, $r^2 = 0.80$

5. (a) $Y_c = -0.25 + 0.6085X$ (b) $t_{test} = 25.46$, Reject H_0; Reject H_0 if $t_{test} < -2.132$ or $t_{test} > 2.132$
   (c) $Y_c = 4.618$ (d) $r = 0.996$

7. (a) $Y_c = 1.395 + 0.0872X$ (c) $t_{test} = 8.9$, Reject H_0; Reject H_0 if $t_{test} < -2.306$ or $t_{test} > 2.306$ (d) $r^2 = 0.91$

9. (b) $Y_c = -0.389 + 0.08466X$ (c) $t_{test} = 11.76$, Reject H_0; Reject H_0 if $t_{test} < -1.86$ or $t_{test} > 1.86$
   (d) $r^2 = 0.94$

11. (a) $Y_c = -07262 + 0.1439X$ (b) $syx = 0.4723$
    (c) $t_{test} = 8.22$, Reject H_0; Reject H_0 if $t_{test} < -2.447$ or $t_{test} > 2.447$ (d) $r^2 = 0.918$

13. $t_{test} = -5.3761$, Reject H_0; conclude that there is a relationship between floor number and number of complaints; Reject H_0 if $t_{test} < -2.776$ or $t_{test} > 2.776$

15. (a) $r = 0.9354$ (b) $syx = 3.1944$

17. (b) $Y_c = -7.5533 + 2.3355X$ (c) $t_{test} = 3.4125$, Reject H_0; Reject H_0 if $t_{test} < -2.306$ or $t_{test} > 2.306$
    (d) $r^2 = 0.5927$ (e) The relationship is significant at alpha = 0.05, and change in years of education accounts for 59.27% of the change in income

19. (a) $t_{test} = 15.314$, Reject H_0; Reject H_0 if $t_{test} < -2.306$ or $t_{test} > 2.306$ (b) $r^2 = 0.9671$ (c) 95% C.I. for $Y_c = 13.9884 \pm 1.0606$ (d) 95% C.I. $Y = 7.8964 \leq Y \leq 12.6684$

21. (a) $\log Y_c = -1.67821 + 0.03936X$ (b) $Y_c = 28.15$
    (c) $r = 0.993$ (d) 9.5% increase for each additional service call

Chapter 13

1. (a) $Y_c = 32.516 + 3.027X_1 - 0.360X_2$ X_1 = Dog licenses; X_2 = Supermarkets; $X_1 t_{test} = 3.40$; $X_2 t_{test} = -0.434$
   (b) No

3. (a) $Y_c = -1077.76 + 26.28072X_1 + 3.24836X_2$
   (b) $X_1 t_{test} = 3.725$; $X_2 t_{test} = 4.26669$ (c) $R_2 = 0.85439$

5. (a) $Y_c = -12408.61 + 1.16X_1 + 104.35X_2 + 188.68X_3$; X_1 = Median income; X_2 = Population; X_3 = Unemployment Rate   (b) $X_1 t_{test} = 2.47$; $p > T = 0.0281$; $X_2 t_{test} = 0.57$; $p > T = 0.5776$; $X_3 t_{test} = 2.92$; $p > T = 0.0120$ (c) $R^2 = 0.980536$

7. A significantly high correlation exists between the median income variable and the population variable. Therefore, there is definitely a reason to be concerned with the impact of multicolinearity. Serial correlation frequently occurs when you fail to include important independent variables. Thus, the exclusion of variables such as the amount of financial and available and the average cost of attending college are reasons for concern.

9. $X_1 t_{test} = 7.31$, Reject H_0; $X_2 t_{test} = 0.1369$, Accept (do not Reject) H_0; At the 0.10 level of significance, advertising expenditures and population density should not be considered together to predict sales.

Chapter 14

1. (b) $Y_c = 212.79 + 1.68X$ (c) $Y_{1981} = 223.71$
3. (a) $Y_c = 139.54 + 17.61X$ (b) $Y_{1982} = 227.59$
5. (a) $Y_c = 656.26 + 61.18X$ (b) $Y_{1982} = 1023.34$; The estimate for 1982 is consistent with the upward trend in debt for the past years.
7. (a) $Y_c = 106 + 5.96$ (b) $\log Y_c = 1.72627 + 0.07100X$
   (c) $Y_{1983} = 102.40$ thousand
9. (a) 17 years   (b) $Y_{1985} = \$442.4$ thousand   (c) $\$14.4$ thousand
11. (a) 10 years   (b) $12 thousand   (c) $116 thousand
13. $Y_c = 5.557 + 1.193X$
15. $Y_c = 1.28 + 0.279X + 0.536X^2$
17. (a) $Y_c = 20.44 + 0.49X$ (b) $Y_c = 5.11 + 0.031X$
    (c) $Y_c = 5.1255 + 0.031X$
19. (a) $Y_{1982} = 135$ million   (b) $120 + 10X$
21. $Y_c = 132.6 + 3.6X$ (b) $10 + 0.025X$

Chapter 15
1.

	J	F	M	A	M	J	J	A	S	O	N	D
1976	—	—	—	—	—	—	51.61	126.33	98.96	72.73	95.06	139.79
1977	68.57	88.89	129.73	148.68	104.34	61.54	60.51	138.83	97.56	76.80	94.48	130.23
1978	73.29	90.22	124.44	140.15	103.59	68.58	68.09	134.27	99.30	81.63	96.65	111.25
1979	78.43	92.91	122.29	135.85	104.35	73.17	71.86	127.80	98.25	83.24	96.00	122.03
1980	80.45	92.81	118.03	129.74	102.67	76.19	75.40	124.35	98.46	85.28	96.48	119.40
1981	82.76	93.65	115.94	126.32	102.37	78.87	78.14	121.65	98.63	86.88	96.86	117.33
1982	84.58	94.32	114.29	123.61	102.12	81.01						
Modified total	314.93	369.59	480.7	532.06	412.97	296.81	275.86	512.75	394.3	326.95	384.19	488.99
Average specific seasonal (unadjusted)	78.73	92.4	120.18	138.02	103.24	74.2	68.97	128.19	93.58	81.74	96.05	122.25	1197.55
Seasonal index (adjusted)	78.89	92.59	120.43	133.29	103.45	74.35	69.11	128.45	98.78	81.91	96.25	122.50	1200

Multiplier for Adjustment $= \dfrac{1200}{1197.55} = 1.002045844$

3.

	J	F	M	A	M	J	J	A	S	O	N	D
1975	—	—	—	—	—	—	129.73	148.68	124.15	114.00	93.76	71.11
1976	68.09	48.98	77.92	104.99	116.36	112.95	124.86	136.37	133.33	103.79	100.00	72.36
1977	58.82	57.97	80.77	114.84	137.14	125.11	123.94	123.26	112.68	103.84	96.48	87.46
1978	77.42	65.57	79.12	78.69	90.81	102.13	111.41	120.60	116.51	111.63	95.57	90.75
1979	86.40	73.28	87.28	91.67	103.65	107.01	111.11	115.32	119.30	109.72	107.50	99.17
1980	84.78	84.10	90.32	96.77	103.78	111.78	113.65	115.58	117.24	111.95	99.70	86.75
1981	73.17	73.84	88.44	90.28	97.80	111.80	109.75	114.63	105.57	103.75	102.56	95.45
1982	95.19	87.64	99.45	97.30	89.13	93.75						
Modified total	389.96	354.76	426.93	481.01	512.4	545.67	585.47	611.13	589.88	540.93	494.31	433.27
Average specific seasonal (unadjusted)	77.47	70.95	85.39	96.20	102.49	109.13	117.09	123.23	117.98	108.19	98.86	86.65	1194.12
Seasonal index	78.35	71.30	85.81	96.67	102.98	109.67	117.67	123.84	118.56	108.72	99.35	87.08	1200

$$\text{Multiplier for Adjustment} = \frac{1200}{1194.12} = 1.004924128$$

5.

Year	I	II	III	IV
1978			120.75	83.48
1979	72.73	114.29	133.33	82.42
1980	73.14	112.43	126.53	85.02
1981	81.11	109.25	122.03	88.16
1982	81.25	107.87		
Modified total	154.25	221.68	255.36	168.50
Unadjusted index	77.13	110.84	127.68	84.25
Adjusted index	77.15	110.87	127.71	84.27

$$\text{Multiplier for Adjustment} = \frac{400}{399.895} = 1.000262569$$

7.

Year	I	II	III	IV
1974			114.29	139.13
1975	64.00	88.89	110.34	129.03
1976	72.73	91.43	108.11	123.08
1977	78.05	93.02	106.67	119.15
1978	81.63	94.12		
Modified total	150.78	184.45	218.45	252.11
Unadjusted index	75.39	92.23	109.23	126.06
Adjusted index	74.85	91.57	108.45	125.15

$$\text{Multiplier for Adjustment} = \frac{400}{402.895} = 0.992814505$$

9. (a)

Month	Seasonally adjusted sales
January	$7857
February	$6500
March	$6600
April	$7727

(b) $8400

11. $0.36 billion

13. I = 50.12 thousand; II = 58.24 thousand; III = 103.60 thousand; IV = 82.72 thousand

15.

Year	I	II	III	IV
1978	0.893	1.011	1.086	1.019
1979	0.891	0.986	1.080	1.013
1980	0.889	1.040	1.073	1.028
1981	0.887	1.016	1.067	1.023
1982	0.866	0.992	1.079	1.055
Modified total	2.667	3.019	3.232	3.07
Unadjusted index	88.90	100.63	107.73	102.33
Adjusted index	88.99	100.73	107.84	102.43

$$\text{Multiplier for Adjustment} = \frac{400}{399.59} = 1.001026052$$

17. Cyclical index

Year	I	II	III	IV
1978		100.49	100.19	100.10
1979	99.14	99.36	98.99	99.64
1980	100.70	100.89	101.04	99.83
1981	100.27	99.80	99.86	98.69
1982	98.55	98.63	100.52	

Chapter 16

1. (a) 1980 = 100; 1981 = 105; 1982 = 110   (b) 5 points
(c) 4.76%   (d) 1980 = 116.7; 1981 = 133.3;
1982 = 100   (e) 1980 = 100; 1981 = 116.7;
1982 = 133.3

3. 1980 = 100; 1981 = 110.3; 1982 = 123.5

5. 1980 = 100; 1981 = 112.5; 1982 = 127.1

7. (a) 1980 = 100; 1981 = 108.64; 1982 = 100.98
(b) 1980 = 100; 1981 = 108.69; 1982 = 100.7   (c) The answers are the same. (Rounding differences may cause slight difference.)

9. (a) 1980 = 100; 1981 = 108.7; 1982 = 121.11

(b) 1980 = 100; 1981 = 108.75; 1982 = 120.78   (c) The answers are the same.

11. (a) 1980 = 100; 1981 = 123.39; 1982 = 137.1   (b) The percentage increase from 1980 to 1982 is 37.1   (c) 11.11%

13. (a) 1980 = 100; 1981 = 92.86; 1982 = 89.29
(b) Laspeyres: 1980 = 100; 1981 = 103.65; 1982 = 118.47   (c) Laspeyres: 1980 = 100; 1981 = 103.65; 1982 = 118.47

15. (a) 1976 = 96.92; 1977 = 102.67   (b) 1979 = 157.04; 1980 = 169.21; 1981 = 177.73

17. No, his real earnings are higher in present job

19. $3.29

21. $74,100

Chapter 17

1. 4.1

3. (a)

	Actions			
	100	110	120	130
100	300	260	220	180
110	300	330	290	250
120	300	330	360	320
130	300	330	360	390

(Events)

(b) EMV (100) = $300, EMV (110) = $309, EMV (120) = $290, EMV(130) = $257   (c) Stock 110   (d) EPUC = $333   (e) EVPI = $24

5. (a)

	Actions		
	10	20	30
10	20	−80	−180
20	20	40	−60
30	20	40	60

(Events)

(b) EMV (10) = $20, EMV (20) = $4, EMV (30) = −$72; Stock 10 books   (c) EPUC = $38   (d) EVPI = $18

7.

	Actions		
	10	20	30
10	0	100	200
20	20	0	100
30	40	20	0

(Events)

9. (a)

	Actions		
	15	20	25
15	45	20	−5
20	45	60	35
25	45	60	75

(Events)

(b) EMV (15) = $45, EMV (20) = $50, EMV(25) = $37; Stock 20 calendars   (c) EVPI = $10.75

11.

		Actions		
		15	20	25
Events	15	0	15	30
	20	15	0	15
	25	30	15	0

13. (a)

		Actions			
		200	210	220	230
Events	200	1000	950	900	850
	210	1000	1050	1000	950
	220	1000	1050	1100	1050
	230	1000	1050	1100	1150

(b) EMV (200) = \$1,000; EMV (210) = \$1,040; EMV (220) = \$1,040; EMV (230) = \$1,010; Stock 210 or 220 bouquets   (c) EPUC = \$1080   (d) EVPI = \$40
(e) 216

15. (a)

Prob. of repeat	(P(X = 2, n = 2)	Prior	Joint	Revised
0.05	0.0025	0.10	0.00025	0.00223
0.25	0.0625	0.40	0.02500	0.22321
0.35	0.1225	0.30	0.03675	0.32813
0.50	0.2500	0.20	0.05000	0.44643
			0.11200	1.00000

(b) EMV (200) = \$1,000; EMV (210) = \$1,049.777; EMV (220) = \$1077.233; EMV (230) = \$1071.876; Stock 220

17. (a)

Fraction defective	100% Inspect	Accept without inspection
0.05	400	250
0.10	400	500
0.15	400	750

(b) EMV (100% Inspect) = \$400, EMV (accept without inspection) = \$350; Accept without inspection.
(c) ECUC = \$302.50   (d) EVPI = \$47.50

19.

          w/o sampling   with sampling
    EVSI = Min (EMV)   −   E (min EMV)
         = \$350 − [0.86576(341.035) + 0.12850 (400) + 0.00576 (400)]
         = \$350 − [\$295.25 + \$51.40 + \$2.30]
    EVSI = \$1.05
              at \$0.20 per unit, cost of sampling = \$0.40, thus
              sampling is worth \$1.05 − 0.40 = \$0.65

21. (a)

	Actions			
	0	5	10	15
Events 0	0	−15	−30	−45
5	0	15	0	−15
10	0	15	30	15
15	0	15	30	45

(b) EMV (0) = 0, EMV (5) = \$10.50, EMV (10) = \$9.00, EMV (15) = −\$3.00; Stock 5 planes   (c) \$10.50   (d) 7

23. (a)

	Actions (1000)				
	75	100	125	150	175
Events 75	22500	11250	0	−11250	−22500
100	22500	30000	18750	7500	−3750
125	22500	30000	37500	26250	15000
150	22500	30000	37500	45000	33750
175	22500	30000	37500	45000	52500

(b) EMV (75,000) = \$22,500; EMV (100,000) = \$26437.50; EMV (125,000) = \$25125; EMV (150,000) = \$19125; EMV (175,000) = \$9937.50; Stock 100,000 boxes   (c) EVPI = \$9037.50

Index

Student *t* Distribution

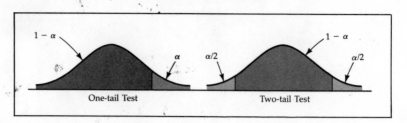

	Level of confidence (1-α)								
	0.20	0.50	0.80	0.90	0.95	0.98	0.99	0.999	Two-tail test
	Level of significance (α)								
n-1 Degrees of freedom	0.4	0.25	0.1	0.05	0.025	0.01	0.005	0.0005	One-tail test
	0.8	0.5	0.2	0.10	0.05	0.02	0.01	0.001	Two-tail test
1	0.325	1.000	3.078	6.314	12.706	31.821	63.657	636.619	
2	0.289	0.816	1.886	2.920	4.303	6.965	9.925	31.598	
3	0.277	0.765	1.638	2.353	3.182	4.541	5.841	12.924	
4	0.271	0.741	1.533	2.132	2.776	3.747	4.604	8.610	
5	0.267	0.727	1.476	2.015	2.571	3.365	4.032	6.869	
6	0.265	0.718	1.440	1.943	2.447	3.143	3.707	5.959	
7	0.263	0.711	1.415	1.895	2.365	2.998	3.499	5.408	
8	0.262	0.706	1.397	1.860	2.306	2.896	3.355	5.041	
9	0.261	0.703	1.383	1.833	2.262	2.821	3.250	4.781	
10	0.260	0.700	1.372	1.812	2.228	2.764	3.169	4.587	
11	0.260	0.697	1.363	1.796	2.201	2.718	3.106	4.437	
12	0.259	0.695	1.356	1.782	2.179	2.681	3.055	4.318	
13	0.259	0.694	1.350	1.771	2.160	2.650	3.012	4.221	
14	0.258	0.692	1.345	1.761	2.145	2.624	2.977	4.140	
15	0.258	0.691	1.341	1.753	2.131	2.602	2.947	4.073	
16	0.258	0.690	1.337	1.746	2.120	2.583	2.921	4.015	
17	0.257	0.689	1.333	1.740	2.110	2.567	2.898	3.965	
18	0.257	0.688	1.330	1.734	2.101	2.552	2.878	3.922	
19	0.257	0.688	1.328	1.729	2.093	2.539	2.861	3.883	
20	0.257	0.687	1.325	1.725	2.086	2.528	2.845	3.850	
21	0.257	0.686	1.323	1.721	2.080	2.518	2.831	3.819	
22	0.256	0.686	1.321	1.717	2.074	2.508	2.819	3.792	
23	0.256	0.685	1.319	1.714	2.069	2.500	2.807	3.767	
24	0.256	0.685	1.318	1.711	2.064	2.492	2.797	3.745	
25	0.256	0.684	1.316	1.708	2.060	2.485	2.787	3.725	
26	0.256	0.684	1.315	1.706	2.056	2.479	2.779	3.707	
27	0.256	0.684	1.314	1.703	2.052	2.473	2.771	3.690	
28	0.256	0.683	1.313	1.701	2.048	2.467	2.763	3.674	
29	0.256	0.683	1.311	1.699	2.045	2.462	2.756	3.659	
30	0.256	0.683	1.310	1.697	2.042	2.457	2.750	3.646	
40	0.255	0.681	1.303	1.684	2.021	2.423	2.704	3.551	
60	0.254	0.679	1.296	1.671	2.000	2.390	2.660	3.460	
120	0.254	0.677	1.289	1.658	1.980	2.358	2.617	3.373	
∞	0.253	0.674	1.282	1.645	1.960	2.326	2.576	3.291	

Source: Adapted from Table 12, Percentage points of the t-distribution, in E. S. Pearson and H. O. Hartley, *Biometrika Tables for Statisticians*, Vol. I, third edition, 1970, p. 146. Used with permission of Biometrika Trustees